Dredging

Dredging

A Handbook for Engineers

Second Edition

R. N. Bray
A. D. Bates
J. M. Land

BUTTERWORTH HEINEMANN

OXFORD AUCKLAND BOSTON JOHANNESBURG MELBOURNE NEW DELHI

Butterworth-Heinemann
Linacre House, Jordan Hill, Oxford OX2 8DP
225 Wildwood Avenue, Woburn, MA 01801-2041
A division of Reed Educational and Professional Publishing Ltd

A member of the Reed Elsevier plc group

First published by Arnold 1979
Second edition published by Arnold 1997
Second impression by Arnold 1998
Reprinted by Butterworth-Heinemann 2001

Transferred to digital printing, 2005

British Library Cataloguing in Publication Data
A catalogue record for this book is available from the British Library

Library of Congress Cataloging in Publication Data
A catalogue record for this book is available from the Library of Congress

ISBN 0 340 54524 0

Typeset in Times by GreenGate Publishing Services, Tonbridge, Kent
Printed & bound by Antony Rowe Ltd, Eastbourne

Contents

Foreword

It is a privilege to prepare a foreword for the interesting and informative book prepared by R. N. Bray, A. D. Bates and J. M. Land; *Dredging: A Handbook for Engineers.*

Dredging technology has resulted in many significant contributions over an extended period; including such notable achievements as the Suez and Panama Canals as well as providing access from the sea to the major ports of the world. While the majority of the world's population are probably aware of these and similar achievements, it is likely that only a limited number of people realize that these accomplishments would not have been possible without progressive improvements in dredging technology.

With rare exceptions, the navigation channels serving the major ports of the world require periodic dredging to provide navigable depths for the safe and efficient transportation of international maritime commerce. It is not an exaggeration to state that without access to the sea and the ports of the world, the world economy would experience a significant negative impact.

Dredging technology is applicable to a wide variety of functions in addition to being essential to maritime commerce. They include 1) improvement of water flow to provide flood control benefits; 2) removal of polluted sediments for processing or capping with clean material; 3) construction materials for concrete mix, landfills and dikes and levees; 4) beach nourishment and restoration; 5) mineral and gem mining; and 6) hurricane protection.

In spite of the importance of dredging technology and its major achievements over time, the literature on the subject was quite limited until the evolution of the non-profit making World Organization of Dredging Associations in 1967. Since then, there has been a significant increase in the number of technical papers presented at seminars and conferences and in trade journals. However, the number of published manuscripts which fully address the entire scope of dredging technology continues to be limited.

This volume addresses all aspects of dredging technology. The book is written by individuals who are expert in the field so its contents can be easily understood by management and ownership personnel as well as those engaged in dredging activities on a day to day basis.

It includes detailed discussions of the improvements which have occurred over the years, the design and construction of a wide variety of dredging equipment, dredging procedures and operations, contractual arrangements and a thorough coverage of topics relative to the protection and enhancement of the environment.

Due to its timeliness, comprehensive nature and accurate coverage of all the aspects of dredging technology, this book should be included in the libraries of colleges and universities as well as the offices of those individuals engaged in maritime activities including port authorities, ship owners and operators, and design and construction engineers.

William R. Murden, Jr.
Member of the National Academy of
Engineering
Alexandria, Virginia (USA)

Preface to the First Edition

Over the past ten years there has been a considerable improvement in the amount of information published on the subject of dredging. In spite of this, the practicability gap – that sparsely populated area between case history and theoretical report – has remained remarkably barren. With a few notable exceptions, there appears to have been little effort directed towards extracting the essence of the subject from the mass of accumulated technical information from the site and the laboratory.

Although the technology of dredging is complex it should not be necessary for the engineer to concern himself with its intricate details. This book, therefore, has been written for those engineers and students who have not the time, nor perhaps the inclination, to delve deeply into the abundance of highly specialised papers on each aspect of the subject. The object is to provide a readable and useful guide to the users of dredging equipment, their methods of operation, capabilities and place in maritime and fluvial civil engineering.

One of the greatest problems in tackling an unfamiliar branch of engineering is to get the feel of the subject and, in this respect, dredging is no exception. The intangibility of the work and the obfuscation which surrounds the prowess of the major exponents of the science do little to aid the outsider. However, the technical secrecy which, it is claimed, is necessary to maintain commercial competitiveness is instrumental in fostering a degree of misunderstanding in those not acquainted with the dredging profession. For, without an adequate amount of knowledge, the engineer is unable to discover whether he is being overcharged for relatively simple work or whether the work is really difficult and justifies a high unit cost. In order to try and overcome this problem I have resorted to discussing average characteristics, outputs, etc., and it has been necessary to make many generalisations. For similar reasons the book is virtually devoid of mathematics, with the exception of some simple algebraic statements, and even these have not been developed in the text. In short this is intended to be a practical handbook which, if it does not provide the answer directly, points out the types of problem which may be encountered and suggests a number of ways by which they can be overcome.

Many of the subjects mentioned in the text are worthy of detailed study and it is hoped that, by referring to the sources mentioned, the engineer will be encouraged to research them further. Indeed some aspects, such as estimating output, ship behaviour and reclamation, could well be expanded into treatises in their own right. It is hoped that the condensed coverage given here will be more manageable for general use.

A considerable time has elapsed since the inception of this book during which my everyday work and my writing have been allowed to mingle with considerable freedom. My sincere thanks are, therefore, due to the Partners of Livesey and Henderson for permitting me to work in this manner and also to my colleagues for putting up with it. My appreciation is also due for the assistance, facilities and helpful comments which have been forthcoming from them all. Apart from the numerous organisations who, unknowingly, have assisted me and those which are listed in the acknowledgements, I would also like to record my special thanks to Captain Cornelius J. Wennink for his comments on Chapter 9. Finally, my grateful thanks to Caroline Pontin and Rosemary Lemon for battling with my handwriting and corrections respectively and Nigel Wright for transforming my rough sketches into illustrations.

R. N. Bray
1979

Preface to the Second Edition

Since the original publication date of this book in 1979 there has been a considerable amount of technical development in the dredging industry. During the same period the legislative and environmental framework within which dredging is carried out has become better defined and more rigorously imposed. For these reasons a minor updating of the text would have been inadequate and the authors have resorted to a major re-write of much of the book. However, some chapters which proved to be particularly popular in the First Edition, such as that covering production estimating, have been retained because the basic principles remain unaltered.

Apart from the overall updating of most of the subjects which has been undertaken, the book is now more comprehensive. It covers more types of dredger, includes greater detail of ancillary equipment, such as instrumentation and pipelines, and contains new chapters on project implementation and costing. Many of the illustrations and all the photographs are new. We hope that this will make the book more interesting and useful.

Although the pace of technological development in most industries seems to accelerate continually, and dredging is no exception, there is always a need for a basic balanced readable treatise on the fundamentals of the field. After a somewhat lengthy gestation period, we believe that we have now produced the most comprehensive primer on the subject. To have extended the text into the realms of higher technology and greater detail would have almost certainly ensured that parts of the work would have been obsolete, prior to publication, so fast does the scene now change.

Welcome to a fascinating field of engineering. You will soon discover that what at first appears to be a narrow, specialist aspect of maritime engineering will emerge as a hugely varied subject, encompassing many of the major scientific and engineering disciplines.

R. N. Bray
A. D. Bates
J. M. Land
December 1995

Acknowledgements

The authors gratefully acknowledge the assistance of the dredging industry during the preparation of this book. In particular, special thanks are due to the British Standards Institution (BSI) and the Permanent International Association of Navigation Congress (PIANC) for their permissions to use or copy illustrations from their publications. Thanks are also due to:

Applied Explosives Technology Ltd
A. W. Wakefield
Construction Industry Research & Information Association
Conbar International Ltd
D. Blankenvoort International Dredging Co.
Damen Shipyards
Demex Explosives Technology
Dredging International nv
Ellicott Machine Corporation International
HAM Dredging and Marine Contractors
Holland Dredging
IHC Holland nv
Jan de Nul nv
Pneuma
Rhode Neilsen A/S
Royal BosKalis Westminster nv
Shipyard De Donge Ltd
Van Oord – Utrecht nv
Van Rees nv
Wartsila Turku Shipyard

1 Introduction

1.1 Historical

Dredging is an ancient art but a relatively young science. Traces of man's work involving primitive dredging techniques have been discovered in many places, dating back to thousands of years BC (Linssen, 1975). In such instances, the vessel was probably little more than a raft and the excavating means a man with a bucket. The development of this method of excavation into the spoon and bag dredger and the subsequent proliferation of dredging machines has been well described (Gower, 1967).

With the development of powerful dredging machinery there was a corresponding increase in the scope and complexity of engineering projects which could be executed by dredging. For a considerable time the art of using this equipment was known only to small groups of men who passed their knowledge and experience on from one generation to the next. With the advent of the Industrial Revolution, which transformed many arts into sciences, the dredging process was subjected to greater scientific analysis.

The development of dredging methods has been influenced by geographical characteristics. Whereas the major developments in mechanical excavating methods in the 19th century probably occurred in the lowlands of Europe, the development of hydraulic dredging techniques, using the centrifugal pump, was pioneered by the Americans. The latter had numerous sites suitable for reclamation by pipeline dredgers, whilst the European sites were more often confined, necessitating transport by barge to open sea. In due course, the slow mechanical methods were largely replaced by suction dredgers and hydraulic dredging became commonplace worldwide.

Other notable trends in dredging technology have been the development of rock cutter suction dredgers in the 1970s to tackle the many rock dredging projects in the Middle East, and the recent introduction of dredging machines specifically designed to be used within strict environmental constraints. No doubt this latter trend will continue into the foreseeable future as the dredging industry continues to adapt to the demands of the market.

1.2 Modern dredging practice

Dredging today is usually categorised as either capital or maintenance dredging. Capital dredging normally denotes projects which involve dredging as a one-off operation, whereas maintenance dredging is used to describe dredging which is of a recurrent nature.

Dredging is carried out for a variety of reasons and by a number of different types of dredger. However, the basic reason for dredging is usually to achieve one, or a combination, of the following:

- to excavate material from underwater, thereby creating a greater depth in a river, lake or sea;
- to fill a void, either underwater or on land, with material excavated from underwater;
- to replace material underwater, (i.e. to substitute poor quality material with good quality);
- to win material for construction;
- to win minerals;
- to improve the environment.

Examples of the types of work where these results are achieved are:

Material excavation

For construction:	to form new harbours, basins, entrance channels, canals; to excavate trenches for foundations, immersed tube tunnels and for laying pipelines and cables.
For navigation:	to deepen or widen harbours, channels and rivers; to remove obstructions such as boulders, rock outcrops or sand banks.
For maintenance:	to remove sediment from harbours, rivers and canals; to remove material from lakes and reservoirs to restore water capacity.
Others:	to alter seabed contours to modify the wave climate; to excavate around wrecks to sink them into the seabed; to improve the hydraulic efficiency of rivers and storage reservoirs.

Filling

For construction:	to form new or improved land for ports, industry, agriculture, roads, etc.; to place fill-in trenches for the foundations of breakwaters, caissons, immersed tube tunnels and pipelines; to backfill around the same.
For coast defence:	to form dykes and new beaches; to recharge beaches with additional sand or shingle.

Material replacement

For construction:	to excavate and remove unsuitable material for foundations, or reclamation, and replace with suitable material.

Material winning

For construction:	to obtain aggregate for making concrete; to obtain sand for reclamation fill.

Mineral winning

For mining:	to excavate minerals from the sea or river bed (manganese nodules, tin, gold, etc.); to excavate, process and re-deposit materials containing minerals and mineral ores (gold, tin, rutile, etc.).

Environmental improvement

For contaminants:	to excavate and remove contaminated materials to a safe disposal site, or for processing.
For capping:	to excavate clean, inert material and place over contaminated material as a sealing layer.
For habitat:	to excavate and place appropriate materials to create or improve habitats, such as wetlands.

In this book, with the exception of the mining processes, all these subjects are described in more detail.

1.3 Definitions

In the text the following definitions are used:

Dredger:	(in the USA, a dredge) a vessel, or item of floating plant, equipped with means to move or excavate soil or rock underwater.
Dredge:	to excavate or move soil or rock with a dredger.
Dredging:	excavating or moving soil or rock by dredger.
Dredgemaster:	the man in control of the dredging operation, not necessarily the master of the vessel.
Dredging equipment:	any piece of equipment (dredger or ancillary) which is required to perform the act of dredging.

1.4 Environmental aspects

By its very nature every act of dredging has an environmental impact. The rationale for carrying out dredging work is to cause a permanent or temporary change in the topography of the sea, lake or river bed. The same may be said of the site where the dredged material is to be relocated.

When the act of dredging takes place, other effects are apparent. Some turbidity is normally seen near the dredger. In some types of dredging work turbidity is an unavoidable effect of the dredging process. Noise and other environmental disturbances may also occur. If the dredged material is contaminated there is the possibility of pollutants being released to the surrounding environment. It is necessary to evaluate the impact of these effects.

The beneficial effects of dredging are numerous (see Section 1.2) and the uses to which the dredged material may be put have been studied extensively (see PIANC, 1992). It is important, therefore, that these benefits should be balanced with the environmental concerns described above. There are many aspects of the environmental effects of dredging being researched. Many of these have shown that the alarmist reactions to dredging activities are at best misplaced. However, these reactions do indicate an ignorance of the processes involved and their significance in the natural world. The need to understand and quantify natural changes in the environment is a prerequisite to justifying the use of dredging equipment.

In this revised edition of the book, environmental matters have been addressed in Chapters 2, 5, 6, 13 and 14.

1.5 Soils information

The dredging process is essentially concerned with the excavation and relocation of soils and rocks. The behaviour of these materials during the process has a profound influence on the ease with which a dredger can perform its task, the environmental effects of the work and the characteristics of the relocated material. The need for comprehensive soils information for dredging work cannot be too highly stressed.

In particular, the following geotechnical aspects of dredging work are fundamental to the understanding of the dredging process and the evaluation of dredging projects:

- the *in situ* characteristics of the material to be dredged;
- the change in strength and volume of the material during the dredging and relocation processes;
- the behaviour of the material during pipeline and hopper transport;
- the potential for change in material grading;
- the behaviour of the material during and after placement in a relocation site.

In this revised edition of the book soils aspects are covered in Chapters 3, 4, 5, 6, 8 and 13. Rock aspects are covered in Chapters 6 and 14.

1.6 Use of the book

This text is intended for two types of usage; as a general primer, for those unfamiliar with the subject, and as a basic *aide-mémoire* or handbook for practitioners. For the former it is suggested that Chapter 2 should be used as the basic guide to understanding the subject and implementing a project. The reader will find that he is referred to the other chapters as necessary. The supporting Chapters 3 to 14 will provide a useful reference for those more familiar with this type of work.

References

Gower, G.L. 1967. *A history of dredging*, Dredging Symposium, Institution of Civil Engineers, London.

Linssen, J.G.Th. 1975. The performance and the future development of dredging equipment, *Proceedings of the 1st International Symposium on Dredging Technology*, Canterbury, England.

PIANC 1992. Beneficial uses of dredged material, *Report of Working Group 19, Permanent International Association of Navigation Congresses (PIANC)*, Brussels.

2
Project Implementation

2.1 Introduction

A dredging project, in common with most other works of civil engineering, evolves in a series of stages. The stages are required to optimise expenditure during data gathering and analytical periods, and to provide points at which the project may be aborted if studies indicate it to be lacking viability. This staged process applies to both capital and maintenance dredging projects, although there may be times during either of these when, due to the exigencies of a particular situation, a number of stages will merge into one.

In this chapter the overall philosophy of this staged approach is outlined. The reader may therefore use this chapter as a general guide to the implementation of any dredging work, referring where necessary to the more detailed guidance given in the following supporting chapters. Although the chapter is written from the viewpoint of the client who intends to engage the services of a dredging contractor to carry out his dredging work, much of it is also applicable to an organisation that owns dredging equipment.

The staged approach is illustrated in simple flow-chart form in Figure 2.1. Essentially there are five stages:

- preliminary appraisal;
- pre-contract planning;
- preparation of contract documents;
- tender procedures;
- project execution.

In practice the preparation of contract documents is often carried out simultaneously with some of the tender procedures.

Before considering these five stages it is prudent to review the background against which any dredging project must be examined; the environmental balance of the proposed dredging site and any other site affected by the proposed works, and the legislation controlling these works.

2.2 The balance of nature

The act of dredging is primarily associated with change of the natural environment, the sea, lake or river bed, or even the coastline. Man-made changes of this type are an interference with, and will have an effect on, the balance of nature. It is essential, therefore, that any

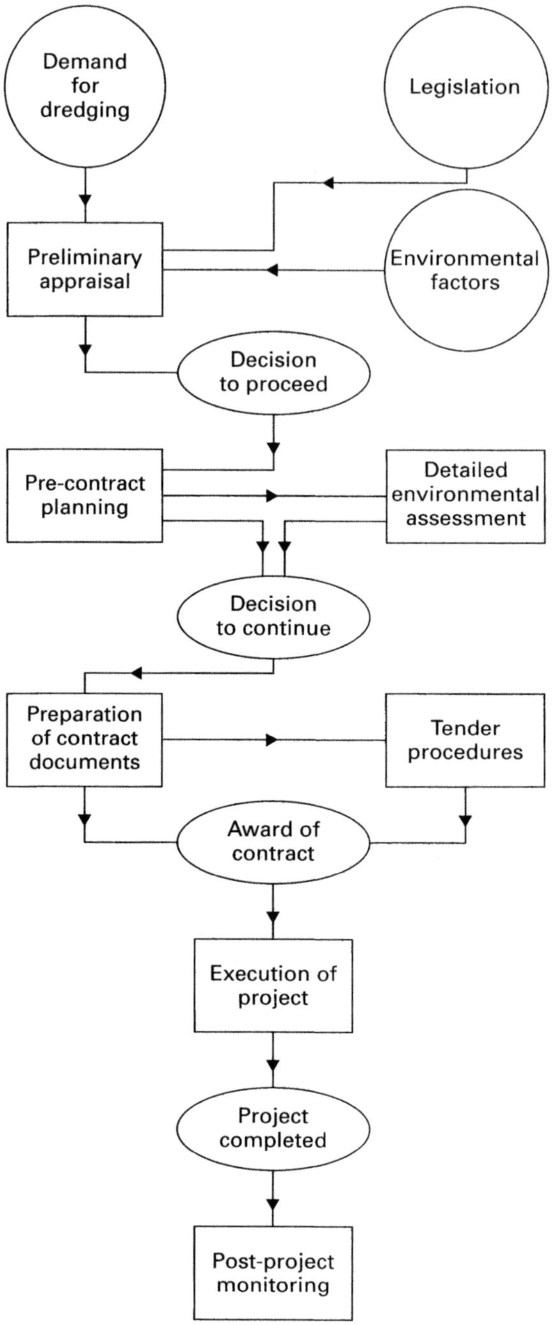

Figure 2.1 The staged approach to project implementation

proposed dredging scheme should be viewed in the context of its effect on the environment and in this respect the environmentally-sensitive approach is to:

- investigate the environmental balance;
- examine the need for dredging and whether it can be justified on socio-economic grounds;
- assess the effects of dredging if it is considered to be justifiable;
- investigate the mitigating measures that can be taken to reduce the environmental impact, where this is necessary and possible.

2.2.1 THE ENVIRONMENTAL BALANCE

It is particularly important, when considering the balance of nature, to recognise the differences between the on-land and underwater environments.

Above water a number of forces are at work. Sun, wind, rain and frost are all combined in an environment which generally acts as a leveller. Erosion and deposition are usually slow. Thus, if a hole is excavated it may fill with water but is unlikely to fill with earth. Similarly, if a pile of earth is placed in a field it will not disperse overnight, nor necessarily for a number of years. Relative equilibrium is thus generally maintained.

Below water level almost all material movement is caused by forces exerted by the water. A large proportion of the surface of the bed of seas or rivers is in motion, especially in shallow water or in the near-shore zone. Equilibrium is often characterised by a balance of material transported into and out of a site. If a hole is excavated it often fills quickly with material. A heap of deposited material tends to flatten out and disperse.

There is, therefore, a considerable difference between the effects of excavation above and below water. Dredging, and its effects, are thus not a simple extension of land excavation.

Sediment movement underwater consists essentially of three stages; erosion, transport and deposition. The state of any site can be defined by the net transport of material into and out of it, taking account of seasonal variations. To understand the regime of any site, it is necessary to examine the natural mechanisms of sediment movement in the area. Once these are understood the effect of dredging or depositing material in a site may be assessed.

2.2.2 THE BASIC MECHANISMS OF SEDIMENT MOVEMENT

The mechanics of sedimentation are complex and the study of sedimentary mechanisms is a subject which is already well-developed. Readers who wish to study the subject in detail should resort to standard textbooks or consult specialist organisations. The summary of mechanisms given here is intended only as a guide. For convenience the areas of interest are split into three groups, those relating to coastal, fluvial and estuarial sites. Methods for carrying out the measurements suggested below are given in Chapter 6.

Coastal sites

Sediment movement at coastal sites is known as littoral transport. Littoral transport is caused by the action of waves and currents. It is divided into two classes; longshore transport (parallel to the shore) and onshore-offshore transport (perpendicular to the shore). The material moved by either of these processes is called 'littoral drift'.

Onshore-offshore transport is primarily a function of beach slope, sediment particle size and wave climate. It is caused by the water motion at the seabed due to the passing of each wave. Sediment is moved if the water motion has sufficient velocity to move particles on the seabed. The threshold velocity for the movement of the most easily moved particles is around 0.15 metres per second, whilst that for movement of a coarse beach sand of around two millimetres diameter is about 0.30 metres per second. Table 2.1 shows depths of water in which a wave, one metre high, causes those velocities to occur at the seabed. It can be seen from the table that movement will occur at considerable depths when wave periods are large.

Table 2.1 Depths of water at which velocities on bed exceed 0.15 and 0.30 m s^{-1}

Wave period (seconds)	Depth of water (metres) for maximum seabed orbital velocity of: 0.15 m s^{-1}	0.30 m s^{-1}
4	9	6
6	17	10
8	25	14
10	33.5	16
12	40	18
14	49	20
16	56	22
18	61	22
20	65.5	22

Longshore transport is a result of the turbulence caused by the breaking wave, which stirs up the sediment and transports it in the wave direction. The movement along the shore is due to the component of wave direction which is parallel to the shore and the longshore current generated by the breaking wave, together with any other influencing currents. Since wave direction and energy vary throughout the year, the direction and amount of longshore transport also vary. However, over a long period for a given site, there is usually a net drift in one direction.

In order to estimate the seasonal or net annual longshore drift it is necessary to obtain high quality wave records of the area in question and accurate surveys of the beach and seabed. As a supplement to theoretical estimation of drift movement, tracer experiments are sometimes used to quantify drift volumes.

In most coastal zones there are also unidirectional currents. These may be oceanic or tidal in nature and, hence, may vary seasonally, tidally or both. Current velocities are generally low, less than 0.15 metres per second, and are then not capable of moving sediment unless it is already in suspension. In dynamic areas where wave action causes significant sediment motion, bed particle sizes are usually 0.2 millimetres diameter and upwards. In such places unidirectional currents will not have an appreciable effect unless their velocities are in excess of 0.5 metres per second.

When there are weak currents and wave action is light, the silt and clay fractions of material, which have been transported into the coastal zone, will tend to settle out of suspension. This is usual in well-protected bays, harbours and estuaries which are not subjected to significant tidal flushing action.

When the littoral drift is interrupted by a natural or artificial barrier, such as an inlet or jetty, the drift material eventually finds its way past the barrier in order to preserve continuity. The action of sediment being moved from one side to the other side of the barrier is known as bypassing. At inlets, the bypassing action consists of a combination of tidal flushing and wave action which transfers the drift material to an offshore, or inshore, bar and back to the beach again.

If the obstruction to drift material is artificial and there is no appreciable tidal action, the material will build up on one side of the barrier and eventually continue round the outer end of the barrier. Often this will result in the shoaling of the harbour or entrance channel which the barrier was protecting. One method of avoiding this problem is to dredge material on the updrift side of the barrier and discharge it on the downdrift side, thus restoring the balance of nature and helping to eliminate erosion of the coast on the downdrift side. Whilst this concept appears simple, in reality it is often difficult to arrange.

An investigation into the sedimentary mechanisms at work in a coastal site should include the following:

- measurement of the wave climate, (i.e. wave height, direction and period distribution on a seasonal and annual basis);
- current measurement on a tidal basis, and seasonally if it is considered that seasonal weather patterns could affect the regime;
- seabed and beach material sampling;
- bathymetric surveying of the seabed and beach profiling (the measurement of beach levels down to low water level, at regular intervals, to ascertain how the cross-section of the beach changes throughout the year);
- measurement of suspended sediment concentrations on a tidal and seasonal basis.

Fluvial sites

In fluvial sites, that is in rivers above the region which is influenced by tidal flows, sediment movement is caused by the flow of water in the downstream direction. Sediment movement occurs in three forms: suspension, saltation and rolling. Light particles are carried permanently in suspension; larger particles tend to move in a jumping motion off the river bed and back on to it again, which is called saltation; and the largest particles rarely leave the bed but roll along on the surface. The factors that determine the manner in which a specific size of particle behaves are the water velocity distribution in the river, the degree of turbulence and the river bed profile.

Most rivers behave in a seasonal fashion, that is the flow of water down the river and the quantity of sediment being supplied to the river are related to the season. Often a high proportion of the annual sediment load is transported down the river in one major flood. In other cases there is a gradual increase and decrease of flow throughout the season, augmented by irregular floods of short duration.

Whatever the pattern of river flow and sediment supply, it is known that rivers alter their bed profiles according to the flow. This has the effect of changing the bed roughness and affecting the flow characteristics. In this manner rivers naturally reduce their frictional resistance when overloaded by floods and increase it during seasons of low flow.

An investigation of the sedimentary mechanisms at work in any particular fluvial site must involve the following measurements:

- stage-discharge relationships, (i.e. the relationship between river depth, current velocities and flow across a river section);
- the quantity of material in suspension, and the river flow, at various times of the year;
- the sediment particle sizes occurring in the river;

- the height, wave length and rate of movement of the ripples or dunes of material on the river bed.

Estuarial sites

Estuarial sites combine the features of coastal and fluvial sites but are largely characterised by the tidal flow into and out of the site. The rise and fall of the tide at the mouth of the estuary causes a volume of sea water to enter the estuary during the flood and to leave during the ebb. The total volume of sea water moved into and out of the estuary during a tidal cycle is known as the tidal prism. The relationship between the volume of the tidal prism and the volume of fresh water entering the estuary during a tidal cycle serves as a classification for the various types of estuary, (i.e. well-mixed, etc).

Apart from the relationship between fresh and sea water volumes, the rate at which the fresh water mixes with the sea water is of great importance in an analysis of sedimentary mechanisms in the estuary. When the fresh water volume is high compared with the tidal prism it is common for diffusion to take place very slowly, and the mixing of the two types of water is largely incomplete at the mouth. In such circumstances the fresh, less dense, water tends to flow over the saline water and continue out to sea. A wedge-shaped volume of saline water remains under the fresh water and this wedge, which is called a saline wedge, projects into the estuary. The existence and position of the saline wedge is known to be largely responsible for silting patterns observed in estuaries. Well-mixed estuaries do not have a saline wedge.

Thus, for sites situated at the mouths of estuaries, an examination must be made of any coastal effects, the ebb and flood currents, the freshwater flow and sediment load and the existence of the saline wedge. Further up the estuary, coastal effects can be ignored and the mechanisms become dependent on tidal currents, sediment load and saline wedges. In some estuaries considerable volumes of silt are found in the form of tidal mudflats. Movement of silt onto and off these mudflats is affected by meteorological conditions (heavy rain, ice/snow and waves) as well as the other parameters mentioned above.

In very wide and shallow estuaries, the scour and siltation which occur are variable and can be affected by small variations in tidal currents. Some estuaries remain relatively stable for many years and then suddenly change to assume a new stability in a different form. Complete analysis of the sedimentation pattern of a whole estuary would be very complex and would not usually be attempted. However, physical and numerical models of estuaries are able to help in determining the overall patterns.

The following measurements, which should cover seasonal variations, are necessary to investigate sedimentation in estuarial regions:

- tidal and fluvial currents, directions and distributions with water depth;
- water temperature and salinity, and their distribution with water depth and tidal state;
- bed and suspended sediment samples;
- wave records, if the region is exposed to wave action;
- measurement of the density and bed shear of surface deposits;
- examination of mudflat areas and characteristics.

2.2.3 THE NEED FOR DREDGING

From the various sedimentary processes described in Section 2.2.2 it can be seen that in most sites the various forces of nature are delicately balanced. Any artificial disturbance, such as that caused by the dredging or disposal of dredged material, is likely to alter the balance in some manner. In some instances the balance of nature may have already been altered by some other artificial means such as the construction of a breakwater or jetty. Thus, the need for dredging may arise because:

- the required bed configuration does not exist, and must be formed;
- the required bed configuration is unstable and needs to be maintained or;
- the original bed stability has been upset by other artificial means and needs to be maintained.

Dredging works are classed in one of two categories: capital or maintenance. The formation of a new bed configuration by dredging, whether the configuration is stable or not, is known as capital dredging, the implication being that the work involves the payment of a single capital sum. Any other dredging work would be recurrent and, since it is performed to maintain the desired bed configuration, it is known as maintenance dredging.

The desirability of dredging in a particular location depends on economic and environmental factors, and must be viewed in the light of the total development being planned. From most environmental, and in many cases economic, aspects the possibilities, in order of preference, are:

- no dredging at all;
- capital dredging only;
- capital and maintenance dredging.

The first two of these are discussed below. The third is the result of the failure to achieve one of the others. Figure 2.2 illustrates the various possibilities.

Development without dredging

There are three ways to avoid maintenance dredging when adequate water depths are available. These are:

1. develop at a site where there is no sediment movement;
2. develop at a site where sediment movement occurs but is not affected by the development;
3. develop at a site where sediment movement occurs but where the development assists in preventing siltation at the site.

Examples of (1) can be found on rocky coastlines and where water close to the shore is deep. Structures can also be built on predominantly sandy coastlines at locations where the gross littoral drift is negligible. This occurs at some headlands where the drift on either side is always away from the headland. Another location often chosen for port development without dredging is inside sheltered bays where there is already a sufficient depth of water and virtually no sediment movement.

A type (2) development can be carried out by use of structures which do not impede water movement and, hence, do not interrupt the movement of sediment, such as open pile jetties. An alternative to this is to construct a berthing face using a natural feature which has adequate water along side.

Type (3) developments include structures in rivers, which tend to channel the main flow of water and, hence, maintain adequate water depth, and breakwaters, which deflect the littoral drift out from the coast into some natural sink, such as a submarine canyon. (This latter type may, however, cause erosion problems on the downdrift side of the development.) Also included in category (3) are harbours whose entrance configurations are designed to be self-cleansing.

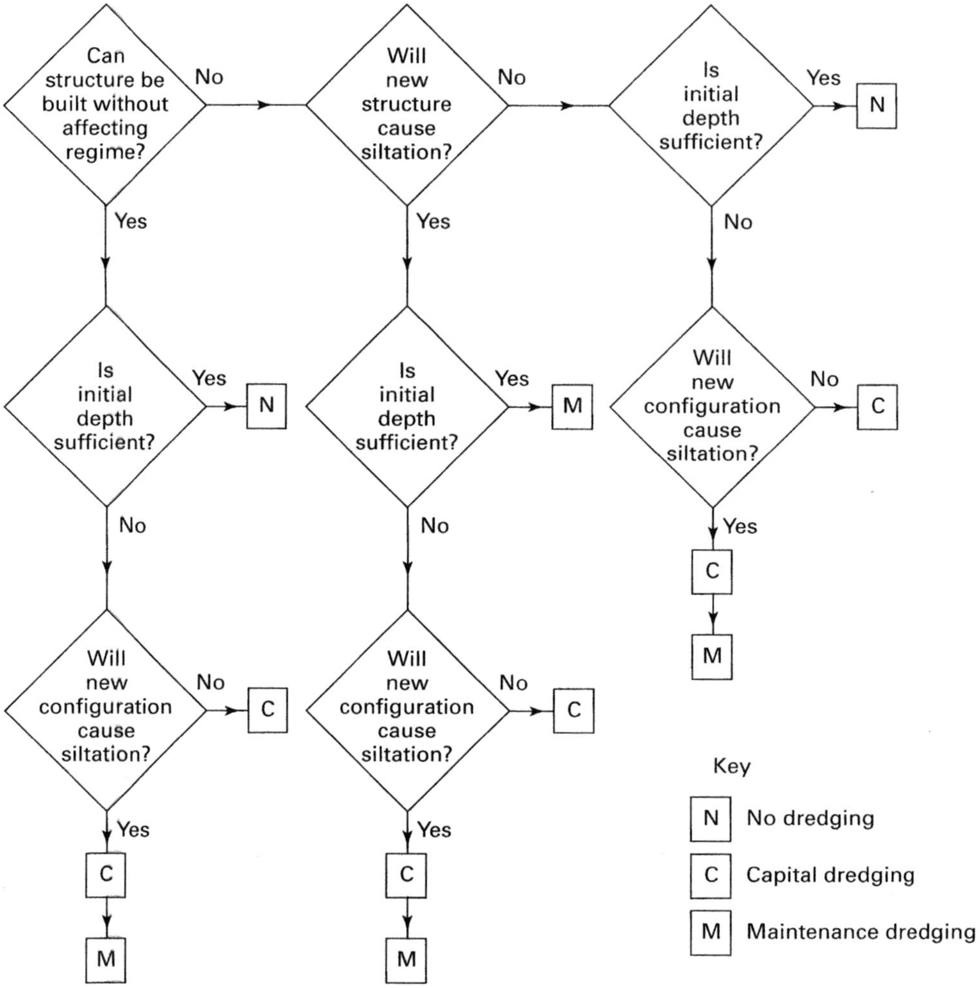

Figure 2.2 The dredging need

Development with capital dredging only

There are two main ways to prevent maintenance dredging arising as a result of capital dredging:

(a) dredge at a site where there is no sediment movement;
(b) dredge at a site which will be stable in its developed state.

Examples of (a) are generally the same as those in (1).

Examples of type (b) dredging occur in rivers where the bed is composed of hard material and where an enlargement of the river cross-section does not sufficiently reduce current velocities to cause siltation.

Where dredging is carried out in coarse-grained soil, the dredged area will remain unchanged if the water currents are too weak to move any more coarse material but sufficiently high to maintain fine material in suspension.

When channels are dredged in open water at an angle of 15° or less to the predominant current direction, the mean velocity of flow tends to increase due to an increase in hydraulic radius. In areas where there is no appreciable sediment movement due to wave action, this tends to lead to a self-cleansing situation.

2.3 Legislation affecting dredging

Another of the key factors which will influence a development involving dredging works is the legislation relating to dredging operations and the projects for which they are required. Much of the need for this legislation has arisen due to the pressure applied to governments by groups concerned with environmental protection, particularly where contaminated materials are to be handled. However, in the context of modern thought, the requirement for environmental protection has been extended to most construction activities, including dredging. The legislation may apply to:

- the site where the dredging is to take place;
- the site where dredged material is to be deposited.

In the UK much of the legislation relating to projects in the maritime field is derived from European and International conventions. It is probable therefore that other countries will have similar legislation relating to these points. An interesting review of the institutions and the conventions they have developed is given by Lammers and Neervoort (1988), and the American view of the effects of legislation is given by Murden (1989).

The main items in the UK for 1995 are as shown below. However, it is possible that current efforts to integrate coastal zone management will result in the future formation of a single authority responsible for developments in the coastal zone.

The Town and Country Planning Act 1971

This Act covers all major development proposals within a specific local authority's jurisdiction. Any proposed development must be the subject of an application for planning approval and no works may go ahead without the consent of the planning authority. This would apply to a port extension involving dredging and the reclamation of land, but only extends to low water.

One of the duties of the local authority is to administer EEC Directive 85/338 on Environmental Assessment. This directive applies to any project which is likely to have a significant effect on the environment. For certain circumstances these assessments are obligatory (see Annex 1 of the Directive) and for others they are discretionary (see Annex 2 of the Directive). Environmental assessments are submitted, where appropriate, with the relevant planning applications.

Food and Environment Protection Act (Part II) 1985

This act embodies the requirements of the Oslo and London Conventions. It is necessary for those disposing of materials at sea to obtain, and renew annually, licences from the regulatory authority. Applicants for a licence must provide detailed information about the nature of the material to be deposited in the disposal ground. The regulatory authority in England and Wales is the Ministry of Agriculture Fisheries & Food (MAFF). In Scotland and Northern Ireland, the Act is administered by the Departments of Agriculture & Fisheries (DAF), and Environment (DOE), respectively.

This Act affects most dredging projects in so far as any dredged material to be deposited permanently in the sea comes under the Act.

Control of Pollution Act (Part II) 1984

Under this Act (commonly referred to as COPA II) standards are set relating to the discharge of wastes into rivers and coastal waters. The Act derives from the various EEC Directives on pollution and is currently administered by the Environment Agency on behalf of the Department of the Environment. In order to obtain consent for a discharge under COPA II the effects of the discharge on water quality and the environment have to be fully investigated and approval obtained from MAFF and the Department of the Environment.

The main effect of this Act will be on projects involving reclamation where the discharge of tail waters from the reclamation area may contain high levels of turbidity which could be unacceptable in sites having a low ambient turbidity level.

Coast Protection Act 1949

Under this Act, which is currently administered by the Marine Directorate of the Department of Transport, permission is required for any dredging works proposed in navigable waters. The Act is even applied to authorities which have specific parliamentary powers to carry out dredging works. Applications for permission under this Act may also bring in the Harbour Works Environmental Regulations 1988 and 1989.

Other permissions

Apart from those permissions mentioned above, which arise from specific Acts, it is also necessary to obtain permissions from those who own or control the land, waters or seabed where the project is to take place. In many cases the works will be carried out within the boundaries of a local port, harbour or conservancy authority. It is also usually necessary to obtain permission from the Crown Estates Commissioners, who administer all crown land, which includes much of the foreshore around the UK and the seabed out to the limits of territorial waters.

Since the Crown Estates usually also hold the mineral rights out to the limits of the continental shelf they are frequently involved in matters relating to the dredging of sand and gravel for reclamation or other engineering uses.

2.4 Preliminary appraisal

In the previous sections of this chapter the background to any dredging project has been set out in some detail. This background is of fundamental importance in a preliminary appraisal of dredging works.

The preliminary appraisal is a coarse screening process. Its objective is to evaluate the technical feasibility of carrying out dredging works and to provide a framework in which pre-contract planning may be carried out in a cost-effective way. The preliminary appraisal would normally include:

- definition of the proposed project in outline;
- preliminary determination of conditions at the dredging site, the dredged material relocation site and the transport route for dredged material;

- determination of the probable effects of dredging;
- determination of potential sites for the relocation of dredged material and the probable effects of relocation at these sites;
- preliminary determination of the type of dredger to be used on the site and the method of transporting material to the relocation site;
- budget estimates and programme times for carrying out the work;
- identification of data requirements for the pre-contract planning stage.

2.4.1 DEFINITION OF THE PROPOSED PROJECT

At this stage the project would need to be defined in terms of the outline configuration of the dredging site, dredging depths and quantities, reclamation details (if any), relocation sites and the interaction of the dredging with other civil engineering construction works on the site.

Guidance on the design of a dredging site is given in Chapter 3 and matters relating to the use of dredged material, such as for reclamation, are given in Chapter 4, or for disposal are discussed in Chapter 5.

The interaction of the dredging with other civil engineering works on the site is of importance in that it often defines the programme time available for the dredging. Other constraints may also be applied in the form of confined working, sequencing of operations and special tolerances or bed formations required. A number of these aspects are discussed in Chapter 3.

2.4.2 SITE CONDITIONS

Knowledge of site conditions is required to make a preliminary assessment of the effect of the dredging, the relocation of dredged material and the probable types of equipment to be employed. It is necessary, as a minimum, to have some data on the following:

- The sea (or river) conditions at the dredging and relocation sites, and the transport route; tides, wave heights and water current velocities are particularly important.
- Dimensional details of the sites in terms of maximum and minimum water depths, thickness of material to be dredged, constraints imposed by adjacent services and facilities, and access restrictions.
- Any possible contamination of the materials to be dredged and any environmentally sensitive aspect of the proposed works or adjacent areas.
- Details of the soils and rocks to be dredged in terms of their geotechnical characteristics, their distribution and volumes.

2.4.3 PROBABLE EFFECTS OF DREDGING SCHEME

The proposed dredging is likely to affect the environment in some way, particularly with respect to the balance of nature outlined in Section 2.2. It is thus important in the preliminary stages to determine what these effects might be. These might include:

- in-filling of the dredged area, and thus a future maintenance dredging commitment;
- removal of littoral material resulting in coastal erosion on the downdrift side;
- draw-down of beach material due to dredging close to the shore;

- increase of wave energy at the shore due to dredging of offshore bar, or increased water depth, leading to alterations to coastal regime;
- refraction of waves passing over dredged area, leading to focussing or dissipation of wave energy;
- generation of turbidity, noise or excessive marine traffic from the dredging activity.

The quantitative evaluation of effects such as those mentioned above may be highly complex and computations of likely effects are often left to specialist organisations to carry out. What is important at this stage is that the engineer should be aware of the possible effects.

2.4.4 PROBABLE EFFECTS OF RELOCATION

The satisfactory relocation of dredged material is often a greater problem than its dredging. Relocation may take a number of forms, such as use of the material as a resource, placing the material onshore for reclamation or placing material in the near-shore zone to create new wetlands. These alternatives are discussed in some detail in Chapter 4. The removal of the material to offshore disposal grounds is discussed in Chapter 5.

In some forms of dredging the material is put into suspension at the dredging site and the removal of the material from the site is affected by the natural currents in the area. For guidance on whether this form of disposal is feasible and to evaluate its effects one must be aware of the nature of the local hydraulic and sedimentary regime, as described earlier in this chapter.

2.4.5 PRELIMINARY ASSESSMENT OF SUITABLE DREDGERS

In Chapter 7 the different types of dredging plant are described in detail and it will be seen that there is an enormous variety in types and sizes. For any proposed project there may be a number of dredgers which are potentially able to carry out the work. There will also be a number which are most unlikely to carry out the work and it is this latter group which it is important to identify at this stage.

The reason for eliminating certain dredger types from the project at this stage is that the detailed site investigations, which will be planned in the next stage of the work, will be tailored to some extent to match the possible dredgers. It is thus important to identify the suitable dredgers and thereby avoid carrying out unnecessary investigations, or worse, failing to make appropriate investigations.

Because of the great number of variables affecting the performance of dredging plant, it is not possible to be categorical about the choice of plant for all cases. However it is essential that some general guidance is given, otherwise the engineer is left with no idea how to start his appraisal.

In the British Standard Code of Practice for Maritime Structures (Part 5) Recommendations for Dredging and Land Reclamation (BSI, 1991) guidance on this subject has been given in a series of tables relating to various categories of work; maintenance dredging, capital dredging, land reclamation and beach replenishment, rock pretreatment and dredging. Tables 2.2 to 2.5 are based on the BSI tables.

Tables such as these should be used with care, particularly when the designation is 'marginal' or 'not usually suitable', since it is these categories which sometimes prove to be feasible in exceptional circumstances. For example, a dredger which is not usually suitable may be used on a site because the dredger happens to be near by and its use saves the cost of mobilising a more suitable piece of equipment.

Table 2.2 Guidance on the selection of plant for maintenance dredging (After BS 6349: Part 5: 1991)

Site conditions	Standard trailer	Light trailer	Cutter suction	Bucket wheel	Grab hopper	Grab pontoon	Bucket	Backhoe	Dipper	Barge unloader
Bed material										
Loose silt	1	1	1	1	2	2	2	2	N	1
Cohesive silt	1	2	1	1	1	1	1	2	N	1
Fine sand	1	1	1	1	2	2	2	2	N	1
Medium sand	1	1	1	1	2	2	2	2	N	1
Coarse sand	1	2	1	1	2	2	2	1	N	1
Sea conditions										
Impounded water	3	2	1	1	1	2	2	2	N	1
Sheltered water	1	1	1	1	1	1	1	1	N	2
Exposed water	1	2	3	3	3	N	3	3	N	N
Disposal to:										
Shore	2	2	1	1	N	2	2	2	N	1
Tide	1	1	1	1	N	N	N	N	N	N
Sea	1	1	N	N	1	1	1	1	N	N
Quantities										
<100 000 m^3	2	1	1	1	1	1	2	1	N	1
<250 000 m^3	1	2	1	1	1	2	1	2	N	1
<500 000 m^3	1	2	1	1	3	2	1	2	N	1
>500 000 m^3	1	2	1	1	3	3	1	3	N	1
Heavy traffic	1	1	3	3	2	2	3	1	N	2
Confined working	N	3	2	3	2	1	3	1	N	2

Key
1 = Suitable; 2 = Acceptable; 3 = Marginal; N = Not usually suitable
Note: Other factors not referred to may influence the choice of dredger. The table provides only a general guide.

Table 2.3 Guidance on the selection of plant for capital dredging (After BS 6349: Part 5: 1991)

Site conditions	Standard trailer	Light trailer	Cutter suction	Bucket wheel	Grab hopper	Grab pontoon	Bucket	Backhoe	Dipper	Barge unloader
Bed material										
Loose silt	1	1	1	1	2	2	2	2	3	1
Cohesive silt	1	1	1	1	1	1	1	2	3	1
Fine sand	1	1	1	1	2	2	1	2	3	1
Medium sand	1	1	1	1	2	2	1	2	2	1
Coarse sand	1	1	1	1	1	1	2	2	1	1
Gravel	1	2	1	1	1	1	1	1	1	1
Soft clay	1	2	3	1	1	2	1	2	2	1
Medium clay	2	3	3	2	2	2	1	1	2	2
Stiff clay	3	N	3	2	3	3	1	1	1	3
Boulders	N	N	3	3	3	2	2	1	1	N
Very weak rock	3	N	1	2	3	3	2	1	1	3
Weak rock	N	N	1	3	N	N	3	1	1	N
Moderately weak rock	N	N	2	N	N	N	N	2	2	N
Pretreated rock	2	N	3	N	3	2	2	1	1	N
Sea conditions										
Impounded water	N	3	1	1	1	1	1	1	1	1
Sheltered water	1	1	1	2	2	1	2	1	2	3
Exposed water	1	2	3	3	3	3	3	2	3	N
Disposal to:										
Shore	1	2	1	1	N	N	N	N	N	1
Tide	1	1	2	1	N	N	N	N	N	N
Sea	1	1	3	3	1	1	1	1	1	N
Quantities										
<100 000 m^3	2	1	1	1	1	1	2	1	1	1
<250 000 m^3	1	2	1	1	2	2	1	1	2	1
<500 000 m^3	1	3	1	1	3	2	1	2	3	1
>500 000 m^3	1	3	1	1	3	3	1	3	3	1
Heavy traffic	1	1	2	2	2	3	3	2	2	2
Confined working	3	3	3	3	1	1	3	1	2	2

Key 1 = Suitable; 2 = Acceptable; 3 = Marginal; N = Not usually suitable

Note: Other factors not referred to may influence the choice of dredger. The table provides only a preliminary engineering guide.

Table 2.4 Guidance on the selection of plant for land reclamation and beach replenishment (After BS 6349: Part 5: 1991)

Site conditions	Standard trailer	Light trailer	Cutter suction	Bucket wheel	Grab hopper	Grab pontoon	Bucket	Backhoe	Dipper	Barge unloader
Bed material										
Fine sand	1	1	1	1	N	2	1	1	3	1
Medium sand	1	1	1	1	N	2	1	1	3	1
Coarse sand	1	2	1	1	N	2	1	1	3	1
Gravel	1	3	1	1	N	2	1	1	3	1
Cobbles	2	N	2	2	N	2	2	1	2	3
Very weak rock	3	N	1	3	N	3	2	1	1	3
Weak rock	N	N	2	N	N	N	N	1	2	N
Sea conditions										
Impounded water	N	3	1	1	N	1	2	2	2	2
Sheltered water	1	1	1	1	N	1	1	1	1	1
Exposed water	1	3	3	3	N	N	3	3	3	N
Placed by:										
Direct dumping	3	2	N	N	N	1	1	1	1	N
Direct pumping	N	N	1	1	N	N	N	N	N	1
Transport and pump	1	2	N	N	N	2	2	2	3	N
Dump and pump	1	1	N	N	N	1	1	1	1	N
Quantities										
<100 000 m^3	2	1	1	1	N	1	2	1	2	2
<250 000 m^3	1	2	1	1	N	2	1	1	2	1
<500 000 m^3	1	2	1	1	N	2	1	1	3	1
>500 000 m^3	1	3	1	1	N	3	2	2	3	2
Heavy traffic	1	1	3	3	N	2	3	2	2	2
Confined working	N	3	3	3	N	1	3	2	2	2

Key 1 = Suitable; 2 = Acceptable; 3 = Marginal; N = Not usually suitable
Note: Other factors not referred to may influence the choice of dredger. The table provides only a preliminary engineering guide.

Table 2.5 Guidance on the selection of plant for rock pretreatment and dredging (After BS 6349: Part 5: 1991)

Site conditions	Diver driller	Floating pontoon	Jack-up pontoon	Standard trailer	Cutter suction	Bucket wheel	Grab pontoon	Bucket	Backhoe	Dipper
Bed material										
Very weak rock	N	N	N	N	1	2	2	1	1	1
Weak rock	2	N	N	N	1	3	3	2	1	1
Moderately weak rock	2	3	3	N	1	N	N	3	3	2
Pretreated rock	N	1	1	2	2	3	2	2	1	1
Sea conditions										
Impounded water	1	1	3	N	2	2	1	2	1	1
Sheltered water	2	1	1	1	1	1	1	1	1	1
Exposed water	3	3	1	1	2	3	3	2	3	2
Disposal to:										
Shore	N	N	N	N	1	1	N	N	N	N
Tide	N	N	N	N	N	N	N	N	N	N
Sea	N	N	N	1	2	2	1	1	1	1
Quantities										
<10 000 m^3	1	1	2	3	2	2	1	2	1	2
<50 000 m^3	3	1	1	2	1	1	2	1	2	1
<100 000 m^3	N	1	1	1	1	1	3	1	2	2
>300 000 m^3	N	1	1	1	1	1	3	1	3	2
Heavy traffic	N	2	1	1	2	2	2	3	2	2
Confined working	1	1	3	3	3	3	1	3	2	2

Key 1 = Suitable; 2 = Acceptable; 3 = Marginal; N = Not usually suitable/necessary/applicable

Note: Other factors not referred to may influence the choice of dredger. The table provides only a preliminary engineering guide.

2.4.6 BUDGET ESTIMATES AND PROGRAMMES

It may now be possible to prepare some budget cost estimates and programme times, either by using a simplified version of the methods outlined in Chapters 9, 10 and 14, or by obtaining guide prices from a specialist consultant or contractor. Extreme caution must be exercised if deriving budget costs by comparison with similar projects elsewhere. When doing this, a small difference in the site conditions, programme and contract can make a substantial difference to costs.

Generally, if there is insufficient information available to make a reasonable first attempt at costing, it is better not to develop any costs at this stage, but to identify where the unknowns lie.

2.4.7 INFORMATION REQUIRED

One of the most important reasons for carrying out the preliminary appraisal is to identify areas where information is lacking so that the pre-contract planning stage may be made as effective as possible. There is thus good reason to attempt to carry out the evaluations outlined in Sections 2.4.1 to 2.4.6, since any failure to complete an evaluation is an indication that information in some form is lacking.

2.5 Pre-contract planning

Pre-contract planning is defined here as that stage in the development of the project which, after the preliminary appraisal has been undertaken, takes the project through the exploratory and feasibility stages to the point at which contract documents can be drawn up. Depending on the size of the project, pre-contract planning may be executed in one or two stages and parts of the process may be iterative.

For the purposes of explanation the process is described here in two stages and is illustrated as such in Figure 2.3.

2.5.1 PRE-CONTRACT PLANNING: STAGE 1

The first stage is normally only required for major projects. It serves to define the scope of the detailed investigations which are required. This is achieved by carrying out preliminary investigations and using the results from these to establish the optimum type and necessary scope of the detailed investigations.

Details of the types of investigation often required in dredging and reclamation works are described in Chapter 6. Of these the types most likely to be used in Stage 1 would be as follows:

- general desk research to identify existing data relating to the site, its geology and geomorphological characteristics;
- current measurements in a few locations over the tidal cycle;
- a selection of boreholes, vibrocores and/or probes to obtain initial details of the subsoil and an assessment of the likelihood of discovering rock, or some other distinctly different material, during more detailed investigations;
- hydrographic surveys over the full extent of the possible dredging site and, if necessary, the relocation site and occasionally the route between.

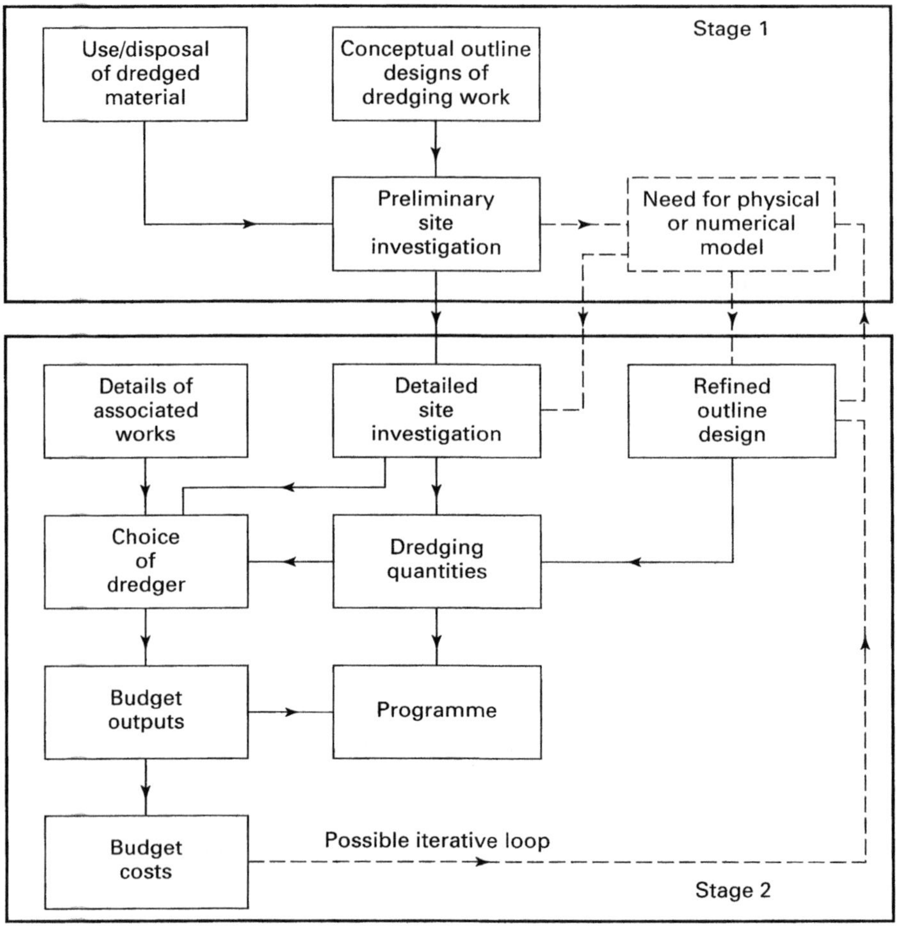

Figure 2.3 Pre-contract planning

With the results of these investigations it is then possible to identify those characteristics of the site which are likely to be of over-riding significance in the determination of optimum design, dredger choice and least cost solutions. This will then assist the engineer in deciding on the merits of introducing various more sophisticated and wide ranging studies in Stage 2. In particular it will be important to ascertain the value of carrying out investigations such as the following:

- environmental baseline surveys to establish flora and fauna populations, turbidity levels or the presence of contaminants on the site;
- geophysical surveys to define more accurately the interfaces between different substrata or to locate rock;
- magnetometer surveys to locate wrecks or other magnetic anomalies caused by debris or ordnance;
- numerical or physical modelling studies to identify, simulate and predict water flow, wave climate and sedimentary regimes over the site, before and after the works.

2.5.2 PRE-CONTRACT PLANNING: STAGE 2

This is the main investigative stage of the project during which all necessary site data will be collected and the scope of the works will be more clearly defined and costed. During this stage one would expect to carry out most of the following activities (for a project which is predominantly dredging):

- Undertake all relevant surveys and investigations giving information on the sea/river and soil/rock conditions, dimensional and environmental constraints, siltation and soil behaviour, relating to the dredging site; these are fully explained in Chapter 6.
- Refine the alternative outline designs of the dredging site, as originally proposed in the preliminary proposal or Stage 1, in the light of the investigations and other studies carried out (see Chapter 3).
- Establish dredging quantities for each alternative outline design and separate these quantities into appropriate categories for costing and programming purposes.
- Identify suitable dredger types to carry out the work and match their characteristics with the dredging quantities to arrive at some indicative outputs and programme times. Chapters 7, 8, 9 and 14 give guidance on these aspects.
- Develop a basic cost model for each dredger type to enable a cost estimate for the whole works to be formulated; the procedure for this is explained in Chapter 10.

Where a project contains a significant element of reclamation, the following items would need to be added to the above list:

- Additional surveys and investigations in the reclamation area to assess ground stability under load and the necessity, or otherwise, of removing poor material prior to filling.
- Surveys to assess dimensional and environmental constraints.
- Identification of sources and types of material for filling the reclamation area and for constructing any reclamation bunds.

It is possible that the initial attempts at costing will indicate that the proposed schemes are too expensive and it will then be necessary to go through the process again using reduced quantities or some other method to reduce the cost of the overall project.

During Stage 2 it may be prudent to make contact with a specialist consultant or contractor to discuss the proposed works. The advantages of doing this are that comments may be made on the proposed schemes and suggestions made which may result in savings. In particular:

- the availability of various types of plant, especially if they are limited in number;
- the choice of dredger for a particular site;
- global costs for various types of plant.

Normally dredging companies are receptive to this kind of approach being made since it assists them in keeping abreast of the projects in the pipeline. The natural bias of a company to favour its own type of plant may be offset by talking to a number of companies. However, this type of request should not be made frequently and should not be a substitute for reliable design and cost estimation, which should be entrusted to a specialist.

Another reason why one should be wary of asking one contracting company to spend an appreciable time considering a project and getting involved in detailed design is that it may affect the contractor's position during the tender stage. (See Section 2.7.1.)

2.6 Design and documentation

Once the pre-contract design stage has been completed there should be sufficient information available to formulate the contract documents and develop the detailed design. The procedure for carrying out this phase of the project is illustrated in Figure 2.4.

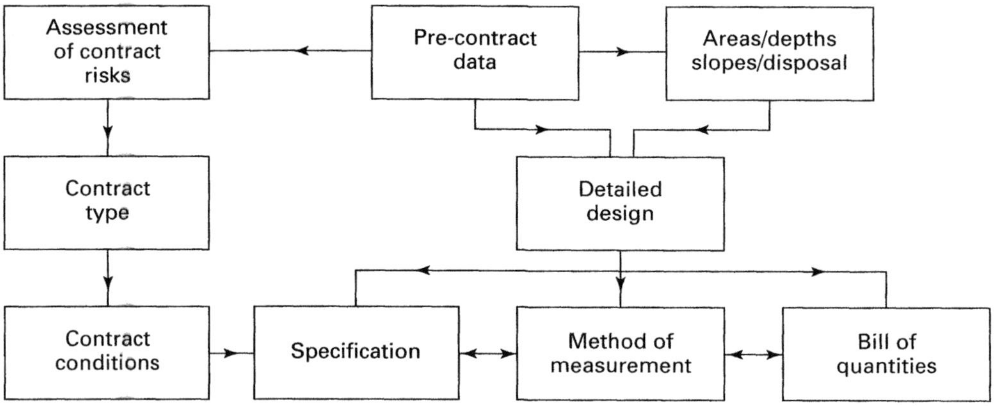

Figure 2.4 Preparation of design and contract documentation

A full description of the contract types applicable to dredging works is given in Chapter 11 together with details of the other sections of a contract document, such as the conditions of contract, specification and Bill of Quantities. (Those relating to rock pre-treatment are given in Chapter 14.) There are a few points concerning these which it is important to note:

- Contract type and apportionment of risk are inextricably linked and thus the choice of contract will probably have a large influence on the cost of the works to be carried out.
- Inappropriate methods of measurement encourage submission of unbalanced tenders and may result in very expensive work if unexpected ground conditions are found.
- Specifications, particularly tolerances, should be matched to the type and size of plant to be used on the site and the prevailing physical conditions. Too often the standards which are specified, and easily achieved, on dry land are transferred to the marine environment, resulting in unnecessary cost.
- Ambiguities in the contract wording should be avoided at all costs since alternative interpretations of a specific clause frequently lead to dispute. Specialist advice should be sought if there is any doubt on this point.

2.7 Tender procedures

Tender procedures normally run in parallel with pre-contract planning and production of contract documents as illustrated in Figure 2.5. Essentially, tender procedures relate to the contact between the client/engineer and the contractor. There are five distinct periods when this contact is significant as follows:

- Informal contact with the contractor during the pre-contract planning stages to discuss working methods, plant availability and budget costs.
- Pre-qualification, when the contractor has to demonstrate that he has the experience,

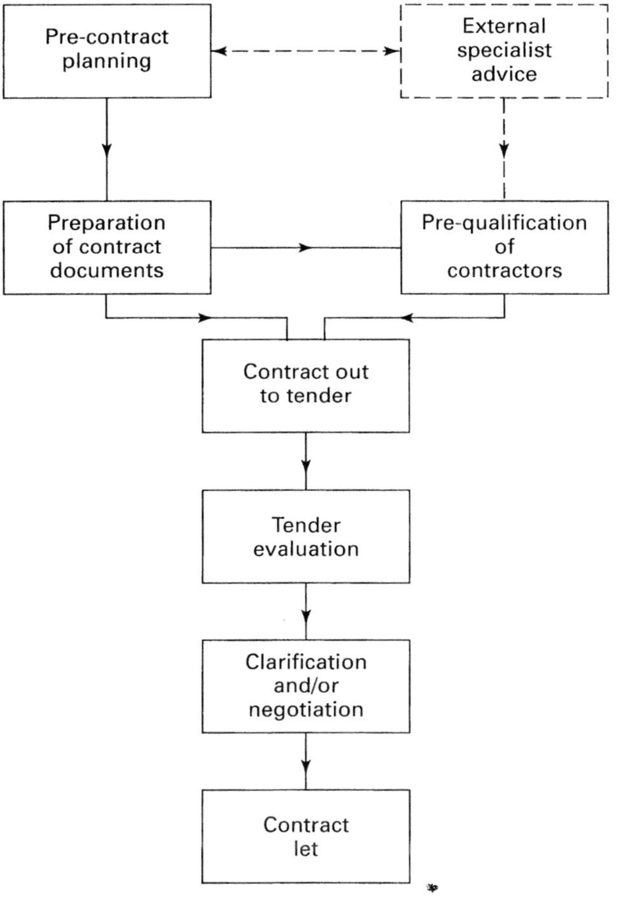

Figure 2.5 Tender procedures

resources and proven track record to undertake the proposed works (or, in the case of an open tender, when the client/engineer advertises for contractors).

- During the tender period when it may be necessary for amendments to the contract to be made, when contractor's queries require to be answered and when contractors may wish to discuss alternative working methods and design.
- Tender evaluation, when the client/engineer may wish to obtain clarifications from the contractors.
- Final negotiations, when the details of the contract to be signed are discussed and agreed.

2.7.1 CONTACT DURING PLANNING STAGES

As briefly mentioned in Section 2.5.2 there may be some advantage in talking to dredging contractors during the planning stages. The client/engineer may be able to learn of new or improved working methods. The contractor will be able to advise the client/engineer of his plant availability and capability, and may be able to give an indication of guide prices for the work.

However, it is suggested that the following principles should be adhered to when making this type of contact:

- Contact should be made with more than one contractor and each contractor should be told that this is the case.
- The scope of the project should be described and the chances of it coming to fruition should be indicated. This enables the contractor to decide for himself how much time he wishes to devote to the discussions.
- Do not contact contractors who are unlikely to be pre-qualified for the work.
- Do not press contractors for budget prices when insufficient information is available for estimating them.
- Do not offer to pay for advice, as this prejudices the contractor's subsequent position as a tenderer.

If these principles are not observed there is every chance that contractors will soon become less enthusiastic about providing assistance during this stage and the opportunity for information interchange may be lost.

2.7.2 PRE-QUALIFICATION

Rules for pre-qualification will vary depending on the size and location of the works and the type of organisation wishing to have the work carried out. Some local authorities in the UK have standing rules which control the minimum number of tenderers for each type and size of contract. As European integration proceeds so the EC regulations will have a greater influence on the procedures to be adopted.

Since national and international regulations are prone to frequent alteration it would be imprudent to try to define them here. However, it should be pointed out that dredging is already an international business and thus it is common for dredging plant to be mobilised from one country to another, particularly when the project demands the use of an unusual size or type of equipment. The following notes relating to pre-qualification may be helpful:

- Check on the international, national and client's own regulations relating to advertising for tenderers, prior to initiating pre-qualification procedures.
- Do not make the pre-qualification documents too complicated. The size of the organisation is less important generally than the fact that it has suitable equipment and is capable of using it effectively.
- If the project demands a particular, fairly scarce, item of plant, ensure that those organisations with suitable plant are alerted to the call for pre-qualification.
- Do not pre-qualify too early. If the scope of the project changes during final design stages, early pre-qualification may exclude contractors who are actually well suited to the final design.
- If the dredging works are a sub-contract to the main contract, it may be possible for a main contractor to pre-qualify with a sub-contractor whose performance is doubtful. If necessary ensure that the importance of dredging to the project is emphasised in the main contract pre-qualification documents and indicate that by proposing an inappropriate sub-contractor pre-qualification of the main contractor may fail.

2.7.3 CONTACT DURING THE TENDER PERIOD

There are a number of reasons for the contractor and engineer to be in contact during the tender period, apart from the usual arrangements associated with the contractor inspecting the site. The client may wish to advise the contractor of some additional information about the site or the design of the works. The contractor may wish to receive answers to queries and may also wish to put forward alternative proposals.

In the case where the contractor finds some point in the documents unclear, ambiguous or lacking in some other fashion, the normal procedure is for the engineer to respond to the question and then send a copy of both question and answer to all tenderers. In this way one can ensure that all contractors are submitting their tenders on a similar basis and it is thus easier to compare them during the evaluation stage.

If, on the other hand, a contractor wishes to ask questions relating to a tender for an alternative method or design the procedure is not so clear. The contractor may wish to keep details of his alternative proposal confidential or, in some cases it may be to his advantage for the competing contractors to be told of his questions and the answers given. One solution to this dilemma is to ask the contractor whether he wishes the exchanges to be copied to the other bidders or not.

Alternative tenders are usually to be encouraged since they may result in considerable savings to the client, but it is normal practice to insist that all contractors must submit a tender in accordance with the original invitation documents.

The length of the tender period is important and should vary in accordance with the size and complexity of the project, the quality of information given to the contractor and the exigencies of the project. If contractors are required to obtain much local information or carry out their own site investigations, the tender period should be extended to reflect this. Short tender periods which coincide with national holidays (such as Christmas, Easter, fasting months or the height of the summer vacations) are not recommended since this is unlikely to improve the contractor's mood or concentration! It may well increase the contingency sum in his tender to cover him for the risk in those items which he has not had adequate time to consider.

2.7.4 TENDER EVALUATION

The process of tender evaluation, as applied to offshore works, may be defined as a requirement to:

- clarify any errors or omissions in the tenderers' submissions;
- ascertain that the tender complies with the specification and that the works will be completed within an acceptable programme;
- examine any qualifying conditions made by the contractor;
- assess how sensitive the tenderer's working method is to ground or weather conditions;
- assess the risk of cost escalation due to the method and programme proposed and the consequential apportionment of risk between client and contractor;
- examine, if acceptable, any proposals the contractor has made for alternative designs;
- examine the sensitivity of all tenders to variations in the quantities to be dredged and variations in the proportions of different materials to be dredged;
- examine the financial implications of the contractor's pricing of the Bill of Quantities.

Of particular importance during the evaluation for dredging works is the working method proposed. Contractors use large items of site-specific plant for dredging and there is always the risk that unforeseen working conditions may render a major item of plant unsuitable for a large proportion of the work.

With respect to the Bill of Quantities there are a number of points which should be checked. These are listed below.

Range of tenders

The range of tenders should reflect upon the accuracy and scope of the specification and the assessment of risk. A wide range of prices on a tight specification and a low-risk project should be investigated further.

Level of tenders

The overall level of tenders should compare with the client's estimate. If there is a wide divergence the reason should be ascertained.

Unusually low tender

A tender well below the client's estimate and the other tenders may indicate an error or omission in the build up of the total price, or that he has failed to read the specification, or has not appreciated the risks which must be taken into account in his tender price. The submission of an unrealistically low tender may result in a claim later.

Cover tenders

These may be difficult to detect but are the result of collusion amongst the tenderers and are fabricated to make competition look real when in fact it is only token.

Unbalanced elements

These should be detectable if a good estimate has been made. They consist of the following.

- Front-loading: achieved by putting a high price on the items which will be started first. This is done to increase the cash flow and will result in serious consequences in the event of the contractor defaulting.
- Cross-subsidisation: where lower rates are quoted for more costly items and subsidised by slightly higher rates for less costly items. This can be used to encourage certain types of work and discourage others.
- Unbalancing: achieved by quoting higher rates for items the quantity for which the contractor knows, or suspects, have been underestimated, and vice versa. This allows the contractor to make substantial profits at very little risk to himself and great expense to the client.

2.7.5 PRE-AWARD DISCUSSIONS

In many cases the evaluation procedures suggested above will indicate that one tender is to be preferred from all others. Sometimes this will not be the case and more than one tender may be regarded as potentially acceptable. In both such instances it is probable that some pre-award discussions will be beneficial to establish the following:

- that any tender qualifications still remaining are acceptable to the client;
- that any ambiguities in the documents have been resolved;
- whether there are any minor design changes which could be made to tailor the design to the contractor's plant and working method, thereby achieving some cost savings (such items as relaxation of working tolerances, modification to fill material specification and the like often having an appreciable effect on the overall cost of the work);
- that the contractor has fully understood the contract requirements.

There are a number of difficulties associated with these types of discussions.

Firstly, if there is more than one contractor and the negotiations involve alternative designs, there may be reluctance on the part of the contractor to divulge details of his proposals for fear that they will be passed on to the other tenderers. There is little that can be done about this, apart from keeping the discussions confidential and assuring the contractors of this confidentiality. However there must be doubts whether this type of 'Chinese wall' attitude can be truly effective, even with the most ethical of negotiators.

Secondly, discussion with the contractor concerning his working methods, and acceptance of them, tends to weaken the client's case when disputes arise. For this reason it is imperative that discussions held in negotiation are recorded and agreed by both parties prior to contract signature. This allows the client or his representative to raise doubts about the proposed method or point out difficulties which might arise and have these points minuted, thereby providing evidence of the views of the various parties prior to the commencement of work, which might prove to be helpful in the event of a subsequent dispute.

2.8 Project control

After contract signature there are four distinct sets of operations to be undertaken to ensure that the works on site proceed as smoothly as possible. These are as follows:

- Ensuring that the public are aware of the work to be carried out and that a public relations system is in place to keep them informed throughout the duration of the project.
- Arranging for all matters concerning third parties to be covered. This would include checking that licences, insurances and consents had been obtained, warning local organisations that works are about to proceed (including the issuing of Notices to Mariners), setting up and carrying out baseline surveys where necessary and so on.
- Agreeing the contractor's operational work plan, including methods for quality control, progress monitoring, measurement for payment and environmental monitoring.
- Controlling the site works, evaluating progress reports, identifying problem areas, issuing instructions to the contractor and dealing with contractor's enquiries.

These activities are shown diagrammatically in Figure 2.6 and are described in some detail in Chapter 12.

Where the works are to be undertaken in an environmentally sensitive location, such as in close proximity to aquatic breeding grounds, it will be necessary to ensure that mitigation measures incorporated in the contract documents are enforced. Guidance on these matters is given in Chapters 13 and 14.

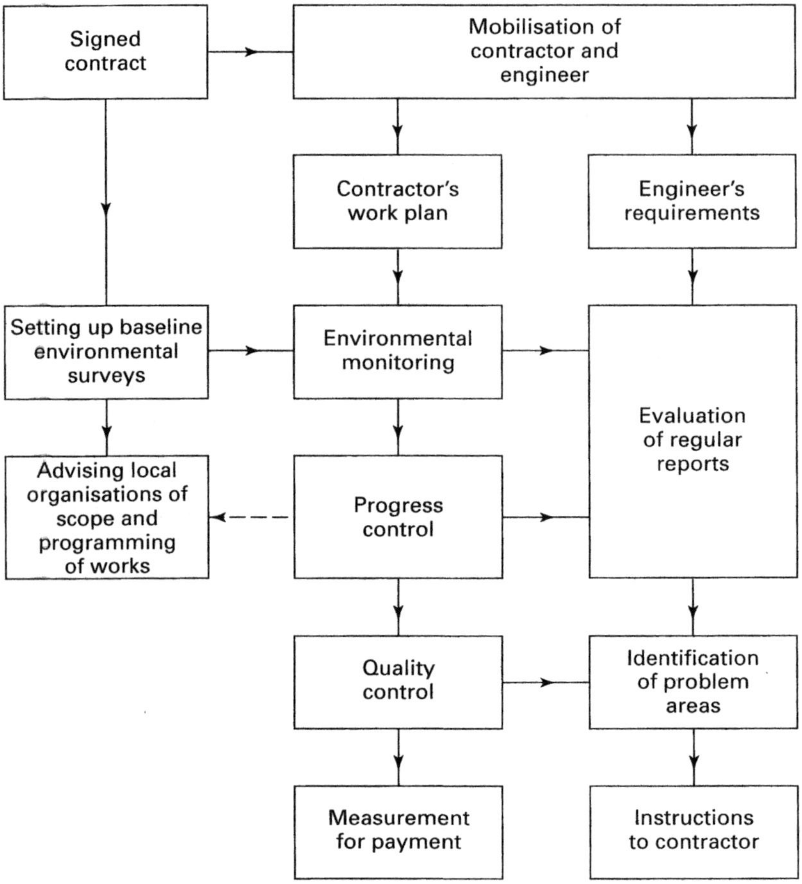

Figure 2.6 Project control

References

BSI 1991. *Maritime structures, code of practice for dredging and land reclamation*, BS 6349: Part 5, British Standards Institution, London.

Lammers, J.G. & Neervoort, F.C.H. 1988. Legal aspects of the environmental consequences of dredging – A survey of law-making institutions, *Proceedings of the CEDA Dredging Days*, Hamburg.

Murden, W.R. 1989. The meeting of the twain, *Terra et Aqua*, No. 40, August, International Association of Dredging Companies (IADC), The Netherlands.

3
Design of Dredging Works

3.1 Introduction

Dredging is a construction process and a dredged area may be considered to be a structure. In this respect it may be a temporary structure, formed during the works and subsequently filled in, or a permanent structure such as a navigation channel or basin. As such, the design of the structure and the method of construction, or the ability to construct it, should be examined as carefully as that for a wharf, jetty or other similar maritime structure.

The fact that the dredged area is hidden and not normally liable to collapse often results in too little attention being paid to the economy of its design. The cost of the dredging work in a port development or sea outfall can often amount to 50 per cent or more of the total cost of the work. However, it would be unusual if any more than five or six per cent of the total design time had been spent investigating, designing, costing and specifying this aspect of the work.

This chapter is concerned with the design of the dredged area and its relationship to the method of dredging, and the use to which the dredged area is to be put. The chapter is divided into a number of sections broadly covering the following subjects:

- dredging for navigational improvement;
- dredging for construction purposes;
- dredging to maintain or improve hydraulic efficiency or for environmental reasons;
- dredging to obtain fill materials;
- design of details, such as slopes and tolerances.

The design of reclamation areas and beach nourishment are covered in Chapter 4.

3.2 Dredging for navigation improvement

Dredging is often carried out to enable a vessel, or vessels to navigate a particular stretch of water. In such cases there is normally a direct relationship between the vessel's characteristics and the design of the dredged area. For smaller vessels the compatibility between dredged area and vessel will depend to a large extent on vessel type (e.g. powered or sailing) and the nature of the facilities being provided. Some useful information on this subject is given by PIANC (1991) and British Standard 6349: Part 1.

For larger vessels it is often necessary to investigate manoeuvrability and hydrodynamic effects connected with such factors as vessel speed, and the proximity of the hull to the bottom and sides of the channel. The design of the dredged area should be preceded by a study of transport economics, where the benefits of dredging will be compared with the dredging costs.

The various components of dredged area are discussed below.

3.2.1 DEPTH

The minimum depth of water that a vessel requires for navigation is made up of a number of components. These are shown in Figure 3.1, which is based on those given by the International Oil Tanker Commission (PIANC, 1973). Although some of the factors, such as squat, may only be appreciable for large vessels, they should all be checked for relevance for each different portion of the waterway which makes up the navigable area of the port.

Whether the navigable waters will be dredged to accommodate all vessels, for all stages of the tide, in all weathers, will be determined by the frequency of entry of the various vessel types and sizes. Often, economy will dictate that this is too generous an allowance and depths will be reduced accordingly. In ports that receive large vessels the entrance channels are often designed to allow laden vessels to enter, or leave, only for a short period during high water. However, for traffic which works to a fixed timetable, such as ferry services, the navigable channels and areas must be designed to provide safe access at all times.

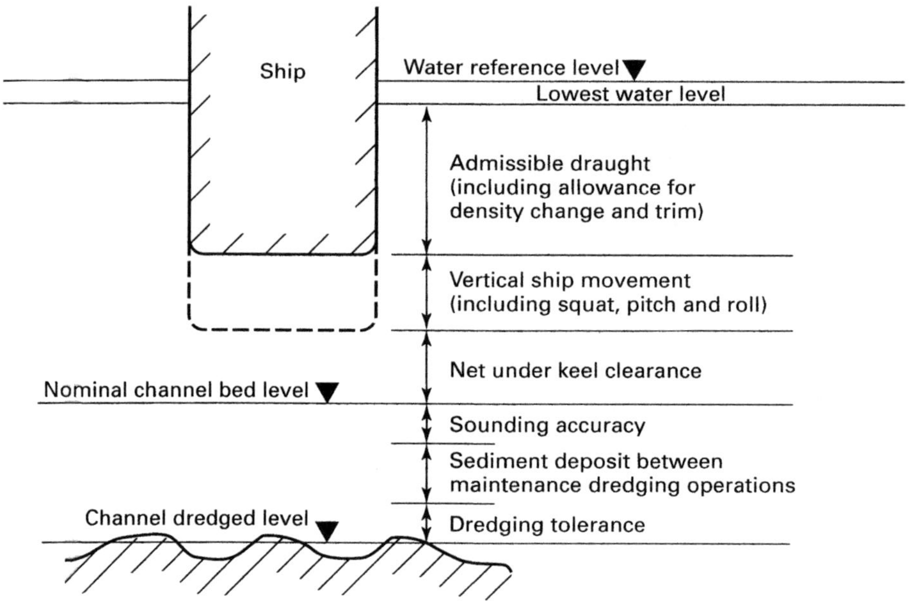

Figure 3.1 Components of depth (after PIANC 1973)

Water reference level

This will usually be chart datum in a port or estuary and low river level in a river. Generally the latter is fixed at some arbitrary low level which is within the lowest five per cent of all low levels recorded. They are not necessarily the lowest level to which the water will fall. Some ports retain a local datum, often related to a dock sill level.

Lowest water level

The lowest water level is the lowest astronomical tide, or lowest recorded river level, less, where appropriate, an amount to account for meteorological conditions, that is high pressure, wind and so on.

Admissible draught

If the vessel to be accommodated has not been specified, but is hypothetical, it will be necessary to estimate its maximum draught when using the port. This could be the maximum summer draught (in the northern hemisphere), the tropical draught, the applicable load line zone for the port in question or any other condition dictated by the vessel's operation.

For small vessels the draught varies according to the length and type of craft (see Figures 3.2 and 3.3). Larger vessels are usually classified according to their capacity or deadweight tonnage (DWT). In Figure 3.4 the average dimensions of container ships are shown; the bigger dimensions for a ship of any capacity are associated with higher speeds (the average speed is around 26 knots) and the smaller with lower speeds. The draughts of tankers, bulk and general cargo vessels are shown in Figure 3.5. The draughts shown in the figures are average draughts and considerable variations occur due to the variety of ship characteristics. They should thus be used as a first approximation only.

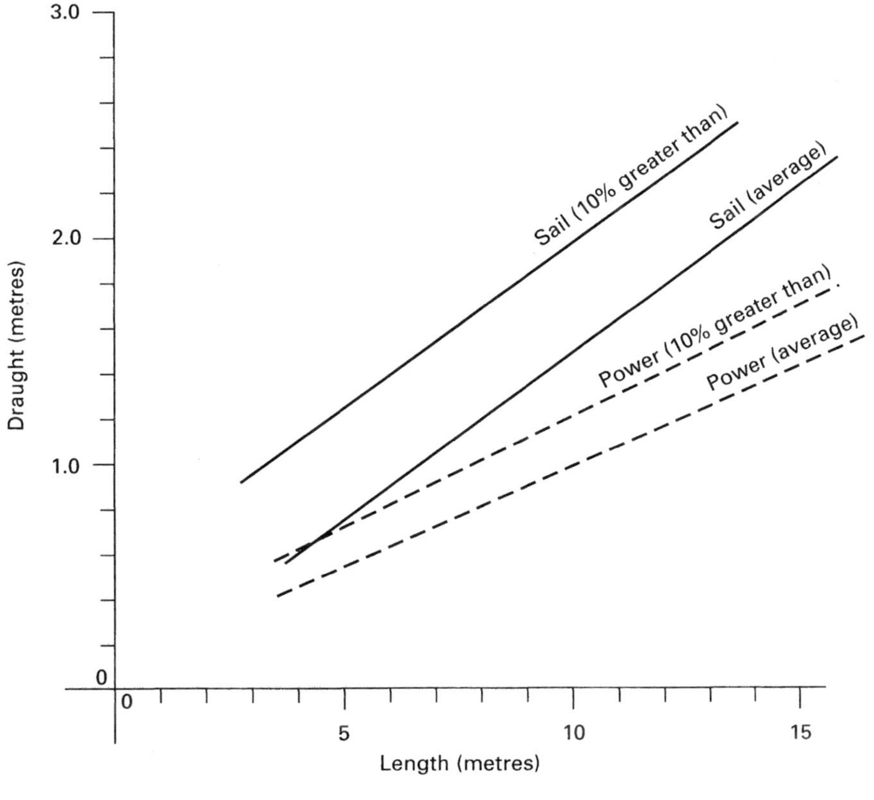

Figure 3.2 Variation in draught with length for recreational sailing and power boats (after de Ridder & de Visser 1986)

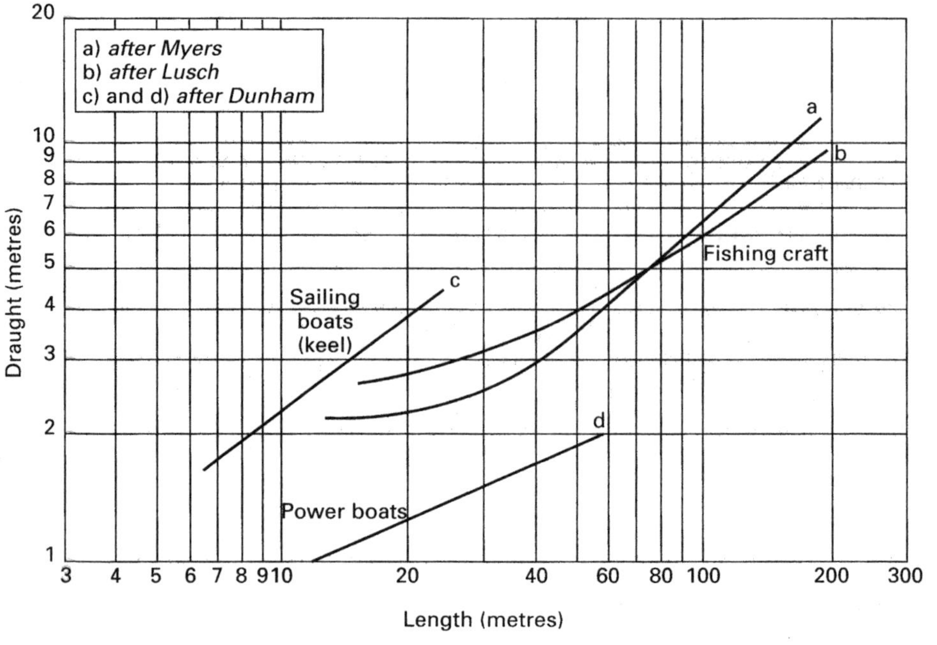

Figure 3.3 Draughts of small vessels (after Myers 1969, Lusch 1970 and Dunham & Finn 1974)

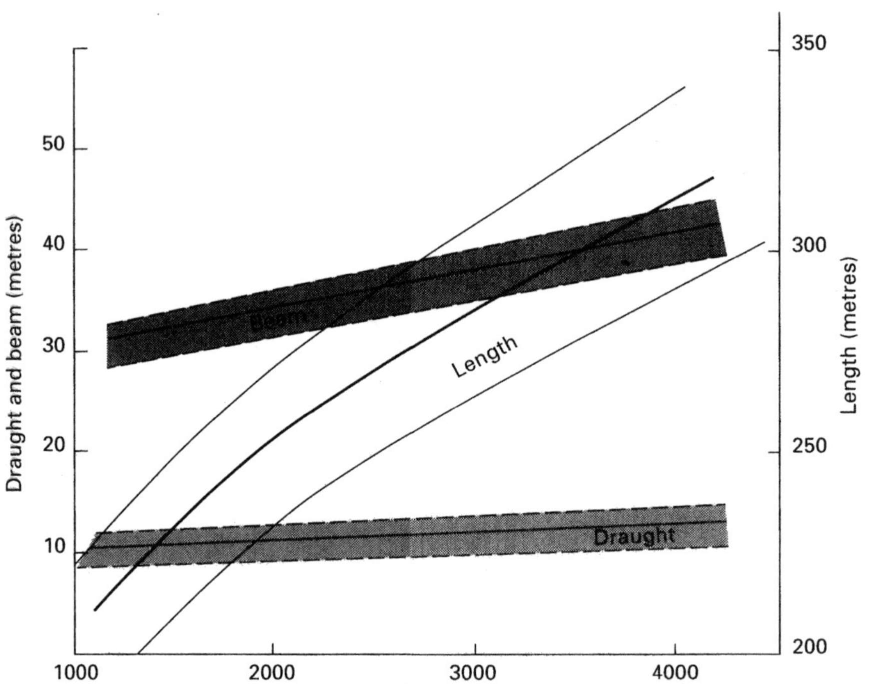

Figure 3.4 Containerships: average dimensions (after NPC 1970)

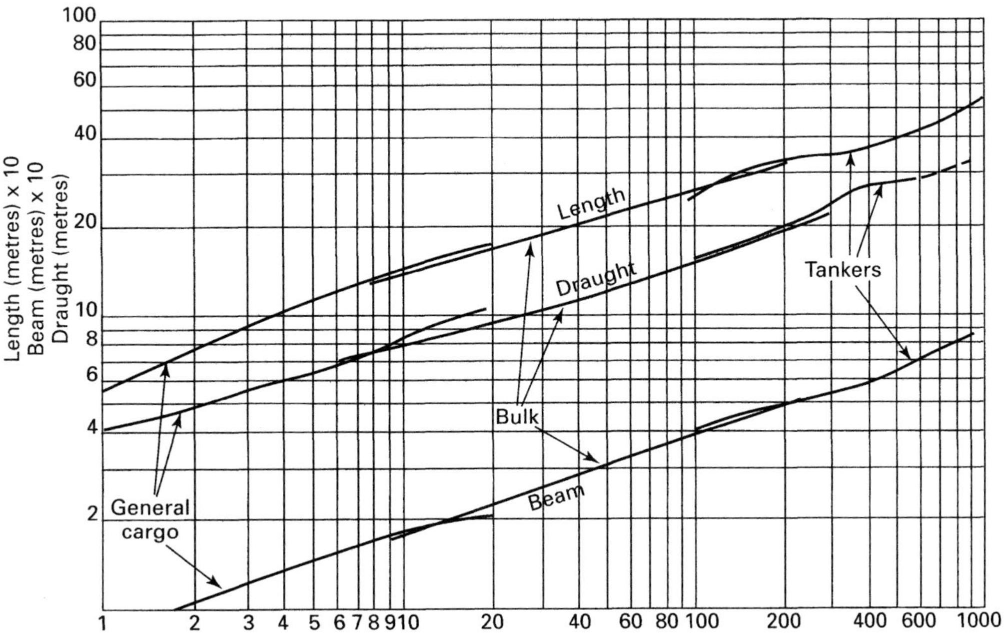

Figure 3.5 Average dimensions of tankers, bulk and general cargo vessels

These draughts apply to salt water conditions and an allowance must be made if the vessel is to enter fresh water in an estuary or river. The increase in draught due to fresh water can be as much as three per cent. Draught will also alter according to the trim of the ship. Vessels are trimmed for a number of reasons. The ship's master will often trim his vessel in order that its draught will be the minimum under the effects of squat. Oil tankers are trimmed at the berth during unloading and it is possible for the draught of the vessel to increase by some two to three per cent as the first hold is discharged.

Vertical ship movement

The position of the keel of a vessel with respect to the sea bed will vary according to the sea state, the vessel's speed and orientation, and the relative dimensions of draught, depth, channel width and so on. On entering shallow water vessels produce waves which are bigger than those produced in open, deep water. Consequently a surface depression is produced due to an average decrease in the water surface level along the ship's profile. This depression, and the drop in pressure around the vessel due to return flows of water, causes the vessel to sink lower into the water and to change its trim. This phenomenon is known as 'squat'.

Squat may be estimated if the characteristics of the channel and the vessels entering it are known, or can be assumed. Barrass (1979) gives a simple formula for this estimation derived from an analysis of about 300 squat observations. The special effects relating to the squat of vessels navigating in muddy waters at low speeds has been studied by Vantorre and Coen (1988).

Vessels will also move vertically due to the influence of waves and swell. The movements involved are heaving, rolling and pitching. The amount of vertical movement will depend on the vessel size, speed and direction of sailing with respect to swell, the ratio of water depth to draught and the amplitude and period of the swell. In severe cases, the allowance for this type of movement can be considerable. For instance at Richard's Bay, South Africa, an overdepth of 35 per cent of the vessel's draught is considered necessary.

Net underkeel clearance

Net underkeel clearance is the safety margin between the lowest calculated position of the vessel's hull and the highest probable elevation of channel bed. When the channel bed is composed of soft material and a slight touch between vessel and bed is not likely to prove damaging, the net underkeel clearance will normally be taken between 0.3 and 0.5 metre. For rocky seabeds this allowance is usually doubled.

Sounding accuracy

In Chapter 6 it is noted that the accuracy of sounding depends on a number of factors such as tidal range, instrument accuracy and movement of the survey vessel. No matter how recently a bathymetric survey has been carried out it is advisable to assume that the actual bed level is at least 0.15 metre above the shallowest depth shown by the survey.

Sediment deposit, dredging tolerances and navigable depth

These are controlled by the site characteristics and the method of dredging, and are discussed in Chapters 2, 3 (Section 3.6.2) and 14.

3.2.2 WIDTH

The minimum required width of a straight channel depends on the size and manoeuvrability of the vessel navigating the channel, the type of channel bank, the effect of other vessels in the channel and the effects of wind and currents. Many channels are in existence which have widths considerably less than those given below and which do not appear to lead to serious problems. The dimensions suggested below should, thus, be considered as ideal rather than mandatory.

The width required may be split into three basic zones: manoeuvring, ship clearance and bank clearance lanes, see Figure 3.6 (Kray, 1970).

The manoeuvring lane

Each vessel in the channel requires a manoeuvring lane which makes allowance for the oscillating track produced by the vessel. The oscillations occur due to the vessel's inherent directional instability and the manner in which a course is steered. Generally the width of the manoeuvring lane will be between 1.6 and 2.0 times the beam of the vessel depending on its controllability.

When there are cross-currents or winds, particularly at the entrances of approach channels, allowance must be made for the yaw of the vessel. The angle of yaw can commonly be 5° and sometimes as much as 10°. For larger vessels a yaw of 5° can add an extra width to the manoeuvring lane equivalent to half the beam. Some vessels, such as liquefied natural gas (LNG) carriers and tankers in ballast, have a considerable windage and the manoeuvring lane for these can sometimes exceed twice the vessel's beam. Bends also increase the manoeuvring lane requirement.

It is common to find that ship simulation techniques are used in the detailed design of shipping channels (see Chapter 6).

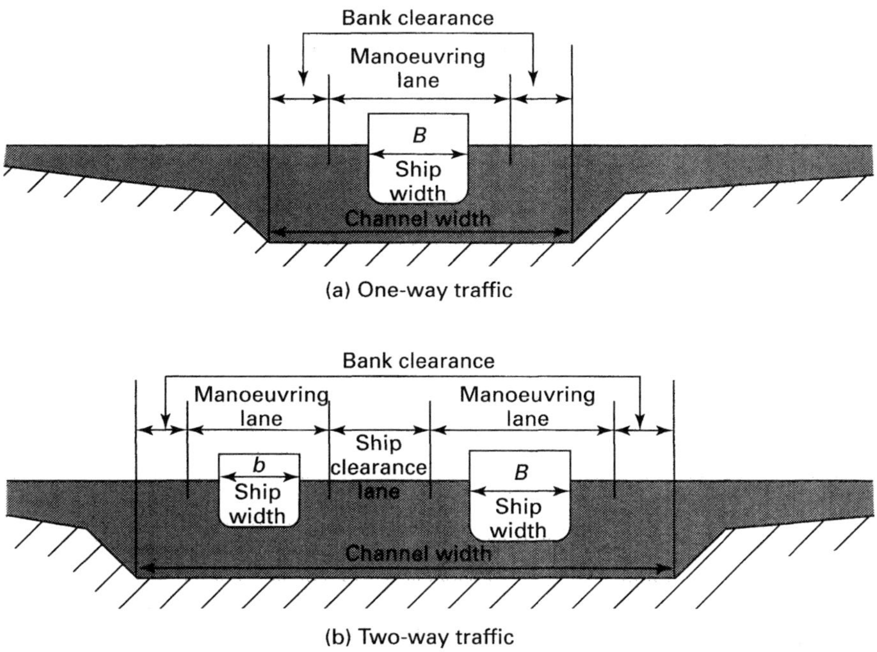

Figure 3.6 Components of channel width (after Kray 1970)

The ship clearance lane

In a multi-lane channel it is necessary to separate manoeuvring lanes by a ship clearance lane. This is to avoid excessive interaction between vessels travelling past one another, either in the same or opposite directions. The degree of interaction depends on a number of variables, including channel shape, channel depth, vessel speed, currents, swell and wind. It is generally considered that an adequate ship clearance lane is 30 m or the beam of the larger vessel, whichever is greater. An upper limit of half the combined beam widths of the vessels has been suggested. Both of these are likely to be excessive at low speeds.

Bank clearance

When a vessel sails too close to the bank of a channel there is a bank suction effect due to the asymmetrical flow of water round the vessel. To counteract this effect, helm must be applied to the vessel and additional width is required to allow this, usually of between 1.5 and 2.0 times the beam of the vessel, depending on:

- the type of channel – an open, trapezoidal section produces less suction than a steep-sided canal section;
- underkeel clearance – bank suction increases as underkeel clearance decreases;
- the vessel's speed – an increase in speed may slightly increase rudder requirement.

The sum of the widths of the above lanes and clearances gives the total channel width requirement. For large vessels in a single lane channel, PIANC (1973) recommend that the total width should vary between five to seven times the vessel beam depending on the sea and wind conditions. A similar method of analysis would give the width requirement for a double lane channel as about nine times the width of the largest vessel. However, as mentioned before, many ports operate safely with much narrower channels.

Additional guidance on the design of channel widths is given in PIANC (1995).

3.2.3 BENDS

The minimum width of a channel will be considerably larger in a bend than a straight channel due to the additional width of manoeuvring lane required. The ease with which a vessel can negotiate a bend depends on a number of factors, including bend radius; length and beam of vessel; velocity of vessel; depth of water and deflection angle. Figure 3.7 defines bend radius and deflection angle.

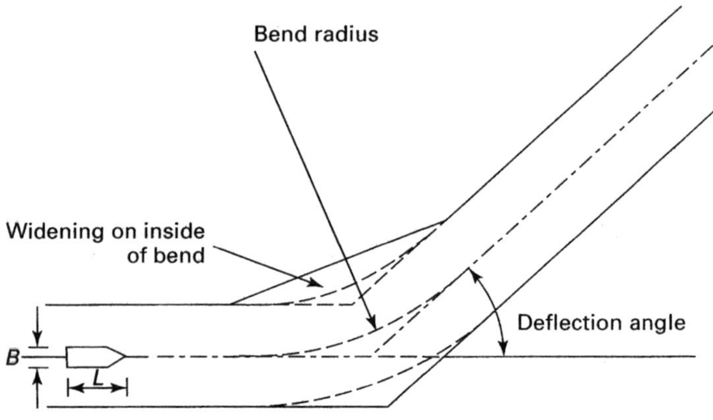

Figure 3.7 Channel bend: definition sketch

Within reason the velocity of a vessel will be controlled to suit navigation of a bend and the other factors must be adjusted until they are mutually compatible. There is a minimum bend radius which must be satisfied and which is usually related to vessel length. It has been suggested (Carvalho, 1926) that the following relationship should be applied:

Angle of deflection	Minimum bend radius
25° or less	3*L*
25° to 35°	5*L*
35° or more	10*L*

where *L* is the length of the largest vessel passing through the channel.

More recent proposals for minimum bend radius have tended to be in the 8*L* to 10*L* range, without being related to angle of deflection.

A number of formulae have been proposed for estimating the required manoeuvring lane width in a bend which do not contain an adjustment for angle of deflection. Model tests (Wicker, 1965) have suggested the following average relationships for a 3810 m radius bend.

Controllability of vessel	Width of manoeuvring lane	
	Deflection angle 26°	Deflection angle 40°
Very good	3.25*B*	3.85*B*
Good	3.70*B*	4.40*B*
Poor	4.15*B*	4.90*B*

where *B* is the beam of the vessel.

There are various configurations which produce the necessary channel widening. Although channels which have gradually widening cross-sections can cause problems due to the variation of bank suction experienced throughout the bend, it is found that one of the most important factors is the ease with which the additional width at the bend can be marked and used. A widening of the inside of the bend (see Figure 3.7) is now generally accepted to be the most suitable manner by which to improve the navigability of the bend.

3.2.4 BASINS AND MANOEUVRING AREAS

The shape and size of basins and manoeuvring areas depend to a large extent on the type of facility being constructed, the sea and weather conditions, the size and number of vessels, and the use of tugs. The use of numerical models in computer simulations of ship entry and berthing is now fairly common in the development of designs for new facilities (see Chapter 6). However, the following points are of use in forming preliminary estimates of the areas required.

Stopping distances

When a vessel enters a port from the open sea, or from a curved approach channel, it will have a certain minimum velocity which is sufficient for it to maintain steerage. Before manoeuvring to the berth can take place, the speed of the vessel must be reduced to virtually zero. The distance required to perform this manoeuvre depends on the speed and size of the vessel.

The speed of a vessel entering a port varies considerably, depending on the conditions at the port entrance. Physical or numerical model testing can be carried out to determine the entry speed and stopping distance required. For moderate entry conditions, an average stopping distance of five times the length of the longest vessel is the minimum which should be provided (PIANC, 1973). This distance includes the diameter of the turning basin, if it forms part of the stopping area.

The action of putting the engines astern to reduce speed tends to cause lateral movement, or sheer, in the vessel, for which an allowance must be made. The shorter the stopping distance provided, the greater must be the width of the channel.

Turning basins

Most vessels are turned either just before berthing or when leaving the berth. The minimum area required to perform this manoeuvre will depend on whether the vessel has tug assistance. The following minimum diameters of turning basin are generally accepted (Hay, 1968):

	Diameter of turning basin
With tug assistance	$2L$
Without tug assistance	$4L$

where L is the length of the largest vessel.

In good conditions these diameters might be reduced to $1.6L$ and $3L$ respectively.

Berths

Where a vessel has to manoeuvre to a berth whose dredged basin is adjacent to shallower water, allowance must be made for overshoot whilst berthing. In good conditions the following should be sufficient.

	Length of berth pocket
Tug assisted berthing	1.25L
Berthing without tugs	1.50L

where L is the length of the largest vessel to use the berth.

The width of a dredged tidal berth, which is deeper than the entrance channel, should be at least 1.25B, where B is the beam of the largest vessel to use the berth. For non-tidal berths the width will vary according to the angle of ship approach and departure, use of tugs, and number of berths along the quay.

Underkeel clearances at berth will depend on the soil and sea conditions. In sheltered areas the clearances will normally be 0.3 to 0.5 metre for soft soils and possibly double that when the bottom is hard. When berths are constructed in exposed locations, such as offshore berths for bulk carriers, an additional clearance is necessary to take account of vessel movement at berth.

In cases where an entrance channel has been designed to accommodate the largest vessels at high water only, the berth will need to be dredged to an increased depth to allow the vessel to float fully laden at the berth at low water. If this method of accommodating large vessels is planned, the siltation in the proposed berth should be studied carefully to ensure that adequate allowances are made for depth reduction between maintenance dredging campaigns. Siltation of deep pockets may be rapid.

3.3 Dredging for pipelines and cables

It is frequently necessary to lay pipelines and cables underwater in depths of water varying from zero to hundreds of metres. In deep water they may be laid on the sea bed with little or no preparation and with no protective covering of soil or stone. In shallow and/or in heavily trafficked waters it is usual to bury the pipe or cable to prevent it being damaged or moved by the currents, or by anchoring, fishing or sinking vessels.

A common method of burying pipelines or cables is to excavate an underwater trench, lay the item in the trench and backfill the trench, either by replacing the dredged soil or rock, or by importing other soil, rock or armour units. Dredging equipment is used to excavate underwater trenches as well as to backfill them. However, since trenches are by necessity narrow and often need to be constructed in shallow waters, they pose particular problems for dredging equipment. For this reason it is necessary to design trenches with care, bearing in mind the type of dredging equipment which is likely to be carrying out the excavation work.

As a trench approaches the shore, or a river bank, the water depth becomes very shallow and finally it becomes easier to excavate the trench by dry-land methods. Prior to this stage the water depths are severely limiting to most items of dredging equipment and the method of construction of the landfall is likely to dominate the design of the trench.

Backfilling of trenches is by no means a straightforward operation and poor backfill design can add considerable costs to a pipeline or cable laying operation. It is prudent, therefore, to review the methods of backfilling and to ensure that those proposed are practicable. In addition it is also worthwhile to examine the whole dredging, pipelaying and backfilling sequence to ensure that the total cost of the combined operations is a minimum. This aspect is addressed in the final part of this section.

3.3.1 TRENCH SHAPE AND DIMENSIONS

A trench may be defined by reference to its depth, bottom width and side slopes (see Figure 3.8).

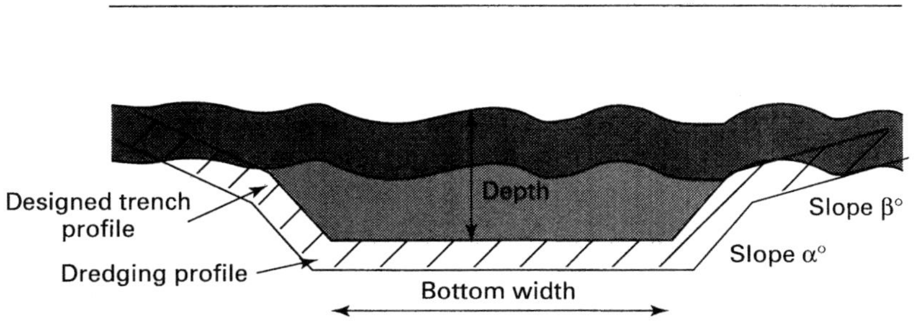

Figure 3.8 Dredged trench: definition sketch

Depth

The depth of the trench is normally defined by its function and by the site conditions. For instance, where the bed material consists of potentially mobile soft silt or sand, overlying firm clay, it may be necessary to excavate to a sufficient depth in the clay for the pipe to be buried and covered by backfill, irrespective of whether the soft material is replaced or not. To the maximum depth of the trench must be added the working tolerance of the dredger (see Section 3.6.2). If very tight tolerances are required, it is likely that the cost of executing the work will increase dramatically.

Dredging depth is not in itself necessarily a problem. However, for an outfall trench or any trench which runs from deep water into the shore or bank, there may well be a problem since the large dredger needed to reach the deepest parts of the trench is often the least suitable for the shallowest parts. When one adds the tidal range to the maximum depth, the difference in working in the deepest and shallowest areas may become extreme and special measures may have to be taken (see Section 3.3.2). Table 3.1 shows the maximum and minimum working depths for a variety of dredger types.

Width

Apart from satisfying the demands of its function, by providing sufficient width for a pipe or cable to be laid (and for larger items, the added width of the device which lays it), the bottom width of the trench must be sufficiently large for the dredger bucket or cutting head to operate. It is thus necessary to determine the type of dredger which is likely to undertake the work and to check that it is possible for it to construct the trench as designed. This becomes increasingly important as the depth of water decreases (see Section 3.3.2).

Very few large dredgers would be able to dredge a trench of bottom width less than two metres and most would require much more than this.

Side slopes

Side slopes may be designed for either permanent or temporary stability using the techniques described in Section 3.6.1. However, in locations where the soils are reasonably

Table 3.1 Maximum and minimum working depths of dredgers

Dredger	Size	Min. depth (m)	Max. depth (m)
a) Trailer	Hopper capacity		
(i) small	Under 1500 m^3	5	15
(ii) medium	1500 to 4000 m^3	7	20
(iii) large	Over 4000 m^3	9	35/80
b) Cutter	Disch. pipe diam		
(i) small	Under 400 mm	1	9
(ii) medium	400 to 800 mm	1.5	17
(iii) large	Over 800 mm	2	30
c) Bucket	Bucket capacity		
(i) small	Under 300 litres	5	12
(ii) medium	300 to 600 litres	7.5	19
(iii) large	Over 600 litres	10	30
d) Backhoe	Bucket capacity		
(i) small	Under 2 m^3	2	8
(ii) medium	2 to 5m^3	2.5	11
(iii) large	Over 5 m^3	3	25
e) Grab (pontoon)	Grab capacity		
(i) small	Under 3 m^3	1	No limit
(ii) medium	3 to 6 m^3	1.5	No limit
(iii) large	Over 6 m^3	2	No limit
f) Drilling pontoons			
(i) modular		2	30
(ii) bulkheaded		3	30
(iii) spudded		5	30

Notes:

1. Minimum depths for spudded dredgers may be controlled by the spuds which may increase the minimum depth to three or four metres.
2. The working of dredgers which load into barges may be restricted by the laden draught of the barges rather than the draught of the dredger.
3. The minimum depth of bucket dredgers may be determined by the depth of the underwater fairleads.
4. 80 metres dredging depth possible only with specially configured large trailers.

strong it is possible that the steepest stable slope would not, in fact, be constructable. The reason for this is that most dredgers have dredging ladders which are wider than the dredging head and to enable them to cut a trench down to a specified depth it is necessary to ensure that the sides of the trench actually allow the ladder to move an adequate amount. It may also be necessary to apply this criterion to the hull of the vessel. These effects are much more pronounced in shallow waters.

3.3.2 LANDFALLS

Landfalls occur in the inshore zone and need to be given special attention when considering trench design. In this area it is the type of dredger which can operate on the site which will determine to a great extent the design of the trench.

The inshore zone, which is defined as the zone in which the dredger in question becomes depth limited due to shallow water, has an important bearing on the selection of the optimum dredger type. The degree of importance will depend on the length of the inshore zone and this will in turn depend on the tidal range and beach slope; the flatter slopes and larger tidal ranges producing the longer inshore zones.

The mobility of the seabed in the inshore zone is often high. This will affect the dredging operation since high sediment mobility will favour dredgers with high production rates. These can excavate a trench quickly in a period of good weather so that a pipe can be pulled, or cable laid, before the trench is infilled by natural accretion.

Many dredgers that are capable of working in shallow water areas are mounted on wide pontoons. For the dredger to operate, usually it must be able to move or swing from side to side in its normal dredging mode which may include having a barge alongside. Considerations such as these will have a significant effect on the size of trench to be excavated in shallow areas.

In addition to the trench size, it is necessary to take account of the accuracy of dredging in shallow areas.

These points are discussed with respect to specific dredger types below.

Trailer dredgers

Trailer dredgers operate without mooring wires and by sailing over the dredging area at around two to four knots. They tend to have appreciable draughts when laden. Since they must manoeuvre like a ship they are not well suited to working in inshore zones. A trailer might be used to remove weak superficial deposits or to clean an already dredged trench just prior to pipe pulling. In soft material the trailer may overdredge to an unacceptable degree and this would tend to affect its ability to work to the specified tolerances. However, this effect can be controlled by careful operations.

Cutter suction dredgers

Cutter dredgers have to swing the cutterhead from side to side to make a cut and this usually involves swinging the whole pontoon. The width of trench needed depends on the dredger characteristics (see Section 7.6, pages 183–4). A trench in shallow water often will need to be around 30 metres wide at the top of the side slope. Modern cutter dredgers are highly automated and, when working in reasonable site conditions, are capable of accurately dredging a trench section.

Bucket dredgers

Bucket dredgers swing from side to side when dredging and in shallow water must also have space for a barge to be moored alongside. They therefore require a fairly wide working area which is approximately equal to 1.5 times the length of the dredger. Although bucket dredgers are able to make a fairly even trench bottom when working in deep water, they suffer from two disadvantages in shallow water:

- the bucket chain tends to hang below the level of the bottom tumbler thereby dredging a deeper trench than intended if appropriate adjustment is not made;
- there is a loss of production due to the fact that the buckets are tilted backwards and cannot be filled to capacity.

There is also a problem in trying to dredge trenches which are too narrow where there is the risk of the bucket chain actually coming off the bottom tumbler. The minimum trench bottom width is normally ten metres.

Backhoe and grab dredgers

The pontoons of backhoe and grab dredging vessels do not swing during their normal production cycle. However, unless side-casting, they both load dredged material to barges moored alongside and thus require sufficient space for these barges to approach and leave the dredger's position. A top width approximating to the beam of the dredger and one barge plus another 10 metres is usually sufficient.

The bottom width of the trench may be smaller than for other dredgers but should still be at least five metres to allow for horizontal inaccuracies.

When using backhoe and grab dredgers it is difficult to be sure that the whole of the trench bottom has been dredged and for this reason the bottom profile left by these dredgers is inclined to be uneven. This may result in unsupported lengths of pipeline.

Working periods in the inshore zone

In the inshore zone the combination of shallow water and tidal movement may restrict periods of access. There are various ways in which the working time in the inshore zone may be increased. These include the following:

- Moving the dredger shorewards during periods of high water and backwards during low water thereby taking maximum advantage of the water depth available at any state of the tide. This method is only feasible with dredgers which may be moved with relative ease and when the tidal range is appreciable. It does help considerably in areas where the beach slope is very flat.
- Using dredgers which are capable of supporting themselves on spuds. These dredgers are relatively rare. Due to the depth of water required to manoeuvre the dredger into shallow areas, a fairly large tidal range may be required to gain the initial access. It must, however, be remembered that, even if the dredger is able to stay on location throughout the tidal cycle, it may not be possible for a barge to come alongside to take the dredged material away.
- Using amphibious dredgers in the shallowest areas.

The landfall of a trench being constructed by a combination of land-based backhoes and a bucket dredger is shown in Plate 3.1.

3.3.3 BACKFILLING

Pipeline trenches are frequently backfilled after pipe installation to secure the pipe and to provide adequate protection from hydrodynamic forces and accidental damage from dragging anchors, sinking vessels and other such hazards.

Uncontrolled bottom-dumping from barges or from trailer dredgers is the simplest and cheapest method of placing material but it can easily give rise to very considerable losses, particularly in cases where the trench is narrow and the water deep. Bottom-dumped material descends very rapidly to the seabed, and after impact, spreads laterally as a density current, often eroding a considerable amount of material from the impact area. The radial spread of material depends on the material type, the water depth, the size of discharge openings of the dumping vessel and the speed of discharge. In some cases, dumped material can be carried a distance of over 100 metres from the dumping area.

To prevent such losses, backfilling must be carried out in a carefully controlled manner, either by dumping slowly or by delivering the backfill material into the trench through a

Plate 3.1 A landfall being constructed with land-based backhoes and a bucket dredger

pipe or by grab. Slow dumping can be achieved with some modern trailer dredgers and by side dumping or split barges. The nature of the material must also be considered when specifying placement methods in the case of backfilling pipeline trenches in order to prevent damage to the pipe. Damage is unlikely to result from the dumping of sand or gravel, but cobbles or boulders, which may be required as a protective layer, may need to be placed with special care.

3.4 Dredging for soil replacement

There are many occasions in maritime works when dredging is carried out for soil replacement purposes. Typical examples are as follows:

(a) When a breakwater is to be built in an area where the bed material is too weak to bear the load of the structure.
(b) Where reclamation is needed over an area of soft soil and the period required for consolidation of the weak underlying materials would be too long and would delay the commencement of construction on the reclaimed land.
(c) Where soft marine clays overlie very hard material into which sheet pile penetration is insufficient for the purposes of lateral stability the clays may be replaced by more suitable material.

In the case of the removal of large areas of soft material, where the volume of dredging in the side slopes is a small proportion of the total volume to be dredged, the design of the dredging works does not pose any particular problems. However, the same does not apply to relatively narrow trenches, such as those which might be required for (a) and (c). The reason for this is that it is desirable to remove the minimum amount of soft material in order to keep the costs of dredging and subsequent filling to a minimum.

3.4.1 DREDGED SLOPES

It is important that the slope angle is specified. There have been instances in the past when the designer, in the absence of adequate ground information, has specified the slope to be 'as steep as possible' in order to minimise the amount of material dredged and filled. Clearly this presents the dredging contractor with a problem, since without more information he is unable to determine how steep the slope should be. In fact he has a number of dilemmas:

- he does not know how deep the trench is going to be and therefore cannot determine where the top of the slope is;
- he cannot estimate the maximum stable slope angle since he has no soils information;
- he does not know how long the trench is to be kept open and thus how prone any temporary slope will be to degradation;
- he can excavate to a steep face but will then risk a major slope failure which might result in a slope being formed which is flatter than a slope which could have been achieved by design and careful dredging.

A typical soil replacement design is shown in Figure 3.9. A trench is to be excavated down to a firm stratum. The bottom width has been specified as a fixed dimension, B. However, the depth of the trench, d, is a variable since the level at which the firm material will be encountered is not known. The slope angle, α, has been specified, but without knowledge of the required depth of the trench it is not possible to determine the top width. If the trench is

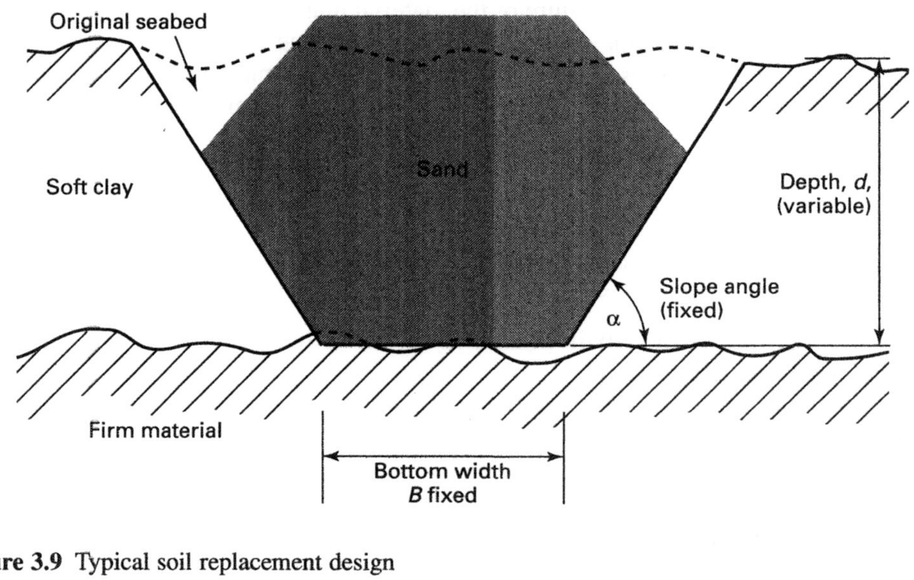

Figure 3.9 Typical soil replacement design

deeper than expected, the slopes will have to be re-dredged in order to achieve the specified bottom width and slope angle. If the firm stratum is shallower than expected, then more material will have been dredged than necessary to achieve the specified bottom width. Contract documents should recognise this uncertainty, particularly when ground investigation data are limited.

Because of these difficulties it is preferable and less contentious for the client, or his engineer, to design the whole trench after having carried out a suitable site investigation.

3.4.2 DREDGED LEVEL

Another detail of the design which can cause problems is the dredged level. Naturally the designer does not wish excessive overdredging below the specified level because it will mean that additional fill material will be required. It is, therefore, imperative to show clearly in the design and specification:

- the level to which the contractor is to work;
- how much tolerance is allowed;
- what is to be done if this tolerance is exceeded (see Chapter 11).

3.4.3 BACKFILLING

Most of the comments given in Section 3.3.3 apply. It is important to ensure that the contract for the backfilling encourages the contractor to use the minimum amount of material necessary by accurate placing.

3.5 Dredging for fill material

Dredging may be required to obtain granular fill material. These sources are usually superficial sandy deposits, lying at, or just below, the sea or river bed. Although one might assume that little design is required to use these deposits, the layout of the borrow area may need to be delineated, and potential volumes of fill estimated, before any dredging takes place. If this is the case the following points should be noted:

- The limits of the excavation should be defined (i.e. the top of the excavated slope) by reference not only to the location of the sandy deposits but also bearing in mind that a borrow pit needs to be of a size and shape appropriate to the type of dredger which is to work in it.
- Any exclusion zones due to adjacent power or telephone lines, pipelines and so on need to be taken into account.
- The side slopes of the borrow pit must be flat enough to be stable and for the dredger to be able to follow the profile of the pit.
- Water depth over the area and its approaches must be sufficient to allow efficient operation of the most appropriate type of dredging plant.
- Pockets of material will inevitably be left undredged in corners and on undulating surfaces of hard material. Due allowance for this must be made when defining the size of pit required to produce a given volume of sand.
- Similarly, if the deposit is to be worked using trailing suction hopper dredgers, or cutter suction dredgers loading into barges, due allowance must be made for the loss of material through overflow. Further losses may also be incurred if the material is rehandled at the reclamation site by a cutter suction dredger.

- Overburden and interburden may need to be removed to allow access to the granular deposits. This material will need to be relocated elsewhere, although some may be returned to the borrow pit if it can be suitably divided into sections. Interburdens of stiff clays are particularly difficult to deal with and areas with numerous lenses of this material may be unworkable.

3.6 Designed details

3.6.1 SIDE SLOPES

In principle, the design procedure for dredged slopes is little different from the design of any other excavated slope. Conventional soil or rock mechanics methods are used to arrive at a slope with an adequate factor of safety including, where appropriate, allowances for earthquake loads. However, in the case of dredged slopes, some additional factors may need to be considered. In exposed areas, for example, it may be necessary to allow for loads imposed by waves. If the slope is subject to currents, it may degrade as a result of erosion and this aspect should also be considered. The passage of vessels close to the slope may also give rise to erosive currents due to propeller wash and return water, that is the water flowing around the vessel in a channel.

If the slope is only partially submerged, it acts as a beach and is therefore likely to assume a beach profile.

The method of construction, or excavation, imposes limitations with respect to the design tolerances which can be specified and may also influence the design slope which is selected. When working with land-based plant, it is relatively easy to form smooth, continuous slopes with close tolerances. Slopes formed with most dredging plant generally have to be formed as a series of box cuts resulting in a stepped profile, approximating to the design slope. If accuracy is required, it is necessary to form the slope as a series of many, very shallow cuts and this method of working inevitably raises costs. If great accuracy is not required, the slope can be formed with a few deep cuts but there will inevitably be minor slope failures and material will fall down the slope as it stabilises. Some modern cutter suction dredgers and bucket dredgers are equipped with sophisticated computer controlled excavation systems which enable slopes to be cut with great accuracy and with only limited cost consequences.

3.6.2 FUNCTIONAL AND OPERATIONAL TOLERANCES

Tolerances in dredging work may be required for a number of reasons:

- to ensure that the completed works are functional in all respects;
- in some cases, to minimise costs;
- for the protection of nearby structures or services;
- to account for inaccuracies in the various methods of dredger positioning.

The ideal tolerance is one which ensures that the minimum amount of material is dredged, without an increase in the unit cost of dredging due to the difficulty of dredging to within the tolerance. Naturally, if the functional tolerance demanded for the work is smaller than the normal operating tolerance for the dredger this optimum balance cannot be achieved.

Functional tolerances are set by the nature of the work and its technical requirements; horizontal tolerances being important for trenches, narrow channels, canals, quay walls and slopes in harbour basins, whilst vertical tolerances are important in prepared surfaces for foundations or alongside structures such as quay walls. Tolerances for slopes should recognise the difficulty of slope dredging and are often less stringent than those for channel depth, for example. One exception to this may be when the slope is to be protected by stones, asphalt or proprietary blocks.

The operational tolerances demanded by dredging plant vary considerably according to the type of dredger and the working conditions. De Koning (1968) suggests that the following factors are important:

- the type and size of equipment;
- the type and properties of soil – hardness, compaction, cementation, cohesion, and so on;
- the prevailing waves – height, length and frequency;
- the prevailing tidal conditions;
- the dredging depth in relation to the optimum position of the ladder;
- the experience of the personnel;
- the instrumentation on board the dredger;
- the method of position fixing;
- the degree of automation of the dredger;
- the quality of survey, sounding work and the equipment used.

A rough guide to the vertical and horizontal operating tolerances for dredgers in different site conditions is given in Tables 3.2 and 3.3. It should be noted that these tables do not take account of all the factors mentioned above and considerable judgement is needed to arrive at realistic operating tolerances for a specific dredger at a particular location. Greater accuracy may be possible under special circumstances, at reduced output and increased cost.

3.7 Optimisation of channel design and maintenance

Many optimisation processes consist of four basic steps:

- choice of a suitable project life;
- identification of potential future demand and developments which may favour a more extensive project;
- estimation of alternative development costs and benefits;
- economic cost comparison and selection of the most cost-effective solution.

Since the estimation of development costs and benefits is an iterative process which usually deals with a considerable amount of basic data, it is often desirable to use a combination of statistical analysis and computer techniques to produce the results. Although not every engineer is likely to have acquired the skills of complex statistical analysis and computer programming, the principles of analysis are very simple and should be understood if the engineer is to be able to use the results in a sensible manner. The same applies to economic cost comparisons. Benjamin and Cornell (1970) and the ICE (1969) give basic guidance on the subjects of statistics and economics.

The optimisation of navigation channel design is described in two stages:

- the main optimisation procedures involved in determining the optimum amount of channel deepening and widening;
- the sub-optimisation procedures for minimising capital and maintenance dredging costs.

The sub-optimisation procedures can also be used independently in cases where the required dredged depth has already been defined.

Table 3.2 Vertical working tolerances (After BS 6349: Part 5: 1991)

Site conditions	Standard trailer	Light trailer	Cutter suction	Bucket wheel	Grab hopper	Grab pontoon	Bucket	Backhoe	Dipper
Bed material	mm	mm	mm	mm	mm	mm	mm	mm	mm
Loose silt	200	200	200	200	200	200	200	150	200
Cohesive silt	300	300	150	150	250	250	150	150	150
Fine sand	200	200	150	150	200	200	150	150	150
Medium sand	200	200	150	150	200	200	150	150	150
Gravel	200	200	150	150	200	200	150	150	150
Soft clay	250	250	150	150	250	250	150	150	150
Medium clay	300	300	150	150	300	300	150	150	150
Stiff clay	250	250	150	150	250	250	200	150	200
Very weak rock	300	N	300	250	N	300	300	350	300
Weak rock	N	N	300	250	N	350	300	350	300
Moderately weak rock	N	N	300	N	N	N	N	350	350
Pretreated rock	350	N	350	350	350	350	350	350	375
Adjustments for site conditions									
Sea conditions									
Sheltered water									
Small plant	125	150	150	150	175	175	100	100	100
Medium plant	100	150	125	125	150	150	100	100	100
Large plant	75	150	100	150	150	150	75	75	
Exposed water									
Small plant	300	350	N	N	500	N	N	N	N
Medium plant	250	350	350	350	400	400	350	300	300
Large plant	200	350	300	300	350	300	300	250	250
Currents									
Moderate (0.5 m/s)	0	0	0	0	100	100	0	0	0
Strong (1.0 m/s)	100	100	50	0	200	200	100	0	0

Notes 1. Accuracies are plus or minus.
2. Site condition adjustments should be added to 'Bed material' figures.
3. None of the figures are absolute limits, but difficulties may arise where lower limits are specified.

N = Not usually appropriate

Table 3.3 Horizontal working tolerances (After BS 6349: Part 5: 1991)

Site conditions	Standard trailer	Light trailer	Cutter suction	Bucket wheel	Grab hopper	Grab pontoon	Bucket	Backhoe	Dipper
Bed material	mm	mm	mm	mm	mm	mm	mm	mm	mm
Loose silt	2500	2000	500	500	500	500	500	250	500
Cohesive silt	2500	2500	500	500	500	500	500	250	500
Fine sand	2500	2000	500	500	500	500	500	250	500
Medium sand	2500	2000	500	500	500	500	500	250	500
Gravel	2500	2000	500	500	500	500	500	250	500
Soft clay	2500	2500	500	500	700	700	500	250	500
Medium clay	2500	2500	500	500	700	700	500	250	500
Stiff clay	2500	2500	500	500	700	500	500	250	500
Very weak rock	2500	N	500	500	N	700	700	700	700
Weak rock	N	N	500	500	N	800	600	600	600
Moderately weak rock	N	N	600	600	N	N	700	800	700
Pretreated rock	2500	N	1000	800	1000	1000	1000	700	800
Adjustments for site conditions									
Sea conditions									
Sheltered water									
Small plant	500	700	700	700	1000	700	700	400	400
Medium plant	500	700	500	500	1000	500	700	350	350
Large plant	500	700	500	500	1000	500	700	300	300
Exposed water									
Small plant	2000	2500	N	N	2000	N	N	N	N
Medium plant	1750	2500	1000	1000	1750	1500	1500	700	1000
Large plant	1500	2500	1000	1000	1750	1500	1500	700	700
Currents									
Moderate (0.5 m/s)	1000	1500	500	500	1500	1000	1000	200	300
Strong (1.0 m/s)	2500	3000	1500	1800	3000	2000	2000	700	700

Notes 1. Accuracies are plus or minus.
2. Site condition adjustments should be added to 'Bed material' figures.
3. None of the figures are absolute limits, but difficulties may arise where lower limits are specified.

N = Not usually appropriate

3.7.1 PORT ENTRANCE CHANNELS

Choice of project life

In most economic appraisals the evaluation period, or life of the project, is no more than 20 years. In exceptional circumstances, such as when a specific cargo is likely to increase dramatically in later years, a longer period may be chosen. However, with prevailing discount rates and the difficulty of predicting traffic growth and trends in ship sizes over long periods, the analysis of operations over periods in excess of 20 years is almost meaningless and, in any case, hardly appreciable in terms of cost comparison.

Identification of demand and possible developments

The deepening of an approach channel brings at least three major identifiable benefits (Stonham, 1977):

- it increases the number of ships that can sail at any time free of tidal restrictions, and it reduces the time constraints on movement of deeper draught ships;
- it reduces the delays to ships otherwise able only to enter harbour at spring tides;
- it raises the maximum size of ship that can be accommodated.

Thus, the channel deepening not only improves the existing operations but may actually lead to an increased demand. It is therefore necessary to examine the developments and the potential future demand together.

The entrance channel can be improved in the following ways:

- The channel can be straightened and widened – this allows the larger vessels to enter port unrestricted by bends, if there is sufficient depth. It can also change the channel from one-way to two-way.
- A one-way channel can be deepened – this can be carried out to various depths.
- One half of a two-way channel can be deepened – this produces a stepped channel in which there is always one deep water channel. The other channel is available for smaller craft or large vessels in ballast or light-loaded.
- The full two-way channel can be deepened to various depths.

For each of these possible alternatives there will be an annual traffic flow consisting of a range of ship sizes. The distribution of vessel sizes and types will be a function of the type of port, future traffic demand, and degree of port development, and can best be forecast by analysing present traffic and projecting on the basis of future developments.

Estimation of development costs and benefits

Development costs will include:

- the capital cost of dredging to the new configuration;
- the cost of future maintenance dredging to maintain the new configuration;
- the cost and future maintenance of new navigational aids for the improved channel;
- the cost of shipping delays suffered during the improvements;
- the cost of improving port facilities, handling methods and storage to cope with larger and more numerous vessels.

The question of how far the channel development affects the port operations depends on many factors. Great care should be taken in defining the limit of the cost/benefit comparison.

Benefits will include reductions in shipping delays, both in queueing to enter the port and in queueing to leave; increased throughput over the existing, or new, berths and increased utilisation of other port facilities.

Ship waiting times will be a function of ship size distribution, ship arrival distribution, variation in available water depth and berth service times. Variation in water depth is determined by the tidal characteristics of the area and the degree of channel improvement carried out. In order to take account of all these variables it is often best to resort to a simulation of the annual port operations.

Simulation can be carried out using computer techniques and consists essentially of programming the computer to behave in a manner related to the known variables. In effect, a model of the port is set up and shipping movements are followed through as they would occur in real life. The random nature of such variables as ship arrivals, types and sizes is accommodated by feeding the model sub-random figures, whose distribution is similar to that determined from statistical records of actual events. This method of simulation is called the Monte Carlo method.

Economic cost comparison

The economic cost comparison is carried out by determining the actual capital and running costs of the improvements proposed, together with the values of all the benefits accruing. These are are then expressed in present values by discounting them back from their year of occurrence to the beginning of the evaluation period. If the present value of the benefits exceeds the present value of the costs, the scheme is viable. If a number of alternatives are viable, then the best is that which has the highest net present value (NPV), given by the present value of the benefits less the present value of the costs.

3.7.2 RIVER CHANNELS

The analysis of river channel improvement schemes can be carried out on the same basis as that for port entrance channels, when the river has a significant amount of traffic on it. However, there are cases when a river is to be opened up solely for the purpose of allowing vessels to transport a specific product and in such a situation the optimisation procedure is modified.

For example, in a study in 1977 (Livesey, 1977), the minimum cost of transporting iron ore in barge trains down a river which has relatively shallow and tortuous upper reaches was determined. Amongst the many variables considered were:

- barge train shape, size and power;
- annual water level variations;
- increased traffic levels;
- channel depth and width improvements.

Because the draught and, to some extent, the speed of the barges was limited by the water depth, it was decided that a computer simulation program was desirable. The program which was developed simulated a complete barge train journey up and down the river and, in this manner, determined the journey time, load carried and number of barge trains required each month. Various channel improvements were proposed which included eliminating navigational hazards, improving bends and dredging to obtain various minimum water depths. It was determined that significant savings in transport cost could be obtained by dredging.

3.7.3 SUB-OPTIMISATION

Minimising capital dredging costs

The cost of capital dredging depends on the volume to be dredged and the unit cost of dredging, plus the mobilisation and demobilisation of dredging plant. In any development alternative, the shape of the channel, that is width, depth and bend geometry, will have been fixed, but the absolute position of the channel within the river, estuary or port entrance will still be a variable. By altering the position of the channel, the quantity of material to be dredged can be reduced to a minimum.

When the seabed consists of both hard and soft material there will usually be separate unit prices for dredging the different materials. In such a case the minimum cost of dredging may not be associated with the minimum quantity of soil to be dredged, but by an optimum combination of volumes of hard and soft soil. It might be advantageous, for example, to halve the volume of hard dredging and triple the volume of soft dredging in order to obtain the lowest dredging cost.

Rapid calculation of dredging quantities greatly facilitates this type of optimisation and computer techniques are helpful. By assembling geotechnical models, incorporating the different materials to be dredged, it is possible to program a computer to calculate dredging quantities for any desired channel shape, size and position.

Minimising maintenance dredging costs

The cost of maintenance dredging depends on the quantity of material to be dredged, the unit cost and frequency of the dredging. Often, the quantity of material and the frequency of dredging are determined by the rate of siltation and the extra depth which was dredged previously to accomodate siltation. It is a fairly simple calculation to determine the optimum extra depth, or overdredging, which gives the minimum maintenance dredging cost over a period of a few years.

However, in locations where the bed material comprises fine silt, the consolidation of the silt must be taken into account in the optimisation procedure. The cost of dredging silt is related, amongst other factors, to the silt density. If overdredging is increased, the time between dredging campaigns is increased and the silt has more time to consolidate to a higher density. Because of this additional consolidation, and depending on the dredging method, the optimum amount of overdredging may be greater for fine silts than for those soils which do not consolidate over time.

3.8 Dredging and quay design

Although the choice of quay structure may not be directly affected by the dredged area, the point at which the dredged area and the quay structure meet is of significance. The following points should be checked:

- whether dredging should occur before or after the construction of the quay;
- whether the dredged area is to be deepened at a later date and whether the quay design facilitates this;
- whether the dredging tolerances are of importance in relation to the method of construction.

Figure 3.10 (a) to (d) shows four common types of quay design.

3.8.1 OPEN PILE STRUCTURES

In open pile structures (see Figure 3.10a), dredging has to be carried out before installation of the piles since the side slopes and the adjacent dredged area could only be formed afterwards with difficulty. This order of construction would not be essential in cases where the seabed under the structure was already of adequate depth, such as on the shore side of a T-head jetty, although it should be borne in mind that dredging close to a structure, as opposed to dredging in open water, may well increase the dredging cost.

Provision for further deepening of the dredged area should be made by ensuring that the leading line of piles is driven, or cast, to sufficient depth to be stable at the deepest dredged level envisaged, inclusive of overdredging. In addition, the side slope should be formed in a position suited to the deepest dredged level.

It should be noted that the relatively still conditions which are created by pile clusters beneath a suspended deck may encourage siltation. The deposited material may be difficult to remove and may eventually become unstable and slip into the berth area. The maintenance of bed areas beneath open pile structures may require special methods because of the restricted access.

3.8.2 CAISSONS

Caissons which are open at the base allow the excavation of soil from within the structure. They are suitable for placing on site before the dredging of any adjacent basin and manoeuvring area.

Box caissons, however, are closed at the base (see Figure 3.10b). It is necessary to prepare a flat bed of suitable material on which to place them. The flat bed may be prepared by dredging or dredging and filling. Dredging in adjacent areas is often carried out after the caissons have been positioned, particularly when this allows the dredging plant to work in calm waters.

A caisson which is to be used as a berthing face must be founded at a level at least as low as the deepest dredged level. Otherwise, subsequent deepening of the berth by dredging, or scour, will undermine the caisson foundation. Excessive overdredging on the berthing line could have a similar result.

3.8.3 SHEET PILES

The sheet pile form of quay construction is normally used when it is wished to retain soil behind the berthing face. The piles are driven before dredging of the berth is started. A typical sheet piled quay is illustrated in Figure 3.10c. It is essential that the design of the sheet pile wall takes account of the maximum dredged depth which is likely to be attained in the berth area, including allowable overdredging, since the stability of the wall depends upon the fixity of the pile toes. Dredging operations alongside sheet pile walls must be carefully supervised and overdredging tolerances not exceeded.

When the original ground level at the toe of the proposed berthing face is too low for sheet piles to be driven, or when the soil is unsuitable, it is sometimes necessary to remove the original soil, fill the area with more suitable material and then drive the sheet piles. In other cases, the ground level behind the sheet piles may not be high enough to allow anchorage of the piles until some of the fill has been placed. Final dredging of the berth area should not be carried out until the anchorages have been installed.

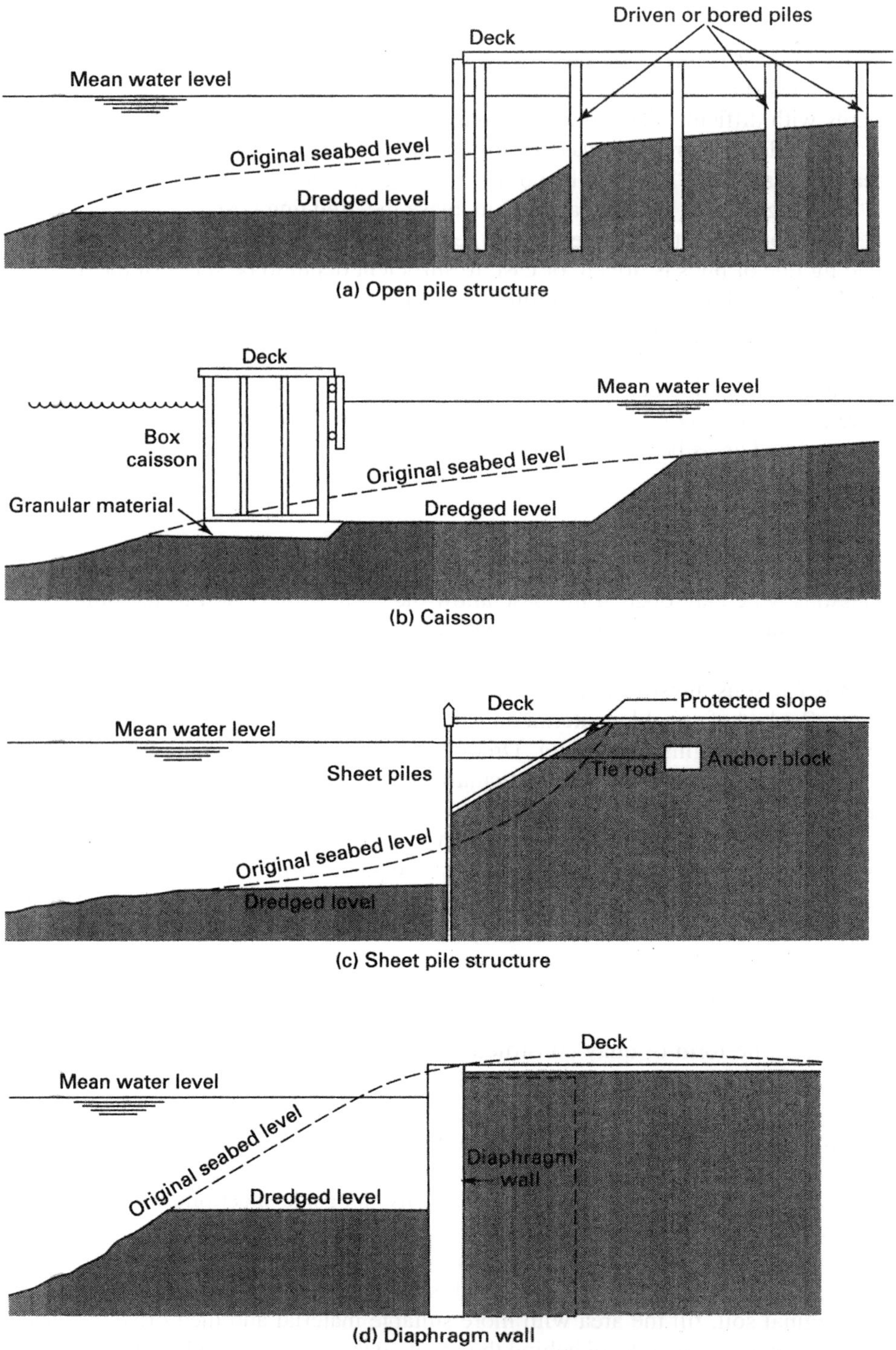

Figure 3.10 Quay designs

When backfilling behind the sheet pile wall is to be carried out with slow-draining soils, dredging along the face of the wall is best carried out after the backfilling. The dredging should be performed in separate lifts, at appropriate time intervals, to allow excess pore water pressures in the fill to dissipate. If this is not possible, the design of the sheet piles should be checked to ensure that they are capable of withstanding any additional imposed load during construction.

3.8.4 DIAPHRAGM WALLS

A quay incorporating a diaphragm wall is illustrated in Figure 3.10d. The method of constructing the diaphragm wall depends on the ability of the soil to stand vertically when supported by a bentonite slurry in the excavation. Since the slurry must be contained it is necessary to retain the soil on either side of the wall until the casting has been completed. When a quay is constructed by this method the dredging of the berth is carried out after the wall has been cast. The diaphragm wall is also a retaining structure and, therefore, excessive excavation below dredged level must not be allowed since this would affect stability. The structure must be designed for the lowest dredged level planned, including allowable dredging tolerances.

In situations where diaphragm wall construction is to be used at a site whose original ground level is below water level, a different technique must be used. Fill material, hydraulic or otherwise, is placed within a temporary bund and the site is reclaimed to normal working level. The diaphragm wall is cast in place and, afterwards, the material on the seaward side of the wall is removed.

References

Barrass, C.B. 1979. The phenomenom of ship squat, *Terra et Aqua* No. 18, December International Association of Dredging Companies (IADC), The Netherlands.

Benjamin, J.R. & Cornell, C.A. 1970. *Probability, Statistics and Decision for Civil Engineers*, McGraw-Hill.

Carvalho, F.V.de M. 1926. Paper of the 14th International Navigation Congress, *Permanent International Association of Navigation Congresses, (PIANC)*, Section 2, Communication 1.

Dunham, J.W. & Finn, A.A. 1974. *Small craft harbours: Design, construction and operation*, Special Report No. 2, US Army Corps of Engineers, Coastal Engineering Research Center.

Hay, D. 1968. Harbour entrances, channels and turning basins, *The Dock and Harbour Authority*, 48, No.567, January.

ICE 1969. *An Introduction to Engineering Economics*, The Institution of Civil Engineers, London.

de Koning J. 1968. *Boundary conditions for the use of dredging equipment,* Paper of the course on dredging operations in coastal waters and estuaries, May, Delft, The Netherlands.

Kray, C.J. 1970. Supership effect on waterway depth and alignments, *Journal of the Waterways and Harbours Division*, ASCE, 96, No. WW2, Paper 7305, May.

Livesey. 1977. *River Paraguay Transport Study, Stage 2*, Livesey & Henderson, Hydraulics Research Station, Sir J H Biles & Co., Maxwell Stamp Associates Ltd.

Lusch, L. 1970. Some aspects of the choice of fishing ports, *Fishing Ports and Markets*, Fishing News Ltd, London.

Myers, J.J. *et al.* 1969. *Handbook of Ocean and Underwater Engineering* McGraw–Hill.
De Nekker, J. *et al.* 1977. Developments in Dutch maintenance dredging, *Proceedings of the 24th International Navigation Congress,* Section II, Subject 5, Permanent International Association of Navigation Congresses, (PIANC), Leningrad.
NPC 1970. Newcastle University: *Transhipment in the seventies,* National Ports Council.
PIANC 1973. *International Oil Tanker Commission, Working Group No. 2 Report,* Bulletin No. 16, Volume III, Permanent International Association of Navigation Congresses, (PIANC).
PIANC 1991. Guidance on facility and management specification for marine yacht harbours and inland waterway marinas with respect to user requirements, *Report of Working Group 5 of the Commission for Sport and Pleasure Navigation.* Supplement to Bulletin No. 75, Permanent International Association of Navigation Congresses, (PIANC).
PIANC 1995 Approach channels – Preliminary guidelines, *First Report of the Joint Working Group PIANC and IAPH,* in cooperation with IMPA and IALA. Supplement to Bulletin No. 87, April 1995.
de Ridder, F. & de Visser, A. 1986. *Trends in recreational boating in the Rhine delta, the Netherlands,* Bulletin No. 55, Permanent International Association of Navigation Congresses, (PIANC).
Stonham, P.E. 1977. Cost-benefit of improving port entrance channels, *Terra et Aqua,* No. 12/13, January, International Association of Dredging Companies (IADC), The Netherlands.
Vantorre, M. & Coen, I. 1988. On sinkage and trim of vessels navigating above a mud layer, *Proceedings of the 9th International Harbour Congress, KVIV*: Vol.II, pp.4., 149-161, Antwerp, Belgium.
Wicker, C.F. 1965. *Evaluation of present state of knowledge of factors affecting tidal hydraulics and related phenomena,* May, US Army, Corps of Engineers, Vicksburg, Mississippi, USA.

4
Use of Dredged Materials

4.1 Introduction

Reclamation is perhaps the best known use of dredged material and, in the past, if the material was not needed for use in reclamation, it was generally disposed of, most frequently at sea. As a result of increased concern for the marine environment the use to which dredged material is put or, more precisely, the manner in which it is disposed, has become contentious. National regulation and international agreements governing the disposal of dredged materials have been made more stringent (see Chapter 5) and one effect of this has been to encourage, and in some cases force, engineers to find uses for dredged materials which formerly would have been considered worthless.

This chapter reviews, under three main headings, the possible uses of dredged material:

- reclamation;
- beach recharge;
- other beneficial uses.

4.2 Reclamation

4.2.1 THE REASONS FOR RECLAMATION

The act of raising the level of land which is either just below or adjacent to water is known as reclamation. The land is effectively being reclaimed from the water. When the level is being raised with fill material which has been dredged from underwater, the material is usually known as hydraulic fill. It should be noted that the term hydraulic fill is usually used only in the context of material which is placed by hydraulic pipeline. It is used here in its broader sense (all types of dredged fill).

Reclamation is by far the most common use of dredged materials and it may be carried out for one or more of the following reasons:

(a) it is cheaper to place the dredged material in a reclamation area than to dispose of it at an acceptable onshore or offshore location;
(b) it is environmentally more acceptable to reclaim with dredged material than to dispose of it at an acceptable onshore or offshore location;
(c) there is a need for the reclaimed land for port, industrial or residential development or for agricultural or recreational use (see Plate 4.1).

Plate 4.1 Sand filling operations on low-lying ground at Tin Tsui Wai, Hong Kong, to make land for a new township

In (a), distance is generally the controlling factor. The distance between the dredging site and the disposal area can vary considerably from one project to another and transport costs increase as distance increases. The apparent cost of transport is also partly determined by restrictions, either practical or environmental, which might be placed on the dredging operation. Restriction of hopper overflow, for example, often means that dredgers have to sail below the normal full hopper load. For any dredger and a given set of dredging conditions, there is an easily definable line which separates those situations where reclaiming is cheaper than dumping (see Figure 4.1). However, it is stressed that Figure 4.1 should be interpreted as a guide. There are considerable differences between the efficiencies of individual dredgers and the actual economic distance for reclamation therefore varies from one vessel to another.

Amongst the many factors which affect the economic and technical feasibility of any reclamation scheme are the following:

- the use and ultimate value of the reclaimed land;
- land ownership;
- the environmental effects of the reclamation;
- the engineering properties of the fill foundation, that is the ground on which the fill will be placed;
- the engineering and/or pedological characteristics of the fill material;
- the availability of material for constructing fill retaining bunds;
- the methods of dredging, transport and placement of the fill material;
- the method of improving the quality of the fill material, if necessary.

These are considered in turn in the next section.

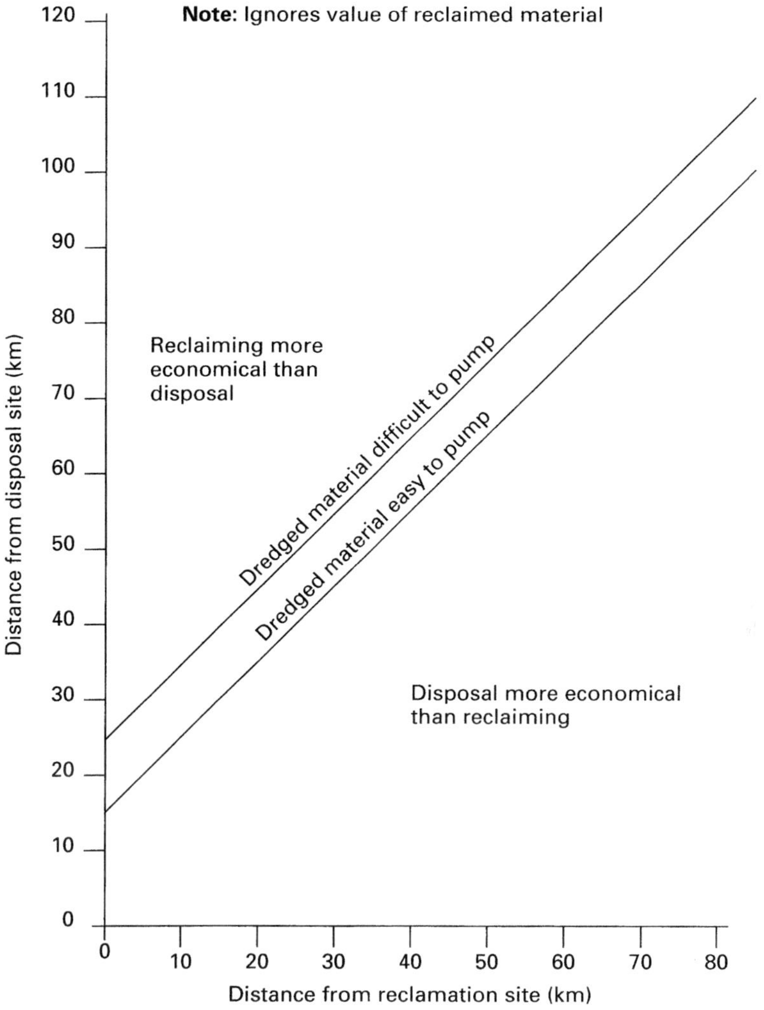

Figure 4.1 Limits of economy for dredging systems

4.2.2 USE AND VALUE OF THE RECLAIMED LAND

Reclaimed land can be used for many purposes. Most often, reclamation has been carried out by port authorities as a means of disposing of material arising from capital and maintenance works, and thus a large proportion of land reclamations have been used to extend the shore facilities of existing ports.

In some countries, such as Singapore and Hong Kong, a shortage of land has prompted large scale reclamations for a wide range of uses including:

- residential, industrial and commercial development;
- transport corridors;
- sites for ports and airports.

Hong Kong, Singapore, Australia, the USA and Japan, amongst others, all have major international airports sited on land formed by dredged reclamation.

Reclamation may also be undertaken as part of sea defence schemes, notably in the Netherlands, or to provide land for agriculture and recreational development.

In fact, reclaimed land can be put to almost any use but the decision as to whether a reclamation should proceed is almost invariably based on economics. Does the value of the land which is created by reclamation justify the cost of the reclamation or is it a cheaper solution than other forms of environmentally acceptable use or disposal?

Irrespective of the purely economic justification for a reclamation scheme, the authors stress the fact that dredged material is a resource. When it is the product of a dredging operation whose primary objective is something other than reclamation, it should always be remembered that the bulk of the cost of obtaining it has already been spent. If it is discarded rather than used to reclaim land (even if that land might not be needed for several years), a valuable resource may have been lost.

4.2.3 LAND OWNERSHIP

Once a suitable location for reclamation has been selected, it is important to discover to whom the existing land belongs, or if there is no land, to whom an area of land would belong if reclaimed at that location. In many countries much of the foreshore is owned by the state. Arrangements may have to be made to lease a reclamation area.

4.2.4 ENVIRONMENTAL EFFECTS OF THE RECLAMATION

The act of reclamation can have a wide variety of environmental effects and merits detailed study during the planning stage. Potential effects include:

- changes to the hydrodynamic and sediment transport regimes of the area;
- changes to the groundwater regime;
- a variety of ecological impacts such as habitat destruction.

The last category is beyond the scope of this book but it should be noted that reclamation has been used with considerable success, particularly in the US (Landin, 1988), to create or enhance wildlife habitats both as a method of disposal and as an end in itself.

Effects on the hydraulic regime

Reclaiming an area of land, that is substituting soil for water, is just as likely to affect the hydraulic regime of a site as dredging a portion of the seabed, that is substituting water for soil. It is therefore essential that the effects of reclamation should be considered at the feasibility stage of a project by means of numerical or physical models (see Chapter 6), including consideration of the various sediment erosion, transport and deposition mechanisms which maintain the balance of the area (see Chapter 2). In many cases, alteration of the shape and alignment of the waterfront will significantly affect the amount of siltation or scour in the nearshore waters and, by influencing the cost of future maintenance, may even be significant in determining the feasibility of the scheme.

Effects on the groundwater regime

Reclamation may also effect the local groundwater regime. During reclamation using marine-dredged materials, for example, the large volumes of saline water which are used to place the material can pollute coastal aquifers if not properly controlled. In some instances, large reclamations may result in the permanent raising of local groundwater levels. This may have an effect on drainage and even endanger the foundations of structures which were designed and built when the water table was at a lower level.

4.2.5 THE FILL FOUNDATION

Irrespective of the nature of the fill material which is used, the engineering quality of the reclamation as a whole is very strongly dependent on the quality of the material on which the fill is placed – the fill foundation. Thus, if a reclamation is to support structures, it is of little benefit to use the best available fill material if the ground or the seabed on which it is placed cannot support the fill, is so weak as to result in failure of structures placed on top of the fill or consolidates excessively when the fill is loaded.

Such situations most frequently arise in the case of foundation soils comprising soft clays and silts. Two factors need to be considered:

- the method by which the fill is placed;
- the need for improvement of the ground, either before or after filling, so that it will adequately support any structures superimposed on the fill.

Filling methods on soft soils

The main objective when filling over very soft soils is to place the fill in a manner which does not lead to instability and the formation of mud waves. This requirement is particularly important when the fill material is granular and of markedly differing engineering characteristics to the substrate, for example dredged sand placed on soft mud.

In order to avoid instability and mixing when filling with dredged sand, it is necessary to place a base layer of fill in such a manner as to ensure that:

- the fill is placed in thin layers of uniform thickness;
- that the leading edge of the fill forms a shallow slope, typically at 8° or less.

It is difficult to achieve these objectives when filling underwater with conventional hydraulic pipeline filling techniques because the fill is brought up to sea level in one lift and because the leading edge of the fill can form a steep slope in the range of 1:10 to 1:3, depending upon grain size of the sand and the hydrodynamic conditions at the filling site (see Table 4.1).

Table 4.1 Indicative slopes of hydraulically placed granular materials (from Athmer & Pycroft 1986)

Grain size, mm	Indicative range of slopes		
	Above water	Below water Calm seas	Below water Rough seas
0.060 – 0.200	1:50 – 1:100	1:6 – 1:8	1:15 – 1:30
0.200 – 0.600	1:25 – 1:50	1:5 – 1:8	1:10 – 1:15
0.600 – 2.000	1:10 – 1:25	1:3 – 1:4	1:4 – 1:10
> 2.000	1:5 – 1:10	1:2	1:3 – 1:6

Uncontrolled bottom-dumping from barges or trailer dredgers is usually too energetic a process to be considered as a method of forming a base layer; it inevitably results in the erosion of considerable volumes of seabed material and mixing with the fill. Thin base layers can be placed by a number of methods, such as the following:

- very slow controlled discharge from a trailer dredger as it passes over the site with part-open bottom doors or valves;
- spraying sand from the bow-nozzle of a trailer dredger as the ship is underway;
- using small barges which are subdivided into several small compartments each containing only small volumes of sand. The compartments are emptied in sequence and, because the volume of each 'dump' is very small, a layer is progressively built up with minimal disturbance of the seabed or of previously-placed sand;
- using a special spreader barge supplied with sand through a floating pipeline. The barge is winched backwards and forwards over the filling area to ensure even placement of fill. In shallow water, this may be the only possible solution;
- discharge by pumping slowly through the suction pipe of a trailer dredger.

When filling on weak soils above water level, it is often possible to use conventional pipeline placement because sub-ariel fill slopes are usually sufficiently shallow to prevent failure of the substrate as long as a reasonable degree of care is taken. Indicative values of fill slopes during placement are given in Table 4.1 (Athmer and Pycroft, 1986).

Geotextiles are often considered in instances where the surface materials are extremely soft. In principle, the geotextile can be used to prevent mixing and, if sufficiently strong, to add strength to the overall reclamation. However, if the surface materials are fluid, or almost so, the use of geotextiles may actually encourage instability because they float on the fluid materials which are immediately displaced as filling commences. In such cases it may be better to omit the geotextile and to permit a degree of mixing when placing the initial thin layers of fill. This will avoid differential loading of the substrate and the mixed material will generally have, or rapidly develop, better engineering characteristics than the original bed materials. This will make the placement of subsequent layers easier.

Improvement of fill foundations

Techniques of improving the quality of weak, or soft, fill foundations are well established. The three most common methods are as follows:

- surcharging;
- artificial drainage using wick drains or sand drains to accelerate consolidation;
- vacuum pre-loading and drainage.

All three methods, described below, require instrumentation and monitoring if they are to be properly controlled. This usually involves the installation of settlement gauges and piezometers to measure the rate and magnitude of consolidation and pore water pressure dissipation. Pore water pressure monitoring may also be necessary during the filling process in order to ensure that the rate of filling does not overstress the underlying soils.

Surcharging

Surcharging, sometimes called pre-loading or pre-compression, is one of the most widely applied methods of improving poor fill foundations. It involves the placement of an additional thickness of fill, beyond that required for reclamation purposes and sometimes in excess of ten metres. The additional load can be used to achieve two objectives:

- It causes the underlying materials to consolidate to a greater degree than would be the case when loaded only with the design fill. It thus reduces the length of time between fill placement and the effective completion of the primary consolidation which would have occurred due to the load imposed only by the design thickness of the fill. When the surcharge is removed, it enables the reclamation to be used earlier than would otherwise be the case.

- It can be used to consolidate the underlying soils to such an extent that settlement, particularly long-term differential settlement, of structures placed on the fill is minimised.

The general principle of surcharging is schematically illustrated in Figure 4.2.

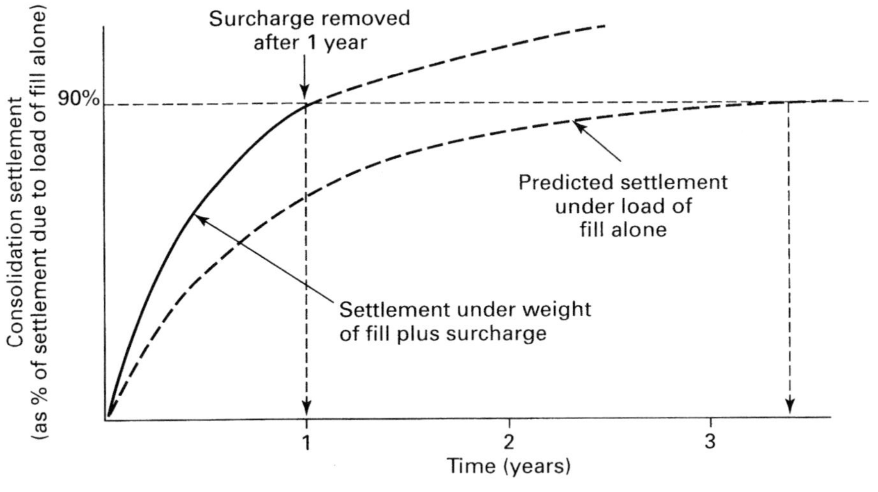

Figure 4.2 Schematic illustration of the principle of surcharging (pre-loading)

Surcharging is an attractive and simple method of improvement if materials for the surcharge are readily available and particularly if they can be put to good use afterwards, perhaps as fill material elsewhere. Surcharges, on their own, are most effective when the layer of compressible material to be improved is relatively thin. In the case of thick layers, surcharges are often combined with artificial drainage to reduce the consolidation time. The volume of material required for surcharging can be reduced by the use of a 'rolling' surcharge. By this method the surcharge material is placed on part of the area and is moved repeatedly until the whole area has been treated.

Artificial drainage

In the case of fine-grained, low-permeability soils, vertical wick drains or sand drains can be installed in order to hasten drainage and thus reduce the amount of time needed to achieve the required degree of consolidation. The time required for a foundation soil to achieve a given degree of consolidation is proportional to the square of the shortest drainage path and inversely proportional to the permeability. In thick layers of low-permeability soils, the installation of vertical drains can therefore considerably reduce the length of the drainage path and thus the consolidation time (see Figure 4.3). This benefit may be magnified by the fact that such soils are often more permeable in the horizontal direction than in the vertical direction.

Wick drains comprise thin strips of highly permeable man-made materials (usually of the order of 5 × 150 millimetres) which are inserted into the soil by special machines at centres typically between one and three metres, depending on the site conditions and on the required rate of consolidation. Sand drains work in a similar manner but are constructed by forming a small hole in the soil and filling it with sand.

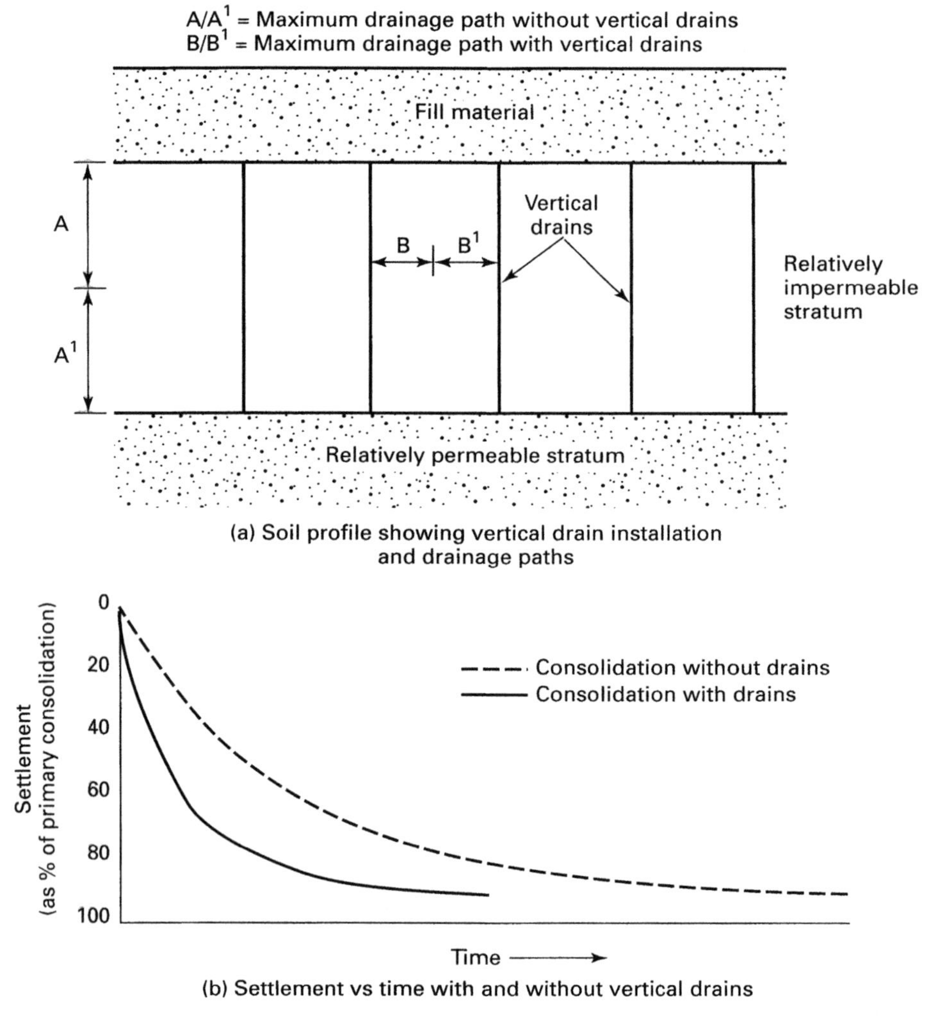

Figure 4.3 The effects of vertical drains on drainage path length and consolidation (settlement) time

The ability of the artificial drainage system to withstand deformation must be checked when consolidation settlements are expected to be very large. Modern wick drains can often tolerate very considerable deformation without significant loss of effectiveness. In this respect, it should be noted that, as consolidation proceeds, the rate of pore water expulsion decreases. It is, therefore, not necessary for the drain to function at its full design capacity throughout the consolidation period. However, in some extreme cases, it may be necessary to install a second set of drains at an appropriate time during consolidation.

Vacuum pre-loadng and drainage

Vacuum pre-loading is a method of effectively surcharging the underlying soil without actually placing any surcharge material. It involves placing an air-tight membrane over a sand blanket placed on the soft soil. The membrane is sealed at the edges using impermeable soil or slurry trenches. Vacuum pumps are then used to draw water from the soil under the membrane. This method is generally only applied in conjunction with vertical drains. The surcharging effect is achieved by increasing the effective stress in the soil by reducing pore water pressures. Choa (1990) gives an account of this method being used at the Port of Tianjin, People's Republic of China.

4.2.6 ENGINEERING CHARACTERISTICS OF FILL MATERIALS

A wide variety of dredged materials have been used as fill. The suitability of different materials for use as fill depends partly on the intended use of the filled area and partly on the methods used to dredge, transport and place the fill. In most cases the basic requirement is either that it is capable of sustaining plant and animal life or that it can support appreciable loads. This section focusses on fills for engineering development.

In order to assess the suitability of the material for reclamation, the following points are usually examined:

- the initial load-bearing capacity of the as-placed material;
- the required load-bearing capacity of the fill;
- the methods, time and expenditure required to achieve the desired degree of improvement, if improvement is necessary;
- the settlement of the fill during improvement.

Measures of fill properties

There are many methods of measuring the quality of fill in geotechnical terms. However, a difficulty encountered in many reclamations is that a large proportion of the fill material may be placed underwater and will always remain below the water table. Such material will always be difficult to reach and test. When assessing fill quality, a basic distinction must be made between granular, relatively free-draining soils, and fine-grained, poorly-draining soils.

Granular soils

Traditionally, non-cohesive, or granular, fills have been assessed on the basis of the relative density, D_r, defined as follows:

$$D_r = \frac{\gamma_m(\gamma - \gamma_0)}{\gamma(\gamma_m - \gamma_0)} \times 100(\%) \tag{4.1}$$

where γ_o = minimum dry density of the soil in laboratory
γ = field dry density of soil
γ_m = maximum dry density of soil when compacted in laboratory

The relative density can be related to other soil properties such as strength and settlement characteristics which are used to design foundations. It can vary from zero to 100 per cent; zero per cent representing the soil in its loosest possible state and 100 per cent representing the soil at its most compact. Soils with a low relative density have a low angle of internal friction and are thus unable to sustain heavy foundation loads. They are also subject to large settlements when loaded and, in some cases, may be prone to liquefaction during earthquakes.

However, there are difficulties inherent in the measurement of relative density and the results are so sensitive to minor variations or inaccuracies in the measurement of the limiting densities, that the use of relative density as a measure of fill quality must be considered, at best, dubious (Tavenas *et al.*, 1973).

Relative compaction is a more useful measure of fill quality and can be determined with a higher degree of confidence and reproducibility. Relative compaction, C_r, is defined as:

$$C_r = \frac{\gamma}{\gamma_m} \times 100\ (\%) \tag{4.2}$$

where γ_m = maximum dry density when compacted in laboratory
γ = field dry density

However, both relative density and relative compaction can only be determined on materials placed above water. Because of these difficulties and limitations, there has recently been a move towards the direct measurement of primary engineering properties of fill.

Two methods, the Standard Penetration Test (SPT) and the Cone Penetration Test (CPT), are now widely used (see Chapter 6). Cone penetration tests, in particular, can be used to establish a wide range of soil properties. Both the SPT and the CPT can be related to relative density and relative compaction by empirical relationships and can be used rapidly to establish the fill properties above and below the water table. Athmer and Pycroft (1986) give a detailed comparison of the various methods of fill testing. They report that, on one site, fill quality control based on relative density determinations above the water table cost approximately 20 times more than testing using the cone penetration test. The latter was used to investigate the materials both above and below water level.

Cohesive soils

The engineering quality of cohesive soil is principally a function of its water content. The higher the water content the lower the strength and the more likely it is to consolidate excessively under load. The strength properties of cohesive fills can be measured *in situ* by a variety of well proven techniques including the CPT, some of which are described in Chapter 6. The consolidation characteristics are usually assessed by standard laboratory tests.

Fills derived from clean sand

Fills derived from fairly clean sand, that is less than 15 per cent fines, are the most common and are generally suitable for industrial and commercial developments because the material can be placed to a medium relative density, and because it will drain and compact naturally in a short time. It may also be compacted further by mechanical means. If reasonable care is taken in the placing of the material, a medium dense sand fill can be obtained which is capable of supporting foundation pressures of 25 to 150 kPa, depending on its thickness, the nature of the material under the fill, the level of the water table.

When the sand is placed above the water table, even higher densities can sometimes be achieved, partly as a result of the compacting effect of water draining through the fill and partly because of compaction by the earthmoving plant used to spread the fill evenly over the reclamation area.

Generally, relative densities in the placed state vary between 45 per cent and 55 per cent, but there are examples of relative densities in excess of 90 per cent being achieved above the water table.

The integrity of non-cohesive fill can be affected by earthquakes. The vibration of the earthquake temporarily increases the pore water pressures in the soil, sometimes leading to liquefaction which can have catastrophic consequences. The liquefaction potential of a soil depends on:

- soil type, that is particle size and grading;
- the relative density or void ratio – less dense material is more susceptible to liquefaction;
- initial confining pressure – the risk of movement caused by liquefaction of material is reduced if the material is confined both laterally and vertically;
- intensity of ground shaking;
- the duration and frequency of ground shaking.

Generally, medium silts to fine sands of a uniform grain size are most susceptible to liquefaction. However, when reclamation is to be carried out in earthquake zones, all types of sand fill should be checked for liquefaction potential. If necessary, the relative density can be improved using one of the techniques described in Section 4.2.9.

Fills derived from mixed soils

Fills derived from silty or clayey sands can either be placed with their original particle size distribution unaltered or attempts can be made to reduce the fines content to improve the quality. It should be noted that some dredging and placing methods have the effect of reducing the fines content (see Section 4.2.9).

If fines reduction is not implemented, it is preferable to place the fill in such a way that the fines are well-distributed in the reclamation area. The result will be a relatively compressible and heterogeneous fill whose relative density may vary from 45 per cent to 85 per cent. However, the fill is likely to have a reasonably high permeability due to the sand content and can be improved simply and quickly by surcharging to the extent where it may support heavy foundation loads. Soils of this type may also be susceptible to liquefaction and this aspect should be checked in earthquake zones.

Fills derived from cohesive soils

Fills derived from stiff cohesive soils which have been dredged by cutter suction dredger generally comprise a mixture of clay balls and slurry. The balling of clay does not occur if only mechanical methods of dredging and transport are used. Whatever the method of dredging employed, the resultant fill usually consists of ridges of clay lumps separated by areas of mixed or fine materials. The engineering properties of the fill are variable.

Laboratory tests are of little use for determining the properties of fill comprising stiff cohesive soils since the compressibility of the fill depends on the rate and degree of deformation of the clay lumps and closure of voids. Fills derived from stiff clays can be used successfully for bearing considerable loads if pre-loading of the fill is carried out at a slow rate and sufficient measurements are made to predict settlements when the final loads are applied.

Fills derived from soft cohesive soils are discharged from the pipelines as a slurry. They have a very high water content and generally remain soft for a long time. The slurry discharged from the pipeline will usually have a solids content of between 10 per cent and 20 per cent by volume and, after the excess water has drained, this may increase to 30 per cent to 40 per cent (Salem and Krizek, 1973). Slurries formed of soft cohesive material have a low permeability and the rate of drainage depends on the methods of placing, pre-loading and draining as well as the extent to which permeable sandy laminations are developed during deposition (Bishop and Vaughan, 1972).

The following engineering relationships hold for the natural consolidation of clay slurries in the reclamation area.

- Maximum settlement can be predicted by examination of the relationship between the compression index and the liquid limit. The relationship is normally linear.
- The shear strength of cohesive soils with high water content is found to vary almost linearly with water content.

Field and laboratory testing have shown that given sufficient care in the methods of sampling and testing these slurries, their behaviour over long periods of slow consolidation may be predicted with some reliabililty by means of existing soil mechanics theory and empirical relationships. However, it should be noted that coefficients of permeability in the field may be several orders of magnitude higher than those measured in laboratory conditions (Krizek and Giger, 1975). The development of slurry consolidation equipment may prove to be of great use in assessing consolidation characteristics when materials are too soft or fluid to test by conventional means.

In order to appreciate the time involved in the natural consolidation of clay slurries, a mixture of organic silts and clays of medium to high plasticity, and inorganic clays of high plasticity with sand, silt and clay proportions in the ratio 1:3:2, was investigated (Krizek and Giger, 1975). It was found that the field strength increased at the rate of 4 kPa per year over a period of ten years, and the dry density increased at a rate of about four per cent per year over an eight year period.

The characteristics described above are generally those of slurries deposited above water level. Where slurries are to be deposited below water level their properties are likely to be far inferior. A shear strength of only 2 to 3 kPa may be achieved after ten years. However, tests recently undertaken in Hong Kong (J.M. Land, personal communication 1992) have shown that dredged, normally consolidated, marine mud placed into redundant marine borrow pits achieved a strength profile of approximately 60 per cent of the original *in situ* material (1.8 kPa per metre depth) after a period of only one to two years.

The rate of placement has a significant influence on the properties of the material. Generally, it is best to place the material slowly and in thin layers so as to maximise the initial consolidation, which is hindered in thick layers. When the slurry is placed on sand which has good drainage characteristics the rate of improvement of shear strength can be considerably accelerated

Fills derived from rock

In many parts of the world, notably in the Middle East, dredged rock has been used as a fill material. The quality of the fill will be largely a function of the properties of the rock material, the method of dredging and placement and the degree to which the rock degrades during dredging and transport.

Sedimentary rocks are the most widely used as fill. Fills formed from weathered mudrocks, dredged using a cutter suction dredger and transported by pipeline, may have properties only slightly better than those of fill formed from stiff cohesive soils. These rocks often degenerate markedly during dredging and pipeline transport. On the other hand, fills formed in a similar manner using sandrocks can often have excellent engineering characteristics.

Metamorphic and igneous rocks are very rarely used as fill because they are generally avoided, if at all possible, as far as dredging is concerned. If strong rocks are to be dredged, usually preceded by drilling and blasting, the resulting material is often reserved for construction of fill retaining bunds or for purposes other than use as general fill.

4.2.7 FILL CONTAINMENT AND THE AVAILABILITY OF MATERIAL FOR BUND CONSTRUCTION

In most reclamation sites it is necessary to contain the fill by some kind of levee or bund. If suitable material for constructing bunds is not available on the site, it will be necessary to import material. The availability and cost of this material will affect the overall economics of reclamation.

The degree of containment required for a hydraulic fill will depend on the soil type and the use to which the land is to be put. When placing coarse granular material total containment may not be essential. Such material may form natural slopes during placement of approximately 1:25 above water and 1:4 below (see Table 4.1).

Fills derived from fine cohesive materials must be contained on all sides and overflow weirs and slots (see Section 4.2.8) must be carefully located and controlled.

Containment bunds, dykes or levees usually need to retain all solids inside the reclamation area. Filter layers may need to be incorporated to prevent loss of fines. The filtering can be achieved by constructing filter layers of selected materials or by using geotextiles.

During filling the height of a bund must be sufficient to contain the fill material in its unconsolidated state with an additional freeboard to take account of impounded water and wind waves inside the reclamation area. For slow draining slurries, a volume must also be included which allows adequate time for material to settle out of suspension. A minimum freeboard of about one metre would be reasonable. When the bund is constructed using the general fill material, it is often built up in stages as the level of the reclamation area rises.

When subject to wave attack or scour, protective armouring must be placed on the bund. For wind waves only, simple rip-rap protection may suffice. The crest level of the bund will be dictated by considerations of wave run up and overtopping. Filter layers will also be required to prevent loss of fines from the bund. In some locations, where rock of sufficient size might not be available, the slope can be protected with a bituminous sand revetment or might even take the form of an artificial beach.

4.2.8 PLACING THE RECLAMATION FILL

Sandy fills

Hydraulic fills of a predominantly sandy nature must be placed in such a way that the small proportions of fine material which they contain are not allowed to segregate and accumulate to form mud pockets in the reclamation area. For this reason the formation of ponds in the reclamation area should be avoided and filling should be carried out from the land towards the sea, where possible, and not vice versa, to facilitate the drainage of the area. It may be advisable to limit the amount of filling at any one point to a certain lift height, which will be determined by the level of the discharge pipe. The higher the lift height, the further the material is being carried before deposition, and the greater is the chance that segregation will occur. A maximum lift height of two or three metres is reasonable for filling areas above water level.

Mixed-soil fills

When fills are derived from silty or clayey sand, it may be desirable to reduce the fines content. However, it can be difficult to remove the fines or environmentally unacceptable to do so. In these circumstances, efforts should be made to ensure that fines are distributed as evenly as possible over the site in order to minimise the variation of fill quality and the consequential risk of unacceptable differential settlements. This can be done by raising overflow heights sufficiently to form a large pond and by moving the discharge point at frequent intervals.

Alternatively, it may be feasible to encourage the formation of silt ponds at particular locations where structures are not planned or their settlements are not important.

Clay fills

Fills consisting of stiff clays cannot be placed by hydraulic means in any form other than a heterogenous mixture of lumps and slurry. If dredged by a machine with a digging bucket, however, they will tend to retain their *in situ* character and can be placed using conventional earthworks techniques, for example in layers which are compacted using sheepsfoot rollers.

Slurries formed by the hydraulic dredging of soft clayey and silty soils must be retained in settling ponds in order to allow the material in suspension to settle out. Layers up to a metre thick are formed before the discharge is moved to another pond and the first pond is left to settle, drain and dry.

Control of discharge waters

In many cases, and particularly when filling using hydraulic transport, a considerable volume of excess water needs to be discharged from the reclamation site. Attention has to be given to providing a suitable means of allowing the water to escape without flooding adjacent areas. A more difficult problem, however, is ensuring that large amounts of sediment do not escape, possibly causing environmental damage.

To overcome this problem, it may be necessary to construct a settling basin alongside the main reclamation area. The basin will need to be enclosed and have dimensions sufficient to ensure settlement of soil particles. The dimensions will be determined by the particle sizes in suspension, and the flow of dredged mixture. Settlement basin design is outside the scope of this book and may be found in text books dealing with water supply and irrigation.

In some situations, it may be possible to reclaim the settling basin at the end of the project using stockpiled material, if there is sufficient time for the fines trapped in the basin to consolidate prior to the area being used or if the area is not required to take heavy loads. Alternatively, ground improvement techniques could be employed.

The outlet from the reclamation area will usually consist of an adjustable weir or a drop inlet (see Figure 4.4). The size of the overflow will vary according to the type of fill and the rate of flow. In general, the width of a weir should be sufficient to limit the velocity over the weir to a minimum.

The plan area of a circular drop inlet overflow should be about five to seven times the area of the pipe which is delivering dredged mixture to the reclamation site. The overflow should be positioned as far from the delivery pipeline as possible in order to allow maximum time for the soil particles to settle. For large reclamations there may be several overflows of substantial construction in steel or concrete.

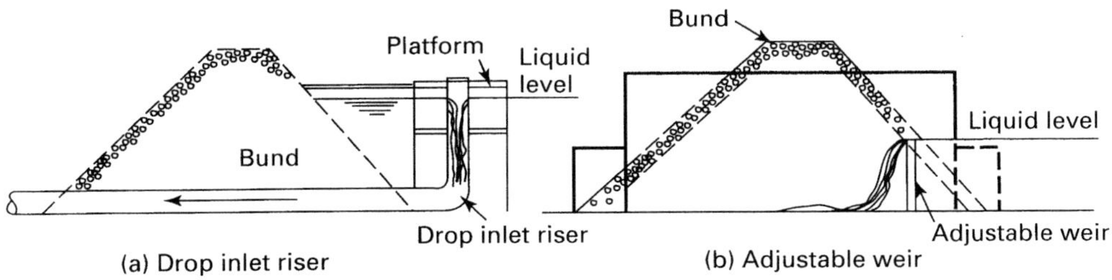

Figure 4.4 Two types of overflow from reclamation areas

4.2.9 IMPROVEMENT OF FILL QUALITY

It is often necessary to improve the quality of the fill material in order to achieve the required engineering properties. There are three stages at which this can be done:

- during dredging, transport and placement by reducing the fines content and thus improving its drainage properties;
- during placement by desiccation or the incorporation of drainage layers;
- after placement by improving the strength and consolidation properties.

Improvement by fines removal

During hydraulic dredging and placement, materials often undergo significant changes of composition, exhibiting a marked reduction of the fines and fine sand content. The process is essentially one of washing and segregation of the component particles of the soil and can be exploited to improve the quality of the material before it is finally incorporated in the fill. In extreme cases, it may be possible to utilise materials which, without such processing, would not be considered suitable for use as fill.

There is, of course, a disadvantage to this process; it may result in the release of considerable amounts of fine material to the surrounding marine or fluvial environment. This potential, and the consequential effects, should be carefully evaluated at the design stage of the project.

The most widely applied method by which the fines content of soil can be reduced is by overflowing from trailer dredgers. The principles governing the loss of material during overflow are considered in Chapter 8. A similar process, and result, can be achieved when loading barges with cutter suction, stationary suction and dustpan dredgers. Figure 4.5 illustrates the changes of particle size distribution which might be achieved by trailer dredgers working in sand. This should be treated as indicative only, in view of the varying performance characteristics of different dredgers.

During hydraulic transport, mixed soils will often become extensively disaggregated and, when discharged into the reclamation area, will easily segregate into coarse and fine fractions. The reclamation cells and the method of filling can be designed to take advantage of this, concentrating the fine material in certain parts of the area where the engineering requirements are less stringent or where it can later be removed, leaving only the relatively clean, high-quality material in the fill.

Rehandling, for example by dumping dredged material into a rehandling basin or an underwater stockpile for pumping ashore with a cutter suction dredger, will also result in some reduction of the fines content.

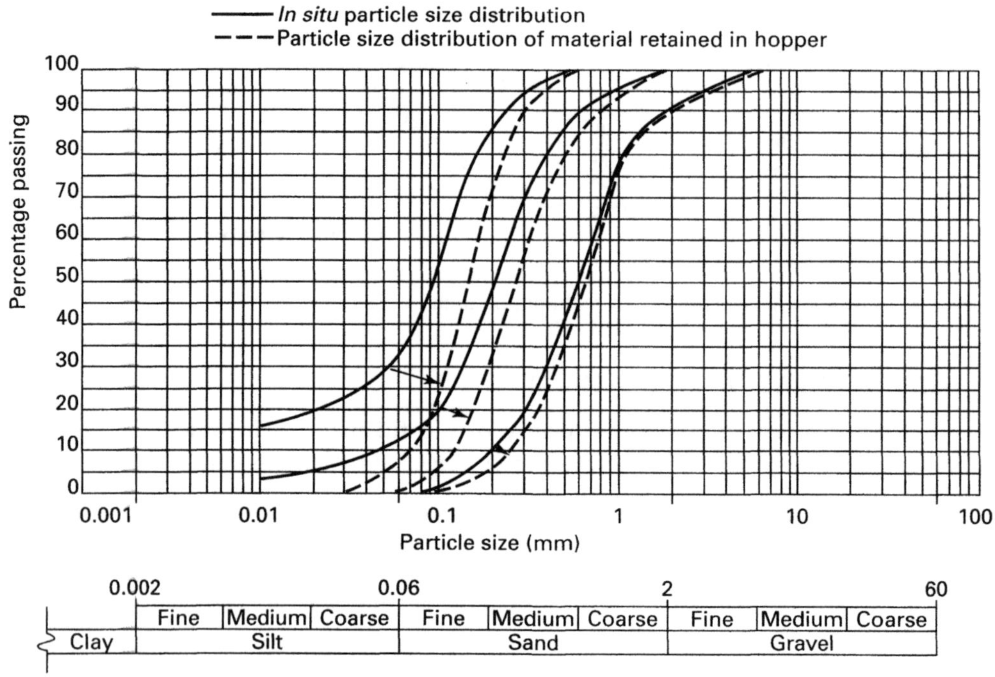

Figure 4.5 Examples of the change of particle size distribution when sand is dredged by a trailing suction hopper dredger

When these methods of fill improvement cannot be applied, or when they are inadequate to achieve the required fill quality, it may be possible to utilise hydrocyclones. Experiments with hydrocyclones (Tiederman and Reischman, 1974) have shown that whilst they are not effective for concentrating dredged materials at high solids contents, the capability of a hydrocyclone to recover medium-size sand from material containing a large percentage of fine silt is excellent. During the construction of the Nakdong Barrage in Korea (van Duin, 1987), hydrocyclones were extensively used to treat more than 6 Mm^3 of dredged material.

Improvement by desiccation

Cohesive sediment reclamations can be improved by desiccation aided by 'progressive trenching', a technique which has been applied in the USA and in the Netherlands primarily as a means of increasing the capacity of disposal areas (see Chapter 5). De Nekker and In't Veld (1975) describe this method being used for treating Rotterdam Harbour mud. The slurry is deposited in ponds in layers of one metre thickness. After two months, sufficient drainage and evaporation has occurred to enable a specially developed amphibious vehicle to work in the mud. The following ditching system is then employed to drain the mud:

Time after deposition, in months	Ditch depth (m)	Ditch spacing (m)
2	0.10	2
4	0.20 – 0.30	2
6	0.50 – 0.60	10

When the soil has ripened sufficiently, another layer is deposited on top and the process is repeated. In this manner a seven-layer deposit can be built up in about ten years. Due to consolidation of the soil, a seven-layer deposit will eventually be about four metres thick. Similar techniques have also been used in onshore disposal areas along the River Thames in the UK.

There are four factors which determine the time required to improve the material:

- the liquid limit of the soil which determines the moisture content at the point when all surface water has been drained-off, typically about 1.8 times the liquid limit;
- the plastic limit of the soil which determines the moisture content at which the soil can be used for reclamation, typically about 1.2 times the plastic limit;
- the climatic conditions;
- the execution of the ditching; drainage ditches must be maintained at a level deeper than the desiccation cracks which form on the surface of the mud.

Haliburton (1978) describes predictive methods to determine the speed with which such reclamations can be formed.

Improvement by the incorporation of drainage layers

Reclamations can be constructed using soft clays placed in layers with intervening drainage blankets of sand. In Singapore, trials of this method (Lee *et al.*, 1987; 1990) showed that hydraulically-placed clay slurries achieved a surface moisture content of 200 to 220 per cent after about five days. A 50 millimetre thick layer of sand was then placed on a one metre thick clay layer and two days allowed for the sand-impregnated surface of the clay to improve in strength. Thereafter, further thin layers of sand were placed, ultimately building up to form a layer about 300 millimetres. Two such sand-clay 'sandwiches' were formed. The layered fill was surcharged with a one metre layer of sand to accelerate consolidation. After 13 weeks of surcharging, the shear strength of the clay had increased to more than 15 kPa and the water content had reduced to less than 70 per cent.

Whilst clearly having some potential for forming reclamations where good quality granular materials are in short supply, this method is difficult and expensive to execute. Choa (1990) lists several problems including segregation of material around the discharge points, excessive losses of sand, and slow placement rates.

Improvement after placement

Methods of improving fills after they have been placed depend mainly on the nature of the fill and fall into two groups:

- those applicable to relatively free-draining materials;
- those applicable to poorly-draining cohesive materials.

Methods which can be used to improve cohesive materials are the same as those which can be employed to improve fill foundations and these have been reviewed in Section 4.2.5.

Fills comprising free-draining, that is sandy or gravelly soils, are relatively easy to improve after placement. Usually, the requirement will be to increase the relative density or relative compaction of the material in order to improve its bearing capacity, to minimise the settlement of superimposed structures and to reduce the liquefaction potential of the fill. Improvement techniques utilise compaction, which is the process whereby the soil particles are packed more closely together by high energy dynamic methods, thus increasing the dry density of the soil. Compaction-based methods fall into three groups depending on the dynamic process involved:

- deep vibration;
- impact;
- rolling.

Deep vibratory methods

Vibroflotation is a patented process in which a vibrating probe is jetted into the soil to the desired depth. After application of vibration for the required period the annular space round the probe is filled with granular material as the probe is withdrawn. The vibrational energy imparted by the probe to the ground has the effect of increasing the density of the soil *in situ*. Relative densities can be improved from between 40 and 50 per cent to around 80 per cent or more by this method.

Vibroflotation is usually carried out at centres of between two and three metres if high relative densities are required. The method becomes slower and less effective as the proportion of silt and clay in the sand increases, due to damping of the vibration and other effects, and the method is generally limited to soils with a fines content of less than 15 per cent. The effectiveness of the technique is seriously impaired if the proportion of fines rises above 20 per cent. Densification by this means is effective both above and below the water table. Figure 4.6 shows the range of soil particle size distributions which are most suited to improvement by this technique.

Impact

Impact methods, commonly called ‘dynamic compaction’, involve dropping heavy weights onto the surface of the fill using a specially adapted crane. It is suitable for use in all granular materials and is the only technique by which the density of fills derived from broken rock, rubble, stones and heterogenous clay/sand mixtures can be improved.

Compaction is achieved by tamping the upper surface of the fill by dropping a large concrete or metal block onto it from a considerable height. The weight of the block is usually between 10 and 20 tonnes and the drop height between 10 and 20 metres. The tamping process is normally carried out in at least two passes separated by an appropriate time interval. The effect of the tamping is to produce high energy impact which transmits shock waves through the ground to a considerable depth.

Rolling

When the fill is to be used to support shallow-founded structures which are sensitive to differential settlement, it is necessary to excavate and replace the fill in well-compacted layers using conventional rolling plant. Rollers may either impose a static load as they pass over the fill or they may impose a vibrating load.

The maximum density achieved by compaction will depend on the moisture content of the soil and the soil characteristics (see Figure 4.7). Uniformly graded sand and gravels will not compact well in the top 100 to 150 millimetres using vibratory methods. Tests for compaction carried out in this region are not, therefore, representative of the whole layer. Densities between 85 per cent and 100 per cent of the modified proctor test (ASTM D-1557-58T) can be achieved by vibratory roller in optimum conditions within the compacted layer.

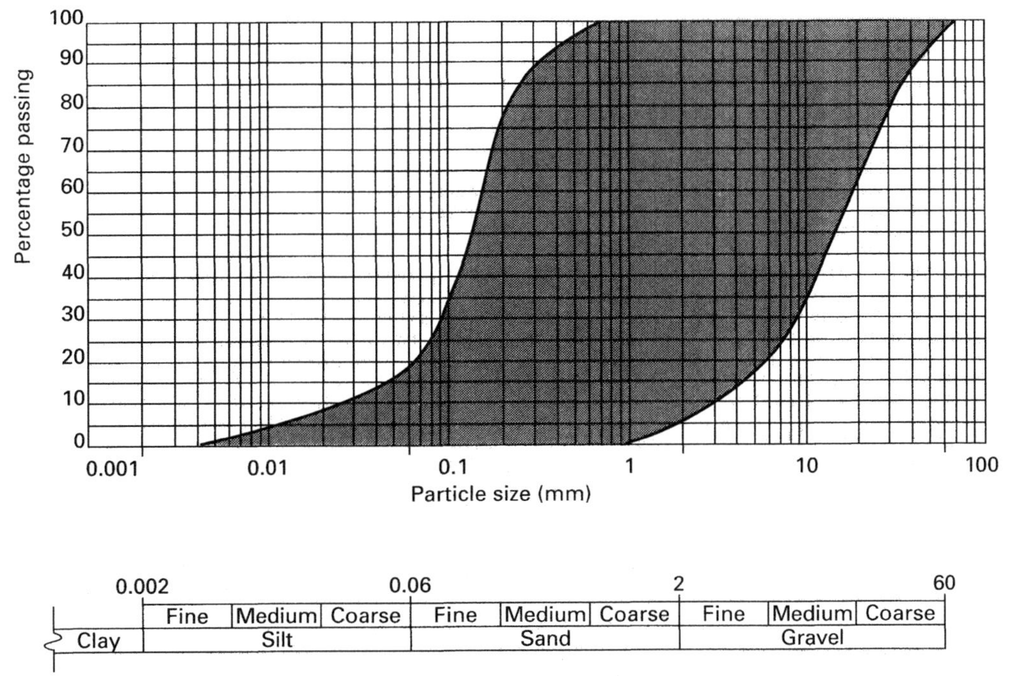

Figure 4.6 Particle size distributions of soils that can be improved by vibroflotation

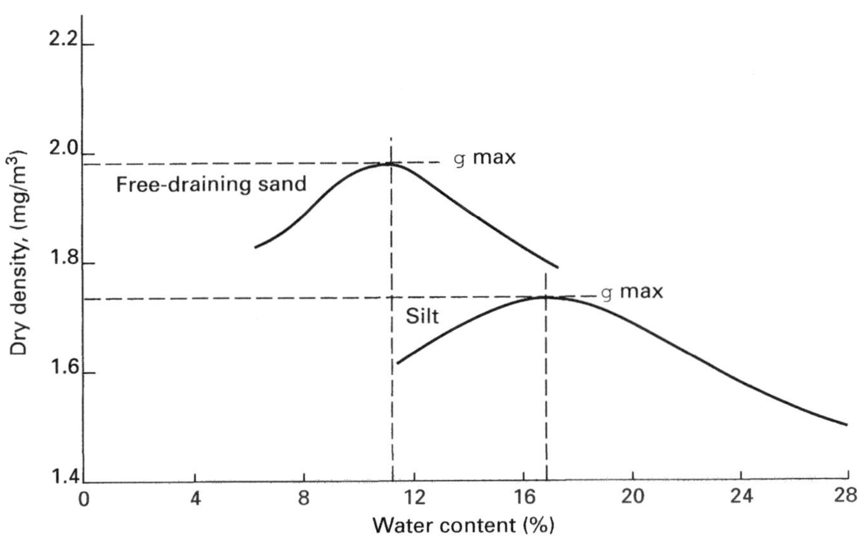

Figure 4.7 Maximum densities of soils versus water content for compacted sands and silts

The passage of heavy earth-moving equipment over hydraulic fill should not be assumed to result in adequate compaction since, not only is this equipment often designed to exert low bearing pressures on the ground, but the resonant vibration frequency necessary for good compaction is unlikely to be present. In addition, it does not cover the fill in a systematic manner.

4.3 Fill specifications

Specifications for hydraulic fill should define the nature of the material which is permitted to be incorporated in the fill and the required engineering properties of the completed fill.

The nature of the material is usually defined by a particle size distribution envelope combined with a sampling and testing schedule which requires a specified rate of testing and degree of compliance. In preparing such specifications, the engineer should be aware of the fact that hydraulically-placed fill is heterogenous on a small scale, even when placed with great care. Individual small samples of material may, therefore, display considerable variation in particle size. The permitted particle-size distribution and compliance schedule must recognise this and should not be unnecessarily restrictive.

Additional clauses are often inserted into specifications to further define the nature of the materials in order to avoid the inclusion of, for example, organic materials, rubbish and debris. An upper particle size may need to be specified depending on the future use of the fill. If driven piles are to be used for foundations, the presence of boulders may be a considerable hindrance to pile driving. In some cases, where service trenches or a large number of shallow foundations are required, a different material will be specified for the uppermost two or three metres of fill.

Specification of the required engineering properties of the fill should preferably be based on a rapid and unequivocal method of establishing those properties. Attention has already been drawn in Section 4.2.6 to the difficulties of using relative density as a measure of fill quality and preference should be given to cone or standard penetration tests which provide direct measurements of relevant engineering properties.

4.4 Beach recharge

Beach recharge is associated with soft coastal defences and with the provision or maintenance of coastal amenities.

For many years engineers have tended to favour hard defences and have created, in concrete and stone, a variety of structures to resist the sea by direct confrontation. Such methods have often proven to be self-defeating because the turbulence which occurs at such sea walls during severe wave attack may result in rapid erosion of the foreshore and consequent undermining of the structure. In addition, concrete coast defences rarely have aesthetic appeal and, with increasing public environmental awareness, the desire for natural solutions now favours the beach form, with its obvious aesthetic and amenity advantages.

4.4.1 BEACH PROCESSES

Beaches comprise a multitude of small particles, usually of sand or gravel size. The beach material is usually the product of erosion of the local, or updrift, coastline. The energy of waves breaking on a beach is dissipated progressively and in the process the beach material is in almost constant motion. In consequence, the form of the beach will adapt to different wave conditions.

If the beach is in equilibrium, there will be little or no loss of beach material from the overall local beach system, only a continuous redistribution. If the beach is not in equilibrium, some loss of material will occur under certain wave conditions. Even though material may be lost from the beach, the beach width and profile may be maintained if the rate of loss is matched by the rate of replacement from adjoining coastal areas. However, if the rate of loss exceeds the rate of replacement, erosion of the beach will result and eventually its effectiveness as a defence against the sea, or as an amenity, will be lost, unless it is restored by artificial beach recharge, sometimes termed 'replenishment' or 'nourishment'.

4.4.2 PLANNING BEACH RECHARGE

Before embarking on the detailed design of a beach recharge, it is necessary to consider two important matters:

- availability of recharge material;
- programming.

Availability of material

The volume of material which is required to fully restore or improve a beach profile may be considerable. If the required volume is large it is unlikely that onshore mining and transport of recharge material will be economic, or environmentally acceptable. Dredging methods usually will be less costly and less disruptive.

It is usual to seek material for recharge which has similar physical characteristics, in terms of particle size and colour, to that which exists naturally on the beach which is to be recharged. Alternatively, a material with greater particle size may be preferred because the resistance to erosion will be greater and hence the life of the recharged beach may be longer. However, there are other considerations and methods which may allow finer materials to be used. These are discussed in Section 4.4.3.

Potential sources of material will need to be identified and investigated in much the same manner as a conventional investigation for dredging works is undertaken. The various stages and methods of investigation are described in Chapter 6. As with all dredging works, the potential environmental effects of removing the material from the source area must be evaluated. In particular, it may not be acceptable to win material from close inshore because it may lead to accelerated erosion of the adjacent shoreline.

On the other hand, good sources of material may be found in the nearshore zone, and there is a body of opinion that favours using these sources because they are linked to the beaches to be recharged, particularly if the beach and the source are in the same coastal system. Normally, it will be legislation which determines whether a usage of this type is possible, rather than purely engineering considerations.

Potential source areas should also be checked to determine the practicality and economics of dredging, bearing in mind the location and the manner in which the material must be placed. Whilst almost any type of dredger might possibly be used to win material, in practice it is likely that the most suitable, from the point of view of transport and placement, will be either the trailing suction hopper dredger or the cutter suction dredger, with the latter being best suited to sheltered nearshore areas.

Generally, it is preferable that the water depth be no more than 30 metres though some dredgers may reach to 50 metres and a few to greater depth. The minimum water depth will usually be about five metres. The distance to the source area is not very critical, unless operations are tidally restricted, in which case there may be an optimum transport distance when dredging and placement operations are synchronised. Distances of up to 50 nautical miles may be economic though shorter distances may be reflected in lower costs.

Guidance on suitable materials for beach recharge and procedures for obtaining them may be found in the CIRIA publications *Beach Management Manual* and *Beach Recharge Materials, Demand, Resources and Availability* (1995).

Programming

Beach recharge requires working close inshore within the surf and intertidal zones. Sea conditions will therefore have a very significant effect on the practicality and cost of the work. Usually, the work will be best undertaken during calm weather seasons.

However, during such seasons, the beach may be in demand for public recreation. The apparent conflict may not be serious. Usually the recharge material will be pumped onto the beach. Only within the immediate area of discharge, spreading and containment, is the risk to the public significant. Hence, the area of public exclusion at any time may be small. Furthermore, where only trailing suction hopper dredgers are employed, discharge will not be continuous. In fact, experience has shown that the public find the progress of beach recharge both interesting and entertaining and often will position themselves to obtain a good view of the works.

Regardless of the constraints described, it should be recognised that the recharge of exposed beaches may be practically or economically impossible during winter, or other periods of increased wave activity. Some conflict of interest may therefore be inevitable.

4.4.3 DESIGN

The detailed design of beach recharge is beyond the scope of this book, which deals only with the outline principles and methods of execution. More detailed treatment can be found in the *Beach Management Manual* published by CIRIA (1995). It is only briefly described here.

The design of beach recharge requires that the following parameters be decided:

- the physical properties of the recharge material;
- the slope of the beach face;
- the elevation of the top of the slope relative to tides and waves;
- the width of the beach-top berm, if any;
- the location and form of associated structures.

Physical properties of the material

Unless it is intended that the wave climate at the beach be modified by the construction of artificial headlands, offshore breakwaters, or offshore bunds, the mean particle size of the recharge material should be similar to, or greater than, that which exists naturally on the beach. A further exception may arise if finer material is available at a significantly lower cost. In most cases sand fill will be cheaper than gravel.

A balance has to be achieved between the life expectancy of the recharged beach, or the recharge interval, and the overall cost of the recharge. It is the average annual cost of maintaining the beach profile within acceptable limits which is the most important cost criterion. By assessing the cost and recharge intervals for a range of fill materials and dividing the total cost by time, the minimum annual cost and hence the optimum material type can be determined (see Figure 4.8). The true optimum is obtained by using discounted cash flow over an appropriate period of time.

Almost without exception, the finer the material which is used to recharge, the greater will be the volume required and the shorter will be the necessary recharge interval. The greater volume arises due to the flatter beach slope of the finer materials.

If fine material is available at a very low cost, as may happen if the materials used arise elsewhere from regular maintenance dredging, then beach life may not be very important, always providing that the beach has the ability to withstand the worst sustained wave attack which might occur between periods of recharge. The use of suitable materials arising from other dredging projects is to be encouraged. Indeed, in the USA and the UK it is now obligatory to investigate beneficial uses, including beach recharge, as a means of disposal of dredged material.

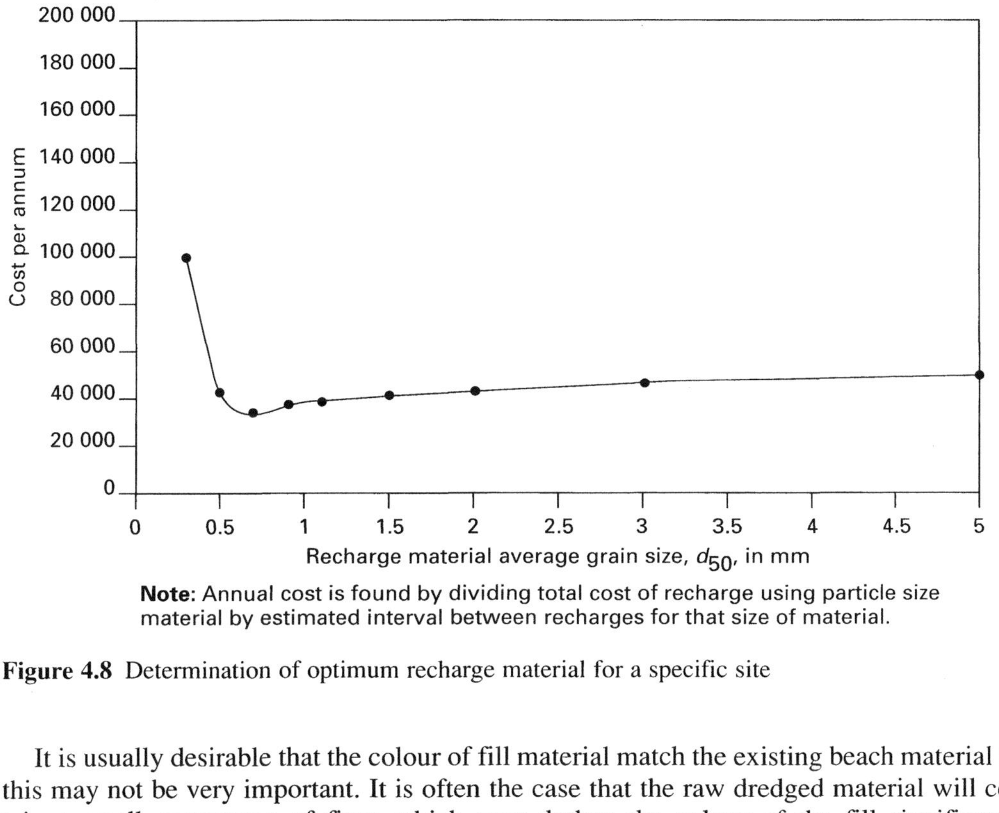

Note: Annual cost is found by dividing total cost of recharge using particle size material by estimated interval between recharges for that size of material.

Figure 4.8 Determination of optimum recharge material for a specific site

It is usually desirable that the colour of fill material match the existing beach material but this may not be very important. It is often the case that the raw dredged material will contain a small percentage of fines which may darken the colour of the fill significantly. However, the fines will be quickly removed from the coarser matrix by wave action. This winnowing, combined with chemical changes due to prolonged exposure to sunlight, will in most cases result in the rapid lightening of colour of the finished beach.

The specific gravity of the fill particles should be similar to that on the existing beach. Particles with a low specific gravity will be less resistant to erosion. Silica sand will have a specific gravity of about 2.65 but shell particles will be significantly less. Hence, it may be necessary to specify an upper limit on shell content.

The fill material should be well graded. Single-sized materials or those in which a particular size dominates, may be less resistant to erosion. However, materials which are not well graded may nevertheless be acceptable if more readily available and hence less expensive.

All materials intended for beach recharge should be free of significant contamination by substances which are potentially harmful (toxins, glass, explosives), a nuisance (oil and grease) or unattractive (clay). Explosives may seem a rather obvious exclusion but it should be recognised that the disposal or loss of ordnance at sea has been common for many decades and hence their recovery during dredging is not unusual.

Slope of the beachface

The design slope used to determine the volume of fill required should be similar to the expected long term average stable slope of the finished beach. It is not essential to produce this slope during the placement of the recharge material. Usually it will be more practical and cheaper to place material at a steeper slope and to rely on natural redistribution by wave action (see Figure 4.9).

The ultimate slope will change in response to the particular wave climate. Typical slopes are in the following range:

fine sand	1 in 75
gravel	1 in 10

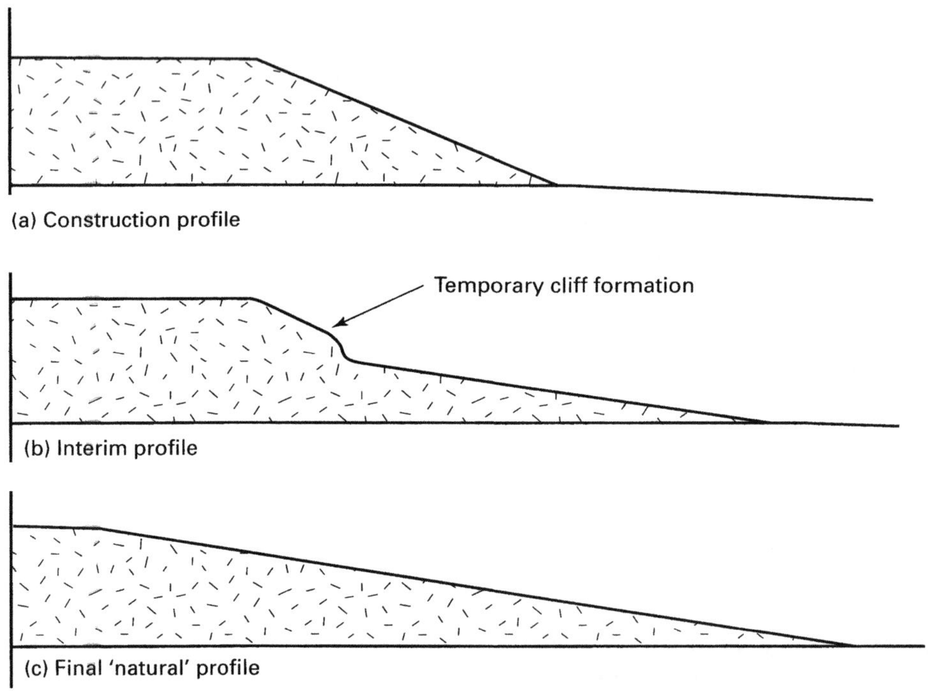

Figure 4.9 Natural redistribution by wave action of a recharged beach

These typical figures should be treated with caution. Actual slopes will be a function of the particle size and distribution of the beach material and the local wave climate, a matter which is covered in detail in a publication of the Construction Industry Research and Information Association, (CIRIA, 1995).

Elevation of the beach-top

The maximum design elevation of the finished beach profile related to tide level is usually dependent on the primary reason for the recharge. If it is for amenity purposes, it may be sufficient to ensure that the beach width is adequate for recreation at all, or most, tidal states. However, if it is for sea defence it may be necessary that the beach elevation be sufficient to prevent overtopping under the most extreme conditions of tide and waves.

It will usually be the case that when a beach is overtopped under storm conditions the rate of movement of beach material will be increased due to the increased height of waves breaking over the beach. Where the beach is backed by cliff or a wave return wall, erosion may be increased due to turbulence at the interface. Thus, it may be desirable that the beach design provide a high level berm such that, even if the overall beach slope is reduced by severe wave attack, a high level margin remains to prevent overtopping.

A disadvantage of high-level berms is that the beach above high tide level is not regularly cleaned by tide and wave action. Neither is the beach material regularly wetted. Sand beaches which remain dry are prone to wind erosion during strong onshore winds. This may not result in an important loss of beach fill volume, but may create very considerable nuisance to the coastal margin infrastructure. The problem may be alleviated by the erection of sand fences close to the seaward crest, but these may be unattractive unless well designed.

The factors which should be considered in determining beach elevation include the following:

- the primary purpose of beach recharge;
- the maximum tide height;
- the maximum wave height and set up;
- the maximum storm surge at the location;
- the maximum frequency and duration of extreme wave events;
- the variation of the beach profile under extreme conditions.

The maximum conditions are the design criteria, which may range from an annual event to a one in ten year event. If the beach forms a major part of a coast defence system, the design criteria may be more stringent. It should be recognised that the beach profile will deteriorate during the interval between recharges and this should be taken into account in the design.

Width of the beach-top berm

Not every beach design will incorporate a top berm but those designed primarily as a sea defence usually will, in order to provide a reserve of beach material which ensures that the defence is maintained throughout the design storm conditions. The primary consideration in deciding berm width is the potential variation of beach profile and reduction in berm width which may occur during storm conditions. For amenity beaches such factors as occupation density, space for beach sports, equipment and sales space may also be important. Those matters discussed under beach elevation apply also to berm width.

Location and form of associated structures

Material may be lost from beaches and foreshore as a result of sediment transport in a seaward direction, or by movement along the beach. The latter, which is probably most significant, is termed littoral drift (see Section 2.2.2). Movement is the result of suspension of beach particles by wave action and transport by wave or tidally induced currents (see Chapter 2).

The movement of beach material can be reduced by changing the pattern or severity of the wave climate at the beach or by the construction of groynes. Change can be effected by a variety of structures. These include the following:

- artificial headlands;
- offshore breakwaters;
- submerged offshore bunds.

With the exception of offshore bunds (see Section 4.5.1), the design and effects of such structures are beyond the scope of this book but they must be taken into account when designing beach recharge schemes. The possibility of constructing them as part of a long-term strategy of beach protection and coastal management should be reviewed.

4.4.4 METHODS OF EXECUTION

It has already been mentioned that the trailing suction hopper dredger and the cutter suction dredger are the tools which are usually best suited to beach recharge. Often the work may be executed using only a trailer for dredging and transport.

The trailer dredger has the advantage of being a ship with the ability to work in open sea. Thus, during the dredging element of the work cycle, work may proceed more or less independently of sea state (see Chapter 7). However, at the point of discharge, sensitivity to sea state and wind becomes critical. In order to unload the hopper a trailer must either bottom-dump, which then requires that the material be rehandled by other plant, or it must discharge by pumping, which requires connection to a pipeline.

Less commonly, the hopper may be pump-discharged over the bow by spraying material onto the beach or into adjoining shallow water. The success of such methods depends on how close the dredger can approach to the beach and the distance through which the material can be sprayed. Usually, the rate of loss of material is higher when using spraying methods but unit costs may be lower. If payment is made only for material retained within the profile, the rate of loss may not be too important to the purchaser. It may, however, have a negative environmental effect.

If material is to be discharged from the hopper to the beach via a pipeline, great importance must be attached to the method of connection between the dredger and the pipeline. This is the weakest link in the system.

Trailer dredgers which are regularly employed in beach recharge and reclamation work are usually fitted with a mechanised pipeline retrieval and connection system which most commonly operates over the bow. Such systems allow connection and discharge during moderately adverse sea conditions. Ideally the pipeline will extend into reasonably deep water, say ten metres or more, thus minimising the risk of grounding the dredger during discharge. However, many trailers are unable to pump material over long distances, especially if the material is coarse grained. Typically, when discharging medium sand, the maximum economic pumping distance is about 1500 metres. This must include the distance to the beach-head plus the distance along the beach to the point of discharge.

If the dredger can not safely moor closer than 1000 metres to the beach then, when discharge has extended to points 500 metres each side of the landfall, the sub-sea pumping main must be relocated. Relocation of the pumping main requires calm conditions. Alternatively, all material can be discharged at a single point on the beach and the material can be rehandled by land based plant. Another solution may be to add a booster pump, but unless the booster is mounted aboard the trailer the advantages of using a trailer in isolation may be lost.

If the limitations of pumping distance or production rates are not acceptable a trailer can be supplemented by a cutter suction dredger. The latter, if of reasonable size, can typically pump medium sand over about 3000 metres. The trailer is employed only to dredge, transport and dump to an offshore stockpile, from whence it is re-dredged and pumped to the beach by the cutter suction dredger. The rate of loss of material may be higher when using such methods. Losses in excess of 20 per cent are not unusual and far higher losses have been reported. The cutter suction dredger and associated pipeline are very vulnerable to wave attack and strong winds. The method is best suited to the placement of large volumes at high rates in reasonably sheltered conditions.

Other types of dredger may be used for the primary dredging. Transport is then by barges which may deposit the cargo at the toe of the beach at high tide, or to a stockpile for rehandling by cutter suction dredger, or be pump discharged by a floating barge unloader. The latter suffers similar shortcomings to the trailer/cutter suction combination.

Material which is pumped on to the beach is most effectively distributed by a combination of regular extension of the discharge pipeline, and pushing and shaping over short distances by bulldozer. Losses may be minimised by using beach fill for the formation of temporary training bunds, parallel to and at an appropriate distance from the shore.

4.5 Other beneficial uses of dredged material

Dredged material can be put to a wide variety of uses, in addition to beach recharge and conventional reclamation. In some cases, these beneficial uses have arisen as a result of a need to find a disposal solution for dredged material which was not required for reclamation. Some uses are obvious extensions to its use as a fill material and include bank stabilisation and dyke construction. Other uses include:

- offshore mounds or berms for coastal protection;
- the creation of wildlife habitats;
- use as soil;
- production of ceramics (artificial gravel).

These are briefly reviewed below.

4.5.1 OFFSHORE MOUNDS AND BUNDS

Offshore bunds have a similar effect to that of offshore breakwaters but the berms are fully submerged and are constructed using dredged sand, gravel, clay, rock or mixtures of these.

Because the bunds are submerged their influence on the inshore wave climate is less than that of an offshore breakwater. Nevertheless, the change in wave climate is beneficial and may be sufficient to reduce the rate of loss of beach material to an acceptable level.

Submerged bunds have several advantages. They are invisible; they create a nearshore wave breaker zone which may be attractive to surfers thus preserving or enhancing amenity value, but most important of all, they may be cheap. Relatively low cost is possible because the construction materials are easily dredged and placed by bottom-discharge from hoppers. Suitable materials may arise from unconnected capital or maintenance dredging, thus providing a beneficial use for materials which would otherwise be wasted. Subject to haul distance, the cost of disposal by submerged bund formation may be less than that of disposal in deep sea (Murden, 1994).

A disadvantage is that submerged bunds which are formed using easily dredged materials are more prone to erosion than structures formed of rock. However, loss of material from the bund may not be a total loss to the near shore sediment system. Eroded materials may move onto the local beaches, or foreshore, or feed beaches on the downdrift coastline, with beneficial effect.

The selection of submerged bunds as an aid to coast protection will be dependent on the capital cost of construction and cost of renewal, or maintenance. Where the material provided arises as a means of disposal of material from recurrent dredging activity the overall economics may be very attractive. This probably represents the optimum overall solution with maximum benefit to the local community.

4.5.2 CREATION OF WILDLIFE HABITATS

Dredged material has been used extensively, particularly in the US, to create or to restore large areas of wildlife habitat. In some respects, this use was aided by the public perception that onshore and coastal dredged material disposal sites were 'waste' disposal sites, leading to a reluctance to use these areas for commercial development or recreation (Landin, 1988). As a result, the sites were often extensively colonised by both plants and wildlife. Habitat formation and enhancement has now become one of the most important uses of dredged material in the USA.

There are numerous accounts of habitat formation in the literature but one example is described here as an illustration of how this type of use serves both the interests of those who require dredging to be undertaken and the preservation or improvement of the natural environment.

Gaillard Island was constructed in Mobile Bay, Alabama in 1980–81 as a disposal site for dredged material (Landin, 1988). The island has an area of 1300 acres and a 700 acre enclosed containment pond will continue to be used as a deposit for maintenance dredging material for many years. Seabirds began nesting on the island before construction was completed and 25 species were regularly nesting there by 1988, including the endangered brown pelican.

4.5.3 USE AS TOPSOIL

Whilst superficially attractive, dredged material has not been widely used as a soil. Potential problems of using dredged material as soil include:

- contamination;
- salinity and acidity;
- uptake of pollutants by plants;
- the cost, and the space required, for conditioning the material.

These aspects have been researched in the UK (Riddell *et al.*, 1989) and elsewhere. In the UK research, it was found that heavy metal concentrations, even at trigger levels, had little or no effect on plant growth and uptake by the plants was less than expected. Concentrations in three year old vegetation were less than trigger levels (see Chapter 13) even when the concentrations in the soil were higher.

Salinity and acidity can be reduced by leaching and the addition of appropriate fertilisers. The main practical problems associated with the use of dredged material in this manner stem from the requirement for dewatering and, in some cases, the need to alter the particle size distribution of the material in order to produce a soil with the right water retention and workability characteristics. The space required for treatment can often be a problem and the overall cost of treatment and unreliability of demand generally render this use unattractive.

4.5.4 PRODUCTION OF CERAMICS

Research in the Netherlands (Veltman *et al.*, 1989) has shown that fine grained dredged materials can be sintered to produce artificial gravel or bricks. This use has particular attractions when contaminated dredged materials are used, as the sintering process permanently seals the contaminants into the end-product.

It is, however, an expensive process and is not yet economic. The economics might improve if the amount of material which was processed in this manner increased substantially. The product cost is likely always to be more expensive than natural gravel and conventional aggregates and ceramics. Furthermore, other methods of disposal are less expensive and better suited to the large volumes commonly arising from dredging.

References

Athmer, J.B.E.M. & Pycroft A.S. 1986. Methods and results of quality control for hydraulically placed sandfill disposal of dredged materials, *Proceedings of the 11th World Dredging Congress 1986*, Brighton, CEDA, Delft, The Netherlands.

Bishop, A.W. & Vaughan, P.R. 1972. Consolidation of fine grained dredged material after hydraulic deposition, *Report to the National Ports Council by Dept of Civil Engineering*, Imperial College, London.

Choa, V. 1990. Soil improvement works at Tianjin East Pier Project, *Proceedings of the 10th Southeast Asian Geotechnical Conference*, Taipei. Vol. 1, pp 47–52.

CIRIA 1995a. *Beach Management Manual,* Research Project 483, Construction Industry Research and Information Association, London.

CIRIA 1995b. *Beach Recharge Materials, Demand, Resources and Availability*, Research Project 482, Construction Industry Research and Information Association, London.

De Nekker, J. & In't Veld J.K. 1975. Dredged Rotterdam Harbour Mud: its qualities and use as soil, *Terra et Aqua,* No. 8/9, International Association of Dredging Companies (IADC), The Netherlands.

van Duin, O.S.R 1987. The treatment of dredged soils in the Nakdong estuary barrage and reclamation project, *Dredging and Port Construction*, February.

Haliburton, T.A. 1987. Guidelines for dewatering/densifying confined dredged material, *Technical Report DS–78–11, USAWES,* Vicksburg, Mississippi, USA.

Krizek, R.J., Giger, M.W. & Jin J.S. 1975. Dewatering of dredged materials by evaporation, *Proceedings of the 1st International Symposium on Dredging Technology*, Canterbury, England.

Landin, M.C. 1988. Dredged material: a recognised resource, *Terra et Aqua*, No. 38, December, International Association of Dredging Companies (IADC), The Netherlands.

Lee, S.L., Karunaratne, G.P., Yong, K.Y., & Ganeshan V. 1987. Layered clay-sand scheme of land reclamation, *Journal of Geotechnical Engineering*, Vol. 113, No. 9, ASCE.

Lee, S.L., Karunaratne, G.P., Yong, K.Y., Tan S.A. Tan, T.S. & Vijiaratnam A. 1990. Developments in land reclamation, *Proceedings of the 1990 Convention of the Institute of Engineers*, Singapore.

Murden, W.R. 1994. An update on submerged berms in the coastal zone utilising dredged material., *Proceedings of the XIV world Dredging Congress 1994*, Amsterdam, Central Dredging Association, Delft, The Netherlands.

Riddell, J.F., Fleming, G. & Smith P.G. 1989. The use of dredged material as topsoil, *Terra et Aqua*, No. 39, April, International Association of Dredging Companies (IADC), The Netherlands.

Salem, A.M. & Krizek R.J. 1973. Consolidation characteristics of dredging slurries, *Journal of the waterways, Harbours and Coastal Engineering Division, ASCE*, Vol. 99, No. WW4.

Tavenas, F.A., Ladd, R.S. & La Rochelle, P. 1973. Accuracy of relative density measurements: results of a comparative test programme, *Evaluation of Relative Density and its Role in Geotechnical Projects Involving Cohesionless Soils*, ASTM STP 523, ASTM, 1973, pp. 18–60.

Tiederman, W.G. & Reischman M.M. 1974. Feasibility of hydrocyclones for dredge spoil, *Journal of the Waterways, Harbours and Coastal Engineering Division, ASCE*, Vol. 99, No. WW4.

Veltman, M., Visser, F. & Fresco J. 1989. Management of contaminated sediments by insulation and recycling, *Terra et Aqua*, No. 40, August, International Association of Dredging Companies (IADC), The Netherlands.

5
Disposal of Dredged Material

5.1 Introduction

Despite increasing pressure to find uses for dredged materials, there are often cases where there is little or no alternative to outright disposal. The decision to dispose of the material, rather than to use it (see Chapter 4), is generally based on economic considerations but, even in these days of environmental concern and awareness, this should not necessarily be regarded as a misguided approach. There are usually sufficient options available for the disposal of both contaminated and uncontaminated materials for economics to be considered a perfectly valid criterion, within limits, on which to decide the most appropriate disposal route.

The manner in which dredged material is disposed of has become perhaps the single most contentious, and also misunderstood, aspect of dredging. This chapter reviews the methods of dredged material disposal and summarises their potential effects under the following main headings:

- disposal at sea;
- disposal in shoreline enclosures;
- onshore disposal.

The degree of contamination is usually a key factor in deciding which disposal route is most appropriate and what special precautions need to be implemented in order to minimise the effect of contamination on the environment. In this connection, there exists the possibility of cleaning contaminated dredged material or separating the contaminated fraction from that which is uncontaminated or only mildly contaminated. This aspect of disposal is considered in Section 5.5.

Disposal of dredged material is frequently governed by international agreement and local legislation. These subjects are considered in Section 5.6 together with a review of decision-making procedures which can be implemented when assessing appropriate or acceptable methods of disposal.

5.2 Disposal at sea

Disposal at sea has become the focus of considerable attention in recent years. In the case of coastal dredging works, it is almost invariably the easiest manner in which unwanted dredged material can be disposed of and it accounts for a very considerable amount of material.

However, this method of disposal may have a wide variety of effects on the environment, including:

- elevated suspended solids concentrations;
- changes to natural seabed sediment regimes;
- destruction of benthic fauna and flora;
- damage to fisheries and fish breeding areas;
- release of pollutants, including heavy metals, to the marine environment.

The release of pollutants is a matter of considerable concern. The amount of dredged material from harbour areas, estuaries and sea channels in several West European countries which was dumped at sea in 1986 is shown in Table 5.1 (Donze, 1990) together with the estimated tonnages of heavy metals that were contained in the material.

Table 5.1 Amounts of dredged material dumped at sea from harbour areas, estuaries and navigation channels in some of the Oslo Convention member states in 1986 (from Donze 1990)

Origin	Country	Amount dumped (tonnes × 1000)	Amount of dumped elements (tonnes)							
			As	Cd	Cr	Cu	Hg	Ni	Pb	Zn
Harbours	Belgium	7 823			258	231	2.5	77	617	767
	Denmark	401	0.5	0.0	2	1	0.0	1	1	6
	France	8 999	275.7	3.1	115	60	22.1	27	129	340
	UK	5 896	0.7	5.7	336	241	2.9	88	301	671
	Ireland	1 651	5.7	<0.3	8	9	<0.0	8	12	47
	Netherlands	16 569	99.4	5.9	330	149	2.2	139	277	1036
	Portugal	394								
Sub total		42 000	382.0	<15.1	1050	692	<29.6	340	1337	2866
Estuary	France	858								
	UK	3 765	0.0	3.8	96	100	1.9	62	150	752
	Ireland	102	1.8	0.1	1	2	0.0	1	3	7
	Portugal	247								
Sea	Belgium	29 168			1467	294	6.1	296	745	5754
	UK	5 572	0.0	1.7	132	125	2.1	97	229	558
	Netherlands	10 007								
Sub total		50 000	1.8	5.5	1697	521	10.1	456	1126	7071
Grand total		92 000	383.8	20.6	2746	1213	39.7	796	2463	9938

In some cases, ill effects have resulted from offshore disposal. However, there is no reason why this method of disposal cannot be continued; the only requirement is that it should not be indiscriminate and it should always be preceded by detailed studies to establish not only the suitability of candidate sites but also whether there is a real need for disposal in preference to one or more beneficial uses.

5.2.1 *OFFSHORE DISPOSAL OPTIONS*

Offshore disposal areas can be either retentive or dispersive depending upon the nature of the material placed there and the hydrodynamic characteristics of the area. Retentive areas are designed, or intended, to confine the dredged material to a well-defined area. As such, they have a limited capacity. Dispersive areas are intended to be sites where material, after placement, is dispersed over a wide area by means of the natural sediment transport processes operating in that area. The capacity of such areas to receive dredged material is only limited by any constraints imposed by the rate and pattern of dispersion.

An offshore location classification was proposed by Holiday (1978) and was summarised in the PIANC report on disposal of dredged material at sea (1986). Four offshore zones were identified, distinguished by the nature and severity of sediment movement processes:

- The deep ocean – considered to be that part of the ocean deeper than 200 metres or beyond the continental shelf break; the general assumption is that material deposited on the deep ocean bed remains there.
- The open shelf – between the continental shelf break and the 40 metre depth contour; this is a relatively high-energy zone characterised by tidal currents and waves.
- The nearshore – from the 40 metre contour up to the breaker zone; this is a high-energy zone dominated by waves, longshore drift and tidal currents; the potential for sediment movement is considerable.
- The inlet zone – the area adjacent to estuaries, rivers and inlets where large-scale sediment movement occurs; the energy levels are similar to those in the nearshore zone except that tidal currents may be strong and coastal irregularities and structures may result in complex, multidirectional wave and current patterns.

There are four main methods of offshore disposal of dredged material:

- unconfined disposal by dumping in licensed disposal areas;
- confined disposal by controlled bottom-dumping into seabed depressions or redundant sand borrow pits;
- confined disposal between underwater dykes constructed on the seabed;
- formation of islands.

These are schematically illustrated in Figure 5.1 (PIANC, 1986). Effective dispersal is generally only likely to take place in the case of unconfined dumping, and then only in the case of dumping in the open shelf, nearshore and inlet zones. The other options listed above imply a desire to retain the material within the confines of the disposal site. All of the above disposal options can be extended, in the case of disposal of contaminated materials, to include capping with inert material in order to isolate the material from the environment.

5.2.2 UNCONFINED DUMPING IN LICENSED DISPOSAL AREAS

This method of disposal has traditionally been used throughout the world. Losses arising from dilution and dispersion can be considerable, depending on the nature of the material, the manner in which it is dredged and placed in the disposal area and on the hydrodynamic regime at the site.

Aside from the obvious impacts on the disposal site itself, unconfined dumping may affect a very wide area around the site. The adoption of this method must, therefore, be preceded by detailed studies to investigate and quantify the potential effects. Assessments of the effects on the disposal site in question need also to be viewed in relation to other disposal sites in the area, particularly in semi-enclosed sea areas. Whilst the effects of one disposal site may be acceptable, the combined effects of several sites may be far from acceptable, particularly in relation to the disposal of contaminated materials. The Black Sea

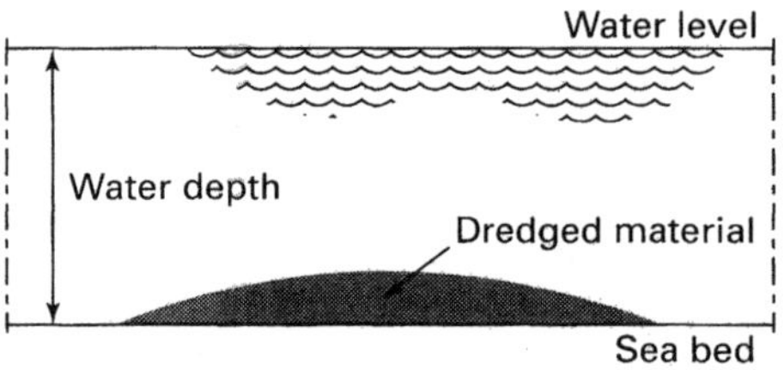

UNCONFINED DISPOSAL

Environment	Rate of dispersion with open dispersal
Deep ocean	Not anticipated
Open shelf	Very limited dispersion anticipated. At the most only until the initial consolidation has been completed. (This takes a few days.) There is little experience in the use of this method. It will depend upon sea current and waves.
Near-shore	Usually considerable. In principle, the fine fraction will be washed out and the coarse material (sand) will remain behind.
Inlet zone	In principle, total dispersion.

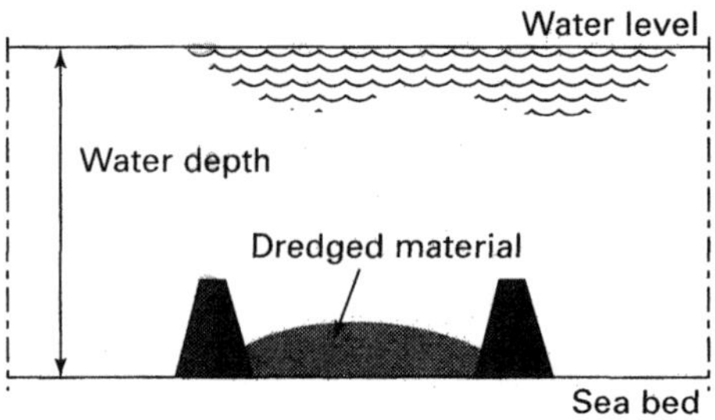

Environment	Rate of dispersion with underwater dams
Deep ocean	Not applicable
Open shelf	Very limited
Near-shore	Dispersion will be reduced if the dimensions are such that the sea currents are sufficiently diverted from the disposal site. Disposal takes place in the calm season. Consolidation, until equilibrium is reached, is completed before the storm season.
Inlet zone	Strong currents and waves will make it almost impossible to maintain the dams. In principle total dispersion is possible.

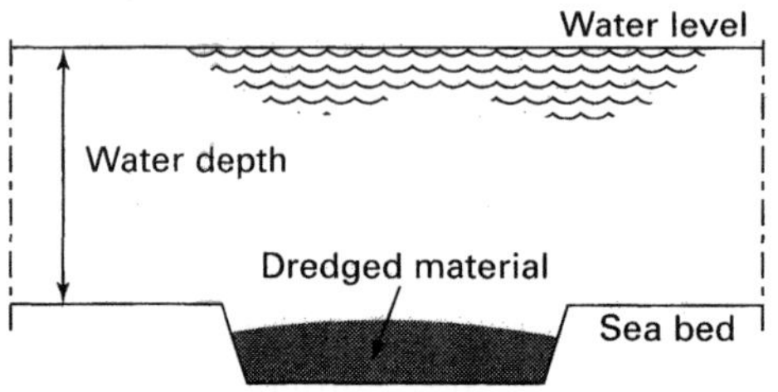

Environment	Rate of dispersion with a pit in the sea-bed
Deep ocean	Only possible if a natural pit is available.
Open shelf	Technically almost impossible, unless a natural pit is available.
Near-shore	Dispersion will be less than that occurring when underwater dams are used.
Inlet zone	Strong currents and waves cause sediment transport. This results in very rapid natural filling of the pit by coarse material. The finer fractions of the material will disappear from the pit.

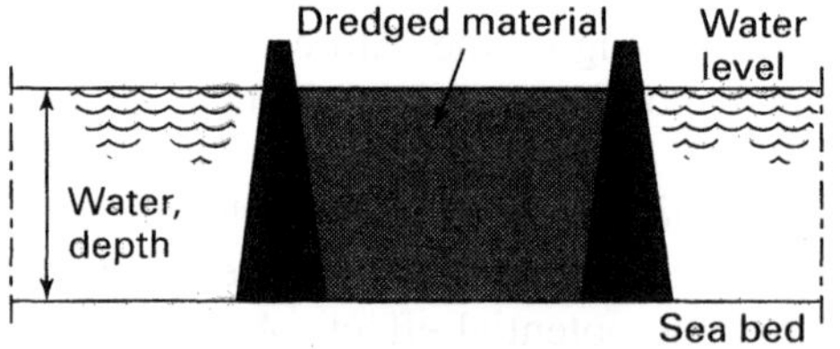

Environment	Rate of dispersion with an artificial island in the sea
Deep ocean and Open shelf	Not applicable
Near-shore	No dispersion of contaminated dredged materials. At most very slow dispersion of contaminants via the groundwater. This is valid for all options.

CONFINED/SEMI-CONFINED DISPOSAL

Figure 5.1 Schematic illustrations of the four main options for offshore disposal (from PIANC 1986) is a case in point, where offshore disposal of dredged material is now prohibited.

Potential effects of unconfined disposal at sea may be temporary or permanent. Temporary adverse effects are principally related to high concentrations of suspended sediment and include:

- increased turbidity;
- reduction of water quality;
- release of contaminants into solution;
- reduction of recreational value, eg. swimming, diving and sailing;
- possible fouling of nearby beaches.

More permanent adverse effects may include:

- detrimental changes to water and sediment movement due to changed seabed morphology;
- possible increased coastal erosion in the case of near-shore disposal sites;
- permanent removal of sediment from a longshore transport system in the case of regular maintenance dredging.

The effects of offshore disposal are generally most pronounced in shallow inshore areas, partly because the potential for material erosion and transport from the dump site is so great and partly because the inshore area is inherently sensitive. Removal of material from the longshore transport system can be at least partly resolved by placing the dredged material in an inshore area downcurrent of the dredging area so that it may re-enter the system. In Alabama in the USA, an offshore disposal site has been designed so that material which is lost from the site nourishes local beaches and reduces the magnitude of coastal erosion (Clarke *et al.*, 1988).

Most unconfined dumping operations involve disposal in a well-defined area. An alternative approach which can be adopted with clean materials is to spread the material in very thin layers over as large an area as possible. 'Thin layer' disposal has been investigated by the Corps of Engineers in Mobile Bay, Alabama (Nester and Rees, 1988). Dredged material was spread in layers of 150 to 600 millimetres within the actual disposal area and 0 to 300 millimetres in the fringe area. No significant impacts to water quality parameters were detected, with the exception of total suspended solids concentrations during disposal activity. No detectable differences in the macrofauna community were noted that could be attributed to the disposal activity. Fluctuations could readily be attributed to natural seasonal and spatial variability. Reworking of the sediments occurred after the disposal by both physical and biological forces. Three weeks after disposal, a broad recolonisation occurred over the study area. The fisheries data did not show any significant variability that could be linked to the presence of the thin cover of dredged material.

The present authors believe that the acceptable layer thickness will vary considerably from one area to another and that in many areas the average layer thickness would need to be significantly less than those described above if damage to benthic communities is to be avoided.

Deep ocean disposal appears to have relatively little environmental impact when compared with disposal in shallower waters. Miller *et al.* (1980) describe studies to investigate the effects of dredged material disposal in 400 metres of water in Hawaii. Detailed baseline investigations followed by monitoring during and after disposal operations involving 640 000 cubic metres of material led to the following conclusions:

- the dumped material spread over a wide area;
- except within the immediate area of the dump site, benthic foraminifera populations appeared to have recovered to pre-disposal levels within six months;

- an increase of zooplankton populations observed during disposal operations returned to baseline level within six months of the end of dumping;
- similarly, heavy metal concentrations returned to background levels within six months;
- no significant differences were found between heavy metal concentrations in shrimps and zooplankton at the disposal site and those at a control site.

Broadly similar conclusions have been reported from disposal monitoring studies elsewhere, but that is not to say that this method of disposal can safely be utilised in all cases. Guidelines for the assessment of deep ocean disposal areas are reviewed in Section 5.6. One of the main disadvantages of deep ocean disposal is cost. There are relatively few areas of the world where water depths in excess of 200 metres exist within a reasonable distance of dredging sites and transport costs are, in most cases, likely to be prohibitive.

5.2.3 DISPOSAL IN SEABED DEPRESSIONS

The semi-confined disposal of dredged material in seabed depressions was suggested by Carpenter (1973). Since then, studies have been undertaken as part of the USACE Dredged Material Research Programme into the use of seabed depressions as disposal sites (Bokuniewicz, 1983). Such depressions may be either natural or artificial, such as those remaining after extraction of fill materials for reclamation projects. DeLoach and Waring (1984) described the disposal of contaminated dredged sediments containing kepone into a deep depression in the James River in Virginia. The material was discharged through a vertical pipeline at a depth of about five metres below sea level. Extensive monitoring demonstrated that the material remained within a small area around the disposal point. The depression was successfully filled without any discernable adverse effects on the surrounding marine environment including nearby oyster beds.

In the Netherlands and in Belgium this method of disposal has been employed successfully at several locations, principally in connection with contaminated materials. In the 1st Petroleumhaven, in Rotterdam, the harbour bottom was dredged to create three pits into which material contaminated with pesticides was placed (Loxham *et al.*, 1988; Kleinbloesem and van der Weijde, 1983). The material was then capped by a layer of clean sand. In Antwerp, approximately 464 000 cubic metres of contaminated sediments were placed into a specially dredged depression in the Doeldok. The material was placed by a diffuser pontoon. In the North Sea, approximately 50 000 cubic metres were successfully placed in a natural seabed depression (Vellinga, 1984).

In Hong Kong contaminated dredged marine mud is placed in specially-dredged pits by simple bottom-dumping from trailer dredgers and barges (Brand *et al.*, 1994). Unpublished work by one of the authors (J.M. Land, personal communication 1992) demonstrated that material can be placed in pits with minimal loss of sediment to the marine environment as long as certain precautions are taken. These relate mainly to the matching of dumping vessel size and discharge speed to the material type, water depth and the dimensions of the target pit. Within these limitations, it has been found that the characteristics of the site are more important than the method of placement in determining the loss of material during and immediately after placement.

Properly controlled disposal into seabed depressions is generally unlikely to have any long-lasting adverse effects. The greatest potential effect may relate to the reason why the depression was there in the first instance. Some natural seabed depressions will be unsuitable for disposal of dredged material, if retention is a requirement, as they are the product of natural forces and attempts to backfill them may be countered by nature. All proposals for such disposal should be supported by detailed studies of potential losses from the depression. These are most likely to arise as a result of erosion by tidal and wave induced currents, particularly if the pit is filled to the level of the adjacent seabed. In shallow waters, propeller scour from passing vessels may also be a consideration.

Disposal into dredged borrow pits restores the original seabed morphology and is unlikely to have adverse effects unless placement is undertaken in a careless fashion leading to large losses of material.

5.2.4 DISPOSAL BETWEEN UNDERWATER BUNDS

Disposal between underwater bunds is, in most aspects, similar to disposal in seabed depressions, the only difference being that the construction of the bunds significantly raises costs. In addition, selection of a suitable site for such a disposal method might be difficult. It is clearly not a practical option in deep water. In shallow waters, wave and current action can lead to some loss of materials and, if severe, would require the construction of formidable enclosing bunds if the integrity of the construction is to be maintained.

In Mobile Bay, Alabama, an underwater bunded disposal facility was constructed in the 1980s in about 10 metres of water (Landin *et al.*, 1995). Approximately 24 million cubic metres of dredged material has been placed in the facility. No long term adverse effects have been detected and several beneficial effects have been recognised including an increase of the local finfish population and the use of the upper slopes and the top of the bunded area by shrimps (Clarke and Kasul, 1994). An additional beneficial effect is that the bund is expected to dampen hurricane-generated storm surges moving into Mobile Bay.

Another example of such a method being applied is the Juzhnuy Harbour project in the Black Sea (Miroshnichenko, 1982). During the late 1970s almost nine million cubic metres of dredged material were dumped within the confines of a U-shaped submarine bund located approximately 1.5 kilometres offshore. Observations indicated that the dumped material remained in the disposal site except during storms when small amounts of fine material were lost.

5.2.5 DISPOSAL TO FORM ISLANDS

The seabed bund option described above can be extended to form offshore islands. Such areas may be of considerable environmental benefit, notably in the provision of wildlife habitats. This solution has been adopted in many places and, if the retaining bunds are adequately designed and constructed, results in minimal loss of material. As with disposal between underwater bunds, it is not a practical option in deep water.

In Tampa Harbour (Rees, 1980), bunds were formed within turbidity-control barriers by pumping dredged material to an initial level of about 2.4 metres above sea level. They were then raised to between six and nine metres above sea level using draglines. Enclosed areas can be infilled immediately and planted, or can be used as and when required for the disposal of material arising from maintenance dredging operations.

A 670 hectare artificial island was constructed in Mobile Bay, Alabama (Fowler *et al.*, 1986) to accommodate dredged material. The enclosing bunds were constructed from a variety of materials, including soft clays, which were placed both hydraulically and by dumping from barges. The island was reclaimed on very soft clays and construction of the retaining bunds had to proceed very carefully to minimise failures. Bund slopes varied between 1:32 and 1:45 when constructed with mixtures of silts, clays and sands and between 1:53 and

1:66 when constructed principally of sands. It was estimated that complete consolidation of the embankments and foundation soils would require more than 800 years. The island has been widely publicised as an outstanding example of a disposal operation which resulted in the creation of valuable wildlife habitats (see Chapter 4).

5.2.6 CAPPING OF CONTAMINATED MATERIALS

Capping contaminated sediments is an attractive method of isolating the material from the marine environment. Truitt *et al.* (1989) give summary accounts of ten capping operations carried out in The Netherlands, Japan and the USA (see Table 5.2). The majority of the USA capping operations involved placement of both contaminated material and capping material by simple bottom-dumping from barges. The capping operations in Rotterdam Harbour utilised a plough (or bed leveller) to move previously dumped material over the contaminated sediment. In Hiroshima Bay (Kikegawa, 1983; Togashi, 1983), placement of the capping layer was achieved using a gravity-fed tremie pipe and by pumping through a tremie fitted with a diffuser bar. It is evident from Table 5.2 that the thickness of the caps placed in the USA by dumping from barges is very variable and in one case coverage was found to be incomplete. It might also be surmised that, because the caps were placed by dumping, mixing of the capping and the contaminated materials occurred.

Capping operations need to be very carefully designed if they are to be effective. Shields and Montgomery (1984) list ten considerations for planning capping operations (see Table 5.3), some of which are reviewed in more detail below.

Site selection

Obviously, the chosen site must be one where dredging is not going to take place in the future. It must also be one which is not going to be used as an anchorage. Even the best-designed and constructed cap is not going to maintain its integrity under the action of falling and dragging anchors unless it is very thick.

Type of capping material

The type of material used to form the cap has a bearing on the required thickness, the stability of the cap itself and on the method of placement. Both mud and sand can provide an effective seal in the sense that they can be placed in uniform layers with a thickness sufficient to prevent the contaminated material being reached by burrowing organisms. Sand is less attractive to burrowers than mud and the cap thickness which is required to ensure that bioturbation does not reduce the integrity of the cap is therefore less.

The capping material must be able to withstand erosive forces of tidal and wave-induced currents. In the case of a sand cap, this would require the selection of material which is sufficiently coarse. In the case of clay, the requirement would be a sufficiently high degree of cohesion. Each site would require separate assessment to determine the nature of the material required to resist erosion.

Thickness of the cap

The thickness of the cap must be sufficient to ensure that it effectively isolates the contaminated material from the water column and from pelagic and benthic biota. The effectiveness of the cap needs therefore to be assessed both chemically and biologically. Other thickness-related factors might also be considered:

- the ability of the cap to resist deformations due to consolidation settlement of the underlying material;
- the weight of the cap which will assist in the strength improvement of the underlying material;
- if erosion is allowed, the thickness of sacrificial cap required to consolidate the underlying material to the extent that, when eventually it is exposed, it will not be eroded.

Factors such as these, whilst being related to some extent to the thickness of the cap, should be considered separately as part of the design process.

Sturgis and Gunnison (1988) describe a laboratory test procedure to estimate the cap thickness required to counteract chemical processes. In summary, it involves the analysis of water drawn from large flasks (see Figure 5.2) which are first partially filled with contaminated sediment and then covered by varying thicknesses of the chosen capping material. Control tests are also carried out on flasks containing only contaminated sediment and only capping material. An example graph showing oxygen depletion rates of Black Rock Harbour sediment, capped with sand of varying thickness, is shown in Figure 5.3. Similar graphs could be produced for any contaminant of concern.

Each potential disposal site should be assessed on the basis of its biological population when deciding on the required thickness of capping material (Shields and Montgomery, 1984). One of the objectives of capping is to isolate the contaminated dredged material from burrowing organisms. Sand tends to attract suspension-feeding organisms which do not burrow deeply, whereas mud will attract deep-burrowing organisms. Sturgis and Gunnison

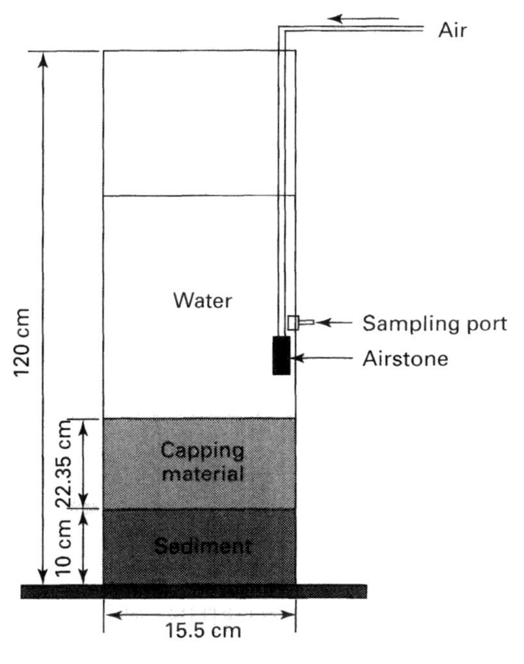

Figure 5.2 Apparatus for determining the thickness of cap material required to achieve chemical isolation of contaminated sediments (from Sturgis and Gunnison 1988)

Table 5.2 Summary account of ten capping operations carried out in the Netherlands, Japan and the USA (Truitt *et al.*, 1989)

Project		Contaminated material			Capping material			
Location (date)	Site characteristics	Volume of material (m)	Dredging method	Placement method	Volume (type) (m)	Thickness of cap (m)	Placement method	Positioning method
Duwamish Waterway, Seattle, Wash. (1984)	Existing subaqueous depression, 21 m deep	840	Clamshell	Scow	2700	0.3–1.0	Sprinkling from scow	Surveying instruments
Rotterdam Harbour, The Netherlands (1981–1983)	Phase I: Botlek Harbour excavated to 29.9 m deep	920 000	Trialing suction hopper	Pumpout submerged diffuser	(clay)	0.6–1.0	Scow, then levelled over site	Surveying instructions
	Phase II: First petroleum harbour excavated to 24.4 m deep	470 000	Matchbox suction	Pipeline submerged diffuser	(clay)	0.6–1.0	Scow, then levelled over site	Automated dredge and suction head equipment
Hiroshima Bay, Japan (1979–1980)	Contaminated bottom sediment overlaid *in situ* with capping material, 21 m deep	N/A	N/A	N/A	(sand and shell)	0.5	Conveyor to gravity-fed submerged tremie Suction/ pumpout through submerged spreader bar	Surveyed grid and winch/ anchor wires
New York Bright (1980)	Generally flat bottom, 24.3–27.4 m deep	660 000 (mounded to 1.8 m thick)	Clamshell	Scows	1 400 000 (majority fine sand)	Average 1–1.2; maximum 1.5–2.7	Scow, hopper dredge	Buoy, real time navigation electronics

Table 5.2 *(Continued)*

Project		Contaminated material			Capping material			
Location (date)	Site characteristics	Volume of material (m)	Dredging method	Placement method	Volume (type) (m)	Thickness of cap (m)	Placement method	Positioning method
Central Long Island Sound disposal area (1979)	Stamford-New Haven, North; generally flat bottom, 19.8 m deep	26 000 (mounded 1–1.8 m thick)	Clamshell	Scows	50 000 (sand)	Up to 2.1–3.0	Hopper dredge	Buoy, Loran-C coupled system
Central Long Island Sound disposal area (1979)	Stamford-New Haven, South; generally flat bottom, 21 m deep	38 000 (mounded 1.2–1.8 m thick)	Clamshell	Scows	76 000 (cohesive silt)	Up to 4.0	Scow	Buoy, Loran-C coupled system
Central Long Island Sound disposal area (1981)	Norwalk, generally flat bottom, 19.8 m deep	70 000 (multiple mounds up to 2.4–3.6 m thick)	Clamshell	Scows	280 000 (silt and sand)	Up to 1.8–2.1	Scow	Buoy
Central Long Island Sound disposal area (1982–1983)	Mill-Quinnipiac generally flat bottom, 19.8 m deep	30 000	Clamshell	Scows	990 000 (silt)	Multiple broad area placement, estimated average 1.8–3.0	Scow	Buoy
Central Long Island Sound disposal area (1983)	Cap site No 1; generally flat; 18.3 m deep	25 000 (mounded 1 m thick)	Clamshell	Scows	60 000 (silt)	Incomplete coverage	Scow	Buoy, Loran-C
Central Long Island Sound disposal area	Cap site No 2; generally flat, 17.0 m deep	30 000 (low mound 0.6 m thick)	Clamshell	Scows	30 000	Irregular; maximum 1.4; minimum 0.2	Scow	Buoy, Loran-C

Table 5.3 Considerations for planning capping operations (from Shields & Montgomery, 1984)

No.	Decision	Influenced by
1	Dredge equipment selection	
2	Selection of capping site	
3	Placement method of contaminated material	1, 2
4	Method of transporting contaminated material to the disposal site	1, 2, 3
5	Selection of capping material	1, 2, 3, 4
6	Placement method for cap	1, 2, 3, 4, 5
7	Dredging plant for obtaining capping material	1, 2, 3, 4, 5, 6
8	Method of transporting capping material to disposal site	1, 2, 3, 4, 5, 6, 7
9	Method for navigation and positioning at site	2, 4, 8
10	Method for monitoring disposal site	2, 9

(1988) recommend that the design cap thickness should be equal to the thickness required to produce an effective chemical seal, plus a thickness equivalent to the maximum depth which burrowing organisms will reach. This depth appears to lie in the general range 0.1 to 0.3 metres (Bokuniewicz, 1983) although there are some organisms which may burrow to depths of 1 metre or more.

In addition to the thickness defined by the need to achieve chemical and biological isolation, the thickness must also be considered from the viewpoint of practicality of placement (Truitt *et al.*, 1989). It is considered that, in view of the difficulties inherent in placing caps

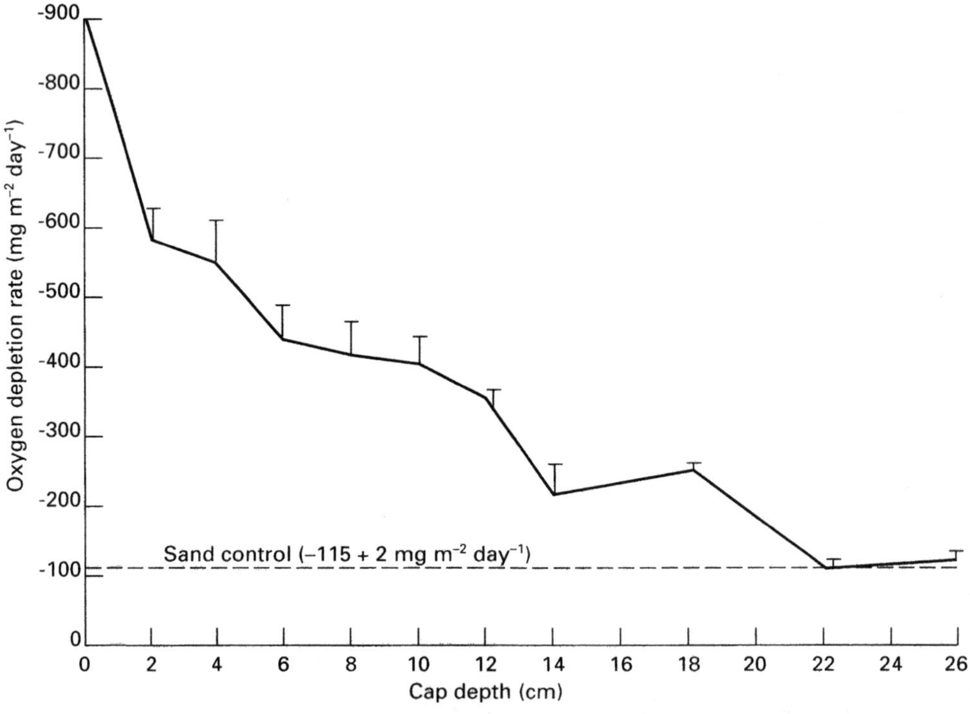

Figure 5.3 Example graph showing oxygen depletion rates of Black Rock Harbour sediment capped with sand of varying thickness (from Sturgis and Gunnison 1988)

evenly over large areas and the inaccuracies inherent in the echo-sounding methods used to monitor and measure levels during placement (plus or minus 0.15 metre), the specified nominal cap thickness should be at least one metre.

Cap stability

Consolidation of the contaminated sediments will result in movement and deformation of the capping material. In the case of caps over mounds of contaminated dredged material, differential settlements could be significant and result in disruption of the cap. If redundant borrow pits are filled and capped, the thickness is less likely to vary significantly over short distances and the cap is less likely to be disrupted in this manner.

Of greater concern is the potential for convective instability which results in the partial or complete inversion of the cap and contaminated sediment layers.

Bokuniewicz and Liu (1989) have described the conditions which may lead to convective instability. These are:

- when the density of the cap is greater than the density of the underlying contaminated sediments;
- when the shear stress along the interface between the two materials is greater than the shear strength of the deposit.

The authors suggest that if the interface between the materials is irregular then, for irregularities of height h, the magnitude of the shear stress along the interface will be $\Delta\rho g\delta h$ where $\Delta\rho$ is the density difference between the materials and δh is the slope of the interface. For motion to start, the authors state that the value of $g\delta h$ must be greater than some critical value. The present authors suggest that this critical value should be the shear strength of the weakest material present at or near the interface.

In practice, it appears that the best means of ensuring cap stability is to eliminate any irregularities. This need not be as difficult as it might first appear. In the case of trailer-dredged sediments, the backfilled material is likely not to form irregular mounds if it has been slurrified during the dredging process. Careful capping could maintain this situation. If the contaminated materials were dredged by mechanical means, it is more likely, even though they may have a tendency to form mounds and irregularities, that they would have sufficient strength to resist large scale deformations under the weight of a relatively thin uniform cap. If necessary, irregularities could be smoothed to some extent using a bed leveller but this would inevitably put some of the contaminated material into suspension.

Method of placement

The method by which the cap is placed is clearly of critical importance to its integrity as a sealing layer. It is evident that uncontrolled bottom-dumping from barges or trailers is far too energetic a process to be considered for cap placement. It would only result in the remobilisation of large quantities of contaminated materials and considerable mixing with the capping layer, and it would be almost impossible to ensure that the cap conformed with the specified requirements. In the USA, most capping has been achieved by controlled, that is relatively slow, bottom dumping, but the results have been caps of variable thickness and, in some cases, incomplete cover. Five methods of capping can be considered, each of which is more sophisticated than simple bottom-dumping:

- allow the cap to be formed naturally;

- place the cap using delivery pipes and diffusers;
- sprinkle the capping material so that it reaches the seabed as a well distributed cloud at the descent velocity of the individual soil particles;
- use bed levellers to move material from the surrounding seabed over the contaminated material;
- place a low-density fluid mud capping material in such a manner that it flows laterally over the contaminated dredged material.

In some cases, part-filled disposal pits will act as sediment sinks and, over a period of time, a natural cap will be formed. The gentle and continuous process of natural sedimentation could be expected to ensure no mixing with, or disturbance of, the underlying contaminated material and an even thickness of cap. Natural capping was achieved in a pit in the North Sea into which dredged material had been placed up to about 75 per cent of its total capacity. It was reported that the dredged material was covered with material which moved naturally into the pit from the surrounding seabed until a balance between erosion and deposition was attained (Vellinga, 1984).

Delivery pipes and diffusers have been used in The Netherlands, Belgium and Japan for the controlled placement of dredged material, including caps. The placement technique and a typical diffuser arrangement are shown in Figure 5.4. There is little doubt that material losses can be reduced to a minimum using this method but in the case of extremely soft, almost fluid trailer-dredged sediments, there remains the question of cap stability. Capping material placed in this manner needs to be spread evenly in thin layers and with a high degree of precision if instability is to be avoided.

The gentle sprinkling of material over the contaminated sediments appears to offer the best chance of overcoming most of the difficulties of capping. The capping material should be sand. In the case of large-scale operations, it could be sprinkled by sailing a trailer dredger slowly over the area to be capped whilst the bottom doors or valves are part open or being very slowly opened. Placement should take place gradually along grid lines and in several stages. Modifications of this method would be to use the self-unloading system of the trailer to spray the sand through a bow-mounted nozzle whilst sailing over the capping site or to discharge slowly through the suction pipe whilst underway.

A plywood diffuser box has successfully been used to place capping material in Tacoma Harbour, USA (Sumeri, 1989). The diffuser box measured 2.4 by 4.9 metres and had holes in the bottom and sides. It was mounted on a spud barge and sand was delivered hydraulically whilst the box was moved from side to side, effectively sprinkling the sand gently over the materials to be capped.

A bed leveller was used in Rotterdam to move clean materials over previously placed contaminated sediments. It is not an attractive solution where large areas are involved, partly because it is slow and partly because it requires a very high degree of precision if the cap is to be placed evenly. The method is only likely to be effective when material is being moved into a depression from adjacent higher areas. It is not considered to be a suitable method for capping a mound of contaminated material.

Expulsion of pore water

It is inevitable that during consolidation of the contaminated sediments, pore water will be expelled which will either move laterally into adjacent uncontaminated bottom sediments or vertically through the capping material, or both. However, a number of studies have concluded that, in the case of fine sediments, loss of contaminants to the marine environment

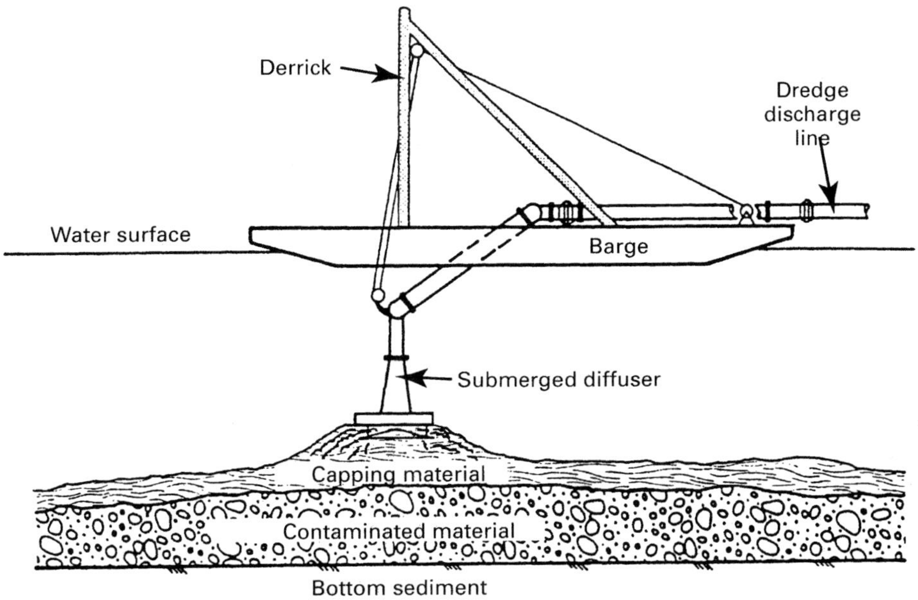

(a) Placement of capping material through a delivery pipe and diffuser (from Shields and Montgomery 1984)

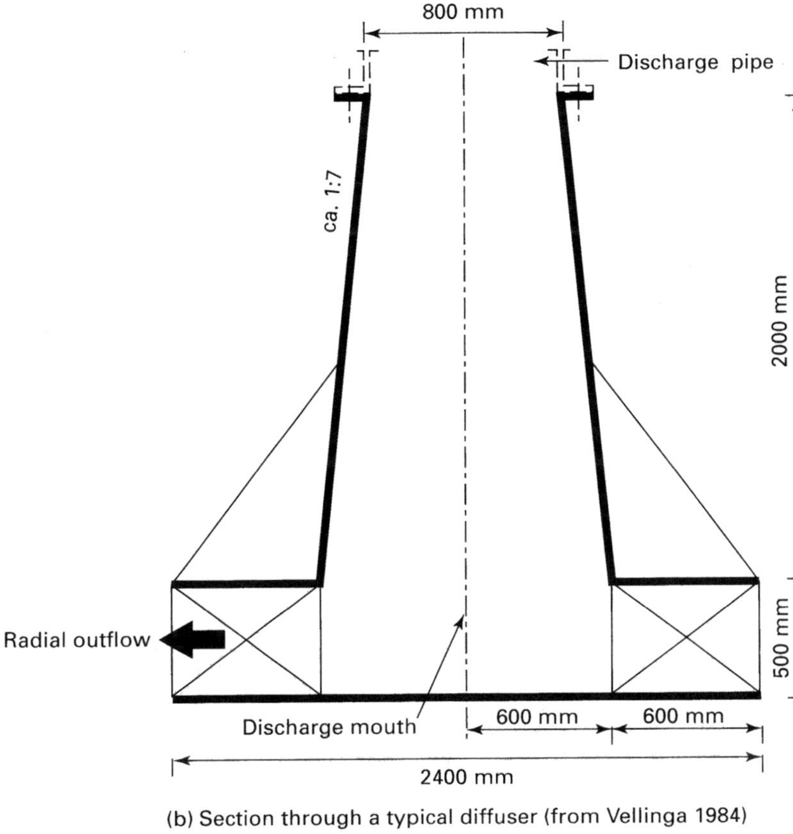

(b) Section through a typical diffuser (from Vellinga 1984)

Figure 5.4 A typical diffuser arrangement

in this manner is minimal, particularly if the dredged material has not been aerated. If loss of pore water is a major concern, one option is to ensure that the volume of the pore spaces in the capping material is sufficient to accommodate the water expelled from the contaminated soils. However, in many cases, this would require the use of an extremely thick cap.

5.3 Disposal in shoreline enclosures

This method has been the subject of very detailed research, particularly in the Netherlands and the USA where it has been put into operation on a large scale. There are two main variants of this type of disposal:

- disposal within low-cost enclosures designed simply to retain solids;
- disposal within a high-integrity enclosure designed to retain all contaminants.

In principle, there is little difference between the two types except for the extent to which the sites and the potential consequences of leakage are investigated prior to design and construction.

The largest of the Dutch shoreline disposal sites is the Slufter Project, located at the mouth of the Rhine. It comprises an eight kilometre long ring dam enclosing an area of 260 hectares with a storage capacity of 150 Mm^3 (van Orden and Witter, 1989). The facility was constructed by dredging approximately 37 Mm^3 of sand to form a pit 28 m deep. The sand was used to construct the ring bund which has a crest elevation of +23 metres above mean sea level (see Figure 5.5). The Slufter accommodates most of the 10 Mm^3 of contaminated sediment which is dredged every year in Rotterdam.

The Slufter facility is designed to accept moderately contaminated materials. No special liners were used in the construction because studies indicated that contaminants would largely be retained within a relatively fresh water lens which will form under the depot as a result of percolation of water from the dredged material into the underlying saline groundwater.

A second facility, the Papegaaiebek (Parrot's Beak) has been constructed nearby to accept 1.5 Mm^3 of very heavily contaminated materials (Vellinga, 1989). It is lined with high density polyethylene plastic sheeting to prevent migration of pore water into the soil. This form of disposal is comparatively expensive but, when properly designed, can be considered as a 'high integrity' solution. In theory, material can be transferred from the transport vessel into the disposal facility with no losses and an absolute guarantee that it will remain there.

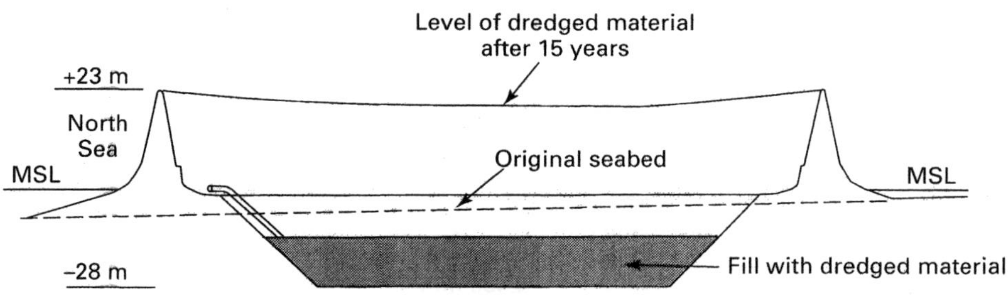

Figure 5.5 The ring bund of the Slufter disposal area, Rotterdam (from van Orden & Witter 1989)

5.4 Onshore disposal

Onshore disposal is an option which might be considered only where there is a large nearby area of ground which has little value in its existing state. In most cases, onshore disposal is achieved by pumping the dredged material directly to the disposal site. Onshore disposal inevitably has some environmental impact but, when well designed and executed, can provide long-term economic benefits even if the engineering properties of the material are poor. For example, it may ultimately provide valuable agricultural or recreational land, or it may be used to create new or improved wildlife habitats.

There are several potential adverse environmental effects, including:

- pollution of groundwater;
- odour;
- visual impact;
- destruction of flora and fauna.

Most adverse effects can either be overcome or made temporary by careful site selection and good placement methodology. In the USA, for example, thin-layer disposal is being used onshore in a similar manner to the offshore thin-layer disposal described earlier (Wilber, 1992). It has been shown that placement in thin layers can reduce the effects of disposal on the environment and, in some cases, prove beneficial by improving low-grade agricultural land or restoring wetlands.

In Hamburg, where a large proportion of the 2 Mm^3 which is dredged each year is heavily contaminated, two 'silt hills' are being constructed on top of an existing onshore disposal area.

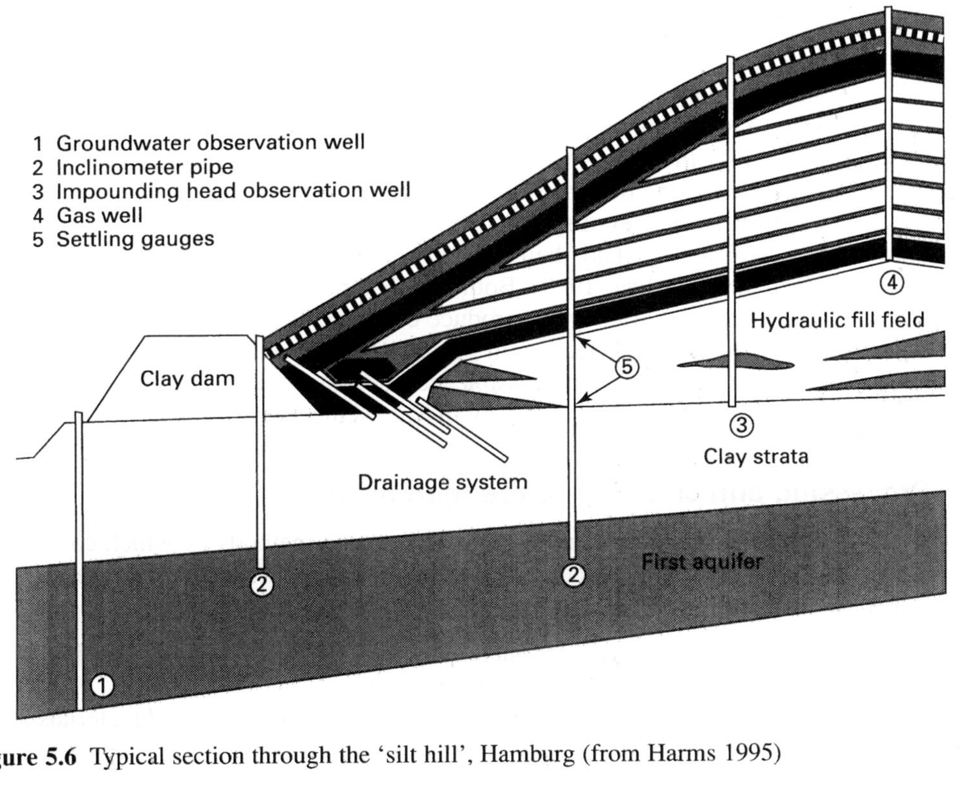

Figure 5.6 Typical section through the 'silt hill', Hamburg (from Harms 1995)

The dredged material is dewatered and the contaminated silt is separated from the less-polluted sand fraction. It is then sealed between layers which incorporate impermeable membranes and which slope towards the perimeter of the hill where drainage water is collected for further treatment. The top of each hill is sealed with a dense binding layer and vegetated (Hensen, 1989). A typical section through a 'silt hill' is shown in Figure 5.6 (Harms, 1995).

The leakage of water from the contaminated soils into the underlying ground can be prevented by incorporating artificial membranes, as at Hamburg. However, research in the Netherlands has shown that natural materials can be used to adsorb heavy metals and organic micro-contaminants from the water passing through them (Veltman *et al.*, 1989). Clay and peat have been shown to be excellent in this respect.

Maximising onshore disposal capacity

In the majority of cases, dredged material is delivered to onshore and shoreline confined disposal facilities as a soil/water slurry with a low solids content, typically in the range 10 to 25% by volume. By matching the delivery rate to the size of disposal area and by installing suitable overflow and outlet structures, a large proportion of the water can be drained from the placed material. However, if the capacity of such disposal facilities is to be maximised, it is necessary to implement measures designed to achieve the highest possible degree of consolidation, and thus volume reduction, of the dredged material.

In the case of sand, such volume reduction is not possible unless one resorts to traditional ground improvement techniques such as dynamic compaction or vibrocompaction. The benefits to be gained are, however, relatively small in volumetric terms and compaction is rarely employed unless required for engineering reasons, for instance foundations.

In the case of silts and clays, considerable benefits may accrue by careful management of the area designed to reduce the water content of the material as much as possible. Such management focuses on the subdivision of the disposal area into cells which are used in sequence with intervening periods of non-deposition when the material is allowed to consolidate and desiccate in a manner similar to that used to improve fills made from cohesive materials (see Chapter 4). Progressive trenching has been shown to be one of the most cost-effective means of accelerating drainage and desiccation.

The North and South Blakeley disposal facilities, Mobile Harbour, Alabama, USA are examples of areas which have had their capacity increased by 'conditioning' involving trenching and dewatering (Walker, 1989). Both sites had been used for some time and were conditioned by trenching and drainage to produce considerable extra capacity. This will be used to accommodate material dredged from Mobile Harbour which will be placed in such a manner as to ensure the maximum possible dewatering between dredging campaigns. Similar techniques have been used extensively in The Netherlands for many years.

5.5 Processing and cleaning of dredged material

In recent years, considerable attention has been focused on methods by which contaminated dredged materials can be processed to reduce their volume and to remove the contaminants. In the case of volume reduction, the objectives are two-fold:

- to reduce the volume of contaminated material which must be disposed of safely, usually in expensive storage facilities of limited capacity;
- to separate the relatively uncontaminated fraction from the contaminated materials and to put it to some beneficial use.

Volumetric reduction is often a key factor in the disposal of contaminated materials. Many of the disposal facilities described in the preceding sections of this chapter are costly to construct and maintain, and they clearly have a limited capacity. Any volume reduction of the dredged contaminated sediments which can be achieved may therefore be of economic benefit. There is a double benefit if such reduction is achieved by separating out an uncontaminated and useful part of the sediment.

Most treatment processes rely on the fact that contaminants tend to be concentrated in the finer fractions of mixed soils. Volumetric reduction thus involves the separation of the fine and coarse components combined with the removal of as much water as possible and usually comprises four stages:

- an initial coarse screening to remove debris which is often a characteristic of those areas where contaminated sediments are found;
- the sediment then passes through a hydrocyclone where most of the fine particles are separated from the sand fraction;
- the sand may then be washed in an elutriator to remove any remaining fine particles;
- the mixture of fine soil and water is usually treated with a flocculant and passed through a settling tank where the bulk of the water is removed from the mixture. More water may be removed by means of belt-filter presses.

Processing the contaminated materials to remove contaminants can also be implemented as an extension to the basic volume-reduction treatment. Heavy metals, for example may be removed by treatment with acid and detergents. Organic solvents may be used to reduce the content of organic compounds. Such processing should always be preceded by extensive trials to prove its efficiency.

Processing plants can either be permanent or mobile. A permanent plant is only justifiable where large volumes require treatment over a long period of time. They can be designed specifically to suit the materials being processed and may therefore be relatively efficient in terms of both throughput and cost. Mobile plants may need to deal with a wide variety of materials and are thus more difficult to design. Whilst being relatively costly, their existence will hopefully encourage the processing of contaminated sediments arising from small projects.

Processing plants are currently (1994) being used for treating the arisings from the dredging of inland waterways in the UK. One advantage of the treatment process is that the dredged material is dried and, thus, becomes acceptable to waste disposal sites whose licence only allows them to accept dry material. Drying dredged material, without further treatment, is beneficial for this reason and also has the effect of reducing volumes and transport costs.

Another method of treating dredged material is to mix it with drying and fixing agents. This is often a cheaper solution than mechanical separation and may, in the future, lead to more beneficial and cost-effective options for dealing with contaminated dredged material.

5.6 Decision-making procedures for dredged material disposal

There are two approaches to the selection of treatment and disposal options:

- a rigid or regulatory approach in which the nature and degree of contamination, if any, serves to define the disposal or treatment method;
- a pragmatic approach in which disposal is assessed on a case-by-case basis in terms of the potential impact on the environment.

Which of these two approaches is preferable is a matter of debate. The pragmatic approach was strongly recommended by the PIANC report on offshore disposal (PIANC, 1986) which stated that: '... there is no single disposal option that presumptively is most suitable for a region, for a type of dredged material or for a group of projects ...' and '... there is no inherent effect or characteristic of a disposal method that rules it out from an environmental stand-point prior to specific on-site evaluation ...'. However, as regulatory approaches are generally based on a detailed examination of the nature of the dredged material, albeit on a regional rather than project-specific scale, and the potential disposal options, it might be argued that there is in fact little difference between the two approaches, the most significant being that evaluation is undertaken more frequently in the case of the pragmatic approach.

5.6.1 THE REGULATORY APPROACH

The regulatory approach is typified by that adopted in Rotterdam where there is a very distinct relationship between the degree of contamination and location. Extensive studies have shown that the sediments in the harbour can be divided into four classes:

Class 1: Relatively clean material which is encountered in the lowest reaches of the Rhine and which is permitted to be dumped at sea.
Class 2: Moderately contaminated material found in the middle section of the harbour and which must be deposited in confined disposal areas such as the Slufter (see above).
Class 3: More contaminated material found in the upstream parts of the harbour. This must also be placed in confined disposal facilities.
Class 4: Highly contaminated materials found concentrated around a limited number of discharge points which must be stored in the high-integrity Papegaaiebek facility.

The distribution of Class 1, 2 and 3 materials reflects the increasing contribution, upstream, of pollutants from the upper reaches of the Rhine. The localised distribution of Class 4 sediments reflects local discharges of polluted materials. Every two years, sampling is undertaken to monitor the contaminant loads and to redefine, if necessary, the material classification (Donze, Ed., 1990).

In Hong Kong, a regulatory approach has also been introduced, largely due to a lack of options stemming from a shortage of land area for onshore disposal or processing, coupled with a very ambitious and condensed programme of dredging and reclamation arising from the Port and Airport Development Strategy. Three classes of contaminated materials have been defined. Class A material is defined in a similar manner as Rotterdam's Class 1 and may be placed in unconfined offshore disposal sites. All material of Classes B and C are placed either in specially dredged disposal pits or in worked-out marine borrow pits and later capped with clean materials.

5.6.2 THE PRAGMATIC APPROACH

The pragmatic approach, which essentially involves studying disposal proposals on a case-by-case basis, is used in a number of countries, including the UK and the USA. This involves very detailed investigations and a complex decision-making procedure to demonstrate the environmental acceptability of the proposals.

In the UK, where the licensing of dredged material disposal comes under the jurisdiction of the Ministry of Agriculture Fisheries and Food (MAFF) in England and Wales; the

Department of Agriculture and Fisheries (DAFS) in Scotland and the Department of the Environment for Northern Ireland (DOE NI) there is a clearly defined process for examining a licence application. This has been described by Campbell (1991). In this process a considerable amount of scientific assessment is involved and this follows closely the guidelines given by the Oslo Convention (see Section 5.6.3).

As well as following the guidelines of the international conventions, in the UK it is also necessary to consider the Environmental Protection Act 1990 and the EC Environmental Assessment Directive (885/337).

5.6.3 INTERNATIONAL REGULATORY FRAMEWORK

All the decision-making procedures described in Sections 5.6.1 and 5.6.2 have been developed within the overall framework of international conventions relating to sea disposal. There are essentially three main conventions:

- The Convention for the Prevention of Pollution by Dumping from Ships and Aircraft, signed in Oslo in 1972, entered into force in 1974 (the Oslo Convention).
- The Convention for the Prevention of Marine Pollution by Dumping of Wastes and other Matter, signed from 1972, entered into force 1975 (the London Convention).
- The Convention on Prevention of Marine Pollution from Land Based Sources, adopted 1974, entered into force 1978 (the Paris Convention).

The Oslo and Paris Conventions are regional conventions, whose signatories are exclusively West European and Scandinavian, whilst the London Convention is global.

The work of the Oslo and Paris Commissions, known now as OSPARCOM, tends to be complementary, both with each other and also with the London Commission (LC). This work, which seeks to establish international regulations for the control of the disposal of materials to the sea, produces a constantly evolving framework within which the signatory nations are to develop their own procedures for controlling and monitoring sea disposal.

At the present time (1995) sea disposal is generally undertaken according to the Guidelines for disposal of dredged materials, which was published by the LC in 1986. However, consideration of dredged materials is constantly evolving and guidelines are now developing along the lines of the Waste Assessment Framework (Campbell, 1993) which, for dredged materials, is now called the Dredged Material Assessment Framework (see Appendix).

References

Brand, E.W., Massey, J.B. & Whiteside, P.G.D. 1994. Environmental aspects of sand dredging and mud disposal in Hong Kong, *Proceedings of the First International Congress on Environmental Geotechnics*, Edmonton, Canada. pp 1–10.

Bokuniewicz, H.J. 1983. *Submarine borrow pits as containment sites for dredged sediment, Wastes in the Ocean.*

Bokuniewicz, H.J. & Liu, J.T. 1989. Stability of layered dredged sediment deposits at subaqueous sites, *Oceans '89.*

Campbell, J.A. 1991. The disposal of capital dredging – the role of the licensing authority, *Proceedings of ICE Conference on Capital Dredging*, Edinburgh, Thomas Telford,

London.

Campbell, J.A. 1993. Dredged material disposal – no longer the bridesmaid, *Proceedings of the CEDA Dredging Days*, Amsterdam.

Carpenter, J.E. 1973. Determining the ultimate capacity of the coastal zone for wastewater and wastewater residuals, *Ultimate Disposal of Wastewaters and their Residuals*, pp 216–224, McJunkin, F.E. & Visiland, P.A. (Eds), Water Resources Institute, North Carolina State University, USA.

Clarke, D., Fredette, T.J. & Imsand, D. 1988. Creation of offshore topographic features with dredged material, *Environmental Effects of Dredging: Information Exchange Bulletin*, Vol. D-88-5. US Army Corps of Engineers, Waterways Experiment Station, Vicksburg, Mississippi, USA.

Clarke, D.G. & Kasul, R. 1994. *Habitat Value of Offshore Dredged Material Berms for Fishery Purposes*, E.C. McNair Jr., Editor, Proc. American Society of Civil Engineers, Dredging '94, Lake Buena Vista, FL, USA. ASCE Publishers, New York, NY, USA.

DeLoach, S.R. & Waring, E.G. 1984. Impacts of an overboard disposal operation, *Dredging and Dredged Material Disposal*, Montgomery, R.L. & Leach, J.W., Eds., Vol. 1, ASCE,

Donze, M., Ed., 1990. *Aquatic Pollution and Dredging in the European Community*, Delwell, The Hague.

Fowler, J., Blaykney, H.N. & Hayen, M. 1986. Verification of design and construction techniques, Gaillard dredged material disposal island, Mobile, Alabama, USA. *Proceedings of the 11th World Dredging Congress 1986*, Brighton, CEDA, Delft, The Netherlands.

Harms, C., 1995. Monitoring the Hamburg silt mound, *Proceedings of 14th World Dredging Congress 1995*, Amsterdam.

Hensen, U. 1989. Disposal of Hamburg's contaminated dredged material, *Terra et Aqua*, No. 40, International Association of Dreding Companies (IADC), The Netherlands.

Holiday, B.W. 1978. Processes affecting the fate of dredged material. *Technical Report No. DS-78-2*, US Army Corps of Engineers Waterways Experiment Station, Vicksburg, Mississippi, USA.

Kikegawa, K. 1983. Sand overlaying for bottom sediment improvement by sand spreader, Management of bottom sediments containing toxic substances, *Proceedings of the 7th US/Japan Experts Meeting*, 79–103, prepared for the US Army Corps of Engineers Water Resources Support Centre by the USAEWES, Vicksburg, Mississippi, USA.

Kleinbloesem, W.C.H. & van der Weijde, R.W. 1983. A special way of dredging and of disposal of heavily polluted silt in Rotterdam, pp 509–526, *Proceedings of the 10th World Dredging Congress*, Singapore. London Commission Publication.

Landin, M.C., Thomas, R.P., Davies, J.E., Palermo, M.R., Clarke, D.G. 1995. Environmental restoration and habitat development using dredged material in U.S. waters, *Proceedings of 14th World Dredging Conference*, Amsterdam.

Loxham, M., Van Meurs, G.A.M., & Taat, J. 1988. Comparison of four strategies for the disposal of contaminated dredged material, *Proceedings; Environmentally Acceptable Methods of Dredging and Handling Harbour and Channel Sediments*. Central Dredging Association, Delft, The Netherlands.

Miller, J.N., Chave, K.E., Allen, M.H., Chave E.H., Moberly, R. & Walters, J.F. 1980. Physical, chemical and biological characteristics of deep water dredge spoil disposal, *Proceedings of the 3rd International Symposium on Dredging Technology*. BHRA Fluid Engineering, UK.

Miroshnichenko, V.G. 1982. *The Use of Marine Canals* (Ekspluatatsiya morskikh lanalov), Transport Publisher, Moscow.

Nester, R.D. & Rees, S.R. 1988. Thin-layer dredged material disposal – Fowl River, Alabama, Test case, environmental effects of dredging, *Information Exchange Bulletin, Vol. D-88-4, USACEWES*, Vicksburg, Mississippi, USA.

PIANC 1986. Disposal of dredged material at sea, *Report of a working group of the Permanent Technical Committee II,* Supplement to Bulletin No. 52, Permanent International Association of Navigation Congresses, Brussels.

Rees, C.P. 1980. Environmental impact of dredging operations, *Proceedings of the 3rd International Symposium on Dredging Technology*, BHRA Fluid Engineering, UK.

Shields, F.D. & Montgomery, R.L. 1984. Fundamentals of capping contaminated dredged material, *Dredging and Dredged Material Disposal*, Montgomery, R.L. & Leach, J.W. (Eds.), Vol. 1, ASCE.

Sturgis, T. & Gunnison, D. 1988. A procedure for determining cap thickness for capping subaqueous dredged material deposits, Environmental Effects of Dredging, *Technical Notes, EEDP–01–9, USACEWES*, Environmental Laboratory, Vicksburg, Mississippi, USA.

Sumeri, A. 1989. Confined aquatic disposal and capping of contaminated bottom sediments in Puget Sound, *Proceedings of the 12th World Dredging Congress, 1989*, Orlando, Florida, USA.

Togashi, H. 1983. Sand overlaying for sea bottom sediment improvement by conveyor barge. Management of bottom sediments containing toxic substances, 50–78, *Proceedings of the 7th US/Japan Experts Meeting, prepared for the US Army Corps of Engineers Water Resources Support Centre by the USAEWES*, Vicksburg, Mississippi, USA.

Truitt, C.L., Clausner, J.E. & McLellan, T.N. 1989. Considerations for capping subaqueous dredged material deposits, *Journal of Waterway, Port, Coastal and Ocean Engineering*, Vol. 15, No. 6, ASCE.

USEPA & USACE (1991) Evaluation of dredged material proposed for ocean disposal. (Testing Manual)

USEPA, 1990. Draft ecological evaluation of proposed discharge of dredged material into ocean waters. U.S. Environmental Protection Agency, EPA-503-8-90/002.

Van Orden, R.G.J. & Witter, A.W. 1989. Some construction aspects of the Slufter, Rotterdam's offshore disposal site for contaminated dredged material, *Proceedings of the 12th World Dredging Congress, 1989*, Orlando, Florida, USA.

Vellinga, T. 1984. Dredging and disposal of contaminated dredged material, *Dredging and Port Construction*, Vol. XI, No. 12.

Vellinga, T. 1989. Environmental effects of dredging and disposal operations, *Proceedings of the 12th World Dredging Congress, 1989*, Orlando, Florida, USA.

Veltman, M., Visser, F. & Fresco, J. 1989. Management and containment of sediments by insulation and recycling, *Terra et Aqua*, No. 40, August, International Association of Dredging Companies (IADC), The Netherlands.

Walker, J.E. 1989. Conditioning and management of mobile harbour CDFs, *Proceedings of the 12th World Dredging Congress, 1989, Orlando,* Florida, USA.

Wilber, P. 1992. Thin-layer disposal: Concepts and terminology, environmental effects of dredging: *Information Exchange Bulletin,* Vol D-92-1. US Army Corps of Engineers, Waterways Experiment Station, Vicksburg, Mississippi, USA.

6
Pre-Contract Investigations

6.1 Introduction

Before undertaking any dredging works, investigations need to be carried out to establish a wide range of parameters essential for design, planning and selection of methods and for the accurate estimation of the cost of undertaking the work. These investigations fall into five main groups:

- Meteorological investigations to establish the wind climate at the site and the incidence of fog, ice and rainfall, all of which can affect dredging operations.
- Hydrographic surveys to measure tides, currents and waves and to define the form and level of the ground and seabed.
- Geological and geotechnical investigations to determine the nature of the materials which are to be dredged, used or disposed of.
- Environmental surveys to identify the potential effects of the works on the environment, both during execution and on completion, and to establish baseline conditions with which the results of later environmental monitoring can be compared.
- General investigations to establish operational, statutory and legal constraints which might affect the works.

The information gathered during these investigations will be vital to the dredging organisation when it is determining the methods which should be used, the rates of production and costs of those methods and the constraints to its operations which may arise from the nature and configuration of the site.

All of these types of investigation should be carried out for dredging works. However, it is rare that dredging is undertaken in isolation; most often dredging and reclamation is a part of a much larger engineering project which may include civil works such as the construction of wharves, jetties, locks, breakwaters and other port facilities. In this chapter, the investigations are described generally in the context of the development of a completely new port complex at a virgin site, in order that most facets of their application and methodology can be covered.

6.2 General principles

It must first be emphasised that comprehensive investigations for dredging works, or any marine works, are fundamental to the success of the project. There have been numerous instances of dredging works which have run into serious difficulties, often leading to very considerable cost increases, as a result of encountering unexpected conditions. This is particularly true of ground conditions and the nature of the materials to be dredged. There is nothing to be gained, and much to be lost, by attempting to limit investigations to a bare minimum.

The five groups of investigation involve different techniques and skills. They may, however, particularly in the case of an undeveloped site, have a significant degree of overlap. In such cases, they should be considered as a single package and planned accordingly.

There are three areas where such a combined approach may be of particular importance. Firstly, offshore geophysical surveys (a method of ground investigation) can often be undertaken at the same time and using the same survey vessel as hydrographic surveys. Secondly, studies of the nature of the seabed, which are of interest to the ground engineer, can often be undertaken by using the hydrographic surveyors' sonar equipment.

Perhaps the greatest area of overlap, however, is in the study of sediment movement and of coastal erosion and accretion, which is equally dependent upon the skills and observations of the hydrographic surveyor and the marine geologist or ground engineer.

In addition, much of the data required by those undertaking the environmental studies can be collected during hydrographic and geotechnical investigations.

There are five basic principles which are critical to the success of pre-contract investigations (PIANC, in press):

- flexibility of approach;
- communication and understanding of objectives;
- use of appropriately qualified staff;
- continuity of staffing;
- adherence to standards.

In addition, the method of procurement may often have a significant bearing on the quality of the investigations.

Flexibility of approach

The design and planning of the investigation should allow changes to the nature and extent of the investigation in the light of the results as they become available. It is rare that the initial design of an investigation will be found to be perfectly suited to the conditions which are actually encountered on site, particularly in the case of ground investigations, and it is therefore essential that the investigation contract be sufficiently flexible to permit changes of method and scope during the investigation.

Communication and understanding of objectives

If the investigations are to achieve their objectives it is necessary to maintain a good degree of communication between the design team and those undertaking the investigations. In addition, both the design team and the investigators must be familiar with the particular needs of the dredging contractor and the nature of the information which he needs in order properly to plan and estimate the works. This implies, for example, that geotechnical engineers and geologists involved in the planning of ground investigations should have a reasonable knowledge of the dredging process and of the manner in which the ground affects that process.

Use of appropriately qualified staff

All staff employed in investigations should be properly qualified and experienced. Hydrographic surveying, geotechnical investigation and environmental impact assessment are all highly specialised fields which can only be properly undertaken by staff holding appropriate formal qualifications and experience. All too often, such investigations are undertaken by persons with inadequate experience and/or qualifications.

Continuity of staffing

It is preferable that the investigations are executed, interpreted and reported by the same staff and that staff changes during investigations are kept to a minimum. This ensures consistency and continuity, the lack of which can often result in errors of interpretation and incomplete reporting.

Adherence to standards

There are a number of national standards and codes of practice which relate to the activities described in this chapter. It is important to specify the standards applied and to ensure that they are observed. In this manner, possible confusion with regard to the methods of undertaking tests and making observations, or the meaning of descriptive terms applied to soil and rock is avoided. This aspect is of particular importance in the case of data which is to be used at a later stage by others who may not have been involved in the surveys, for instance tendering dredging contractors.

Method of procurement

Research in the UK (Rhys and Wood, 1984) has shown that the method of procurement can strongly influence the value of the data obtained during the investigations. It appears that the best results are most likely to be obtained when investigation contracts are negotiated rather than put out to open tender and when specifications evolve as a result of a cooperative approach between the client and the investigation contractor. This approach enables the client to avail himself of the specialist experience of the contractor, particularly with respect to choosing appropriate methods of investigation.

Finally, the authors strongly recommend that, if possible, short-listed or pre-qualified dredging contractors be allowed to participate in pre-contract investigations, particularly the geotechnical investigations. In this manner, they can not only see at first hand the nature of the materials which they have to dredge but they may also provide valuable expert guidance to the investigation teams with respect to the nature and extent of the investigations. Such guidance should not be resisted or resented by the client or his appointed engineer.

6.3 Meteorological investigations

Meteorological conditions; wind, temperature, humidity, fog and ice, are important to dredging operations and to port operations generally. The data required to make a proper and reliable assessment of these conditions is relatively easy to gather, if it is not already available. However, in all cases, it must have been obtained over a long period, usually several years, if it is to be statistically valid.

6.3.1 WIND

The speed, direction and duration of winds are important factors in the determination of sea state and the frequency and intensity of storms. In some cases, wind records are the only data which can be used to predict sea state as wave records are often unavailable. In order to obtain useful data it is necessary to measure wind speed and direction for as long a period as possible. The minimum period is normally considered to be one year but at least ten years' records are required for reliable statistical analysis.

Whilst local meteorological stations may have comprehensive wind records it should be noted that the site topography or location may be such that the data cannot reasonably be applied to the site. It will then be necessary to obtain further site-specific data.

Method

Wind data are obtained using a recording anemometer which samples the wind speed and direction at regular intervals, usually every 15 minutes, and records the observations on magnetic tape or computer disk. These devices can be left unattended for periods of several weeks before the data needs to be collected.

Data presentation

Wind data can be presented in three ways: wind roses, matrices and persistence diagrams. The wind rose is perhaps the most common method and illustrates the percentage distribution of different wind speeds in the eight major directions of the compass. There are a number of ways in which wind roses can be drawn. One type is illustrated in Figure 6.1.

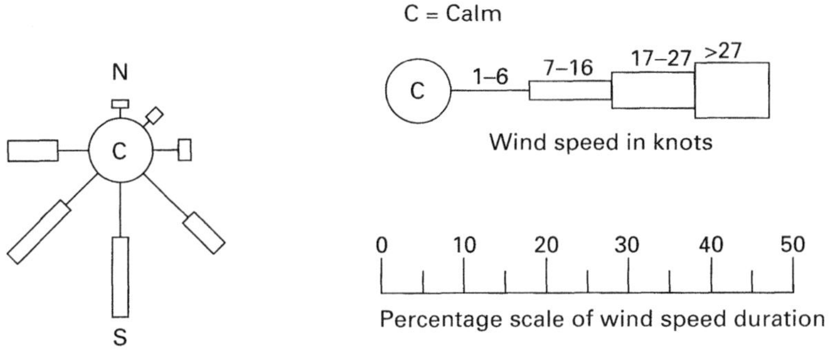

Figure 6.1 A typical wind rose diagram

The wind rose gives a useful indication of the general wind climate but is of little use if data are required for the prediction of waves. In this case, it is necessary to use either high-quality unprocessed data or graphs and tables which have been formulated from such data. Wave climate prediction requires knowledge of wind speed, direction and persistence, as well as fetch, the clear water distance over which the wind has blown. The wind matrix, Table 6.1, tabulates the wind speed and direction and provides the raw data which are used to construct the wind rose. For each of the principle wind directions, a persistence diagram (see Figure 6.2) can be used to illustrate the duration of winds of various speeds. This will be of greatest value if presented on a monthly basis.

Table 6.1 A typical wind frequency matrix

Wind speed (Beaufort scale)	Wind direction								Totals by speed (%)
	N	NE	E	SE	S	SW	W	NW	
8						0.5			0.5
7						0.5			0.5
6						1			1
5			0.5		1	4			5.5
4			0.5		1	5	1		7.5
3		1	2	1	6	7	3		20
2	1	1	2	2	6	9	3	1	25
1	1	4	4	2	10	8	6	2	37
Totals by direction (%)	2	6	9	5	24	35	13	3	Calm 3 100%

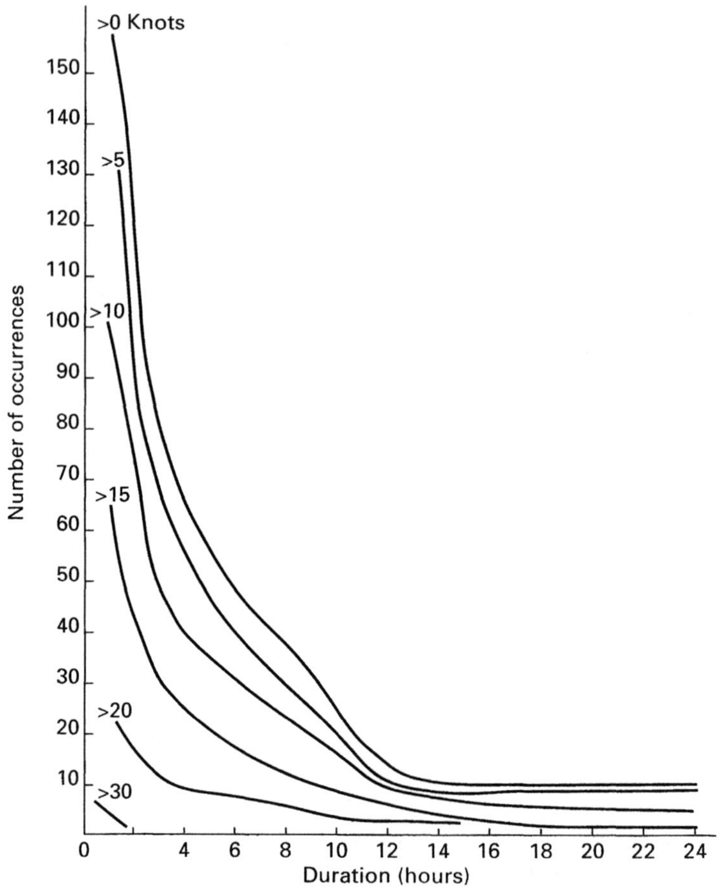

Figure 6.2 A typical wind speed persistence diagram

6.3.2 FOG, ICE, TEMPERATURE, HUMIDITY AND RAINFALL

Wind is generally the most important meteorological feature of most sites with respect to dredging operations. On some sites, however, other factors can be important. Fog, for example, may render navigation hazardous and can prevent accurate positioning if visual methods are being used. In high latitudes, sea and river ice can prevent any dredging operations being carried out. An unusual exception to this is pre-treatment by drilling and blasting which has successfully been undertaken from ice in Finland.

Temperature and humidity can also influence machinery and electronic instruments as well as having an effect on project staff. Rainfall, in some parts of the world, can be so heavy as to seriously hinder visibility.

If sufficient data are not available, it may be necessary to establish a small meteorological station at the site and to take records over as long a period as possible. Data is best presented in the form of monthly groups as most climates exhibit some degree of seasonality. In the case of ice, annual listings of the dates on which ice has formed and melted are most useful.

6.4 Hydrographic investigations

These investigations are undertaken for a variety of purposes:

- to establish the site bathymetry;
- to obtain data relating to tides, currents, waves and sediment movement which can be utilised to optimise the layout of the project, to study the potential effects of the development on the existing environment and to assess the suitability of different types of dredging plant and the delays which may arise from weather and sea state;
- to measure quantities of material dredged during construction and to check that design depths in channels and basins have been achieved;
- to identify seabed features which might hinder dredging operations, for instance pipelines, cables and debris.

Hydrographic surveying involves techniques and equipment which are unique to the marine environment. The simple and traditional methods of lead-lining to establish water depth and the use of a sextant for positioning have largely been superseded as a result of the introduction of electronics and computerisation. These are the most significant of all the modern developments in the field of hydrographic surveying. They enable, in most instances, a greater precision of measurement but, more notably, permit the surveyor to make and process a far greater number of observations in a comparatively short space of time. This latter aspect is particularly important in the case of bathymetric surveys undertaken during dredging when results are often needed quickly to check on progress and to plan operations.

6.4.1 HORIZONTAL AND VERTICAL CONTROL

Horizontal and vertical control, that is positioning, is a vital element in all investigations. The value of observations made during surveys is negated if there is any doubt as to the position where the observation was made and the vertical datum to which it relates.

Horizontal control

There are various methods of position fixing when working in rivers and coastal marine areas. These include visual observation of shore-based transit marks, use of a sextant, use of shore-based theodolites and a wide variety of electronic systems.

If positioning is required for a continuously moving vessel undertaking, for example, bathymetric or geophysical surveys, the optical methods of position fixing require a high degree of skill and speed, and much work at the end of the day plotting the positions. In addition, they are very difficult to use accurately in poor sea conditions or when visibility is poor. The vast majority of surveys are now undertaken using electronic positioning methods. However, the other methods may occasionally be used on small projects.

Whichever method of positioning is being used, it is important that the locations of the reference points from which the position is measured are accurately known. Establishing onshore survey points for the location of transit marks or electronic system reference stations may involve a considerable amount of high quality topographic survey work, particularly in the case of remote areas.

Transits on shore marks

This simple method of positioning (see Figure 6.3a) may be used for small surveys and involves sailing along a line defined by two shore-based marks and establishing the position along the line by reference to pairs of lateral marks. It can only be used when reasonably close inshore and in conditions of good visibility and requires personnel onshore to move the transit marks to new locations after each run.

The accuracy of this method depends on the ability of the helmsman to maintain his course along the transit line, which might be difficult in poor sea conditions or when there is a strong cross-current or wind, and of the surveyor on-board to fix accurately on each lateral transit. Close inshore and in ideal conditions, accuracy is unlikely to be better than one metre and will fall off rapidly as distances increase and conditions deteriorate.

Use of sextant

Using a sextant, once the traditional navigator's method, is a simple and cheap means of positioning which is still occasionally used. Position can be established using shore-based transit marks and measuring a single angle with the sextant or by measuring two angles between three known points onshore.

In the single-angle method (see Figure 6.3b), the vessel is sailed along a transit line defined by two onshore marks. The angle between two more marks, one on either side of the transit line and separated by a known distance, is measured with the sextant by the surveyor on board the vessel who can then calculate the distance from the shore.

The accuracy of this method is again dependent on the abilities of the helmsman and the surveyor and on sea conditions. However, a reasonable degree of accuracy can be maintained for some distance offshore as long as the separation of the two marks being used for angle measurement is sufficient. If the marks are very close and the subtended angle is small, accuracy will be reduced.

The double-angle method (see Figure 6.3c) relies on the fact that a position can be determined by measuring two angles subtended by shore marks at known positions. The marks need not have been set up for the survey; they might be features on prominent buildings or even trees, the positions of which have been accurately surveyed.

This method of positioning is usually used in conjunction with a circle chart on which two families of circles are marked (see Figure 6.4). All of the circles of one group pass through one pair of shore marks whilst the circles of the other group all pass through the second pair; one of the marks may be common to both pairs. Each circle represents a particular angle which will be observed at any point on the circle, between the two marks through which it passes.

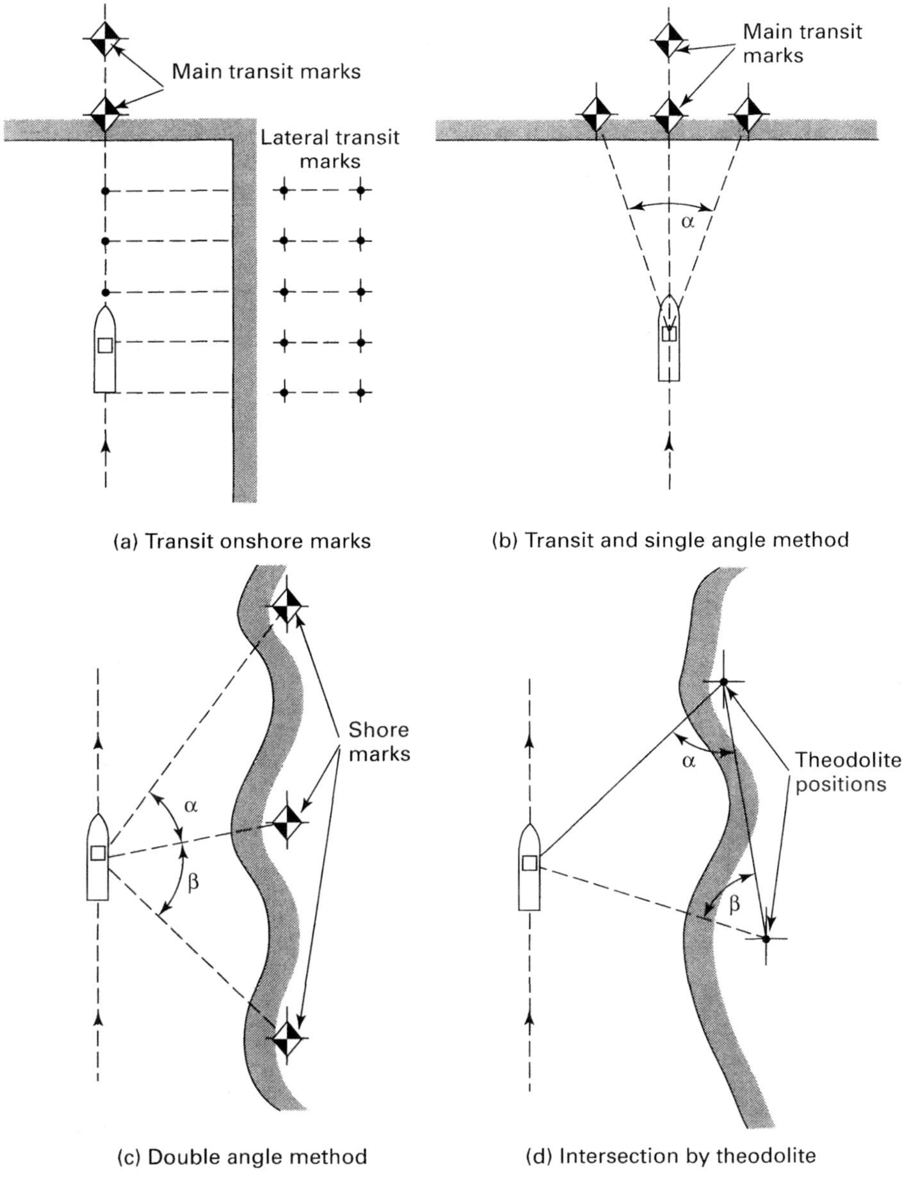

Figure 6.3 Principal optical position fixing methods

The circle charts can be drawn up prior to the commencement of the survey, usually with the aid of a computer. The surveyor measures the angle between one pair of shore stations which places him on one of the circles through those marks. Similarly, he measures the angle between the other pair of marks and identifies the circle on which he is positioned for that pair. The point where the two circles intersect is his position.

When using this method, it is important that the shore stations are located so that the circles within the survey area cross each other at an angle as near 90° as possible in order to ensure the best accuracy. The main advantage of this method of positioning is that it does not

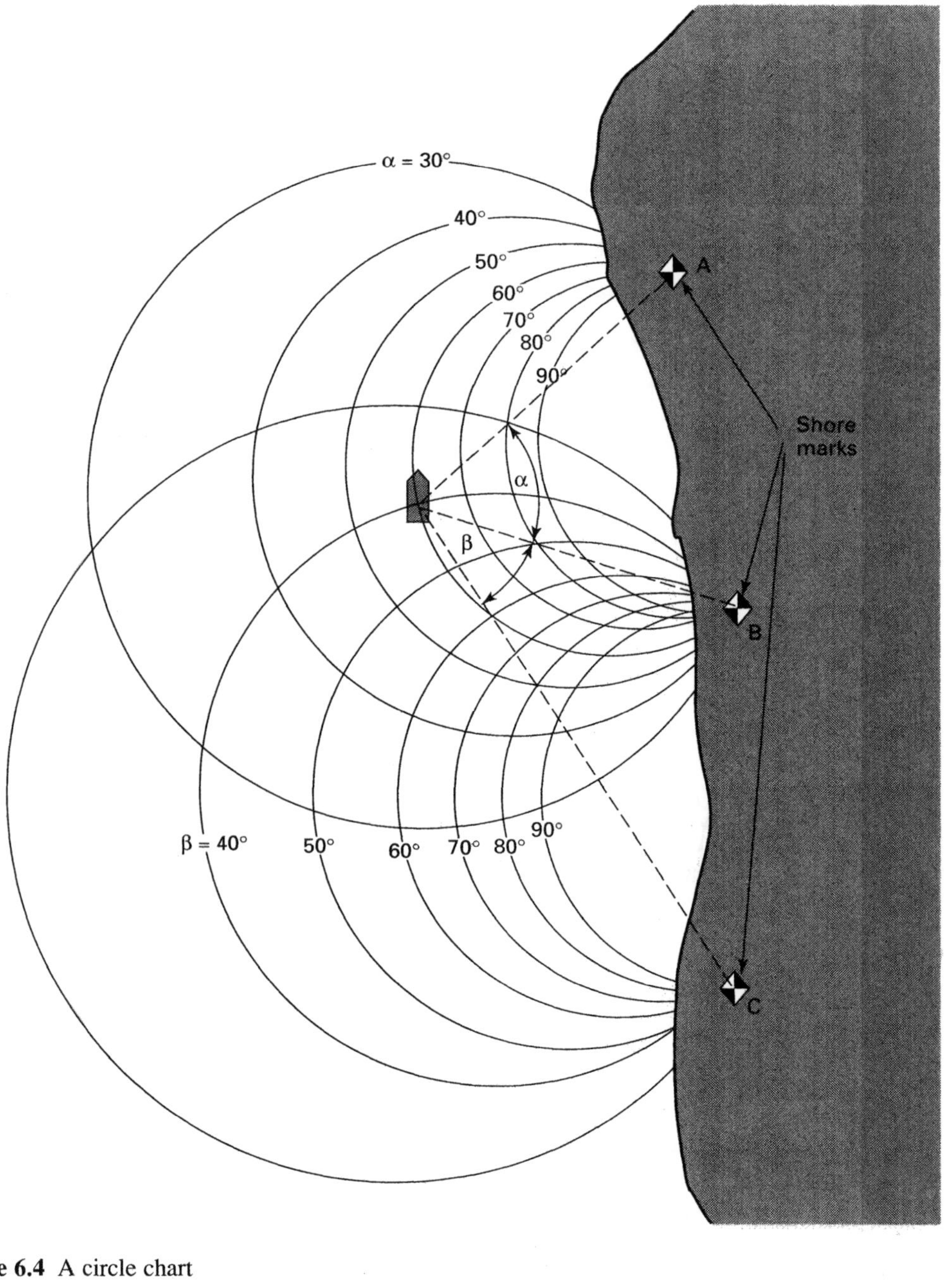

Figure 6.4 A circle chart

require the attendance of shore-based personnel and, if the circle charts are available during the survey, the surveyor can immediately plot his position on the chart without any calculations.

Intersection by theodolite

This method requires two surveyors, each equipped with a theodolite, to be located at two known points onshore. Each measures the angle between a point on the vessel and the other theodolite or another onshore reference point (see Figure 6.3d). Radio contact between the surveyors and the vessel is required because they must measure the angles at the same time as the observations are being made on the vessel. The method can be quite accurate if the vessel is moving slowly and can be used over large distances. In addition, it is not very sensitive to sea conditions as the measurements are made from the shore rather than from the vessel.

Electronic positioning methods

There are four types of system available: satellite, long range navigation, hyperbolic and active ranging systems.

Until recently, the satellite systems were not used for conventional dredging survey work except in very remote areas where there was a need to establish accurately positioned shore survey points against which offshore surveys could be referenced. The surveyor would establish the necessary shore stations using the satellite system and then revert to, most frequently, an active ranging system utilising the newly-established shore stations.

However, with the advent of the global positioning system (GPS), developed by the United States Department of Defence, satellite systems are now in wide use and have supplanted many of the electronic systems which have been used during the last 20 years or more. The GPS comprises a series of 21 satellites giving 24-hour coverage over all parts of the world. Position is established by interrogating three or more satellites each of which provides time and distance data which are integrated by the receiver. The calculated position is accurate to within about 15 to 20 metres.

'Differential GPS' (DGPS) is a much more accurate system which utilises a 'reference' receiver station whose coordinates are precisely known (see Figure 6.5). The reference station can compute adjustments to the data received from the satellites which can then be applied to the mobile receiver being used by the surveyor or the dredger. Two systems are available. Long range DGPS, which utilises permanent differential stations, has a horizontal accuracy of approximately two metres, comparable with other electronic systems. The more recently available short range DGPS, which utilises temporary differential stations, can provide both horizontal and vertical positioning with an accuracy of a few centimetres.

The long range navigation systems such as Loran, Omega and Decca Navigator, and hyperbolic systems such as Hi-Fix Toran and Raydist measure distance by phase comparison of radio signals but are not sufficiently accurate for dredging or dredging surveys.

The active ranging systems include Mini Ranger, Trisponder, Microfix and Hydrodist. They are the methods which until the early to mid 1990s were most widely used by dredging and survey contractors. They measure distance directly between the survey vessel and the shore stations using high frequency radio waves. It is necessary for the shore stations to be in line-of-sight with the survey vessel. These systems produce circular grids and generally have an accuracy of around one to three metres and maximum ranges of 30 to 50 kilometres.

DGPS and active ranging systems are often interfaced with other hydrographic survey equipment using micro- and mini-processors. Computerised systems such as these are able to record several different types of observation simultaneously and greatly facilitate the collection and processing of data.

Vertical control

Vertical control of hydrographic surveys is required to compensate for the effects of tides, river levels and storm surges and to relate bathymetric data to a known datum. Vertical control is obtained by measuring the water level at the time when the depth observation is made

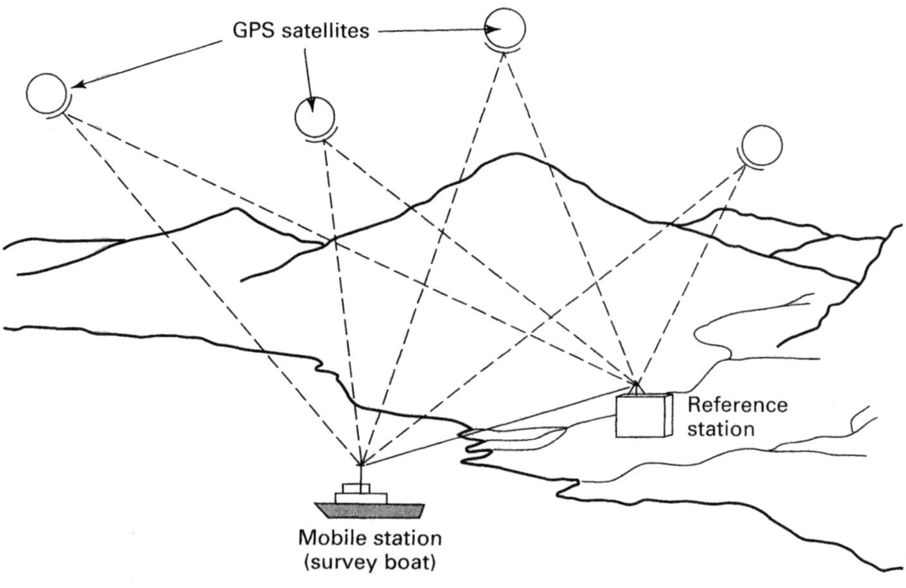

Figure 6.5 The differential global positioning system (DGPS)

and is discussed in more detail in Section 6.4.2. Vertical control can be obtained without measurement of water level using short range DGPS systems. In this case it is only necessary to know the vertical distance between the DGPS receiver and the echo sounder transducer. However, the DGPS receiver must first be referenced against a known datum by placing the receiver at a point whose precise level is known.

6.4.2 WATER LEVEL

There are two reasons for measuring water level; to establish typical levels and fluctuation ranges for the design of port facilities and dredging works and for the vertical control of bathymetric surveys. In both cases it is necessary to measure, over a period of time, the water level fluctuations relative to a known datum. Although most sea ports have published tide data, there are many sites for which no data is available or where tide levels can only be interpolated between major ports. Since every location in the world has a unique tidal regime, it may be necessary to undertake tidal observations in order to establish such a datum. It should be noted that there may be a significant variation between predicted and actual tide levels, and hence predicted levels should not be used as a basis for the reduction of survey soundings.

Water level datums

The various datums which are used and their approximate relationships are shown in Figure 6.6. If water level observations are being undertaken for the purposes of establishing a datum or to facilitate the prediction of tides, they should initially be undertaken over a period of at least 30 days. This is the minimum required for the prediction of future tide levels including the lowest astronomical tide which is often adopted as the local datum.

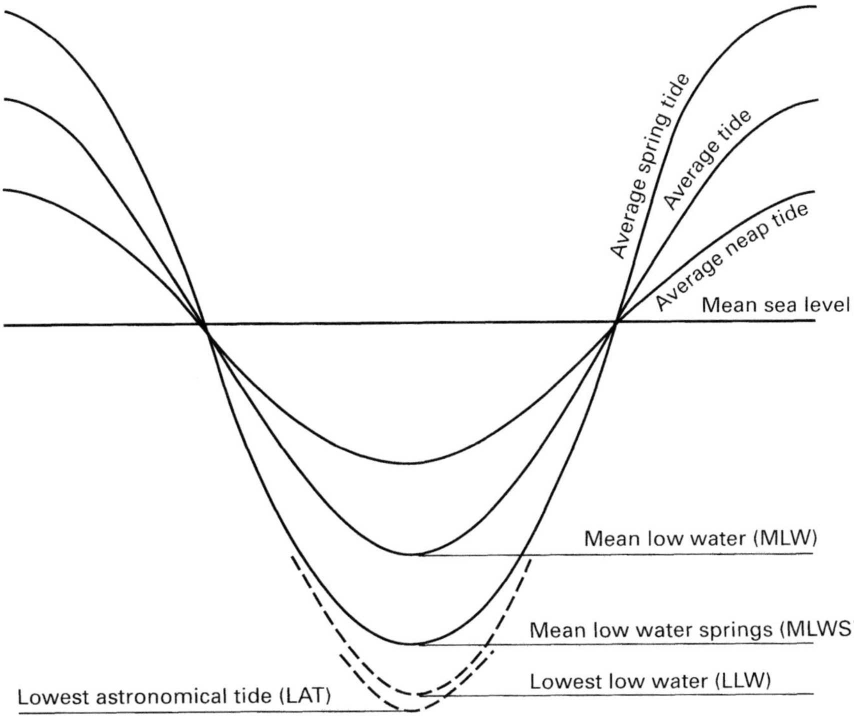

Figure 6.6 Tidal datums

Water level observation methods

There are a number of observational methods which can be used, the simplest being a graduated pole which is read manually, fixed vertically in some sheltered part of the site. In most cases, however, establishing the tidal range and the local datum is undertaken using a recording tide gauge which can be left unattended over an extended period of time.

The three most common types of gauge are as follows:

- the sonar gauge, which measures the distance in air from a transducer set at a known level to the water surface;
- the pressure gauge, which measures tidal movement as water pressure fluctuations;
- the float gauge, in which a float is connected by a fine wire via a series of pulleys and reduction gears to a chart recorder.

The float gauge and the sonic gauge must be installed in a tube or stilling well, which largely eliminates the effects of wave action.

These forms of gauge can be designed to transmit the tide data to a remote point. This is a particularly useful facility for survey vessels undertaking bathymetric surveys. An on-board receiver can be used in conjunction with a micro-computer and interfaced with the echo sounder to produce automatically reduced bathymetric data. Radio tide gauges are also frequently used to transmit tide readings to dredgers so that dredging depths can be adjusted continuously to compensate for the tidal rise and fall.

On large sites, particularly those in estuaries or with complex coastlines, it is necessary to measure tidal level at more than one location in order that adjustments may be made to take account of water surface gradient. In such situations water level at the point of survey should be determined by interpolation.

Water levels in rivers may be measured by the methods described above. However, because the changes in level of rivers are usually slow and are locally affected by floods, it is common to install fairly rudimentary devices at several locations along the river.

For the understanding of river regimes and for the determination of river datums it is necessary that records are built up over long periods, and level prediction is possible only by sophisticated computer techniques combined with detailed knowledge of catchment areas and tributary flows.

Data presentation

Tide data are usually presented in the form of tables giving the predicted levels of high and low tides over a period of a year, when these are available. These are accompanied by listings of mean high and low water springs, mean high and low water neaps, the lowest astronomical tide and the datum to which the levels are related. This information is sufficient for most purposes, but a graph of the tidal height against time before and after high water is useful. Some areas exhibit unusual tidal features such as the double high water in the Solent, UK.

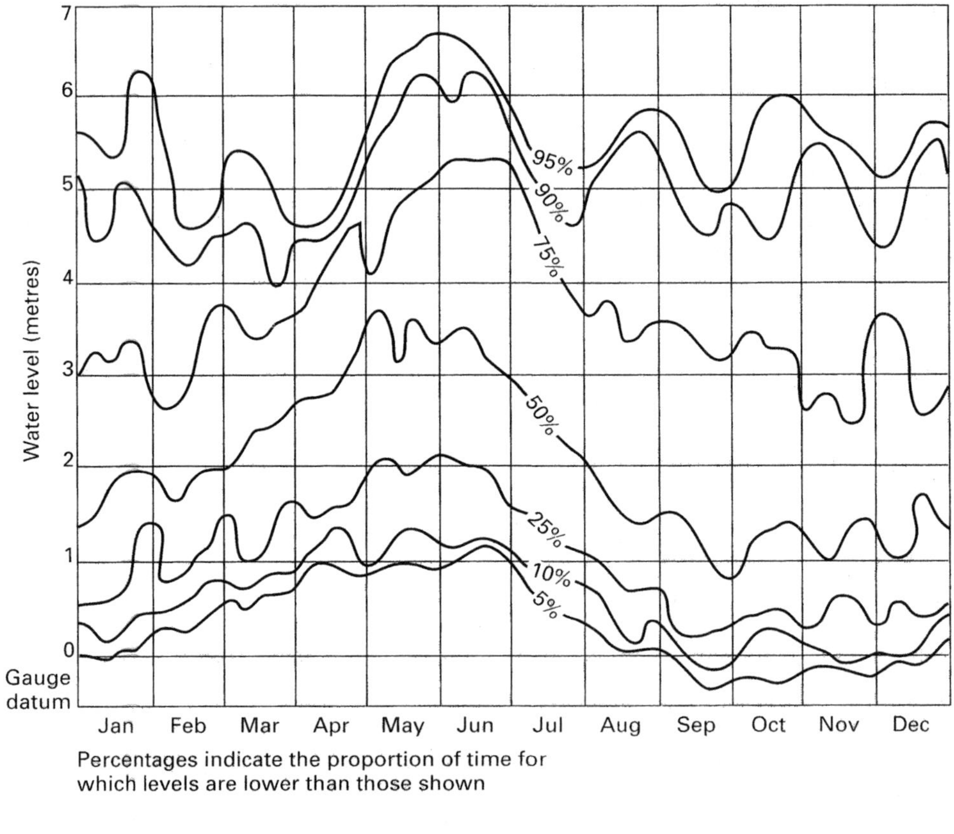

Figure 6.7 River level exceedence graph

In the case of rivers, the water level usually follows an annual cycle although wide variations do occur from one year to another. A level exceedence graph (Figure 6.7 shows an example from South America) can be used which shows the probability of the water level exceeding certain heights during the year. This does not allow accurate prediction of level but it does permit the derivation of average levels and an estimation of the probability of draught restrictions at critical points along the river.

6.4.3 WATER DEPTH

Surveys of water depth, or 'bathymetric' surveys are frequently the largest single component of pre-contract hydrographic surveys. In addition, they are undertaken regularly during dredging works in order to check progress and to measure quantities for payment purposes.

Methods

Simple direct methods of depth measurement such as the use of sounding lines or rods are still occasionally used for small inshore surveys but the vast majority are now undertaken using echo sounders. The echo sounder records the water depth indirectly by measuring the length of time taken for an acoustic pulse to travel to the seabed and back to the transducer (see Figure 6.8). Most echo sounders are equipped with a recorder which produces a permanent trace of the bottom profile. Modern systems often incorporate heave compensators which eliminate from the sounding record the effects of waves and swell on the survey boat.

A comparatively recent development in the field of bathymetric surveying is swathe sounding which is based on a similar principle but which gives complete coverage of a broad strip of the seabed rather than along a discrete line. This is achieved by means of an array of transducers mounted on a boom or by a towed scanning transducer or by one which is hull- or outboard-mounted.

Swathe sounders are, by comparison with echo sounders, expensive but are becoming increasingly widely used because of the speed at which survey areas can be covered and because the complete coverage of the seabed ensures that any obstructions or high spots above the designed dredged level are identified.

Echo sounder calibration

Echo sounding is simple and accurate but the equipment requires regular calibration in order to compensate for sea-water temperature and salinity variations which affect the speed of sound. Calibration can be done by calculation, if the temperature and salinity are known, or directly by measuring to a plate suspended under the instrument at a series of known depths. This latter method, often called a 'bar check', is the most common means of calibration but it should be carried out with care, taking account of the following:

- the bar or plate should be suspended using inelastic chain or wire;
- the bar check should not be undertaken in a strong current – the bar should be suspended directly below the echo sounder transducers;
- the check should be carried out over the full depth range which is to be measured;
- checks should be carried out both before and after each surveying session.

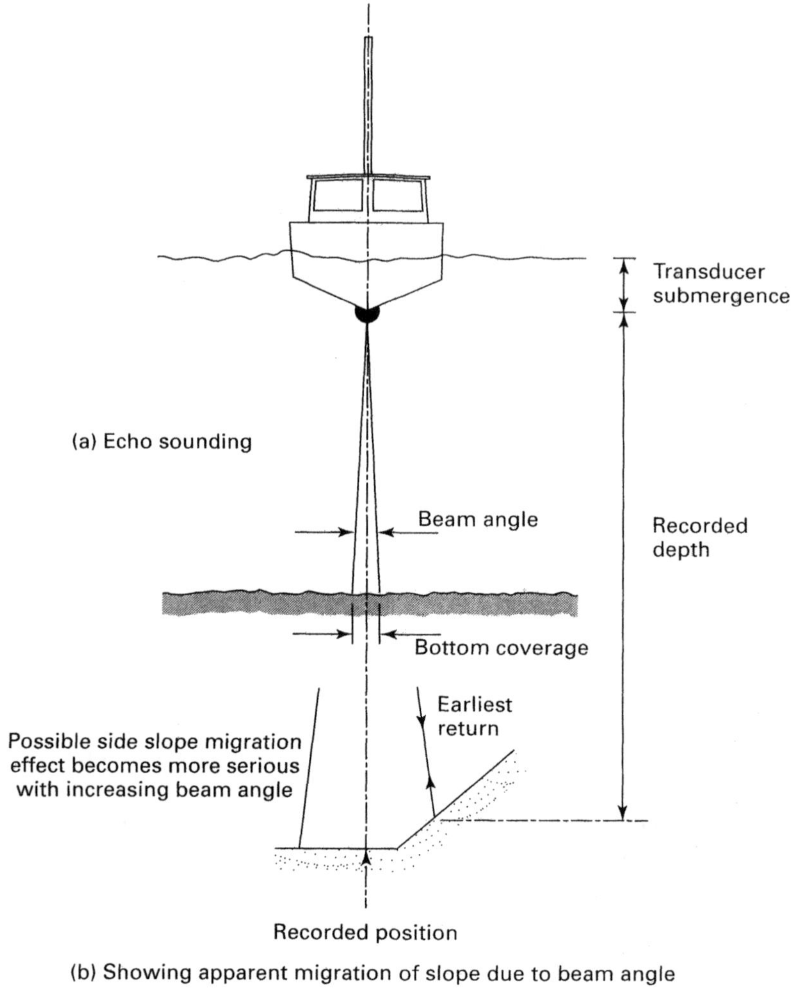

Figure 6.8 Echo sounding

Echo sounder frequency

Echo sounders can work with a variety of acoustic pulse frequencies. The nature of the seabed and the frequency of the transmitted acoustic pulse will influence the echo sounding record. High frequency pulses may be totally reflected by soft clays and silts. Low frequency pulses will also be reflected by such materials but may also penetrate the bed and reflect from layers exhibiting density contrast. Many survey echo sounders therefore operate simultaneously with two pulse frequencies (e.g. 210 and 33 kHz) and the records of such instruments enable the experienced surveyor to interpret the general nature of the seabed.

In areas of active mud sedimentation, layers of low density fluid mud may be present on the seabed. Such layers often exhibit complex variations in density with depth. In such areas, it may be difficult to establish the navigation depth, even with the use of multi-frequency echo sounders, and it may be necessary to undertake *in situ* density measurements.

Beam width

Echo sounder transducers produce fixed acoustic beam widths which may be in the range 3° to 45°. If a single transducer is used, the beam width will influence the interpretation of the record and the manner in which the survey is carried out. The echo sounder will record the first return pulse. If a wide-beam transducer is in use, the first return may be from a comparatively shallow reflector to one side of the vessel and the record will show water depths which are less than the actual depth immediately under the survey vessel. For the same reason, small depressions in the seabed and high spots at the edge of the beam may not be revealed.

The beam width is particularly important when undertaking surveys of steep slopes or dredged trenches where, because the return signal comes from the edge of the cone, the slopes appear from the record to be displaced outwards ('slope migration', see Figure 6.8b). If it is important to establish the position of slopes accurately, it is necessary to use narrow beam transducers. Alternatively, scanning swathe echo sounders, described above, will overcome this problem.

Line spacing

The required survey line spacing will vary according to the purpose of the survey. Surveys for feasibility and design purposes may adopt line spacings of 50 metres or more. Pre- and post-dredging surveys for payment purposes are usually undertaken at much closer spacings, typically ten metres. Check surveys undertaken by the dredging contractor to monitor his progress are undertaken at a spacing of 20 to 30 metres.

Where complete coverage of the dredged area is necessary, for instance post-dredging surveys to check that navigation depths have been obtained, it may be more desirable to utilise swathe sounding. This is particularly important when associated with rock dredging because of the hazard which may be presented by a single high spot.

Presentation of bathymetric data

Bathymetric data are usually presented as charts which may be contoured at appropriate depth intervals to facilitate interpretation. The scale at which the charts are produced will depend on the purpose of the survey and the nature of the features which have been surveyed. Three widely used scales are:

- 1:2500 for approach channels, offshore disposal areas etc.
- 1:1250 for harbour basins, docks and areas of widely varying bed levels and for the measurement of infilling;
- 1:500 for rock outcrops and for post-dredging surveys in rocky or hard areas.

It is traditional in navigational surveys for there to be a fixed relationship between plotting scale and survey density, but this is not the case for engineering work.

6.4.4 WAVES

Accurate wave data is important to the design of port structures such as breakwaters and to the design of fairways and approach channels. It is also of considerable importance to the dredging contractor who needs to estimate the length of time which his floating plant can operate at the dredging and reclamation or disposal sites and along the transport routes between them.

There are three types of waves: locally-generated wind waves with a relatively short period of the order of three to five seconds; remotely-generated swells of 8 to 20 seconds period and very long waves with a period of up to 200 seconds. The latter are difficult to detect and almost impossible to predict. For the dredging contractor, both short and intermediate waves are likely to have some effect on his operations depending on their height and the type of plant he is using.

Observation methods.

Measurements of the wave height and period may be achieved by two methods. The most common is by means of a tethered buoy fitted with a double-integrating accelerometer and the facility either to record the data or to transmit it to a remote recording station onshore. In order to reduce the amount of data to a manageable level, wave movements are usually sampled at intervals rather than continuously measured.

Wave staffs, similar to tide poles, may also be used to make short term visual observations of waves.

Wave direction is difficult to measure except with very sophisticated instruments. Directions are most commonly established visually, supplemented by records of the winds which generate the waves, but such observations are often inaccurate, particularly in confused seas. Surface radar, aerial photography and satellite imagery are being used increasingly often in studies of wave direction. In most cases, the dredging contractor will be more concerned with the height and periodicity of waves than with the direction which is of more importance to the designers of ports and navigation channels.

In the absence of site-specific wave measurements, for example during feasibility studies, useful information can be gained from publications such as the Global Wave Statistics (Hogben *et al.*, 1967) which presents statistics based on observations of swell made from ships. These can be used, with appropriate corrections for refraction and diffraction in coastal areas, to establish an indicative swell climate in most locations. Locally-generated wind waves can be predicted using a number of hindcasting methods based on wind records, if these are available.

Data presentation

There are two simple methods of presenting wave data: scatter diagrams and persistence diagrams. The scatter diagram shows the significant wave height plotted against the zero crossing period (see Figure 6.9). This is useful because these two parameters define the steepness of the waves, that is the ratio between wave height and wave length, which in turn determines the effect that the wave will have on a vessel. The diagram can also be used to compare wave character in different areas. The number of occurrences of waves of specific heights and periods is termed the probability density for those waves. The densities may be expressed in parts per thousand and equal values are joined by contour lines on the diagram which can also be used to illustrate lines of constant steepness.

Persistence diagrams can be used to estimate how much time could be lost during dredging operations. They can be presented in two forms showing either the number of times the wave height exceeds certain values (Figure 6.10a) over given periods of time or the number of times that the wave height remains below certain values (see Figure 6.10b). These diagrams show, in a concise manner, the amount of time which a dredger operator may expect to be delayed by waves and also give some indication of the short term variability of the wave

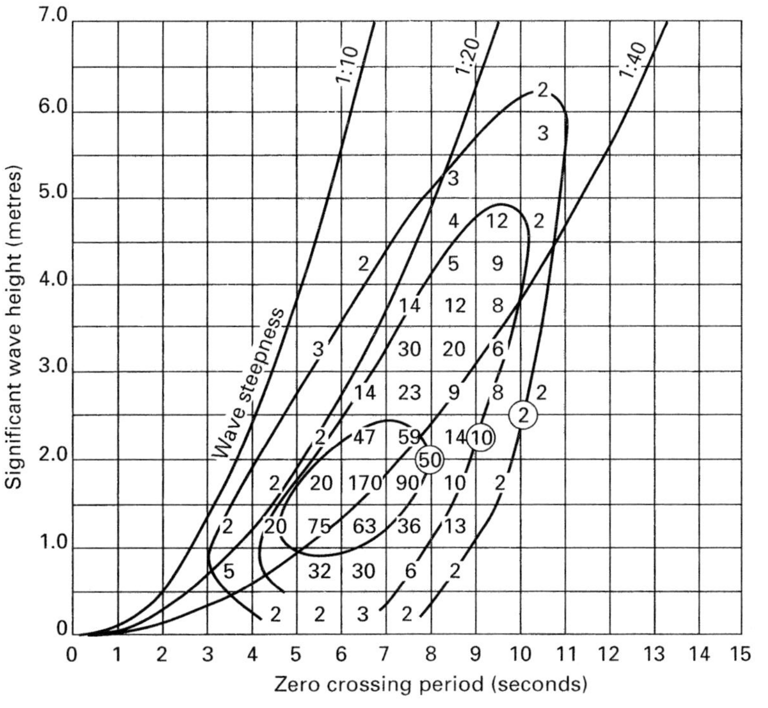

Figure 6.9 A typical wave scatter diagram

climate. These diagrams should only be prepared using good quality data gathered over long uninterrupted periods because they are particularly sensitive to short gaps in the records.

The diagram often used in maritime civil engineering for estimation of design wave heights, the log probability/wave height diagram, is not particularly useful in the context of dredging operations.

6.4.5 CURRENTS

For dredging purposes, the near-surface current is usually the most important with respect to vessel operation. However, in the context of port developments generally and the environmental effects of dredging works, it may be important to measure currents throughout the water column and at several locations in and around the site.

Observation methods

If the main concern is with surface currents, they may be charted by float tracking in which one or more small floats, fitted with drogues to minimise the effects of wind, are released and tracked by the survey vessel. In extremis, oranges have been used for simple float tracking with acceptable results when more sophisticated equipment has been unavailable.

More comprehensive current measurement throughout the water column can be made using a meter comprising a free-running propeller coupled to a mechanical or, more frequently, electronic, revolution counter. The meter incorporates a compass to measure the direction of the current. The most modern forms of current meter are electromagnetic and having no moving parts, give a greater degree of reliability and durability.

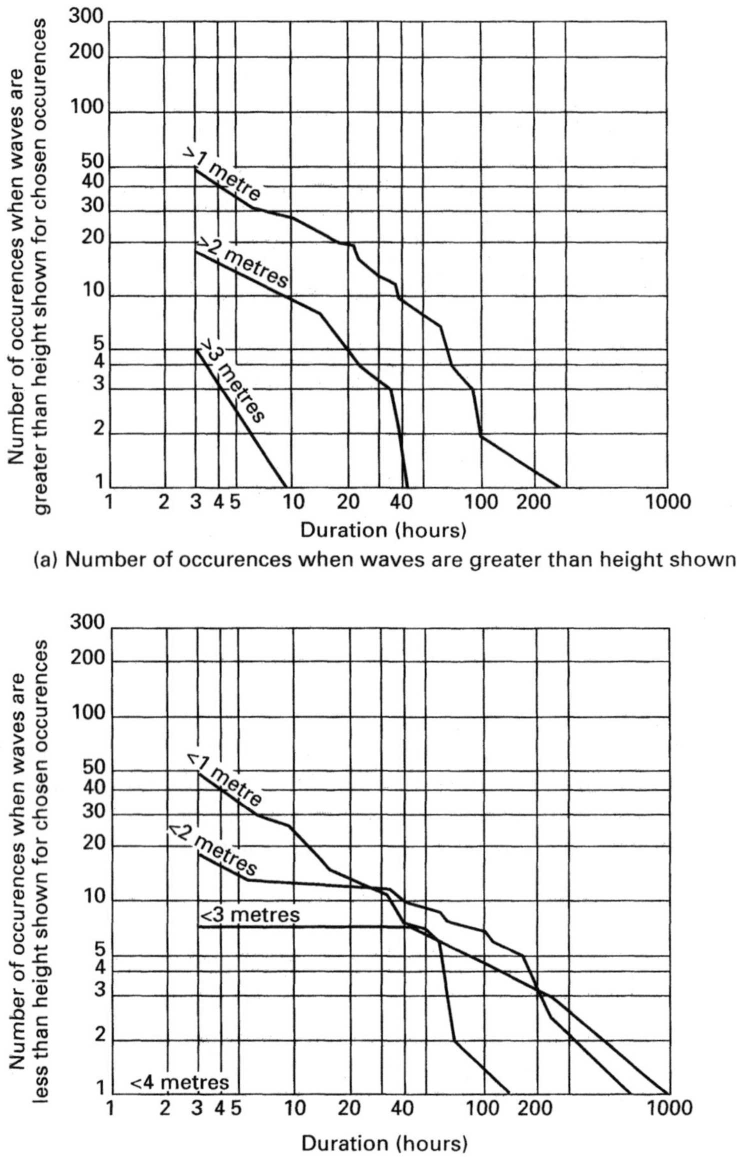

(a) Number of occurences when waves are greater than height shown

(b) Number of occurences when waves are less than height shown

Figure 6.10 Typical wave height persistence diagrams

This type of current meter can be suspended from the survey vessel at the desired depth and observations recorded manually from a deck-mounted read-out device or recorded on magnetic tape or computer disk. As currents should be observed over at least one tidal cycle, this requires the survey vessel to remain on station, which may be expensive or undesirable for other reasons. If this is the case, or when long term observations are needed, the meters can be left in place using a mooring system. In this mode, the meter stores the data on magnetic tape for retrieval at a later date. One drawback of this type of installation is the frequency with which they are lost, either through deliberate theft or as a result of fouling on trawl nets or anchors.

The most advanced and powerful method of current measurement is the acoustic doppler current meter (ADCM) which can be used to measure the current components throughout the water column. The ADCM measures the doppler shift of an acoustic pulse which is reflected from particles suspended in the water column. It utilises four inclined acoustic beams and, by measuring the doppler shift for each beam, can resolve the current components in a series of bins which are typically of the order of one to three metres deep, but can be as little as ten centimetres. ADCMs are usually mounted in frames on the seabed to gather data over an extended period of time. Some can be vessel-mounted and used to gather current data along lines sailed by the survey vessel.

Another method of current measurement, used principally to investigate the dispersion of potential pollutants, is to monitor dyes or chemicals released into the sea.

Data presentation

Current data can be presented in several forms, depending upon the nature of the original observations. In the case of float tracking, the tracks are usually presented on a chart, showing the position of the floats at specified times after release. The wind direction and velocity at the time of observation should also be shown on such plots.

Data from current meters can be plotted graphically using arrows to indicate the current direction; the length of the arrow being scaled to represent the velocity. This method is useful for illustrating variations of current with depth at a single location.

The current state diagram (Bray, 1979) shows the current direction and velocity superimposed on a tidal clock (see Figure 6.11). The current is represented by a ribbon whose variable width indicates the current velocity. The position of the inner edge of the ribbon shows the direction. Although not suited to regimes which have a constantly varying current speed and direction, the diagram does give an instant picture of the general regime in most other situations.

In order to indicate the total current range at the point of measurement, a tidal factor is quoted on the diagram, defined as:

$$\text{Tidal factor} = \frac{\text{Mean tidal range over the measurement cycle}}{\text{Mean spring tide range}}$$

Current state diagrams should be constructed for both spring and neap tidal cycles.

6.4.6 SEABED FEATURES

In any dredging work, it is important to locate features such as wrecks, debris, pipelines and small rock outcrops which present an obstruction to dredging. Often, the general location of such features will be known by the local maritime authority. The owners of cables and pipelines will usually have installation records showing where they were originally placed. In the case of cables, however, it is often found that they have moved some considerable distance from their original positions. The authors know of cases where cables have moved several hundred metres due to seabed movement.

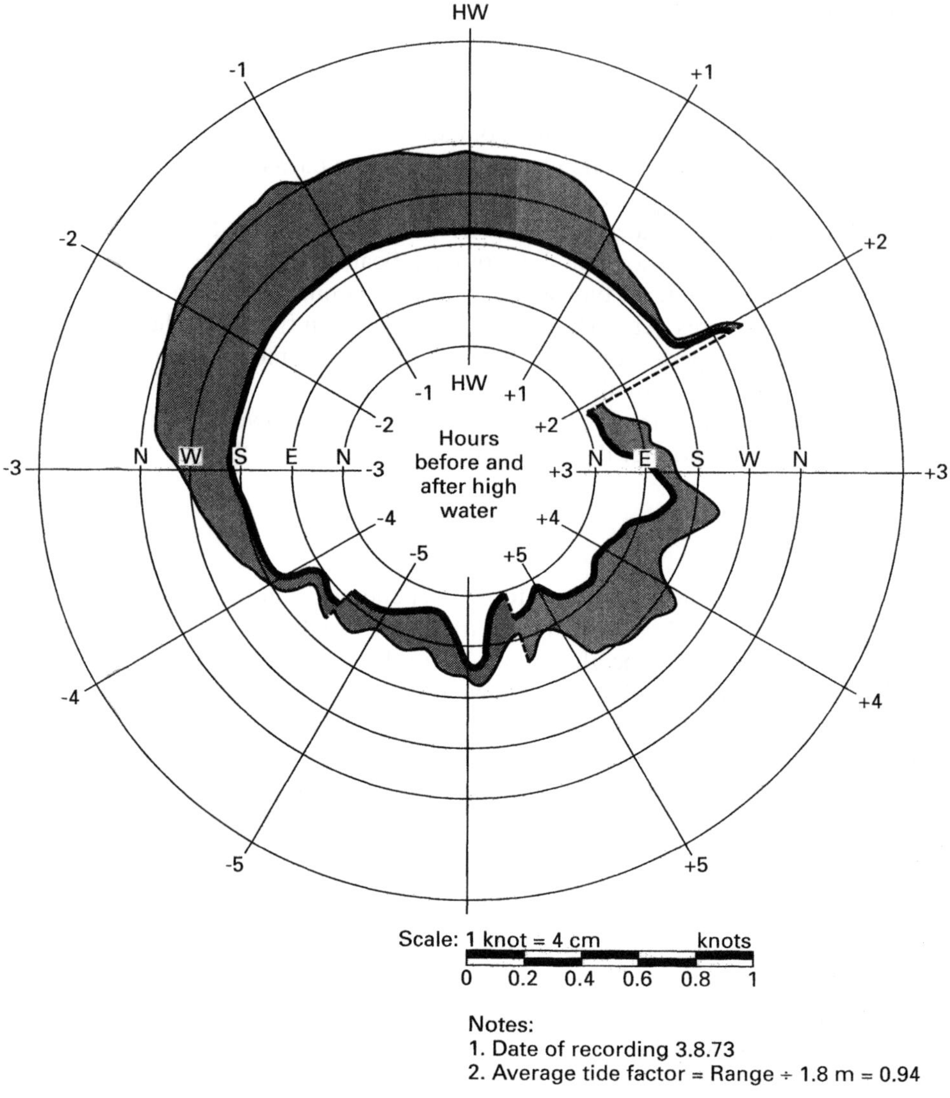

Figure 6.11 A current state diagram

Observation methods

Two techniques are widely used to locate seabed features and obstructions; side-scan sonar, for surface features, and magnetometers, for subsurface features. Swathe echo sounders and shallow geophysical methods may also be of use.

The side-scan sonar comprises a sonar 'fish' which is towed behind the survey vessel and continuously scans the seabed, by means of sonic pulses (see Figure 6.12). The principle of its operation is thus similar to that of an echo sounder except that the on-board recorder converts the return signals received by the tow fish into an oblique picture of the seabed. The general texture of the record and the intensity of the return signals, which are recorded as shades of grey, can be interpreted to determine the accurate location, shape and size of a

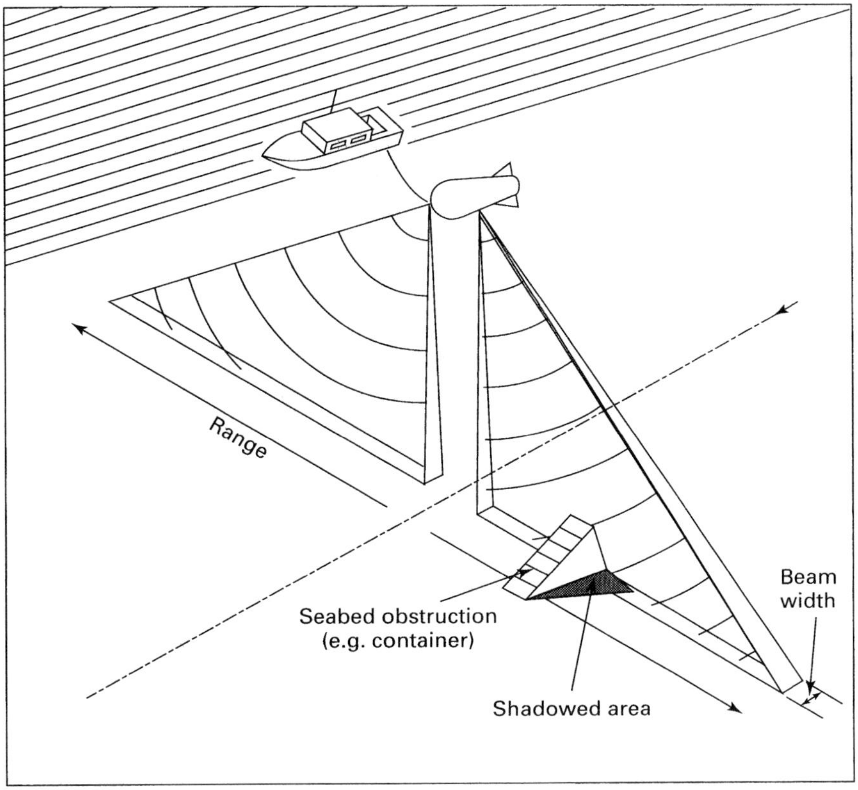

Figure 6.12 Side scan sonar

wide variety of both natural and artificial features. With large objects, a white or pale grey shadow is produced on the record beyond the feature and, using trigonometric methods, the height of the feature above the seabed can be estimated. By experienced interpretation of the texture of side-scan records it is possible to differentiate between different seabed types, (i.e. mud, sand, gravel and so on).

Seabed features which cannot be identified positively from side-scan records will often be further investigated by a diver. High resolution swathe echo sounders (see above) are also increasingly being used for this purpose.

Side-scan sonar cannot be used to locate and identify buried obstructions. For such objects, a magnetometer is used which accurately measures the earth's total magnetic field as it is towed above the seabed. Distortions or anomalies in the field often indicate the presence of buried obstructions. Magnetometers are particularly useful for locating cables with metallic armouring which includes ferrous materials and for locating ordnance. Unfortunately for the engineer concerned with dredging works, many modern cables do not contain ferrous metal and cannot therefore be located using magnetometers. Similarly, small items of ordnance may not disturb the magnetic field sufficiently to be detected. The lack of any apparent anomalies on a magnetometer record cannot therefore be assumed to indicate that an area is clear of obstructions.

Shallow geophysical methods have also in some cases been useful in locating, but not identifying, certain types of subsurface obstructions.

Data presentation

Side-scan records are usually presented in summary form as a chart showing the locations of features which have been identified and accompanied by explanatory notes, either on the chart or in text form as appropriate.

6.4.7 SEDIMENT MOVEMENT

A study of sediment movement is essential in order to design port facilities and shipping channels which require a minimum of maintenance dredging, and to evaluate the potential effects on the marine and coastal environment. In addition, knowledge of sediment movement may be important to the dredging contractor who may wish to estimate rates of sedimentation in his dredging area during the execution of the works.

Sediment movement studies are complex and beyond the scope of this book. They require a full understanding of potential sources of sediment supply, the concentration of sediment in the sea or river water, tides and currents. Observations should be made over as long a period as possible in order to establish seasonal changes of sediment flux, particularly in or adjacent to rivers.

6.4.8 COMPUTERISED SYSTEMS

Many of the survey methods described above can be interfaced with a computer to facilitate data gathering and processing. Computer-based packages come in a variety of configurations and can be tailored to suit the specific requirements of most users.

The simplest system comprises a computer interfaced with the echo sounder and the electronic positioning system. The computer will mark the echo sounder record at pre-determined intervals and record the water depth and the position of the survey vessel at the time; the records can be used to produce vessel track plots and facilitate the production of the bathymetric charts.

More sophisticated systems can be interfaced with a variety of measuring and plotting systems such as telemetric tide gauges, seismic profilers, gyro-compasses, heave compensators and swell filters. They may continuously display the vessel's position to the helmsman, relative to the survey line, enabling him to steer an accurate course and can produce a variety of end products on printers and plotters such as tide and heave-compensated bathymetry, track plots and annotated echo sounding and side-scan sonar records.

With specific regard to surveys undertaken in connection with the progress of dredging works, there are a number of systems now available which include software packages that compare actual bed profiles with the design profiles and which compute quantities to be dredged.

6.5 Geological and geotechnical investigations

6.5.1 THE OBJECTIVES OF GROUND INVESTIGATIONS

Geological and geotechnical investigations, often termed 'ground investigations', are usually required for two reasons:

- to determine the nature of the materials to be dredged and, possibly, used for reclamation purposes;

- to establish the foundation conditions for structures associated with the development such as breakwaters, jetties, land reclamation and shore facilities.

In both cases the basic objectives of the investigations are similar: to identify and characterise in engineering terms the various geological units present at the site. There exist, however, differences between the approaches to the investigations, and the nature of the engineering data required, which are often not fully appreciated by those responsible for their design and execution.

Such lack of appreciation frequently gives rise to difficulties during dredging contracts and, on occasions, to disputes between the contractor and the client. It is not the intention of the authors to treat this aspect in great detail here but a brief discussion is necessary in order to explain the importance of adequate investigation of ground conditions.

There is a common tendency to allocate a certain proportion of a project budget to ground investigation. This results in a higher intensity of investigation for structures, which represent a very high degree of capital investment per unit area, than for dredging.

However, in the majority of construction works, the cost of foundations, which are largely determined by the ground conditions, represent only a part of the overall cost whereas the cost of dredging is very largely dependent upon the nature of the ground being dredged. Dredging contractors are frequently faced with having to interpret dredging conditions, and to estimate costs, on the basis of very widely spaced, and often poorly located, borehole data which may not adequately describe the materials to be dredged. Investigations for dredging should, therefore, within reason, be as thorough as those for structures.

Even in cases where the intensity of investigation is sufficient it is often the case that the nature of the data obtained is relevant to foundation design but not to dredging. Whilst there is a large degree of overlap in the type of data required for both activities, the dredging contractor has particular needs which are frequently overlooked.

A factor which distinguishes geotechnical and geological investigations from others which are undertaken for dredging works is the degree of interpretation which is inherent in the assessment of the results and even in the design of the investigation itself. Rocks and soils are heterogeneous materials and can vary rapidly over very short distances. For each type of material, there are investigation techniques which are more appropriate than others and, therefore, the design of the investigation inevitably involves some degree of speculation. The geologist or geotechnical engineer responsible for the design must ask himself if the methods he proposes to use are appropriate for the expected conditions and must be prepared to vary his methodology to suit the conditions encountered during the investigation.

At the end of the fieldwork, the data which has been obtained will be discontinuous, that is it will have been gathered from boreholes and probings which will be separated by tens or hundreds of metres or will have been obtained by geophysical methods along discrete lines, again separated by tens or hundreds of metres. There is inevitably a high degree of interpolative interpretation involved in the assessment of such data which is not present to the same extent in, for example, hydrographic investigations. Bathymetric surveys, properly undertaken, do not require interpretation; the depth measurements are absolute and subject to known limitations of accuracy. In geological investigations, the observation of similar conditions in two adjacent boreholes may not be taken as a positive indication that similar conditions will be encountered at all points between the holes.

6.5.2 GROUND INVESTIGATION DESIGN

The design and planning of a ground investigation will vary according to the nature of the project in hand. Investigations might be required to establish:

- the nature of materials to be dredged;
- the stability of dredged slopes;
- the foundations of marine structures;
- the foundations of onshore structures (Plate 6.1);
- the sources of natural construction materials such as armour stone, aggregates and reclamation materials and so on.

Whatever the objectives, the investigation should be undertaken in phases, commencing with a desk study and progressing through preliminary field investigations to detailed investigation, laboratory testing and final interpretation. Throughout the investigation, data should be interpreted as it becomes available in order that variations of methodology and intensity of investigation can be introduced where necessary.

Plate 6.1 Land-based rotary drilling rig on a reclamation site

Desk study

All existing information on the area to be investigated should be assembled and studied in order to gain an understanding of the general geology of the area and to facilitate the selection of the most appropriate methods of field investigation. There are few places in the world where it is not possible to obtain some such information but the level of detail will obviously vary. Potential sources of information include:

- records of earlier projects in the area;
- national and regional geological survey departments;
- universities and research organisations.

An additional potential source of information which can be utilised at this stage is aerial photography and multispectral satellite imagery. These might strictly be regarded as indirect methods of field investigation, and are discussed as such in the appropriate section below, but they are most commonly used at the desk study stage.

Having undertaken the desk study, the main phases of investigation can be planned with a greater degree of confidence than would otherwise be the case. There are four main components:

- preliminary field investigation;
- detailed field investigation;
- laboratory testing;
- analysis of results.

For some small, or straightforward projects, the preliminary and detailed field investigations may be merged.

Preliminary field investigation

Depending on the nature of the project, the preliminary field investigation may comprise a campaign of boreholes, geophysical surveying and laboratory testing designed to provide basic information at the feasibility study stage and undertaken quite separately from the main investigation, or it may simply be the initial stages of the detailed investigation during which the general geology is established and methods of investigation are refined.

Detailed field investigation

The preliminary investigation will have provided an overview of the site geology and the engineering properties of the materials present. The detailed investigation should be designed in the light of this valuable information. The suitability of the methods of investigation, which are discussed below, should be reviewed and changes made where the quality of samples or other data appears to be inadequate.

Even during the main investigation, the results should constantly be assessed and reviewed in order that the investigation results in a complete understanding of the site and that all of the required engineering parameters, which will vary from one site to another, have properly been defined.

Laboratory testing

Ideally, laboratory testing of samples of soil and rock should be undertaken concurrently with the field investigation and the results made available to those responsible for the design and progress of the investigation. This will enable them to gain a greater understanding of the nature of the materials they are dealing with and to relate this to the design of the development. In this manner extensions and adjustments to the scope of the fieldwork can be made on an informal basis. In many situations, however, it is not possible to start the laboratory testing until after the fieldwork is completed.

Analysis of results

The analysis of ground investigation results can be divided into two parts: the investigation of the geology of the site and derivation of the engineering properties of the soils and rocks, and the detailed engineering assessment which leads to recommendations for the design of structures, dredged slopes and dredging. The bulk of the analysis generally takes place after the fieldwork and laboratory testing have been completed, but it is essential that some analytical work is undertaken during the field work in order to ensure that sufficient relevant data is collected prior to the demobilisation of the fieldwork teams.

6.5.3 FIELD METHODS OF GROUND INVESTIGATION

Field methods are summarised in Table 6.2 and may be divided into two broad groups: direct methods which involve penetration of the ground or seabed and indirect methods which do not. Direct methods include the sinking of boreholes and pits, which permit sampling and *in situ* testing, together with a variety of probing techniques. Indirect methods are restricted to geophysical techniques and remote sensing.

Table 6.2 Field methods of ground investigation

Direct methods		Indirect methods
Penetration/sampling	*In situ* tests	
Cable-tool (percussion) boring Washboring Vibrocoring Rotary drilling Trial pits Gravity coring Grab sampling Probing	Standard penetration test (SPT) Cone penetration Shear vane Permeability Trial dredging Trial embankments	Satellite imagery Aerial photography Geophysical methods: e.g. seismic reflection, seismic refraction, chirp acoustic, profiling

Direct methods

Direct methods may further be divided into those which provide samples of soil or rock for physical examination or laboratory testing and those which do not. The most common methods which provide samples are the sinking of boreholes by washboring, cable-tool methods and by rotary coring, vibrocoring and excavating trial pits.

Cable-tool boring

Cable-tool boring, often called 'percussion' or 'shell and auger' boring, is a widely used and simple technique of ground exploration. The rig comprises a tripod and winch with a capacity of one or two tonnes (see Figure 6.13). The tools are heavy steel tubes which vary in size and design according to the material being bored, which are raised and lowered at the base of the hole to loosen and retain the soil. The technique can be used in all soil types and in weak rocks. In granular or soft strata, it is usually necessary to line the hole with steel casing to prevent collapse. The sizes of commonly used casing range from 150 to 300 millimetres and, by progressively reducing the casing diameter with depth (i.e. telescoping the casings), the method can be used with considerable effect, even in difficult ground conditions.

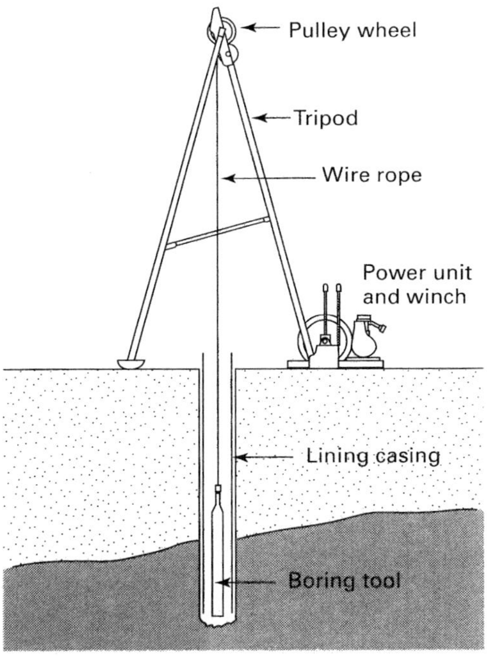

Figure 6.13 The cable-tool boring method

The winch capacity and the length of casing that can be handled is the main limiting factor with regard to the depth capability of this method and boreholes greater than 60 metres deep are unusual, but such depths are rarely required during port investigations.

The method can be used to provide both disturbed samples of the soil for visual examination and classification testing and undisturbed samples for more complex strength and consolidation tests. The latter are obtained by tube samplers which are hammered or pushed into the ground at the bottom of the borehole and which vary in the details of their design according to the nature of the ground being sampled.

Cable-tool boring rigs can be mounted on pontoons and self-elevating platforms or can be operated over the side of large boats from a cantilevered platform. In the case of boreholes put down from floating craft, the sea conditions will affect the quality of the borehole because, as they deteriorate, it becomes progressively more difficult to continue the borehole and to take good quality samples. It is usually possible to continue work in waves and swells of up to 0.75 metres, depending on the size of the vessel being used. Borings undertaken from self-elevating platforms are not affected by sea conditions unless they are severe but delays will be encountered due to the inability of platforms to move from one location to another in bad conditions and because it is often difficult and dangerous to transfer the crew to and from the platform.

Washboring

Washboring can be used in most soil types except for very stiff clays and soils containing appreciable quantities of cobbles or boulders. Using a winch and light tripod, the soil at the base of the hole is loosened by the action of a chisel bit and washed to the surface by water pumped through the hollow rods to which the bit is attached. Lining casing needs to be installed in most cases to support the full length of the hole and to prevent excessive enlargement due to scour of the sides.

Disturbed samples of the soil are obtained by collecting the wash-water returns in a large container and draining off the excess water. This has to be done with considerable care to prevent loss of fine materials and unrepresentative samples. Undisturbed samples can also be obtained by the same methods as those used in cable-tool boring.

Washboring can be undertaken from pontoons, platforms and large boats in the same manner as cable-tool borings and with the same constraints.

Vibrocoring

Vibrocoring is a rapid and cost-effective method of investigating soft soils in the marine environment and is frequently used for port investigations, particularly in connection with the dredging works. A frame carrying a sample tube, typically 75 to 100 millimetres in diameter and up to six metres long, is lowered by crane from a support vessel to the seabed and the tube is vibrated into the seabed. The sampler is then withdrawn, the sample being retained by a basket-shaped spring. Penetration of the sample tube generally takes place within a few minutes and the complete operation, including lowering to the seabed, sampling, recovery and removal of the sample tube can be carried out in 20 to 30 minutes.

In recent years, sophisticated large-diameter vibrocores have been introduced. As with any method of sampling, the size should be chosen so that the equipment is capable of sampling the coarsest material likely to be present.

Care must be taken with the interpretation of the depths from which materials have been recovered when using vibrocores because the sample tube is long and it is not generally possible to allocate core losses to specific depths. Losses can occur at almost any depth due to:

- blockages of the core barrel by coarse material;
- very soft sediments not entering the barrel due to the frictional resistance of material already in the barrel;
- material falling out of the barrel during recovery.

Because there is no rigid connection between the vibrocorer and the seabed, it is possible to work from floating craft in conditions which prohibit cable-tool boring and washboring. As sea conditions deteriorate, however, damage to the equipment can occur as it is placed onto and removed from the seabed due to 'snatching' as the support vessel moves. It will also become progressively more difficult, and dangerous, to move the vibrocore inboard and outboard.

Rotary drilling

Rotary drilling is used to obtain a continuous sample in rock and stiff clays. A tube, called the core barrel, fitted with a diamond or tungsten cutting bit is rotated at the bottom of the hole by means of hollow rods connected to the rig at the surface. As it rotates it cuts an annulus of rock and advances. The remaining central core of rock enters the barrel and is retained by a spring. Cooling and lubricating the cutting bit is achieved by pumping a fluid, usually water, down the hollow rods to the bit. This also serves to remove the cuttings as it returns up the hole between the rods and the wall of the hole.

Core samples are typically between one and three metres in length (see Plate 6.2) and the hole is advanced progressively by drilling the sample, recovering and emptying the barrel and then drilling a further sample. The diameter of cores recovered by this method usually lies in the range 50 to 80 millimetres. There is however, a relationship between the core diameter and the quality of the recovered samples. Equipment designed to produce core diameters at the upper end of that range, or greater, should be used, particularly in the case of weak or friable rock, because it generally produces better and more continuous samples. The depths to which this technique can be applied are, for all practical purposes, unlimited and depend primarily on the size of the rig being used. However, for most jobs, small light-weight rigs are usually found to be adequate (see Plate 6.3).

In order to obtain good cores, rotary drilling needs to be carried out from a stable platform or, when undertaken from a floating vessel, in calm water. Some drilling systems can be used in less-than-ideal conditions from floating craft. One such system uses a drill which is mounted directly on the drill casing that is independent of the movement of the vessel (see Figure 6.14). Other systems are available which incorporate a swell compensation system.

Trial pits

Trial pits are used onshore to investigate soils to a maximum depth of about five metres. Because they are simple hand- or machine-excavated holes, they are cheap and have the advantage that the soils can be examined in far greater detail than is the case with the other methods described above which provide only comparatively small samples.

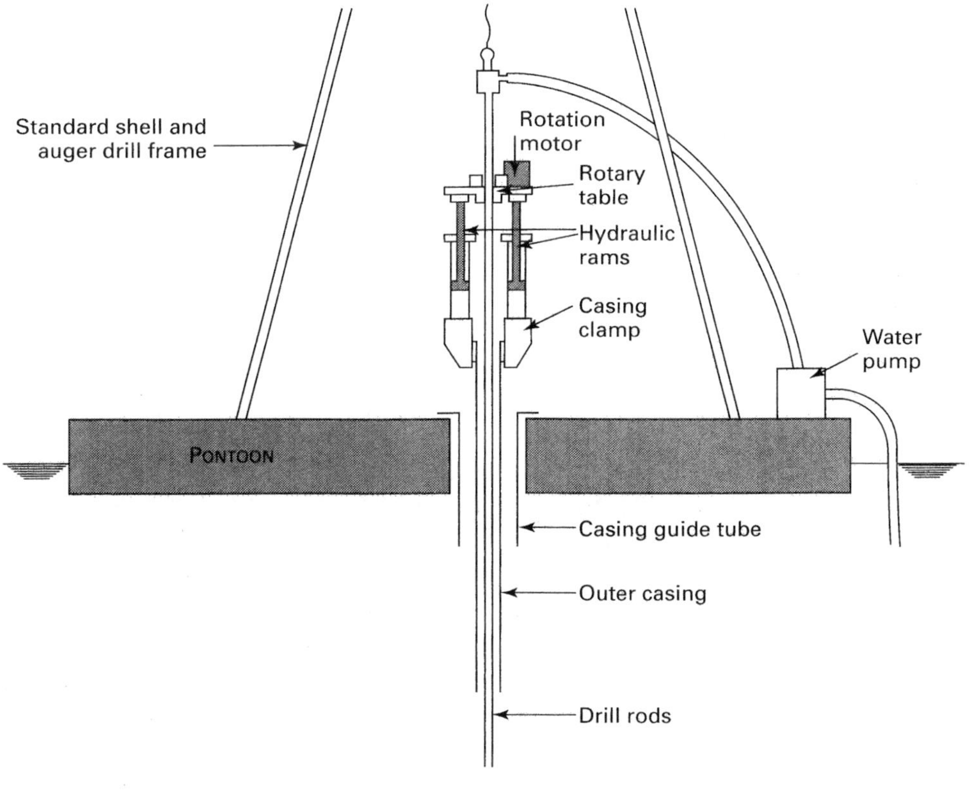

Figure 6.14 Casing mounted marine rotary coring rig (Pilcon Engineering Ltd)

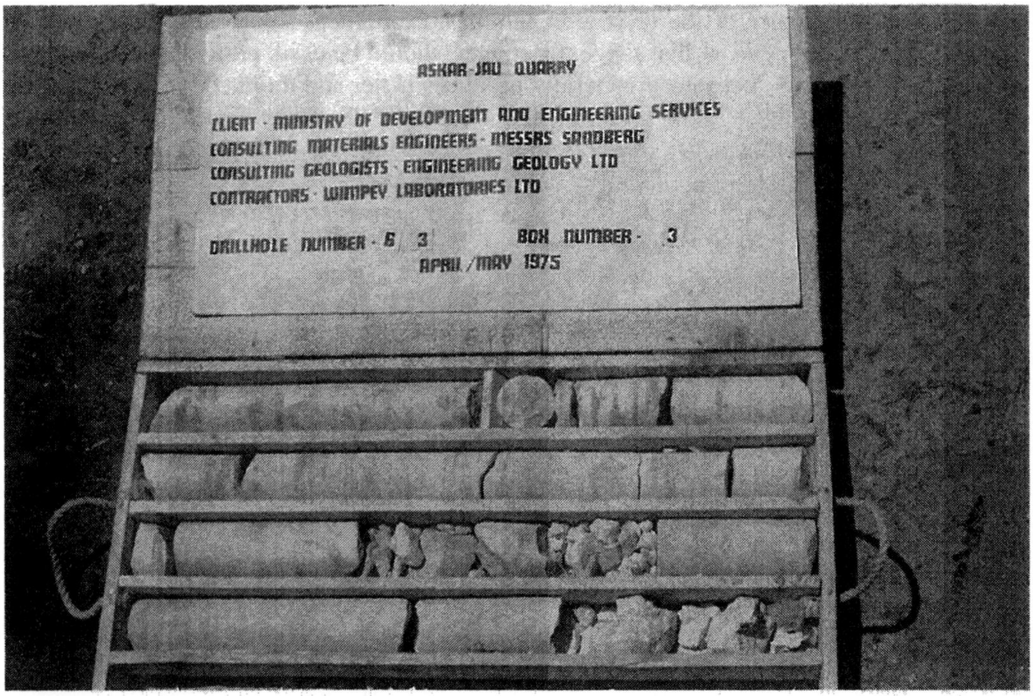

Plate 6.2 Rock cores stored in a core box

Plate 6.3 An overwater site investigation barge with cantilevered drilling rig

Other direct methods of investigation

Other direct methods of investigation are generally somewhat restricted in their use, but they should always be considered as a potential supplement to the techniques just described because they are frequently found to be cost-effective.

Grab sampling is a very simple technique of obtaining small disturbed samples of the seabed material using a small hand-operated grab. The method is often used during preliminary investigations and studies of sediment movement.

Gravity coring is a useful technique of obtaining samples of relatively soft soils close to the seabed. The sampler consists of a streamlined 'bomb' with a sample tube fitted to the base. It is lowered by crane or winch and released at a pre-determined height above the seabed which it penetrates under its own momentum. The size of gravity corers varies widely, the largest weighing several hundred kilograms and capable of penetrating several metres.

Probing is a very simple technique which is often used to investigate variations of the thickness of soft material overlying firmer strata. In its simplest form it involves pushing rods into the ground and recording the degree of effort required. This can be done manually if depths are limited or, for greater depths, using a small tripod and winch mounted on a pontoon. In some instances water may be pumped down the rods to a 'chopping bit' to assist penetration. Interpretation of the records of such probing methods is subjective and they should generally only be used as a cheap method of obtaining additional data to assist interpolation between sampled boreholes or vibrocores.

Indirect methods

Satellite imagery and aerial photography

Satellite imagery and aerial photography are most commonly used during the early stages of investigation. Aerial photography is familiar to most engineers and surveyors as an aid to mapping but pairs of photographs viewed stereographically can also be used by geologists and geomorphologists to gain valuable information concerning for example, geological structure, slope stability and drainage courses.

Multispectral satellite imagery is one of a number of 'remote sensing' techniques which have evolved from conventional aerial photography during the last 20 to 25 years. Today, it is widely used as a valuable and cost-effective aid to ground investigation (Land and Garrard, 1986) and has also been used to produce bathymetric and suspended sediment data (Loman *et al.*, 1992). It involves the detection and measurement, by means of satellite mounted scanners, of reflected or emitted electromagnetic energy which varies in wavelength throughout the electromagnetic spectrum. The various features of the earth's surface each have a 'signature' in terms of their spectral characteristics and, under good conditions and with the aid of computer processing of the images, indicative data can be obtained concerning the geology and geomorphology of the coastline, the nature of shallow seabed sediments, inshore bathymetry and the location of possible offshore sources of reclamation materials.

Geophysical methods

Geophysical surveys are frequently undertaken in connection with port projects, particularly offshore when they may provide a significant amount of information at comparatively modest cost. Their application, however, is often misunderstood and apparent failures of geophysical methods to provide the required information have often been blamed erroneously on the techniques rather than a failure on the part of those who instigated the survey to understand their limitations.

The single most important point concerning the use of geophysical methods for civil engineering purposes is that they invariably require calibration by some means of direct investigation such as boreholes. The techniques described below produce data which are subject to a large degree of interpretation. Such interpretation can only be undertaken with confidence if it can be confirmed, at a number of locations within the survey area, by borehole evidence. Without boreholes the interpreted results of geophysical surveys are reduced to the level of educated speculation. The use of geophysics should therefore always be considered as complementary to direct investigation methods and not a substitute for them. When undertaken in this manner geophysics can provide valuable, cost-effective data between borehole locations.

The second point of importance is that, even with borehole calibration, all geophysical methods are subject to limitations which are dependent upon site conditions. During the planning of geophysical surveys it is strongly advised that the anticipated ground conditions and the investigation methods be discussed in detail with the contractor or with a qualified geophysical consultant in order that the most appropriate methods are selected for the conditions.

Seismic reflection is the technique most commonly used for shallow engineering surveys undertaken offshore. A high-energy acoustic pulse generated just below the surface of the sea passes through the water and the soils under the seabed. It is reflected back by the seabed and by geological interfaces corresponding to changes in density and sonic velocity (see Figure 6.15).

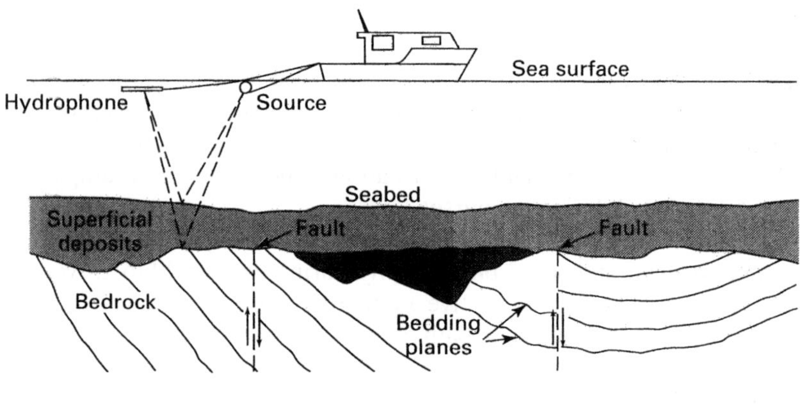

Figure 6.15 Continuous seismic profiling (Hunting Geophysics Ltd)

A profile of the return times from the various reflecting horizons is continuously recorded and, in the hands of a skilled interpreter, can be used to produce a profile of the disposition of the geological units along the survey line. The energy source and the hydrophones, which detect the reflected pulses, are towed behind the survey vessel, which may simultaneously be undertaking bathymetric surveys, and 15 to 20 kilometres of survey line, or more, may be achieved in a working day.

The effectiveness of the seismic reflection method depends partly on the frequency of the acoustic source. Low frequency pulses achieve deep penetration but have a relatively low resolution. High frequency pulses attenuate more quickly and therefore have a shallow penetration but have a better resolution. Three systems are in common use; sparkers, boomers and pingers, the latter two being the most frequently applied to shallow engineering surveys.

Their main characteristics are listed in Table 6.3.

Table 6.3 Characteristics of three seismic reflection energy sources

Type	Frequency, kHz	Maximum penetration, m	Resolution, m
Sparker	0.01 – 1.25	150	0.5 – 1.0
Boomer	0.5 – 2.5	100	0.3 – 0.8
Pinger	2.0 – 12.0	50	0.3

Effectiveness is also dependent upon soil conditions. Interfaces which are sharp and which are characterised by a significant change of density are easier to detect than gradual transitions from one material type to another.

A major limitation of the technique which is sometimes encountered during surveys for port developments is 'masking' due to the presence of biogenic gas in the sediments below the seabed. The presence of even very small quantities of gas in sediments usually obscures the record to the extent that it cannot be interpreted, but such occurrences are generally restricted to recent organic sediments.

The seismic reflection method, whilst providing an interpreted profile of the disposition of the strata along the survey line, gives little direct indication of their nature. Seismic refraction, although less frequently used for offshore engineering surveys provides a more positive indication of the nature of the strata. In this case, the energy source and a 'string' of hydrophones are laid along the seabed, rather than towed behind a moving vessel. The energy source is activated and the hydrophones detect the arrivals of the energy pulses as they are refracted through the various underlying strata. Analysis of the time taken for the energy pulses to travel to successively more distant hydrophones permits the calculation of the seismic velocity and thickness of the various strata. The seismic velocity of soils and rocks tends to increase with increasing strength and competence. The measured velocities provided by this method thus give some indication of the nature of the materials.

'Chirp' acoustic profiling systems which work with a range of transmitted frequencies, (e.g. 1.5 to 10 kHz), are a relatively recent innovation and are becoming widely used for offshore geophysical surveys. These differ from conventional seismic reflection systems such as boomers and pingers in that they provide quantitative data on acoustic reflectivity which permits the identification of relatively subtle and gradational changes of acoustic impedance, in addition to major changes coincident with the interfaces between strata of markedly different characteristics. They are more easily employed than seismic refraction systems. Chirp systems are capable of penetrating up to 40 metres, under the right conditions, and are claimed to have a resolution in the range 0.1 to 0.2 metre. The use of Chirp systems is described by LeBlanc *et al.* (1992) and Schock *et al.* (1989).

Resistivity surveys are only undertaken onshore and are therefore rarely employed during investigations for port development. They are, however, comparatively cheap and should be considered for onshore parts of the development if geological conditions are suitable. Soils and rocks have varying electrical resistivities which depend to a large extent on their porosity, water content and mineralogy. The resistivity is measured by observing the potential drop between electrodes placed in the ground. By progressively increasing the spacing of the electrodes it is possible to generate a vertical profile of resistivity which may, with reference to borehole data, be used to interpolate geological interfaces between boreholes.

In situ tests

One of the main advantages of *in situ* tests is that they provide a measure of the engineering performance or characteristics of the ground without the disturbance which is inevitably induced by sampling, handling and transport to the laboratory, and without the scale effects associated with testing small samples of soil or rock taken from an effectively infinite mass.

The standard penetration test (SPT)

The standard penetration test (SPT) is the most common of the tests undertaken in boreholes. It provides a small sample and gives a qualitative indication of the relative density of granular soils and, to a lesser extent, the consistency of cohesive soils and strength of weak rocks. A small diameter tube fitted with a cutting edge is driven into the soil at the base of the hole by means of a 65 kilogram hammer falling through a distance of 700 millimetres. The sample tube is driven 450 millimetres and the penetration resistance, or N value, is expressed as the number of blows required to drive the last 300 millimetres. In gravels and weak rocks, the cutting shoe is usually replaced with a solid cone.

The shear vane test

The shear vane test is used to determine the shear strength of soft cohesive soils. In some cases, it may be the only manner in which the strength of very soft clays can accurately be measured. The 'vane' comprises two blades fixed at right angles and attached to a thin rod which is pushed into the soil from the surface or from the bottom of a borehole. A calibrated torque head is used to apply a steadily-increasing turning force to the rod until failure occurs. The shear strength obtained in this manner needs to be corrected to allow for the effects of soil friction on the rods between the vane and the torque head, unless the rods are enclosed by small diameter casing.

A special vane testing apparatus has been developed in The Netherlands for measuring the strength of extremely soft muds (Cox *et al.*, 1986). The apparatus works on the seabed and can thus be deployed from a survey boat.

The cone penetration test

The cone penetration test (CPT) is a sophisticated extension of the simple probing technique. It provides a continuous record of the engineering characteristics of the soil profile in contrast to most other *in situ* tests which can usually only be undertaken at discrete depth increments.

Cone penetration testing is the direct measurement of the resistance offered by the ground to the penetration of a thin rod with a conical tip. The resistance can be measured in terms of either or both the end-bearing and side friction components. The Dutch cone penetration test is the most common of a number of variants. The penetrometer is advanced into the ground by a thrust machine at the surface, or mounted on a platform in the case of offshore tests, at a nominal speed of two centimetres per second. Some CPT machines are designed to operate from the seabed, when the reaction force is provided by a heavy, weighted template.

The resistance to penetration can be measured by electric load cells mounted in the penetrometer tip or by pressure gauges at the surface. In the case of electric penetrometers, the end resistance and the friction on the side of the penetrometer are measured and recorded simultaneously. Mechanical penetrometers record the two components of resistance separately at intervals of 200 millimetres. Electric penetrometers can additionally be equipped to measure the pore water pressures induced in the soil as the tip advances, the orientation of the tip, or the temperature of the soil.

The cone penetration test can be used in most soil types except for hard clays and soils containing an appreciable amount of cobbles or boulders. The records of the test can be analysed to provide a remarkably detailed soil profile and to derive a wide variety of engineering parameters which can be used in foundation design and the assessment of soil dredgeability.

Permeability tests

Permeability tests are undertaken in boreholes, usually by the falling-head method in which the rate at which an applied head of water falls can be related to the permeability of the soil or rock. Other more complex methods involving the sealing-off of a section of the borehole with hydraulic or pneumatic 'packers' and injecting water under pressure are also used.

Plate-loading tests

Plate-loading tests undertaken in boreholes or in trial pits are a valuable source of data relating to the bearing capacity and settlement characteristics of the ground. They involve the progressive loading of a steel plate placed on the soil or rock and measurement of the settlement of the plate. Because the applied loads are generally large these tests are only a practical consideration onshore.

Trial dredging

Ground investigation techniques provide only a certain amount of data of a specific nature which must then be interpreted to provide estimates of the performance of dredging plant. In areas where geological conditions are very complex or variable, or when conditions are borderline for the use of certain types of equipment, trial dredging may be undertaken. Trial dredging is expensive but it provides the best type of information if it is properly carried out and recorded. Trial dredging is most frequently undertaken when rock is to be dredged and a decision must be made as to whether or not it can be dredged directly or must first be pre-treated.

If trial dredging is undertaken, it is essential that it be carried out in as wide a range of materials as possible and for a period which is sufficiently long to give a reasonable indication of the range of conditions, and the average condition, to be expected during the main campaign of dredging. Very detailed records should be maintained and the type of plant used should be described in detail when reporting on the results of trial dredging.

Trial embankments

Perhaps the largest and most 'spectacular' of all *in situ* tests is the trial embankment. One of the most difficult things to estimate in geotechnical engineering is the rate of settlement of structures, particularly very large structures, which cover a wide area. Trial embankments are sometimes constructed in areas which are to be reclaimed in order to provide field data concerning the rate and magnitude of settlements which are to be expected during and after placement of the fill material. The embankments are instrumented using a variety of apparatus designed to monitor the progress of the settlement and to observe the change of pore water pressure in the soil. Trial embankments are, of course, expensive and are generally only constructed for investigations undertaken for very large projects.

Laboratory tests

A detailed description of the very wide range of laboratory tests which are undertaken on samples of soil and rock is beyond the scope of this chapter. In general terms, however, most laboratory tests fall into one of five groups:

- index tests;
- strength tests;
- consolidation tests;
- compaction tests;
- materials tests.

Index tests

Index tests are mostly simple and inexpensive tests which provide basic information on the physical characteristics of the material, such as density, porosity, grain size, water content and plasticity. In addition, they can often be used to supplement more sophisticated and relatively expensive tests. For example, for most types of clay there is usually a reasonably well-defined relationship between water content and shear strength. If that relationship has been established by a number of paired tests, then other water content determinations can be used as an indicator of strength.

Strength tests

These are used to determine the strength and elastic properties of soils and rocks in compression, tension and shear. In cohesive soils and in rocks, the unconfined compressive strength is the most widely used measure of strength for dredging investigations. Triaxial compression tests, in which the sample is subjected to an overall confining pressure during the test, are also frequently used, particularly for clays.

Other tests which are used, although generally less frequently, to measure the strength of rock are the Brazilian splitting test (tensile strength), the point load strength test (Broch and Franklin, 1972), the Protodyakonov Test (Protodyakonov, 1962) and the Schmidt Hammer Test. Point load tests are often undertaken because they are very simple and cheap.

Consolidation tests

These are undertaken to determine the magnitude and rate of settlement of structures, including reclamation fill. The standard oedometer test is the one which is most frequently undertaken and utilises samples which are typically 75 millimetres in diameter and 20 millimetres thick. Such small samples are often not wholly representative of the soils from which they are taken because they may not include thin silt and sand partings which are important for drainage during the process of consolidation. For this reason, the rate of consolidation is sometimes underestimated. Large diameter consolidation apparatus such as the Rowe Cell which uses 200 millimetre samples often give more reliable results but they are expensive and require very high quality undisturbed samples.

Compaction tests

Compaction tests are used to determine the compaction requirements and performance of materials used in reclamation, and for roads and container storage areas.

Materials tests

These include all those tests required to establish the strength, durability and chemical properties of natural construction materials such as armour stone and aggregates for concrete manufacture.

Other tests

There are a number of other tests which do not readily fall into any of the above groups and which should be considered when designing a test programme. These include mineralogical analyses to determine the constituents of the soil or rock, particle shape and angularity determinations and calcium carbonate content determinations. These are all useful for the assessment of the abrasiveness of the soils during dredging and hydraulic transport.

Soil and rock description

It can be argued that the most important element of any borehole, washboring, vibrocore or trial pit is the record of the materials which have been encountered. It is therefore of the utmost importance that the description of soils and rocks is both detailed and carried out in accordance with a recognised standard. Many countries have standards for soil and rock description, some of which are very similar. In addition, the Permanent International Association of Navigational Congresses published a report (PIANC, 1984), the Classification of Soils to be Dredged, which sets out descriptive standards, largely based on the British Standard, and which has gained general acceptance by the majority of organisations concerned with dredging.

Soil and rock descriptions are best made by the supervising geotechnical engineer as the borehole is being advanced. He is able to examine all of the material which is recovered from the hole and does not have to rely on discrete samples delivered to a laboratory and which may not enable, for example, the levels of strata interfaces to be accurately identified.

Whilst the authors recommend in principle that the descriptive standards which are used be recognised internationally, it is more important that, whichever standard is used, it is clearly identified in the investigation report and that all personnel involved in soil and rock description during the investigation use the same standard.

Intensity of investigations

The intensity with which a particular site needs to be investigated will depend mainly on the geological complexity of the site and the distances over which conditions change.

In the case of field and laboratory testing, it is relatively simple using standard statistical methods to establish whether or not the number of test results obtained for each geological unit are statistically representative of the properties of that unit. It is much more difficult to know when sufficient borehole, probing and seismic investigation data have been gathered. There are two approaches; the intuitive approach based on detailed interpretation of the data by a qualified geologist or geotechnical engineer and the statistical approach using modern geostatistical methods (e.g. Calle, 1986). Usually, a combined approach will be adopted.

The detail which is required will, to some extent, depend on the properties of the materials present. If, for example, the majority of the materials to be dredged are very similar in dredging terms, there is little point in putting down numerous boreholes in order to provide highly accurate volume estimates for each of the materials. If there is a wide variation of material properties, perhaps sufficient to require two or more different types of plant, or such that rates of production will vary significantly from one material type to another, it becomes more important to quantify accurately the different materials and to undertake very detailed investigations in order to achieve the required accuracy.

Analysis

The analysis of ground investigation data should commence as soon as the first results of the investigation become available and should continue throughout the field and laboratory stages. It is only in this manner that the investigation can be properly directed and modified to suit the conditions as they are revealed.

Presentation of data

Data presentation is an important part of any investigation. There are two types of data which can be presented; purely factual data such as borehole records and laboratory test results, and interpretative data which will include, perhaps, a detailed geotechnical model of the site and an assessment of the properties of the materials present. Similarly, there are two groups of people who will use the data provided by the investigation; the client or his engineer requires the information for design purposes, project monitoring and measurement, and the contractor who requires it in order to be able to cost and programme the work.

Those involved in design obviously need all of the results of the investigation including a detailed interpretation. There are two schools of thought with respect to the amount and type of information which should be made available to tendering dredging contractors. The largest school advocates that only factual data be provided and that the contractor should be forced to undertake his own interpretation of the data. The other school advocates that tendering contractors should be given the benefit of the interpretative effort made by those who actually undertook the investigation and who are therefore best placed to make an accurate interpretation. There are clearly some considerable contractual implications inherent in providing interpretative data to tenderers and these must be weighed against the advantages of providing such data. This is a question best left to lawyers but the authors, as practising engineers, are firmly of the view that any data which might be of use to tenderers should be made available, including interpretative data.

Factual reports

Purely factual reports are relatively simple to prepare but they are often woefully inadequate. More often than not, they fail because the data which is required to make a full interpretation is not provided or the facts are presented in a confusing manner. The following is the minimum which should be included in a factual report:

- a description of the extent of the investigation giving details of the number of investigative locations and the methods used;
- a description of the standards and codes of practice which apply to the fieldwork, the borehole descriptions and the laboratory testing; any deviations from these standards must be clearly stated;
- complete records of all boreholes, washborings, cone penetration tests and so on (including those which have failed or been aborted due to obstructions or lack of penetration); the records should be presented clearly and include details of progress, difficulties encountered (including sea conditions, in the case of investigation carried out from floating plant) and coordinated locations;
- detailed records of all field and laboratory testing;
- clear location plans at an appropriate scale showing the locations of all investigations, preferably in relation to the works which are to be carried out.

Interpretative reports

These should include all of the factual data which is required to make an independent interpretation of the data, as well as the interpretation made by the reporter. Many interpretative reports do not include this information and the user is therefore wholly reliant on the interpretation provided and is unable to make any assessment of, for example, the risk that dredging conditions may turn out to be different from those given by the interpretation.

Interpretative reports should be 'honest' in the sense that, where data is deficient or when two or more alternative interpretations are possible, the user's attention should be drawn to that fact. Geological interpretations, and the geotechnical models which are derived from them, are inevitably subject to some degree of uncertainty and this should be acknowledged where necessary.

6.6 Environmental investigations

Environmental investigations are becoming an increasingly important element in pre-contract investigation. They should be directed at:

- establishing the existing environmental conditions in and around the area of the works;
- identifying the potential impacts of the works on the environment both during construction and, in the longer term, after completion;
- providing baseline data which can be used to monitor the effects of the works.

The importance which is now widely attached to environmental protection, and indeed improvement, means that thorough environmental investigations and impact assessments are often vital to success of a project.

The investigations should address all aspects of the works, including dredging and relocation of dredged materials, transport and placement of reclamation fill, and related construction activities such as piers, seawalls and wharves.

Within this framework, environmental investigations usually address the following subjects:

- soil quality (i.e. level of contamination, if any);
- water quality;
- ecology;
- air quality;
- noise;
- marine traffic;
- employment;
- recreation and amenities;
- sediment transport.

The environmental aspects of dredging works and the types of environmental investigations which might have to be undertaken are discussed in Chapter 13.

6.7 General investigations

Apart from the meteorological, hydrographic, geotechnical and environmental characteristics of the site, there are several characteristics which are relevant to dredging operations and which should be researched during pre-contract investigations.

6.7.1 ACCESS TO SITE

It is surprising how often details of the actual location of the site are overlooked. The following points should be checked:

- Is access to the site restricted by a lock or locks, and if so, how often are the locks open and for how long?
- Is access to the site restricted by the depth or height of either natural or man-made features?

- Is access to the site restricted by width, such as a dock entrance?
- Is access to the relocation site restricted by depth or by times at which it is allowed?

6.7.2 OPERATIONAL CONSTRAINTS

Interference from traffic

If the work is to be carried out in an existing port or waterway there is likely to be a certain amount of water-borne traffic which will cause the dredging or pre-treatment operations to be interrupted. Past records of shipping should be examined and relevant information extracted and passed on to prospective contractors. If possible, it should be ascertained how long a warning period may be given of ship arrival and the effect of the interruption on the operations should be estimated.

Many dredging operations can be temporarily stopped for a few minutes with relative ease, for instance, by slacking the side wires which are crossing the main shipping canal. However, in the case of drilling and blasting operations the slacking of wires may cause sufficient movement of the pontoon to make operations hazardous and, in such circumstances, the round of explosive charges may have to be fired before the vessel passes. Much time may be lost due to this enforced alteration to normal working cycles and an appropriate compensation will have to be allowed in the contract to suit this event (see Chapter 14).

Working areas

Although the main working area will probably be the dredging area, it may be necessary to locate and construct suitable areas for reclamation. If the dredged material is unsuitable for forming the edges of the reclamation area, or if the area is not sufficiently confined, it will be necessary to find suitable local sources of material with which to construct the bunds. Possible quarry sites and borrow pits should be located and their suitability and capacity assessed.

In addition to the dredging and reclamation sites, the contractor and the client's representative will require offices to house their staff. The contractor will also need space to store his materials and spare parts. Adequate space should be located, conveniently close to the site. Much time can be lost travelling to the dredging site over water and the distance to be travelled should be reduced to the minimum since the cost of this travelling time is usually passed on to the client.

Safety

When operations are to be performed in exposed locations, it is essential that there are contingency arrangements made for possible emergencies at sea. Apart from the immediate plans for lifting off injured or sick personnel which are usually set up by the contractor, the engineer who plans the contract should note the location of the nearest lifeboat station and ensure that the nearest sheltered waters are both suitable and capable of harbouring the dredging plant during periods of bad weather.

The transfer of personnel to and from floating plant should be at a safe and proper facility.

Fisheries

Dredging may be prohibited during particular tidal conditions or seasons. For example, during the main migratory period in salmon rivers, or during spawning, when working in close proximity to established spawning grounds.

6.7.3 STATUTORY AND LEGAL CONSTRAINTS

Local and national laws

In most countries there are laws covering the dredging and dumping of material at sea. Also, many shorelines and dump sites are owned by the state and there are additional laws governing operations carried out at these locations. Labour laws often govern the amount of expatriate labour that can be brought into a country or at least fix the ratio of local to imported personnel employed. Further laws may also control when work can be carried out, the number of working days per week, and whether by night or day. Long delays may be incurred at the start of a contract if these laws are not thoroughly investigated and the relevant permission sought. This is particularly true when the use of explosives is planned and temporary explosive stores are required.

In certain countries, for instance Australia and the USA, trade protection laws bar, or severely hinder, the temporary importation of foreign built, registered or owned maritime craft.

Land ownership

Investigations should be made to ascertain the ownership of all lands affected by the dredging and reclamation works. Apart from the dredging site and the contractor's working area and sea access, it may be necessary to gain access to other land for example to erect navigation aids. Landowners should be approached and relevant permission obtained. As mentioned above, many shorelines are state owned and permission may be required for temporary works and so on.

Local harbour regulations

Contractors may be held responsible for dealing with the local harbour authority and paying their pilotage and port dues if necessary. However, it is often advisable to arrange for special moorings to be set aside for dredging craft.

Movement of navigational marks

Navigation buoys may have to be removed to allow dredging operations to be carried out. In this case, arrangements will have to be made in advance with the relevant authority and steps taken for the change in positions to be made known to vessels using the area, usually by means of a Notice to Mariners.

6.8 Ship simulation

During the planning and design of entrance channels and vessel manoeuvring areas it is often difficult for the planner to obtain an indication of how satisfactory or otherwise his concepts are from the point of view of ship handling. This is particularly true in areas where space may be limited, cross winds or currents are severe or channels are of an unusual configuration. To overcome these problems various forms of ship simulation techniques have been devised.

Essentially ship simulation consists of applying a physical or numerical representation of a particular vessel to a known set of port entry or exit conditions. The vessel is then 'driven' by an engineer or master mariner into or out of the port and the ship movement and behaviour is noted. By carrying out a series of runs using different sets of conditions and varying channel/manoeuvring area configurations, it is possible to develop cost-effective but practical solutions.

Simulations may involve actual physical models of vessels, driven along miniature waterways or, more commonly now, computer-based representations of the vessels in a numerical model of the channel, together with the relevant marine and meteorological conditions. Some simulations work in plan on a stop/start basis, others work in real time, whilst the most sophisticated provide a view of the port from the vessel's bridge. Most of the models are developed in-house by specialist maritime organisations, but some may be purchased for general use.

References

Bray, R.N. 1979. *Dredging: A Handbook for Engineers* (1st Edn). Edward Arnold, London.

Broch, E. & Franklin, J.A. 1972. The point load strength test, *International Journal of Rock Mechanics and Mineral Science*, Vol. 9, pp 669–697.

Calle, E.O.F. 1986. A geostatistical approach to the optimisation of pre-dredge surveys, *Proceedings of the 11th World Dredging Conference 1986*, Brighton, UK. Central Dredging Association, Delft, The Netherlands.

Cox, C.M., van Deursen, P.P. & Verhoeven, F.A. 1986. The relevant soil properties for dredging silt, *Proceedings of the 11th World Dredging Conference 1986*, Brighton, UK. Central Dredging Association, Delft, The Netherlands.

Hogben, N., Dacunta, N.M.C. & Olliver, G.F.,1967. *Global Wave Statistics*, British Maritime Technology Limited, Unwin Brothers Ltd.

Land, J.M. & Garrard, G. 1986. The practical application of multispectral satellite imagery to dredging and port development works, *Dredging and Port Construction*, February, 1986.

LeBlanc, L.R., Mayer, L., Rufino, M., Schock, S.G. & King, J., 1992. Marine sediment classification using the chirp sonar, *Journal of the Acoustical Society of America*, 91, pp 107–115.

Loman, J.A., Bezemer, B. & Noorbergen, H.H.S. 1992. Mapping bathymetry and suspended sediment using satellite imagery, *Terra et Aqua*, No. 48, International Association of Dredging Companies (IADC), The Netherlands.

PIANC 1984. Classification of soils to be dredged, *Report of a Working Group of the Permanent Technical Committee II*, Supplement to Bulletin No. 47, Permanent International Association of Navigation Companies, Brussels.

PIANC in press. Site investigation requirements for dredging works, *Report of Working Group No. 23 of the Permanent Technical Committee II,* Permanent International Association of Navigation Companies, Brussels.

Protodyakonov, M.M. Jnr. 1962. Mechanical properties and drillability of rock, *Proceedings of the 5th Symposium on Rock Mechanics*, University of Minnesota, Pergamon Press.

Rhys, L.J. & Wood, I.R. 1984. A question of priority – product before procurement, *Proceedings of the 20th Regional Meeting of the Engineering Group of the Geological Society*, Geological Society, London.

Schock, S.G., LeBlanc, L.R. & May, L.A., 1989. Chirp sub-bottom profiler for quantitative sediment analysis, *Geophysics*, 54, pp 445–450.

7

Dredging Plant and Equipment

7.1 Introduction

There is a wide diversity of dredging plant and methods. Some have evolved to meet a particular requirement, whilst others are more versatile. Some types of dredger may be quite unable to perform particular dredging tasks, or at best may only be able to perform that task at much reduced efficiency.

By the careful design and specification of dredging work it may be possible to reduce the variety of plant required, or to allow plant to operate at greater efficiency. It is important, therefore, that the methods of dredging are properly understood and considered during the design and planning stages of the proposed works.

In special circumstances, such as for narrow waterways or material alongside a berth, land-based plant may be used for dredging, but these are not described here.

Before reviewing the different types of dredging machine it is helpful to examine the physical mechanisms involved in the dredging process.

7.1.1 DREDGING METHODS

The object of dredging is usually to remove material from a particular location and to place it at another. Alternatively, the object may simply be to remove material from a particular place and to allow it to be moved by natural forces to a new location.

The latter condition may be achieved by 'agitation dredging'. This is the simplest of all dredging processes and is most applicable to maintenance dredging in fine-grained sediments. Agitation dredging involves disturbing the seabed materials by forcing them into suspension, after which they may be moved by natural water flow to be redeposited elsewhere.

The suspension of materials may be achieved with water jets, or by raking or pumping. The method works well when the direction and distance travelled by the suspended sediments is sufficient to remove them from the area to be dredged. Thus, where the objective of dredging is to maintain depth in a channel, fine sediments which have settled within the channel may be resuspended and deposited outside the channel area, over the adjoining seabed.

Before such methods are adopted it is important to determine the likely pattern of deposition and to be satisfied that it will be acceptable.

Water injection is a variation of the agitation method. It involves the pressure injection of water into the near surface seabed deposits such that the *in situ* density of the bed material is reduced to the point where the material is induced to flow.

It is more common to place dredged materials within a pre-determined area. In such cases the seabed material must be dislodged, raised and physically confined until arrival at the intended place of deposition. Each element of this process may be achieved in a variety of ways.

The process of dislodgement is often the most uncertain in that it is very dependent on the density, or strength of the seabed materials. This may be difficult to predict with confidence, or may be very variable. If the seabed material is strong, as with rock, it may be necessary to use some method of pre-treatment to break up the bed material before dredging (see Chapter 14). For most ground conditions pre-treatment is unnecessary, but nevertheless, the strength of the seabed material will have an important influence on the choice of method of dredging.

If the seabed materials are loose, or of low strength, the use of mechanical methods of dislodgement may be unnecessary. In such cases, suction methods are usually successful. The eroding force of water drawn into a pump suction intake will entrain sediments in the flow and raise them in the pumped mixture. Such methods may be successful in loose silts, sands and gravels and in very soft clays.

When the density, or strength of the seabed material is too high for economic production to be achieved simply by pumping, mechanical dislodgement is necessary. This may be achieved in conjunction with suction methods, in which case the material is dislodged by cutterhead and raised within the pumped mixture. The cutterhead must be located close to the pump suction intake. The various types of cutterhead are described in detail later, with the particular type of suction dredger to which they relate.

The alternative method of soil dislodgement and raising is by bucket. This may be by single bucket, as in the case of the backhoe, grab (clamshell) and dipper, or may be multi-bucket as in the case of bucket ladder (or bucket chain) dredgers.

The raised materials may be transported hydraulically in pipelines, or by hopper. With minor exceptions, primary hydraulic transport is used only in association with hydraulic dredging methods. However, hydraulic methods may also be used to discharge hoppers which have been loaded with buckets.

7.1.2 TYPES OF DREDGER

In addition to general civil engineering uses, dredging plant may be employed in a variety of special applications, including the following:

- small inland waterways and lakes;
- mining of marine aggregates;
- mining of mineral deposits;
- wreck exploitation, or recovery;
- cleaning around offshore platforms, or within cofferdams;
- shellfish recovery;
- removal of flotsam.

These specialised activities are not examined in detail here, but the plant may have many features in common with dredging plant which is employed in works of civil engineering construction and maintenance, and which is described in the following sections.

7.1.3 STATISTICAL DISTRIBUTION OF DREDGING PLANT

Since development of the various dredging systems has taken place in different parts of the world, predominantly in Western Europe, North America and Japan, and because some dredgers are particularly suited to the site conditions pertaining to specific areas, the world distribution of dredgers is uneven. Table 7.1 shows the worldwide distribution of dredgers as extracted from directories published by the two magazines *World Dredging, Mining and Construction* and *Dredging and Port Construction* The figures for the years 1977 and 1992 have been used, which show how the scene has changed over a 15 year period.

It should be noted that there are difficulties in classifying some dredgers in this form and some areas of the world may not be included comprehensively (i.e. some Eastern European countries). However, it does give an indication of where certain types of dredger originated or are used extensively.

7.2 Trailing suction hopper dredger

7.2.1 GENERAL DESCRIPTION

The trailing suction dredger, frequently abbreviated to 'trailer', is a ship, suited to coastal or deep sea navigation, which has the ability to load a hopper contained within its structure by means of a centrifugal pump or pumps whilst the vessel is moving ahead. Most trailing suction dredgers have twin screw propulsion and a powerful bow thruster, which provide a high degree of manoeuvrability.

Loading takes place with the ship moving slowly ahead. Unloading is normally by means of a bottom-discharge arrangement, or by pump discharge, in the latter case, usually to the shore.

The main advantages of the trailing suction dredger are:

- relative immunity to weather and sea conditions;
- independent operation;
- minimal effect on other shipping;
- the ability to transport dredged material over long distances;
- relatively high rate of production;
- simple, and hence inexpensive, mobilisation procedure.

The main disadvantages are:

- inability to dredge strong materials;
- inability to work in very restricted areas;
- sensitivity to concentrations of debris;
- dilution of dredged materials during the loading process.

The trailer dredger is normally rated according to its maximum hopper capacity, which is typically in the range 750 to 10 000 cubic metres, but exceptionally may be larger. The statistical distribution by size of hopper is shown in Figure 7.1.

7.2.2 METHOD OF OPERATION

The important features of the trailing suction dredger are illustrated in Figure 7.2.

The trailing suction dredger travels between sites under its own power. If the method of dredged material disposal is by deposition at sea, the dredger is usually self-contained and

Table 7.1 Statistical distribution of dredging plant

	Trailing suction hopper	Cutter suction	Suction	Dustpan	Hopper suction	Bucket	Backhoe	Dipper	Grab/ clamshell	Bucket wheel suction
North America	35	406	45	3	5	6	17	13	226	6
	(22)	(394)	(91)	(8)	(3)	(8)	(8)	(30)	(219)	(–)
Central America & Caribbean	–	10	–	–	–	–	–	2	3	–
	(13)	(43)	(–)	(–)	(–)	(2)	(–)	(4)	(4)	(–)
South America	25	118	6	–	7	30	1	2	1	4
	(7)	(55)	(9)	(2)	(1)	(28)	(–)	(–)	(–)	(–)
Western Europe	156	162	56	1	18	49	35	4	114	–
	(94)	(230)	(127)	(1)	(26)	(104)	(4)	(4)	(71)	(–)
Eastern Europe	2	–	1	–	9	3	–	–	15	–
	(10)	(18)	(10)	(–)	(2)	(17)	(–)	(–)	(6)	(–)
Scandinavia	13	16	11	–	15	4	48	2	15	–
	(4)	(28)	(11)	(–)	(7)	(15)	(9)	(7)	(27)	(–)
Middle East	11	15	–	–	2	–	–	1	–	3
	(4)	(12)	(1)	(–)	(5)	(1)	(–)	(–)	(3)	(–)
Indian Sub-continent	18	48	–	–	9	4	2	–	27	1
	(1)	(30	(–)	(–)	(3)	(3)	(–)	(–)	(1)	(–)
Northern Africa	3	26	1	–	2	4	6	–	7	4
	(2)	(11)	(–)	(–)	(–)	(4)	(–)	(1)	(–)	(–)
Southern Africa	6	2	6	–	–	1	2	–	3	–
	(8)	(1)	(–)	(–)	(1)	(2)	(–)	(–)	(5)	(–)
Southeast Asia	36	170	15	–	1	90	2	1	60	–
	(20)	(244)	(18)	(–)	(16)	(54)	(–)	(13)	(94)	(–)
Australasia	7	27	1	–	–	1	2	–	11	7
	(9)	(41)	(16)	(–)	(2)	(13)	(–)	(–)	(15)	(–)
Russia (USSR)	16	1	–	3	7	4	–	–	15	–
	(–)	(–)	(–)	(–)	(8)	(–)	(–)	(–)	(–)	(–)
TOTALS 1992	318	1001	142	7	72	196	115	25	497	25
TOTALS 1977	(194)	(1080)	(283)	(11)	(74)	(251)	(21)	(59)	(445)	(0)

Note: All figures in brackets are for 1977

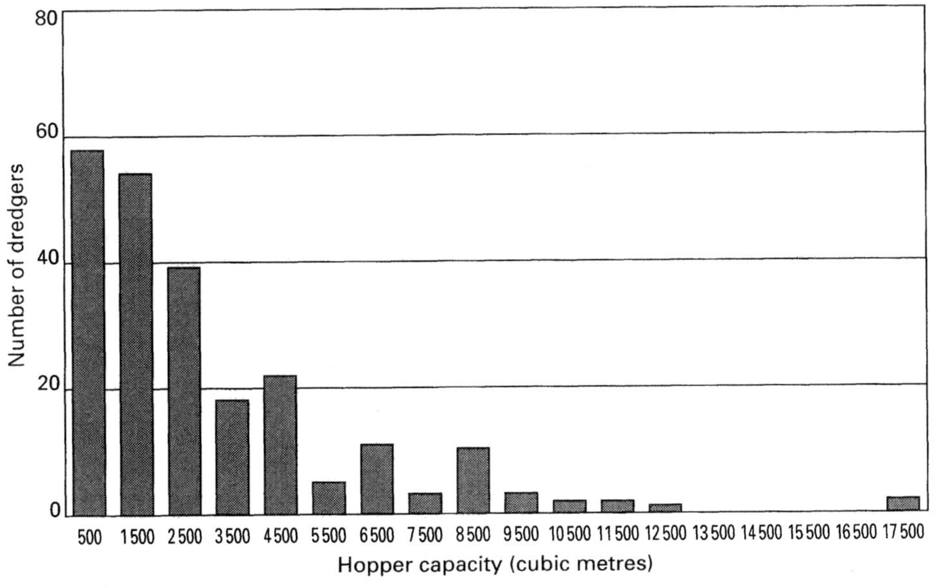

Figure 7.1 Statistical distribution of trailing suction hopper dredgers

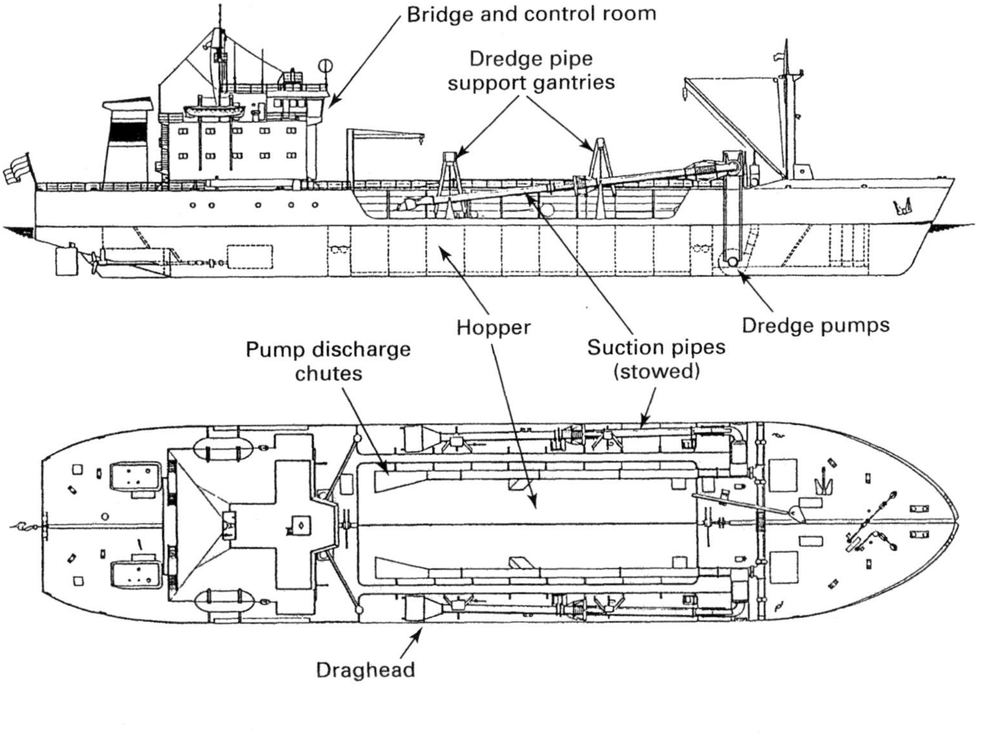

Figure 7.2 Main features of trailing suction hopper dredgers

ready to begin work immediately upon arrival at the work site. In practice, formalities such as customs and immigration controls may be necessary. It may also be necessary to change crew, take on bunkers and perhaps install and calibrate position fixing and tide level measurement instrumentation.

A large proportion of the internal space of the ship is given over to the provision of hopper space into which the dredged material is loaded by one, or two (very occasionally up to four) large centrifugal pumps. The pumps are usually inboard, but may be fitted in the trailing suction pipe (see Figure 7.3 and Plates 7.1 and 7.2). The suction pipe is stowed inboard when the ship is in transit between the dredging site and the unloading site.

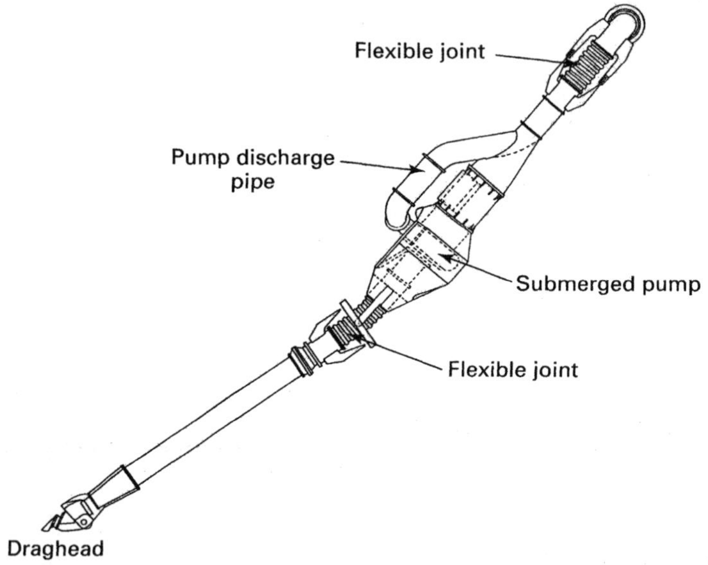

Figure 7.3 Suction pipe of trailing suction hopper dredger, fitted with submerged pump

The maximum depth to which dredging is possible is limited by the vacuum head generated by the dredge pump. If the dredge pump is mounted within the hull the maximum economical dredging depth of a medium size trailer dredger is usually about 35 metres, although, for some larger dredgers, dredging depths of up to 80 metres may be possible with reduced production rates. By mounting the dredge pump externally in the trailing suction pipe greater depth or improved production is possible.

Within the hopper, some, or all of the solids in the pumped mixture will settle out of suspension and the supernatant water is discharged via an overflow arrangement (see Figure 7.4). Adjustment of the overflow level will allow some degree of control over the volume and character of dredged mixture retained in the hopper.

Very fine materials may not readily settle from suspension. When dredging such materials, usually there will not be any significant increase in load achieved by continued pumping after the point of hopper overflow.

Most trailer dredgers are designed to carry a full load of fine grained material. They are not generally able to carry a full load of sand or gravel because of the greater density of the material in the hopper and may only be able to load to 80 per cent of hopper capacity in these materials. The overflow weir level must then be maintained at an appropriate lower-than-maximum level.

Plate 7.1 A trailing suction hopper dredger with a pump fitted in the suction pipe

Plate 7.2 A closer view of the pump fitted in the suction pipe of the trailer dredger shown in Plate 7.1

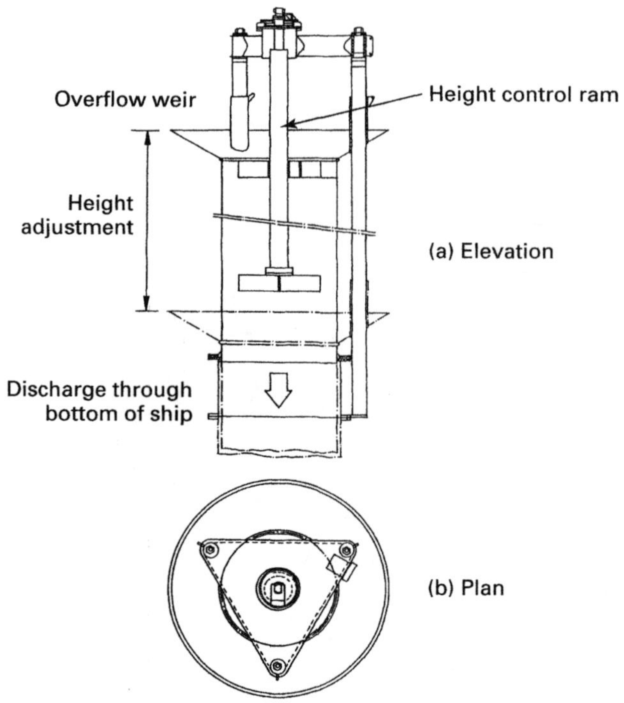

Figure 7.4 Detail of typical hopper overflow for trailing suction hopper dredgers

On approach to the area of dredging, the suction pipe is swung outboard by means of winches and davits. If inboard dredging pumps are installed, the inboard end of the suction pipe is lowered in a fixed track to mate, below the water line, with the pump suction intake, which is open in the side of the hull.

Pipework from the discharge (pressure) side of the pumps is routed to the hopper where discharge is via 'landers' (chutes) to minimise turbulence. If the dredging pumps are located within the trailing suction pipe, then there may be a fixed connection of suction and pressure pipe systems.

Pumping continues until the hopper is full. In fine-grained materials the maximum economic load normally coincides with the commencement of hopper overflow. In coarse-grained, or mixed materials, pumping may continue for some time after hopper overflow, with continued improvement in hopper load.

The intake end of the suction pipe is fitted with a 'draghead', the function of which is to maximise the concentration of solids entrained from the seabed. The draghead is lowered to a level close to, or below the seabed and dredging commences, normally at a speed over the ground ranging from one to five knots. The tow of the draghead can sometimes be used to advantage to improve the manoeuvrability of the ship when turning at low speed in confined areas.

The bearing pressure of the draghead on the seabed is usually controllable by an adjustable pressure compensating system which acts between the draghead and the hoisting winch. The system also serves to alleviate the effects of vertical movement of the ship relative to the seabed due to waves, or swell, and is thus normally known as the 'swell compensator'.

In loose materials the entrainment of solids from the seabed into the suction flow is achieved mainly by the erosive action of the inflowing mixture. Pressure differential also aids the loosening of granular materials. In less dredgeable materials, such as dense sands, or firm and stiff clays, dislodgement of the seabed material may be assisted by high-pressure water jets, scrapers or knives. Dragheads and their attachments are described in Section 7.2.4.

The suction pipes are usually articulated part way along their length by means of a gimble device which supports a rubber hose connection between the adjoining lengths of rigid suction pipe. This articulation permits both vertical and horizontal movement of the draghead relative to the ship.

The hopper design and dimensions have a strong influence on the settlement of soil grains within the hopper. Turbulence within the hopper should be minimised. Modern dredgers have an 'automatic light mixture overboard' (ALMOB) arrangement. This diverts incoming pumped mixtures overboard if the mixture density is less than a pre-set figure. Thus the ratio of solids to water entering the hopper is maximised. This device is principally used in fine-grained soils.

De-gassing equipment may be fitted within the dredge pipework to remove gas from the incoming pumped mixture before it reaches the dredge pump. Without such equipment, when dredging in biogenic gaseous deposits the pump performance may be adversely affected.

When the hopper is loaded, the suction pipes are returned inboard and stowed as the dredger sails to the unloading site. On arrival, the solids are discharged, usually by bottom-dumping (see Section 7.2.4). Alternatively, if the solids are to be discharged onshore, or used in land reclamation, the dredger is moored, connection is made to a pumping main and pump discharge commences. Bottom-dumping will normally require only a few minutes, whereas pump-discharge will usually require about an hour. Occasionally pump-discharge may be via a bow mounted nozzle, whereby the soil/water slurry may be jetted, or sprayed, over a chosen site.

The production cycle

The importance of the component parts of the production cycle will vary according to the site conditions and requirements. Each component of the production cycle must be individually assessed for each new item of work. The cycle components are summarised in Figure 7.5 and are discussed in the following notes.

Loading time

The loading time is heavily dependent on the characteristics of the material to be dredged. It is difficult to calculate with certainty and experienced judgement is very important in assessment. When loading very fine sands, silts and soft clays, it is unlikely that there will be any significant increase in the hopper load achieved by continued pumping beyond the time that hopper overflow commences. In this case, time to load will be a function of total pumped flow rate and the hopper capacity. Continuous pumping for between 10 and 20 minutes will usually fill the hopper. Production will depend on the solids concentration in the pumped mixture. Assessment of optimum loading time is described in Chapter 8.

Turning time

The dimensions of the dredging area and the soil conditions (which determine the optimum trailing speed) may require that more than one turn be made by the dredger during loading.

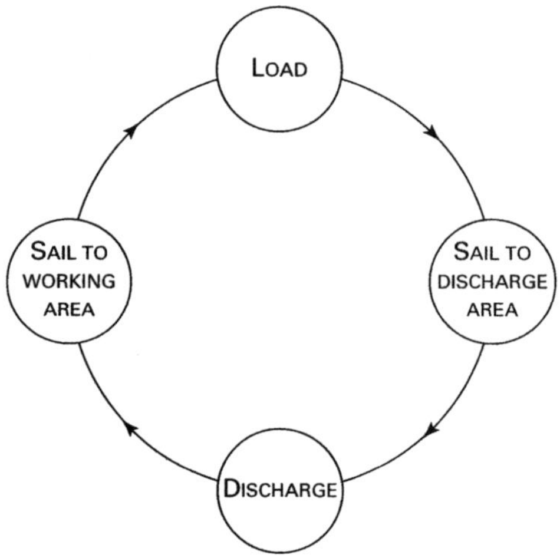

Figure 7.5 Production cycle for trailing suction hopper dredgers

The severity of the turn and the number of turns will increase the non-productive time during loading.

Sailing time loaded

Given by dividing length of navigation route from work site to disposal area by average speed when loaded, under prevailing conditions of weather, waves, current, traffic and navigational restrictions.

Time to discharge (bottom-discharge)

For loose materials and bottom-discharge this may be only a few minutes. For cohesive material, depends on design of hopper, method of discharge, character of material and so on. Difficult to calculate with certainty, experienced judgement very important in assessment.

Time to discharge (pump ashore)

The total time must include time to moor, make connection to the discharge pipeline, disconnect when empty and cast off (normally less than 15 minutes in total, but can be as much as 30). Actual time to pump-discharge hopper depends upon length and terminal elevation of pipeline, particle size distribution of dredged material and the particular characteristics of the dredger. Can be calculated within reasonable limits given adequate data, often of the order 45 to 60 minutes.

Sailing time empty

Given by dividing length of navigation route to work site by average speed of dredger when light, under prevailing conditions of weather, waves, current, traffic and navigational restrictions.

Daily production will be the sum of the hopper contents for each complete cycle achieved. The number of cycles will be given by the total working time divided by the average cycle time. The total working time will be the total hours that the dredger is crewed and available to work, less any lost time (see Chapter 9).

7.2.3 LIMITING FACTORS

The limitations affecting when and where trailers can operate will vary according to the size and characteristics of the particular dredger, but an indication of the extreme limits which will apply to economic operation is given below. These limits apply respectively to the smallest (minimum) and largest (maximum) of this type of dredger in common use. Guidance to the dimensions and other important characteristics of trailer dredgers is given by Figures 7.6 and 7.7.

Minimum water depth to operate	4 metres
Maximum water depth to operate	45 metres (very few to 80/100 metres)
Maximum sailing speed	17 knots
Minimum turning circle	75 metres
Maximum wave height	5 metres
Maximum cross current	3 knots
Maximum ice thickness	200 millimetres
Maximum particle size	300 millimetres
Maximum shear strength (clays)	75 kPa

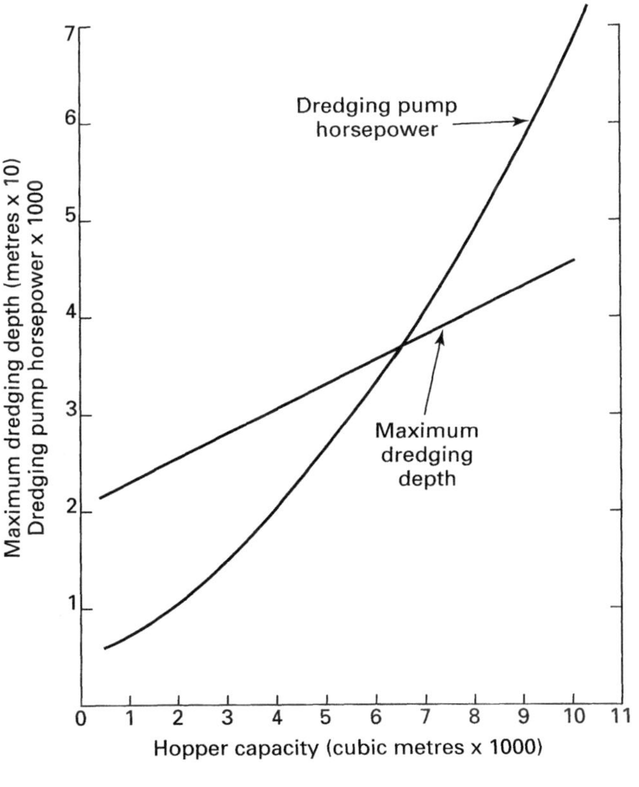

Figure 7.6 Maximum dredging depth and pump power for trailing suction hopper dredgers

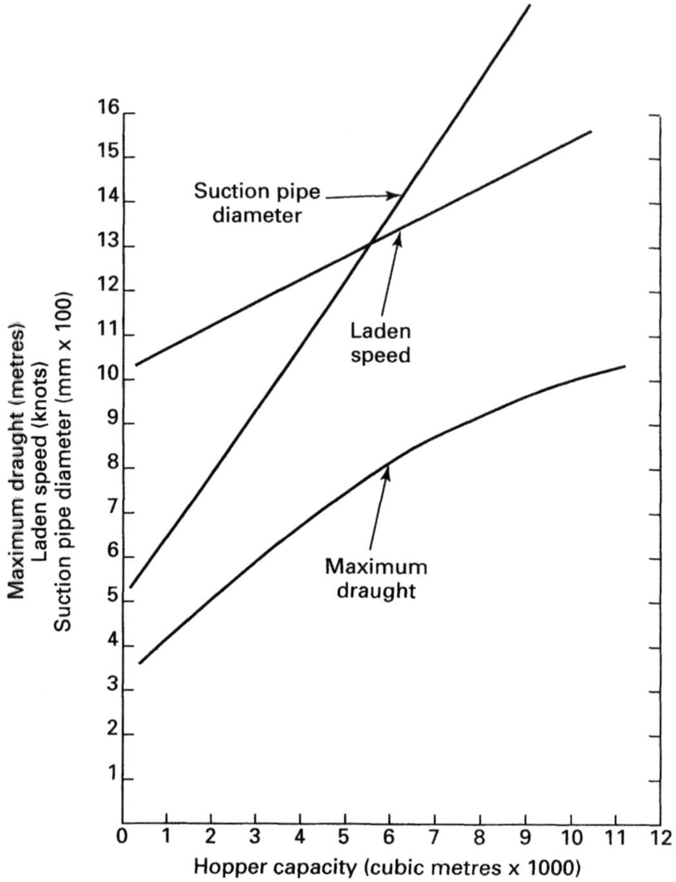

Figure 7.7 Laden speed, maximum draft and suction pipe diameter for trailing suction hopper dredgers

7.2.4 ANCILLARY EQUIPMENT

In independent operation the trailer dredger does not require a wide range of ancillary equipment. The main variations which may occur concern the draghead and the method of discharge. The main options are as follows.

Dragheads

It is usual for part of the draghead assembly to be capable of a limited range of rotation in the vertical plane relative to the lower suction pipe to which it is attached. This allows the angle of the draghead to the seabed to be optimised.

For optimum performance the type of draghead to be employed should be selected according to the type of material to be dredged, and the dredger being used. The more common types and their applications are listed in Table 7.2. The list is not exhaustive. Many contractors and other users have developed dragheads to suit particular requirements.

Table 7.2 Various types of draghead and their common application

Type	General application	See Figure
Fruhling	silts, soft clays and loose sands	7.8a and Plate 7.3
Silt	silts	7.8b
Californian	sands, especially compact sands	7.8c and Plate 7.4
Venturi	sands	7.8d
Waterjet	firm sands and medium clays	7.8e
Active	medium, firm and stiff clays	7.8f

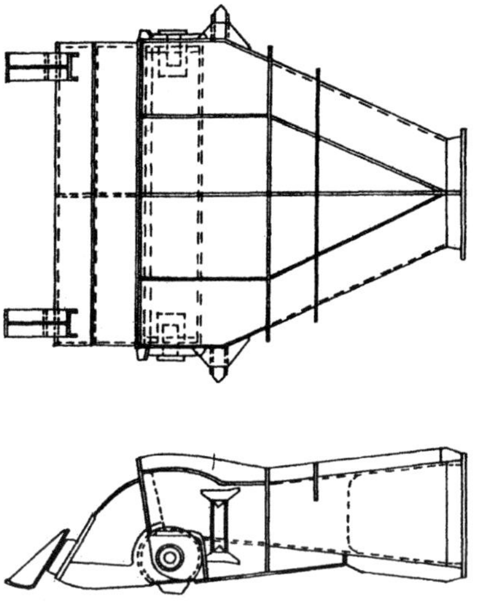

Figure 7.8a Example of Fruhling type draghead

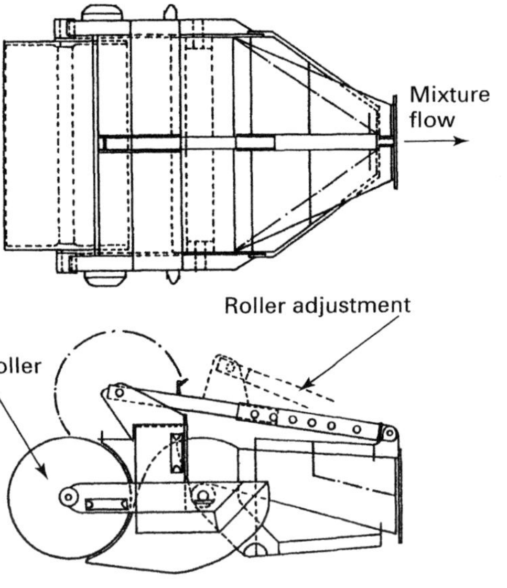

Figure 7.8b Example of silt draghead

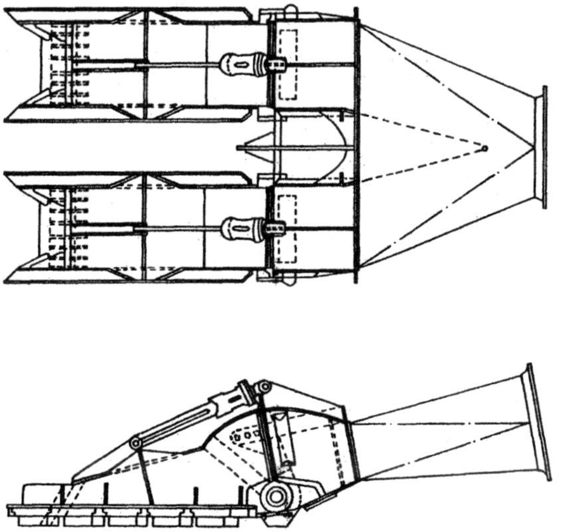

Figure 7.8c Example of Californian draghead

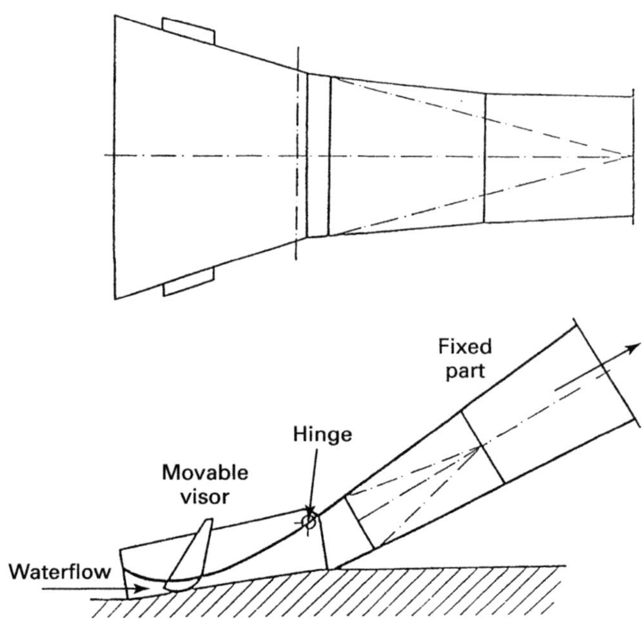

Figure 7.8d Example of venturi draghead

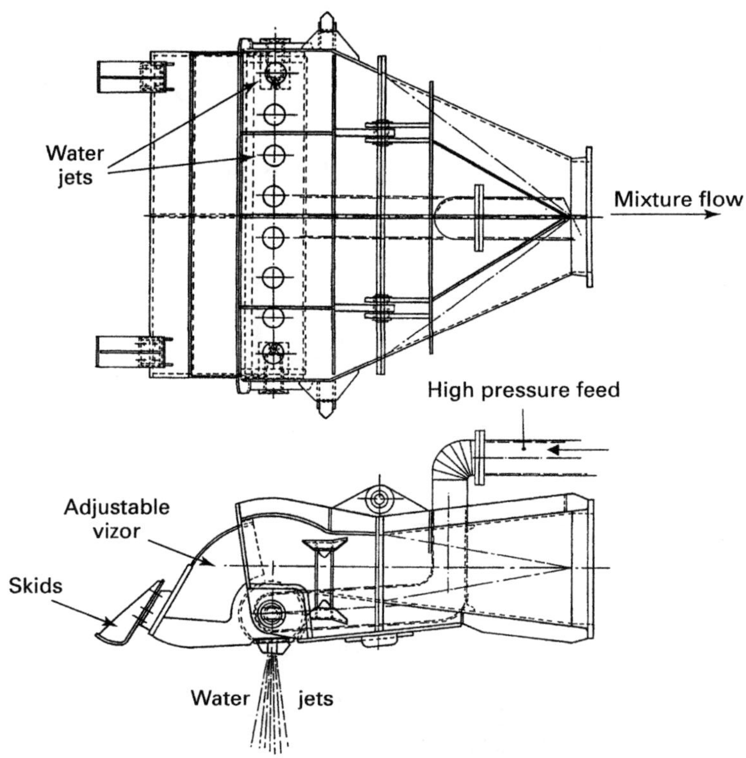

Figure 7.8e Example of draghead fitted with water jets

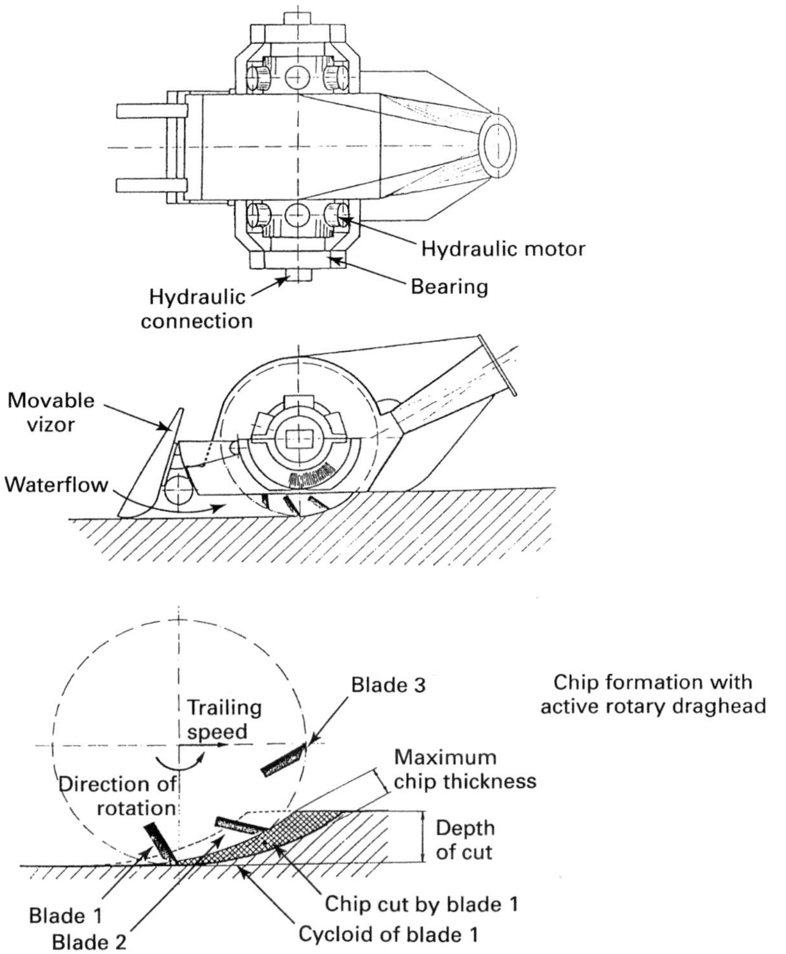

Figure 7.8f Example of active draghead

Teeth, or water jets may be fitted to the underside of the draghead to break up obstinate bed materials. The use of teeth may result in a significant increase in the power required to maintain speed over the ground. The use of water jets may be undesirable in areas which are environmentally sensitive. Active dragheads with a powered rotating mechanical cutter are developed for dredging firm and stiff clays, but these have not been widely adopted.

It is usual for the draghead to be fitted on the top side with adjustable openings to allow some flow of water into the head if the underside becomes buried. This is designed to avoid pump cavitation. These water intakes may be elevated to maintain a flow of water even if the draghead becomes completely buried in soft ground.

Hopper discharge mechanisms

The various methods for discharging hoppers are shown in Table 7.3.

Table 7.3 Various systems of hopper discharge for trailing suction hopper dredgers

Type	Application
Bottom door:	
hinged	Clean silts, sands and soft clays, calm water
sliding	Clean silts, sands and soft clays in shallow water, or rough seas
Bottom valve	Clean silts, sands and soft clays in rough seas
Split hull	Any material, including those containing boulders, or debris, for disposal in shallow water and moderate seas (disposal licence may exclude boulders and debris)
Pump	Silts, sands where disposal is to onshore areas, or material is to be used for land reclamation
Scraper	Shore discharge of dredged aggregates
Grab	Shore discharge of dredged aggregates

Plate 7.3 A Fruhling draghead

Plate 7.4 A Californian draghead

Bottom-discharge

Trailer dredgers usually discharge dredged material from the hopper via a number of doors in the bottom of the hopper. Doors may be hinged on one edge and restrained by chains, or links, attached to hydraulic actuating cylinders. In such cases sealing is achieved by the compression of rubber or synthetic seals. However, hinged doors are susceptible to damage when dumping in shallow water, or rough seas, and may require increased underkeel clearance. These problems are overcome with sliding doors. Sliding doors are also activated by hydraulic cylinders.

As an alternative to bottom doors, bottom valves may be fitted. Bottom valves can be operated in shallower water than hinged doors, but with greater caution than is necessary with sliding doors. Bottom valves are not affected by sea conditions.

All bottom door and valve arrangements present some restriction to the outward flow of the hopper contents. If the material to be discharged is very cohesive, or contains debris, then the restrictions presented by the door or valve surrounding structure may seriously prolong the time to discharge.

The split hull trailing suction dredger has been developed to overcome the problems described above (see Figure 7.2 and Plate 7.5). With this arrangement the entire hull of the dredger is divided longitudinally. Heavy hinges at deck level join the two halves of the hull and large hydraulic cylinders assist the opening and closing of the two halves during discharge. The system allows the rapid discharge of even the most difficult materials and has gained wide acceptance for small and medium sized trailer dredgers, but construction costs are higher and the system is impractical for very large dredgers.

Plate 7.5 A split hull trailing suction hopper dredger

Pump-discharge

When the objective of the work is to reclaim land, it may be necessary to pump-discharge the dredged material directly from the hopper into the reclamation area. Some trailers are specially constructed for this. The main dredge pumps are used in the pump-discharge operation. A system of pipework and valves allow the dredge pumps to draw dredged material directly from the hopper. The hopper contents are diluted by means of an auxiliary pump which draws in water via a submerged valve in the hull. The main dredge pump suction draws from a pipe, or culvert arrangement in the bottom of the hopper well. The flow of material into this cavity is controlled by doors or valves set in a false bottom to the hopper.

On the pressure side of the pumps, discharge pipework, generally over the bow of the dredger, permits connection to an onshore, or floating, pipeline. Because the dredge pumps of a normal trailer are usually low head pumps, they are not usually able to pump-discharge through long pipelines unless aided by intermediate booster pumps.

A variant of this method is to discharge the soil/water mixture through a bow-mounted nozzle, which can jet or spray the material up to 100 metres from the dredger (see Plate 7.6). This is used, for example, when reclamation is taking place immediately behind a seawall, or for beach recharge.

Plate 7.6 A trailing suction hopper dredger discharging over the bows of the vessel onto the beach

All discharge systems may be assisted by a hopper flushing system, designed to speed the flow of dredged material from all areas of the hopper.

Scraper and bucket wheel discharge

Large scraper buckets moved by winches, or travelling bucket wheel excavators, feeding to belt conveyors, are commonly used for the discharge of aggregates to quays. Such systems are confined mainly to marine aggregate mining operations which use purpose-built vessels.

Sidecasting or boom-discharge

The sidecasting dredger is essentially a trailer without a hopper. Instead of the dredged mixture passing into the ship's hopper, it is discharged via an elevated pipe, supported by a boom structure, to one side of the ship (see Plate 7.7). Discharge may be up to 90 metres from the hull centreline. Instead of the conventional one or two trailing suction pipes, up to four pipes may be fitted. One to port, one to starboard and two through a well amidships, or over the stern.

Plate 7.7 The sidecasting dredger Rio Orinoco dredging in the Orinoco River

Dredgers of this type are rare. They are used in the maintenance of long navigation channels which are subject to rapid siltation and where conventional disposal off-site would be uneconomic. The method relies upon material being removed from the channel at a substantially faster rate than it is returned by natural processes.

Because there is no need to sail to a disposal site, production is continuous, the only limits on production being the capacity of the dredge pumps and the solids concentration which can be achieved in the pumped mixture.

The main advantage of sidecasting is the very high rate of production which can be achieved. The main disadvantage is that the dredged material is only removed a short distance from the dredging area and hence may return quickly. There may also be unfavourable environmental effects arising from the high level of induced water turbidity.

7.3 Bed leveller

Bed levellers are often referred to as ploughs. This name is not used here because a plough may be a quite different item of equipment which is used specifically for trenching and placing pipelines.

There are two applications for bed levellers. They may be used independently to move material over short distances or they may be used to improve the efficiency of operations of dredgers.

The main features of the bed leveller are shown in Figure 7.9. Motive power is provided by a tug. The power necessary will depend upon the thickness of material to be removed and the shear strength of the material. As a rough guide, however, for recently deposited silt (the most common type of application), approximately 100 horsepower per one metre width of leveller blade will be required.

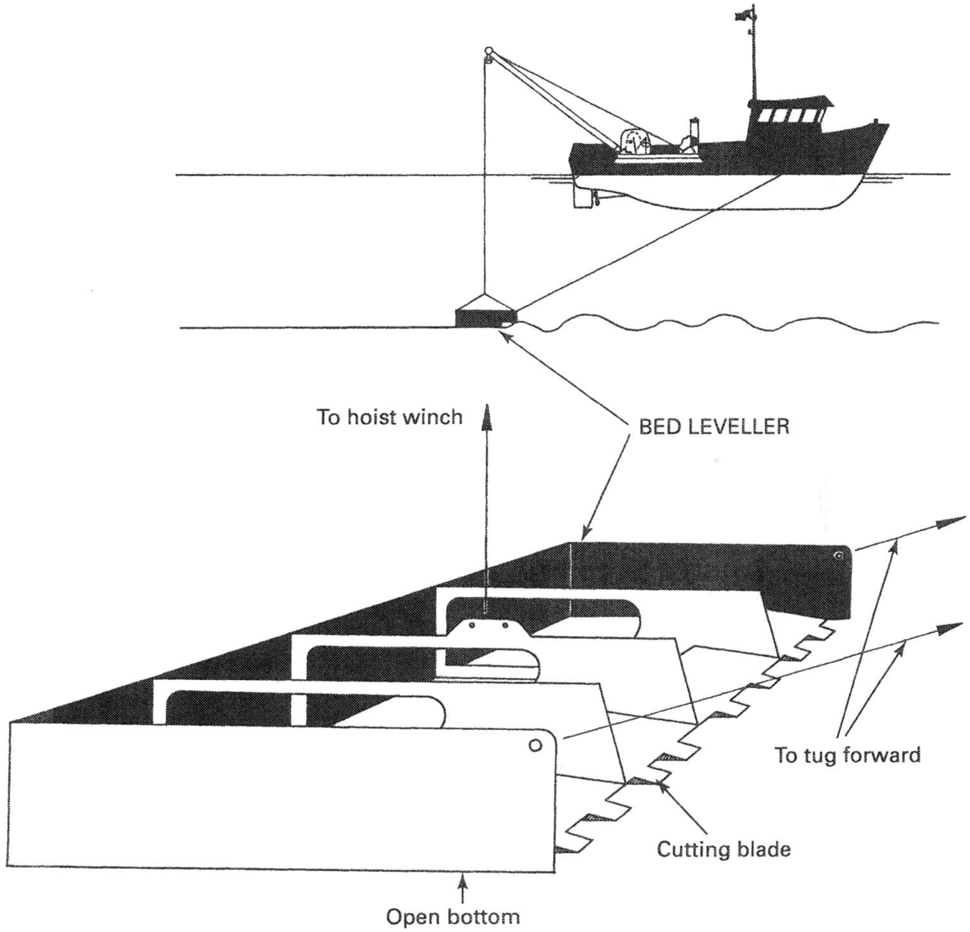

Figure 7.9 Main features of bed leveller

Used independently, the bed leveller moves material from shoal areas over relatively short distances to deeper water.

Some of the limitations from which trailers suffer can be alleviated if the dredger is used in conjunction with a bed leveller. The bed leveller can move material from areas which are inaccessible to the trailer into deeper or more open water where access is possible. It may also be used to level high spots which may remain after bulk dredging by any method. The uses described may significantly improve the efficiency and economics of a dredging operation.

7.3.1 METHOD OF WORKING

When used to move material which is inaccessible to a trailer, the tug carrying the bed leveller is reversed into the area to be cleared and the bed leveller blade is lowered, under winch power, until just biting into the sea bed. The tug is then moved forward into the area where the material is to be deposited. When deeper water is reached the captive material will fall from the scraper box and another cycle begins. In this way, material close to quays, jetties and in entrances or restricted corners can be moved inexpensively without recourse to conventional dredging equipment.

If the required task is to remove high spots, the bed leveller works in a similar way, but merely moves material above the required level into adjacent areas which are below level.

If the bed material is compact, or the object is to force material into suspension, water jets may be fitted to the bed leveller blade to loosen and agitate the material.

7.3.2 PRODUCTION CYCLE

The main elements of the cycle are shown in Figure 7.10.

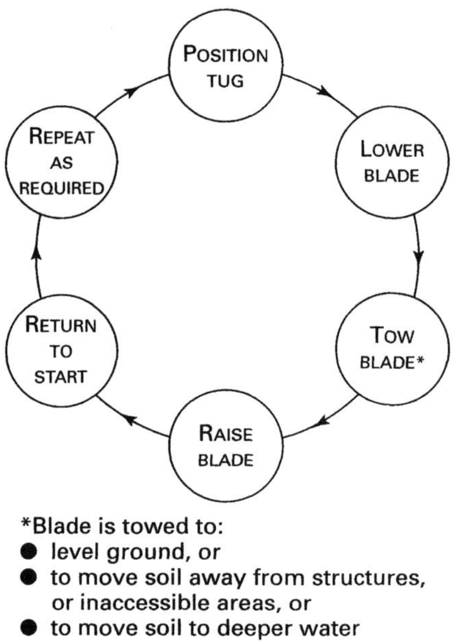

Figure 7.10 Production cycle of bed leveller

- Scrape off material:position tug, lower blade to sea bed, move ahead into area of deposition. Speed ahead, one to two knots reverse to start-point at one to three knots, or turn and sail ahead and turn at three to five knots.
- Repeat until required depth is achieved.
- Move onto adjacent cut, repeat as above.
- Continue until area is satisfactory.

7.3.3 LIMITING FACTORS

The maximum water depth for operation is dependent on the size of tug employed and the method of connection between the blade and the tug. For a rigid connection, the maximum depth is about 15 metres, but for a flexible (wire) connection, which is most common, greater depth is possible.

The limits on economic operation will depend on the size of tug from which the bed leveller is deployed. An indication of the limits for the minimum and maximum sizes in common use is given as follows:

Minimum water depth to operate	3 metres
Maximum depth to operate	30 metres
Maximum wave height	1.0 metre
Maximum swell	1.0 metre
Maximum cross-current	1.5 knots
Soil type	very soft and soft clays and silts, loose and medium dense sands

7.3.4 ANCILLARY EQUIPMENT

Water jets, fed from a high pressure pump on board the tug, may be used to loosen deposits. Teeth may also be used for the same purpose.

Compressed air nozzles, fed from a compressor on board the tug, may be used to increase the suspension of fine sediments where the object is to achieve dredging by agitation.

7.4 Water injection dredger

The principle of water injection systems is that, by the injection of water into certain types of sea bed materials, the *in situ* soil density is reduced to the point where it behaves as a liquid and is mobilised. The method was developed and exploited commercially during the 1980s in the Netherlands.

The method is most successful in low strength fine-grained materials when the opportunity exists for the fluidised material to flow to a lower level. Unlike agitation dredging, the object is not to raise individual sediment grains into the water column, although this can be achieved, intentionally or otherwise, using the same equipment.

7.4.1 METHOD OF WORKING

The water injection unit is self-propelled. When working, a fixed array of water jet nozzles are lowered to penetrate the seabed, water injection at a pre-determined pressure and flow rate is initiated and the vessel moves slowly ahead driving the fluidised material before it. A water injection dredger is illustrated in Figure 7.11.

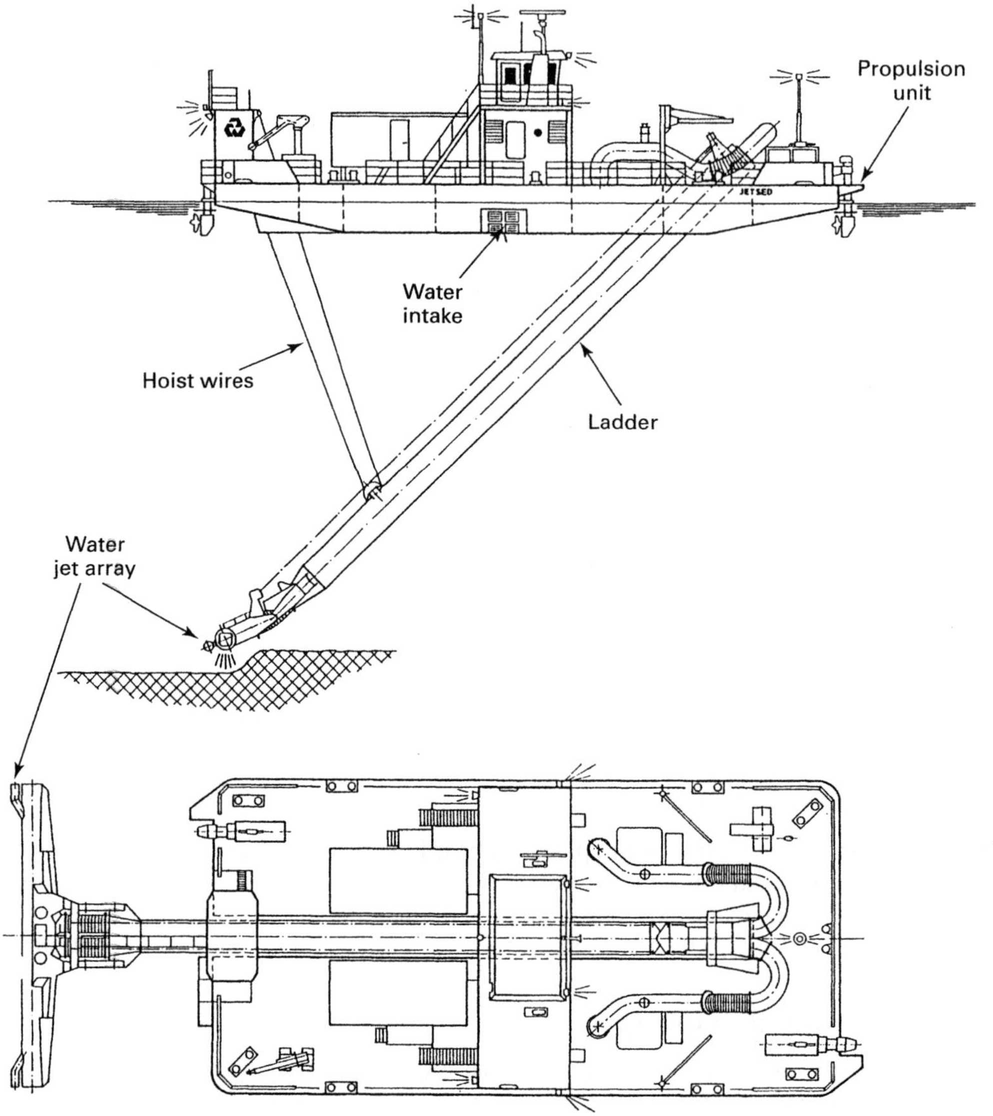

Figure 7.11 A water injection dredging system

If the seabed slopes away from the working area then mass flow of the fluidised material may occur over considerable distances and at very high rates. The method may still find application on level or undulating ground.

There is some evidence that in certain materials, for reasons not fully understood, but perhaps due to a change in particle electrical charge and flocculation processes, consolidation of material which has been subjected to water injection is hindered. This may be beneficial in the sense that it permits natural processes of sediment redistribution to work for an extended period.

Production cycle

The production cycle is similar to that of the bed leveller.

Plate 1 A group of dredgers forming a trench for a gas pipe running through the Wadden Sea, an internationally important nature reserve. The contract specified that no material of any kind was permitted to be discharged to the sea from any of the dredgers or ancillary vessels (courtesy Dredging International; copyright Foto Guido Coolens n.v.)

Plate 2. The cutter section dredger 'Leonardo da Vinci' (27,524 HP) dredging glacial till for reclamation for the Storebaelt bridge in Denmark (courtesy of Jan de Nul; copyright: Ivan Svendsen)

Plate 3 The trailing suction hopper dredger 'Cornelis Zanen'. Hopper capacity 8000 m^3 (courtesy Royal Boskalis Westminster)

Plate 4 The backhoe dredger 'Ijzeren Hein' working in Grimsby (courtesy of Van Oord ACZ; copyright Geoffrey Pass)

Plate 5 Environmental dredging at South Willemsvaart, Holland, using a special grab bucket and treatment plant (courtesy Boskalis Dolman)

Plate 6 Changing the picks on the cutterhead of the 'Zeeland II' (courtesy Van Oord ACZ)

Plate 7 The trailing hopper dredger 'J F J De Nul' (11,750 m³) 'rainbowing' material ashore during reclamation works for Hong Kong's Chek Lap Kok airport (courtesy Jan de Nul)

Plate 8 The HAM water injection dredger 'Jested', working in Harwich Haven (courtesy R N Bray)

Plate 9 The seagoing rock cutter suction dredger 'Marco Polo' (21,913 HP) during reclamation works for the airport coastal road at Mahe, Seychelles. A silt screen is being used as protection for the coral reef (courtesy Jan de Nul)

Plate 10 The suction pipe of the 8,000 m³ trailer dredger 'Delta Queen' owned by D Blankevoort International Dredging company, being lowered into water during trials (courtesy R Boesveld)

Plate 11 The Rhode Neilson bucket dredger 'Ajax', fitted with 700 litre buckets, working in the Port of Belfast, Northern Ireland (courtesy A D Bates)

7.4.2 LIMITING FACTORS

The maximum water depth for operation is dependent on the size of the plant and the method of connection to the jet nozzle array. Existing designs employ a rigid connection and maximum depth is about 15 metres, but by the introduction of a flexible (wire) connection, greater depth may be possible.

The limits on economic operation will depend on the size of the particular plant. An indication of the limits for current designs is given as follows:

Minimum water depth to operate	3 metres
Maximum depth to operate	15 metres
Maximum wave height	0.50 metre
Maximum swell	0.75 metre
Soil type	very soft and soft cohesive soils and very loose and loose fine granular soils

7.5 Stationary suction dredger

The stationary suction dredger shares many of the characteristics of the trailing suction hopper dredger. Unlike the trailer, the stationary suction dredger does not dredge whilst under way, but first anchors and then loads the hopper whilst moored. The result of dredging is usually a pit in the seabed in the form of an inverted cone. In consequence it is not usual for this type of dredger to be used for the formation, or maintenance of channels, or level areas. The usual role of the stationary suction hopper dredger is the winning of granular materials for use in reclamation, or for construction aggregates.

Some stationary suction dredgers are equipped with hoppers and are capable of delivering material to remote sites in a manner similar to trailer dredgers. Indeed, some trailer dredgers are capable of operating in a stationary mode. However, many stationary suction dredgers do not have hoppers and load into barges or pump the dredged material to the delivery point through a pipeline.

Older versions of those dredgers employed exclusively on the winning of concrete aggregates are often converted small coasters, fitted with unsophisticated dredging equipment. The type of dredger which will be described here is the modern purpose-built vessel, which is of greater sophistication. However, the basic principles are similar.

7.5.1 METHOD OF OPERATION

The important features are illustrated in Figure 7.12.

On arrival at the dredging area, the dredger is anchored, the suction pipe intake lowered to the seabed and loading of the hopper commences. The dredger may be equipped to load barges which moor alongside. This has the benefit of increasing the overall productivity and is of particular value where there is a great distance from the dredging area to the point of end use of the dredged material. However, barge loading is usually only possible in sheltered waters.

Fluidisation and entrainment of the soil to be dredged may be assisted by high pressure water jets arranged around the suction intake. This principle may be extended to permit the removal of granular materials from beneath an upper layer of unwanted material, such as clay, or peat. To facilitate this, the suction pipe is articulated and the lower section is maintained in a vertical attitude. The suction pipe is progressively lowered through the overburden and into the desired granular material. The granular deposit in the region of the suction intake

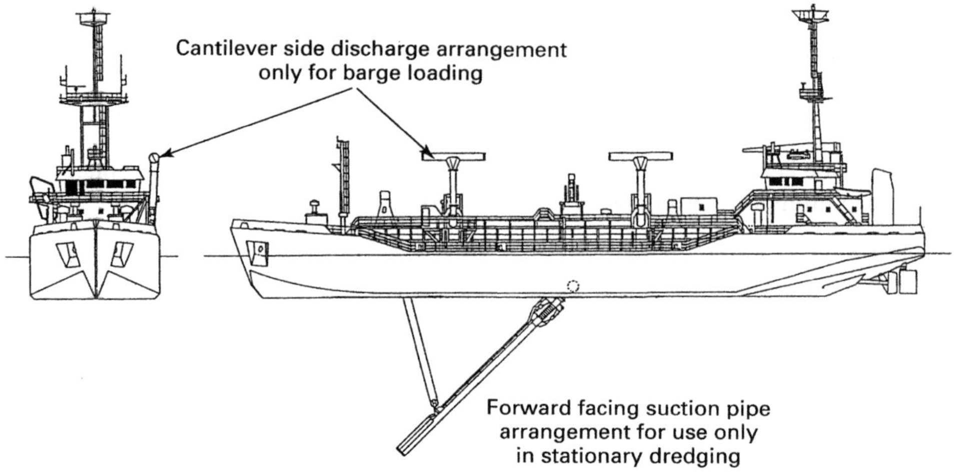

Figure 7.12 Main features of stationary suction hopper dredger

is fluidised by the action of water jets, drawn into the suction intake and thence conveyed to the hopper, or to barges alongside. Subject to the *in situ* properties of the granular deposit, this method may be employed to dredge material from a considerable depth.

For the loading of barges alongside, the dredged material is discharged through moveable outrigger pipes, which terminate in a diffuser arrangement, designed to reduce the velocity of outflow and turbulence within the hopper with the object of minimising the loss of material overboard. Nevertheless, the resulting losses and localised increase in water turbidity may be significant.

The stationary suction dredger is well suited to the winning of fill, or aggregate materials, where the sea bed deposits are of substantial thickness, or in the case of the more sophisticated types of dredger, at considerable depth. If designed for dredging only within sheltered waters, the hull structure may be of a simple pontoon configuration without a hopper, similar to that of the cutter suction dredger. The ship-shaped self-propelled form has the capability for independent operation in fairly exposed sea areas. It is also moved between sites under its own power with relative ease.

7.5.2 THE PRODUCTION CYCLE

Only the loading part of the production cycle is significantly different to that of the trailer dredger (see Figure 7.13).

Loading time

This is dependent on the soil characteristics, the dredging depth, and the particular dredger. If used to load independent hopper barges, loading may be continuous, subject to an adequate supply of barges being available. Loading time is difficult to predict with accuracy and may be best judged from experience. The limiting maximum attainable production, and hence minimum loading time, can be calculated given adequate soils information.

Daily production will be the product of average hourly production and working time. Working time will be influenced by similar factors to those which affect the trailer, but the stationary suction dredger is more sensitive to sea conditions.

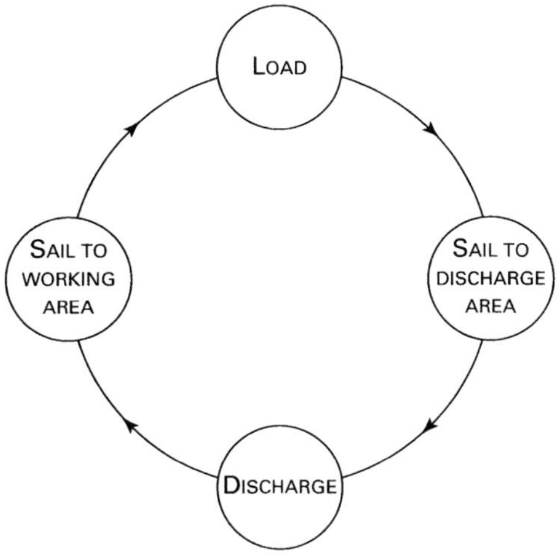

Figure 7.13 Production cycle of stationary suction hopper dredger

7.5.3 LIMITING FACTORS

There are limits on the performance of the stationary suction dredger which will vary according to the size and characteristics of the particular dredger, but an indication of the extreme limits which will apply to economic operation is given below. These limits apply respectively to the smallest (minimum) and largest (maximum) stationary suction dredger in common usage.

Minimum water depth to operate	3 metres
Maximum water depth to operate	50 metres (very few to 80/100 metres)
Maximum wave height	3 metres
Maximum current	3 knots
Soil type	permeable granular soils

7.5.4 ANCILLARY EQUIPMENT

The normally independent operation of the stationary suction hopper dredger does not require a wide range of ancillary equipment. The main variations that may occur relate to the suction head and the method of discharge. (For hopper barges, see Section 7.14.2.)

Suction heads

Various methods may be employed to improve performance, particularly at greater operating depths. For deep dredging, one way in which the limitation of a finite hydraulic suction head is overcome is by mounting the dredge pump part way down the submerged pipe. Another is by the incorporation of a jet pump (see Section 7.13.1) at the suction intake. However, most designs are intended to either improve the flow of mixture within the suction intake, or to fluidise the seabed deposits in the region of the suction intake. The latter effect is achieved by means of high pressure water jets.

Discharge mechanisms

Bottom-discharge

Where fitted, these are the same as the trailer dredger (see Section 7.2).

Side booms

As an alternative to loading the hopper of the dredger, the pumped mixture may be diverted via pressure pipework, to side booms for delivering material to barges.

Pipeline

The dredger may be coupled to floating and land pipelines in the same way as the cutter suction dredger (see Section 7.6).

7.6 Cutter suction dredger

The important features of the cutter suction dredger are shown in Figure 7.14.

In its most common form the cutter suction dredger has a pontoon hull structure without propulsion. However, some, usually large, may have a ship-form hull and be self-propelled. Dredging takes place with the dredger anchored.

The dredger is most often used in works of land reclamation, or other hydraulic filling, but may be employed in a wide range of applications.

The dredging process involves an initial powerful cutting action, followed by suction, and pumped discharge via a pipeline, or occasionally to barges.

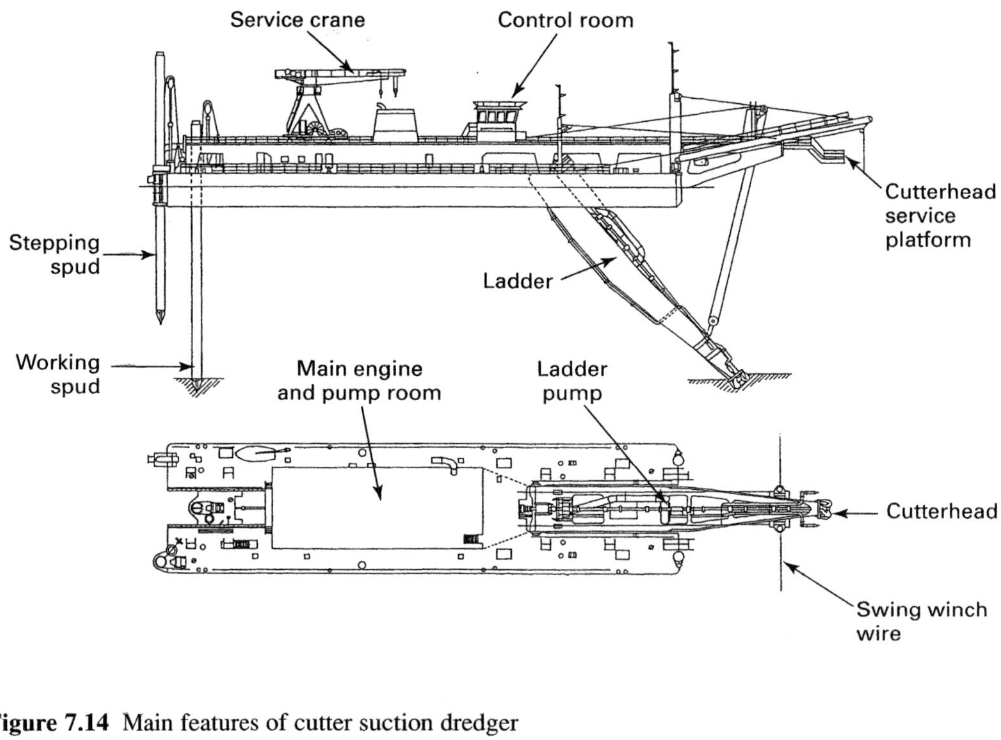

Figure 7.14 Main features of cutter suction dredger

The main advantages of the dredger are:

- the ability to dredge a very wide range of materials, including rock, and to convey the dredged material by pumping with water directly to the disposal, or reclamation area;
- the ability to operate in shallow water and to produce a uniform level bottom with high rates of production;
- the ability, in the case of some modern dredgers to dredge to a pre-defined profile, e.g. in channels.

The disadvantages include the following:

- a sensitivity to sea conditions;
- limited distance through which dredged material can be economically conveyed;
- dilution of the dredged material;
- limited dredging depth;
- high mobilisation costs.

The cutter suction dredger is usually rated according to either the diameter of the discharge pipe, which may range from 150 millimetres to 1100 millimetres, or by the power driving the cutterhead, which may range from 15 kilowatts to 4500 kilowatts. The distribution of cutter suction dredgers by size is shown in Figure 7.15. Typical characteristics are shown in Figure 7.16. Any wide divergence from the average of a characteristic for a given size of dredger is indicative that the dredger has been designed for a special purpose. For instance, dredgers designed to dredge rock may have in excess of three times the power of a conventional dredger.

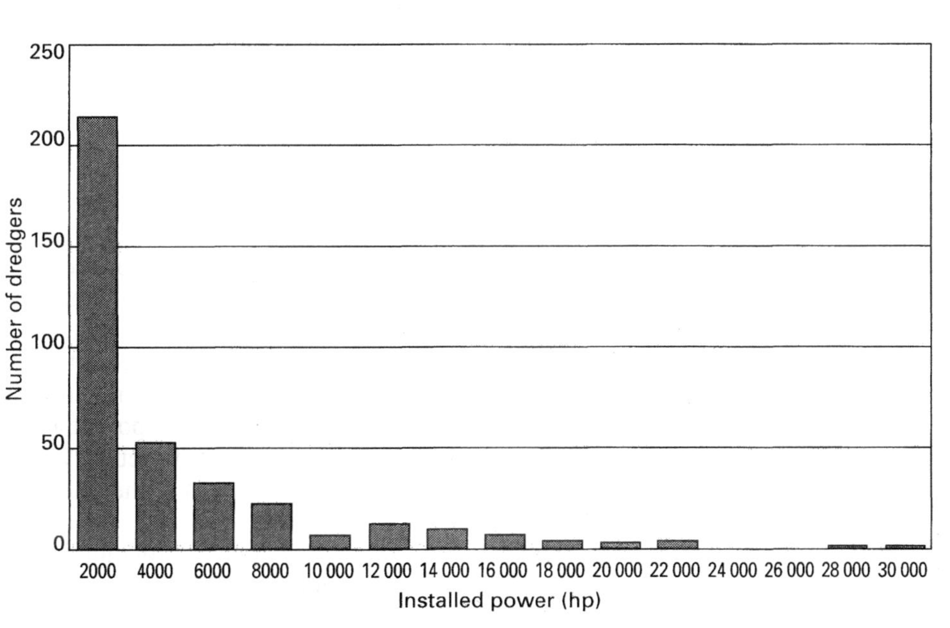

Figure 7.15 Statistical distribution of cutter suction dredgers

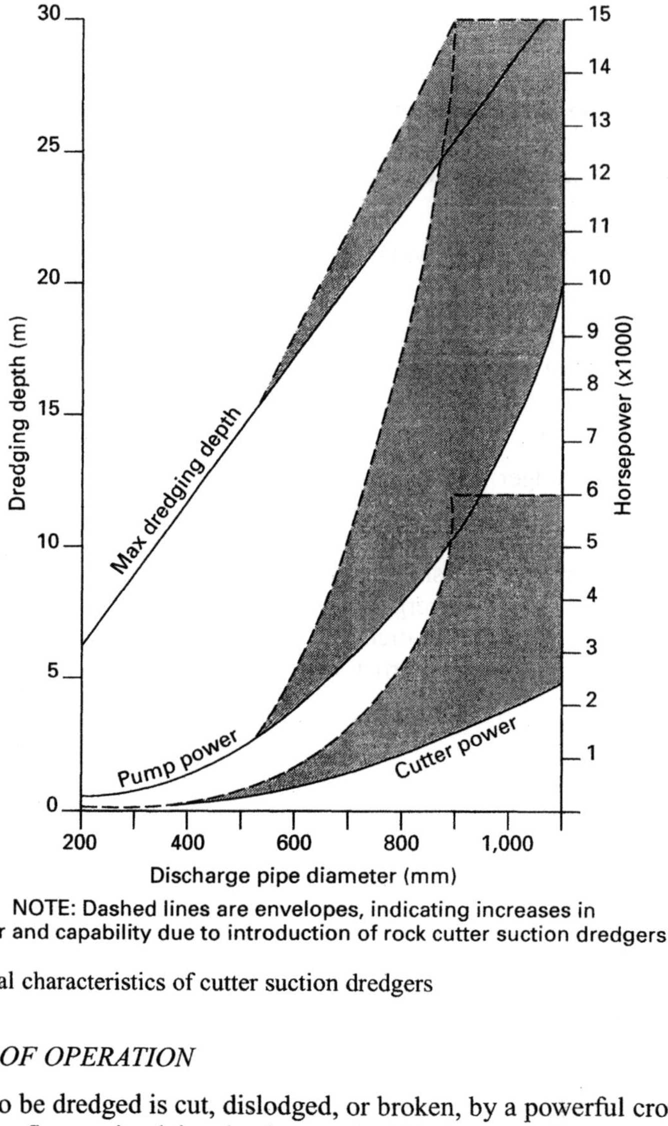

Figure 7.16 Typical characteristics of cutter suction dredgers

7.6.1 METHOD OF OPERATION

The soil or rock to be dredged is cut, dislodged, or broken, by a powerful crown cutter. The cutter typically has five or six plain edged or toothed blades according to the ground conditions (see Figure 7.17).

The cutterhead may be electrically or hydraulically driven. It encloses the suction intake pipework of a centrifugal dredge pump (see Figure 7.18). The cutterhead is mounted at the extremity of a fabricated steel structure, called a 'ladder', which also supports the suction pipe. The cutterhead drive motor may be mounted immediately behind the cutter at the ladder extremity, or on large dredgers may be mounted at the inboard end of the ladder with a long drive shaft.

The ladder is attached to the main hull by heavy hinges which permit rotation in the vertical plane. The suction pipe includes a length of heavy reinforced hose which imparts the flexibility necessary to permit rotation. The hose is located immediately outside the pump-room bulkhead. The ladder assembly is lowered and raised by means of a hoisting winch controlled from the bridge. Alternatively, on small dredgers, the ladder movement may be controlled by hydraulic cylinders.

The main pontoon structure contains the dredge pump (see Figure 7.18 and Plate 7.8) or pumps, the main engines and all ancillary engines, drives and equipment. Generally, small dredgers are diesel hydraulic and large ones diesel electric. For improved performance, or greater dredging depth, a submerged dredge pump may be mounted on the ladder, usually in combination with one or two inboard pressure pumps.

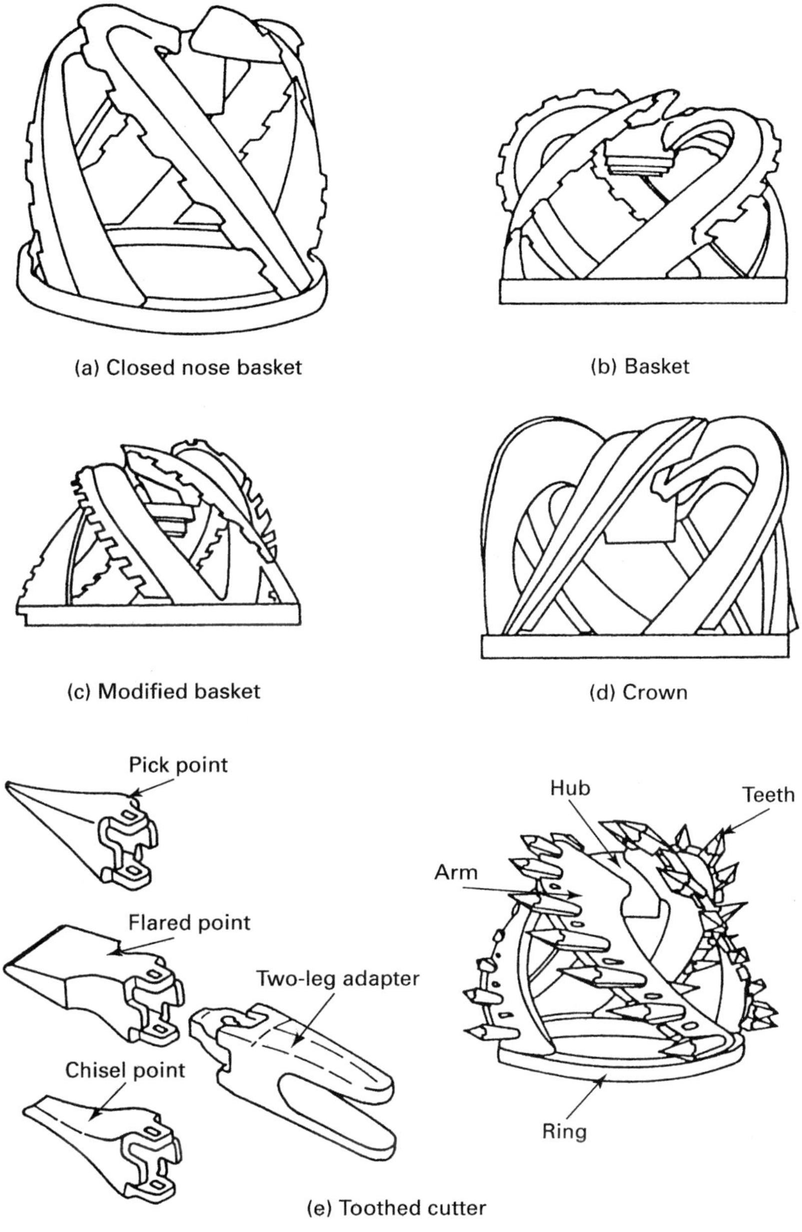

Figure 7.17 Examples of cutterheads for cutter suction dredgers

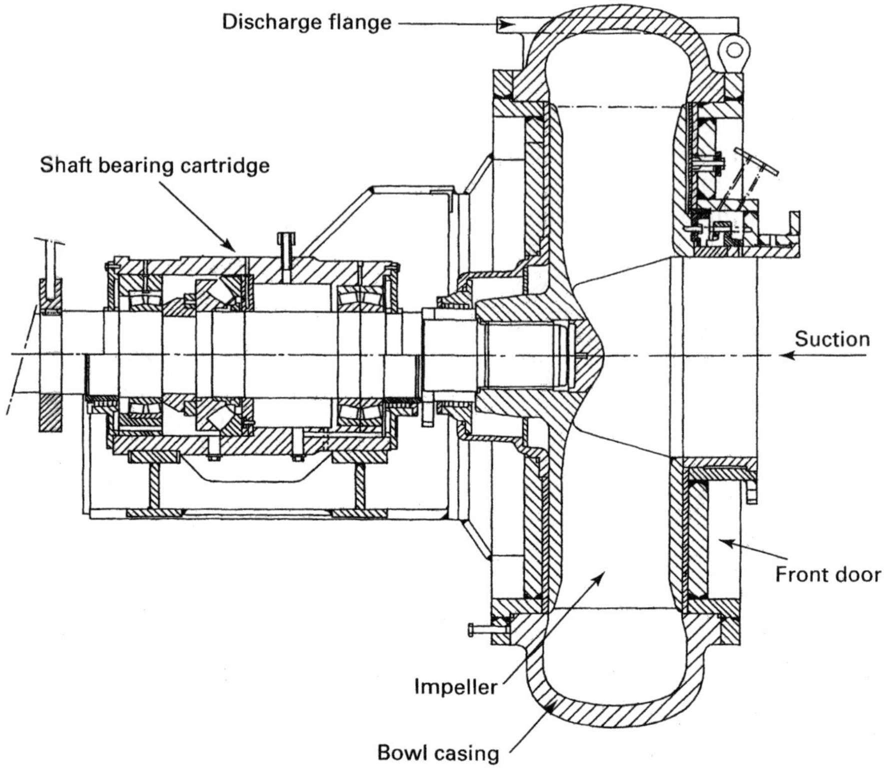

Figure 7.18 Section through typical centrifugal dredge pump for cutter suction dredger

Location and control of the dredger is usually by means of a combination of 'spuds' (see Figure 7.14) and winches. Occasionally, and usually in rough waters, winches only may be employed, the stern of the vessel being located by wires in a 'christmas tree' arrangement. Even more rarely, spuds only may be employed.

The more common combination, of spuds and winches, employs two swing winches and two spuds. One spud is located within fixed gates at the opposite end of the pontoon to the cutter (normally, but not always, the stern). The other 'working' spud may also be fixed at the stern, or may be moveable by tilting, or may be mounted in a carriage which can move longitudinally, or rotationally on the centre line of the dredger. The longitudinal carriage, which is most common, typically has an overall movement of six metres and is powered by a hydraulic cylinder.

The discharge from the dredge pump, or pumps, passes over the stern (or opposite end to the cutter) of the pontoon to a hose, or flexible coupling, to which is connected a floating pipeline (see Section 7.6.4), which in turn is connected to an onshore pipeline.

When dredging, the working spud is 'spudded in' to the seabed. The underside of the cutter is maintained at a level just below the desired finish level and traversed across the arc-shaped face, with the dredger rotating about the working spud, by hauling in on one swing winch and paying out on the other. When a large depth of material is to be removed, several cuts across the face may be necessary before the final level is achieved. In loose materials, the depth (or thickness) of cut may be several times the diameter of the cutterhead. In stiff clays, or rock, the depth of cut is usually less than the diameter of the cutterhead.

Plate 7.8 The main centrifugal dredge pump of a large self-propelled cutter suction dredger

On completion of a cut, or series of cuts to the desired level, a cleaning cut is usually necessary. Thereafter the dredger is advanced into the face a distance approximately equal to the height (front to back) of the cutterhead. This advance may be achieved by stepping with fixed spuds (see Figure 7.19), tilting of the working spud, or advancing the working spud carriage, depending upon the particular system employed.

All spud systems work on the same basic principle that one spud is moved whilst the other is firmly imbedded in the ground. The procedure is repeated in each production cycle. At longer intervals it will be necessary to move anchors as the original positions fall too far astern. Normally the anchor line should not exceed an angle of about 40° to the cut centreline. The cut patterns achieved using different spud systems are shown in Figure 7.19. Working with fixed spud systems is usually less efficient than a spud carriage.

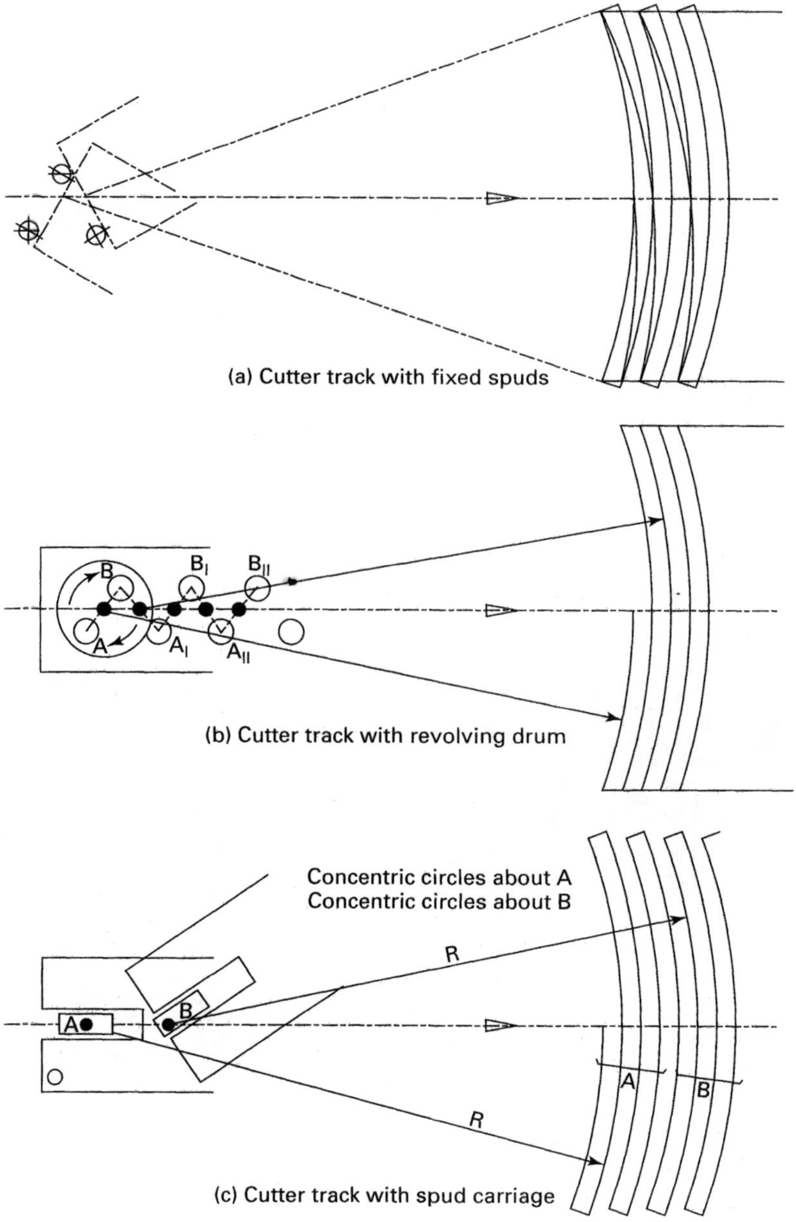

Figure 7.19 Examples of pattern of working using various spud arrangements for cutter suction dredger

7.6.2 THE PRODUCTION CYCLE

The relative importance of the component parts of the production cycle will vary according to the site conditions and requirements. For example, stepping ahead time may be relatively unimportant when dredging to remove a large thickness of hard material, but may be the limiting factor when removing a small thickness of soft material. Ultimately, the production rate will be influenced by the following:

- the pump hydraulic head;
- the applied power available for pumping;
- the ability of the cutter to break down the material;
- the rate at which the dredger can be advanced into the face.

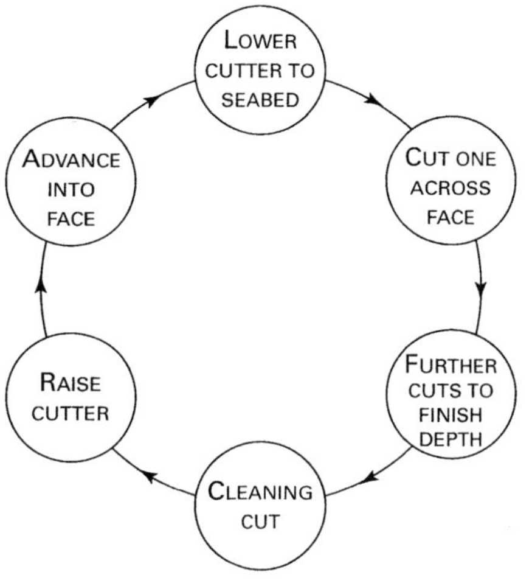

Figure 7.20 Production cycle of cutter suction dredger

The main components of the cycle are illustrated in Figure 7.20 and described as follows.

Traverse face

The speed of traverse is governed by the characteristics of the material to be dredged, the depth of cut, the pumping distance and the dredging depth. These govern hauling winch speed and cutter revolutions.

The time to traverse the face is given approximately by dividing the length of the face (in metres) by winch speed (metres per minute), allowing for acceleration and deceleration at the ends of the swing. The time of traverse is also affected by the direction of traverse, due to the fact that the crown cutter is normally most efficient when undercutting (see Figure 7.21). In hard materials, overcutting may be impractical, in which case a non-productive return swing is necessary.

Number of cuts

The number of cuts across the face to produce a required level is dependent on the thickness of material to be removed and the depth of each cut. The depth of each cut is governed by the characteristics of material to be dredged. Usually one final cleaning cut is required. The number of cuts is given by dividing the depth to be removed (inclusive of any necessary overdredge), by depth of cut and adding one.

Advance into face

Entirely dependent on the spud system employed. The time occupied will be a minimum when using a spud carriage, or tilting spud, and maximum for fixed spuds. If the dredger is located only by winches, the moving time may be reduced, but at the expense of poor positional control and hence irregular contact with the face.

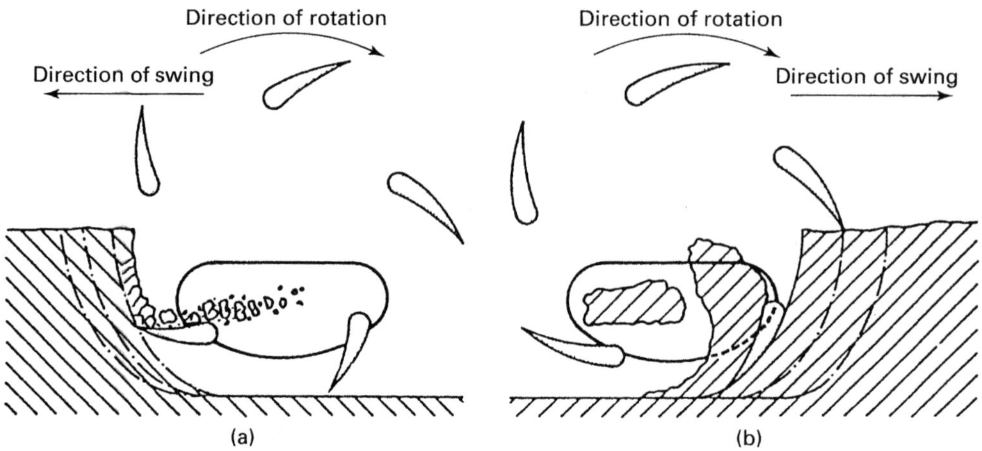

Figure 7.21 Examples of crown cutter action when (a) undercutting and (b) overcutting

Daily production is the product of the average hourly production and the effective working hours per day. The effective working hours will be the total available hours that the dredger is manned, less any lost time.

For a well-managed dredger, under average site conditions, the lost production time may be in the range, 20 to 30 per cent of the theoretically available working time. However, special site conditions will often result in far higher losses.

7.6.3 LIMITING FACTORS

The limits on where the cutter suction dredger can operate and what it can achieve will vary according to the size and characteristics of the particular dredger, but an indication of the extreme limits which will apply to economic operation is given below. These limits apply respectively to the smallest (minimum) and largest (maximum) cutter suction dredgers in common use.

Minimum water depth to operate	0.75 metre
Maximum water depth to dredge	35 metres
Maximum cut width (single pass)	175 metres
Maximum wave height	2.0 metres
Maximum swell	1.0 metre
Maximum cross current	2.0 knots
Maximum ice thickness	200 millimetres
Maximum particle size	500 millimetres
Maximum compressive strength (rocks)	50 MPa

When dredging in enclosed waters, for instance lakes and ponds, where the rate of water inflow is less than the pumping capacity of the dredger, it should be borne in mind that the water level will fall, with potential adverse operational and environmental consequences. Even the smallest dredger may remove water at a rate of 300 cubic metres per hour.

7.6.4 ANCILLARY EQUIPMENT

The cutter suction dredger may be used with a wide variety of ancillary equipment including pipelines, anchors, cutterheads, workboats, lifting equipment, and so on.

Pipelines

The pipeline will have an important effect on both the performance and operational efficiency of the cutter suction dredger. The diameter of the pipeline will directly influence the efficiency of the hydraulic transport process. If the diameter is too small, head losses due to friction may be unacceptably high. If the diameter is too great, the flow velocity may be inadequate to maintain the solids in motion, or the power requirement may become excessive. Typical mixture velocities necessary to transport various types of material without undue risk of pipeline blockage are given in Table 7.4. Hydraulic transport is described in greater detail in Chapter 8.

Table 7.4 Transport velocities necessary for various materials

Material type	Velocity range (m/s)
Silt	2.0 to 3.0
Fine sand	3.0 to 4.0
Medium sand	3.5 to 4.5
Shell	3.5 to 4.5
Very soft clays	4.0 to 5.0
Coarse sand	4.0 to 5.0
Sand with fine gravel	4.5 to 5.0
Sand with medium gravel	4.5 to 5.5
Stiff clays	4.5 to 5.5
Sand with coarse gravel	5.0 to 5.5
Sand, gravel and cobbles	5.5 to 6.5

Note: Higher velocities than those listed above may be possible. This may result in higher production rates but at the expense of increased power demand and pipe wear.

Pipe-wall thickness must be adequate to resist the internal pressure. If, during the work, wear reduces the pipe to an unacceptable thickness, as may happen when pumping abrasive materials, the pipeline must be changed.

Pipelines onshore

Pipelines onshore are most commonly of steel construction with bolted flange connections, including a compressible gasket between flanges to provide a watertight seal. Alternatively pipes may be joined by welding.

Pipes are usually laid on the ground. Over-ramps or pipe bridges may be necessary to maintain vehicular access. Pipelines may branch along their route (see Plate 7.9), either to serve alternative deposition areas, or to discharge at different points within an area. This arrangement permits a better distribution of material, particularly for coarse-grained soils, and also allows the addition of more pipes to a particular line, without the need to cease dredging. Valves at the point of bifurcation (normally gate, or vizor valves) provide for the switching of flow.

At the point of discharge, the pipe may be fitted with a diffuser, or 'spoon', which assists in giving a more even distribution of material.

Plate 7.9 A branch in an onshore pipeline, showing the control valves

When there are significant changes in elevation along the pipeline route, air valves ('snifters') may be necessary to expel trapped air.

Floating pipelines

Floating pipelines can be categorised in terms of their strength and flexibility. Strength is important if the pipe must resist high internal pressures, or high tensile loading. Flexibility is important in confined working areas and, particularly, when working in exposed sea areas with significant wave activity. Usually a good resistance to abrasion, or cutting, is also important.

The most flexible pipeline is formed of reinforced rubber. This will usually incorporate a jacket of buoyant material of such displacement as to provide a positive buoyancy when the pipeline is loaded with a pumped mixture of normal density. Such pipelines are most appropriate for use with large dredgers in exposed coastal locations. However, the capital cost is high.

Where complete flexibility is less important, a composite pipeline construction, which includes floating steel pipes (with a buoyant outer jacket) may be employed. The composite construction will result in a lower capital cost, depending on the relative proportions of steel and rubber pipe. A common compromise consists of alternate lengths of rubber and steel pipe, where each component has an individual length of approximately 12 metres. Occasionally, the proportion of steel may be raised to two to one, but any higher proportion is unlikely to provide sufficient flexibility and the pipe will be more prone to damage.

In calm water, the proportion of steel pipe can be greatly increased, but some flexibility must be retained. This may be provided by rubber sleeve connections between lengths of rigid steel pipe, or by steel ball joints. Ball joints have the advantage of a much greater resistance to wear and a stronger construction, but offer only limited flexibility. An angular movement of 22.5° either side of centre is the normal maximum (see Figure 7.22). Where greater angular movement is required, this may be provided by a swivel assembly (see Figure 7.23).

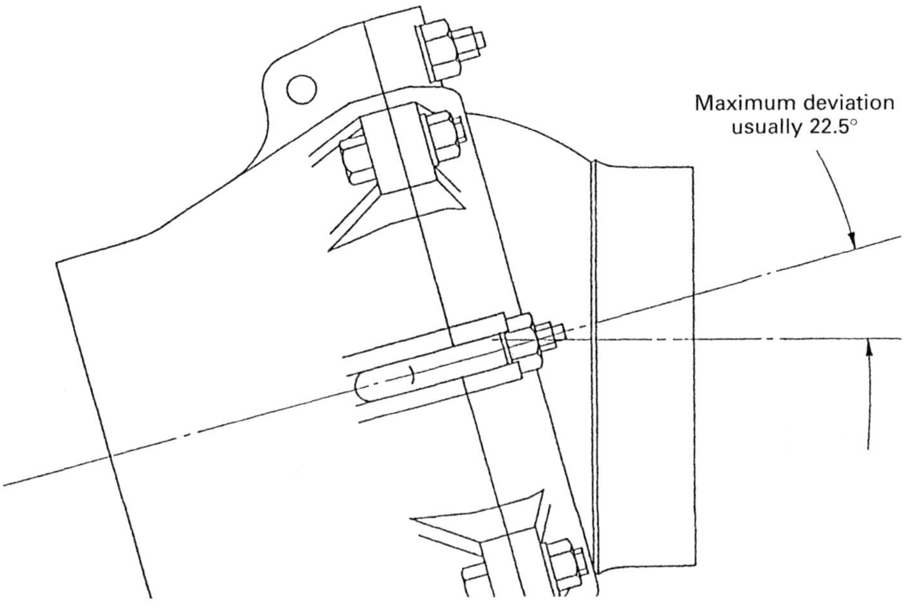

Figure 7.22 Example of ball-joint connector used in discharge floating pipelines

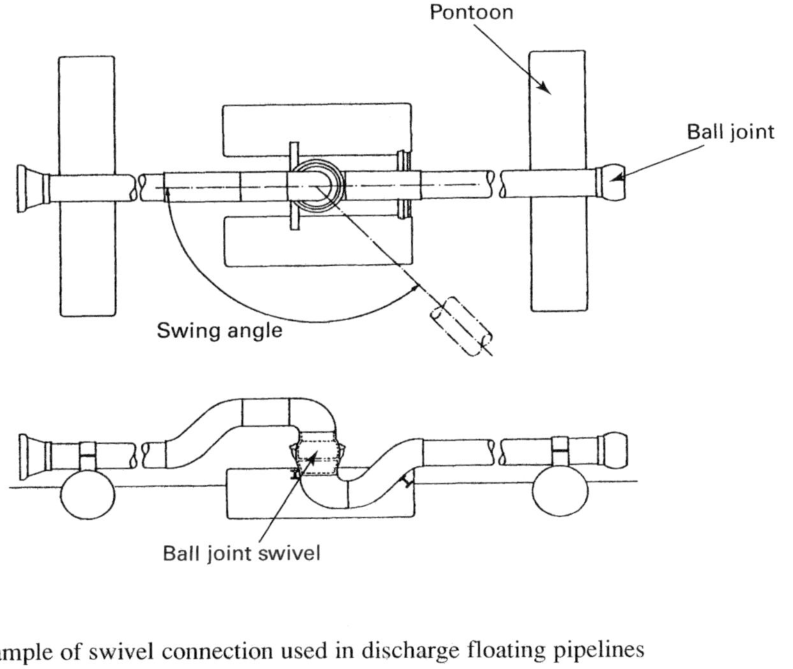

Figure 7.23 Example of swivel connection used in discharge floating pipelines

Pipelines which are of predominately steel construction may be supported by a buoyancy jacket, or by mounting on steel pontoons. Pontoon design may vary from simple rectangular boxes to more elaborate shapes with lower resistance to waves and currents (see Figure 7.24). Oil drums are sometimes used to support very small diameter pipelines, but these deteriorate rapidly and are a poor substitute for purpose-built pontoons. The considerable bulk of pipeline pontoons results in high transport costs between work sites. Some types are designed to permit more efficient stacking to minimise this problem.

Where water level is subject to tidal variation, a specially designed connection is necessary between floating and shore pipelines (see Figure 7.25).

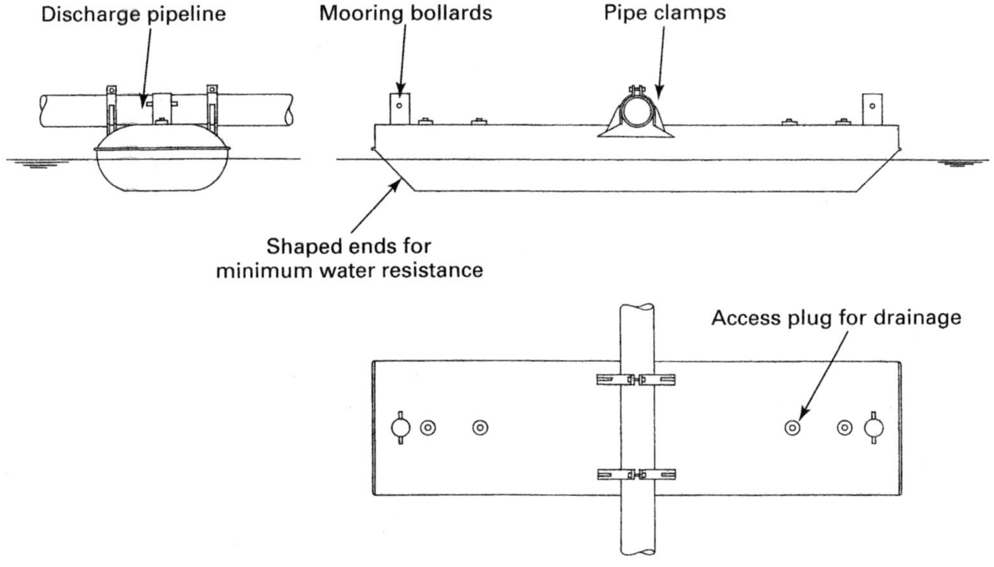

Figure 7.24 Example of floating pipeline supporting pontoon, designed to minimise water resistance

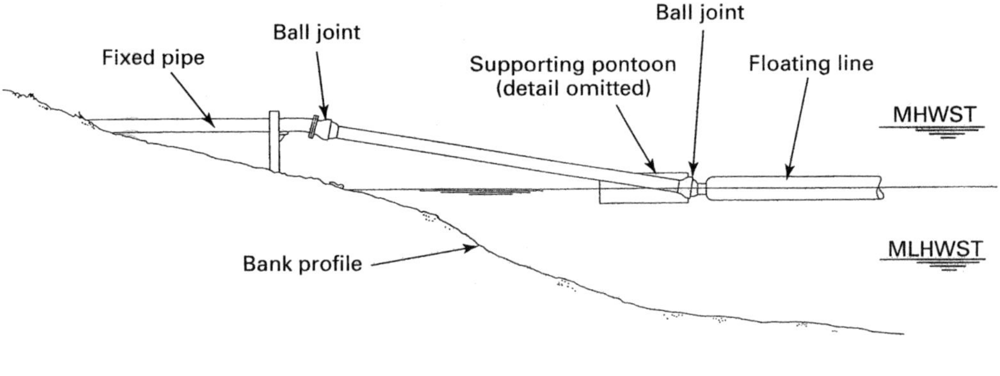

Figure 7.25 Simple shore connection for discharge pipeline

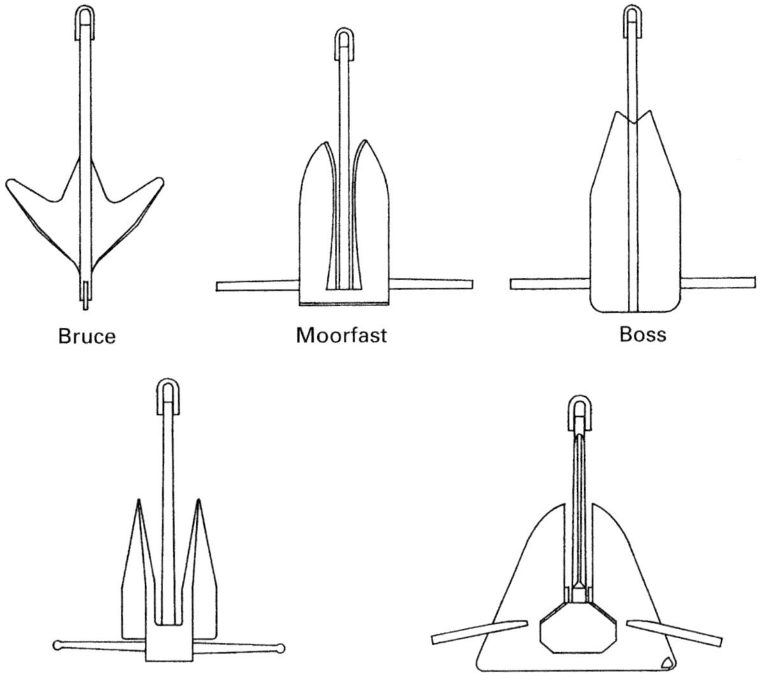

Figure 7.26 Various types of anchors used by cutter suction and other dredgers

Anchors

The choice of anchor should be based upon the ground conditions which exist on the particular site. In many cases, the cutter suction dredger will work with a type of anchor which secures a hold by penetrating the seabed. Occasionally, however, where the seabed is rocky and the winch loading is high, it may be necessary to resort to a gravity anchor, which relies primarily on its mass to resist the winch pull.

Ground anchors

Most anchors rely upon penetration of the seabed to achieve their holding power by acting against the passive resistance of the seabed soils. In order to maximise this effect, the anchors have a large fluke area. Typical holding forces range from 10 to 30 times the mass of the anchor. Various common anchors are illustrated in Figure 7.26.

Gravity anchors

Where conventional anchors cannot penetrate the ground sufficiently to develop the resistance required by the dredger it may be necessary to resort to gravity anchors.

Gravity anchors may consist simply of large concrete blocks which rest on the seabed and are moved by a floating crane, but such arrangements are rarely practical, or economic. More practical designs consist of tripod platform structures which are naturally buoyant, but which can be flooded with sea water to rest on the seabed. Weight can be increased by the addition of further ballast water. The movement of ballast water may be controlled by radio switched submersible pumps. Movement of the de-ballasted anchor is by tug.

Cutterheads

The cutterheads in most common use are of the 'crown' type. A range of different crown cutterheads are shown in Figure 7.27. Each consists of a number (usually five or six) of specially shaped blades designed to cut or break out the seabed material efficiently and convey it towards the suction intake. The main body of the cutterhead is formed from cast alloy steel.

Plain-bladed cutters

Plain-bladed cutters should be used only in soft materials, such as silts, sands and clays. Blade edges are normally replaceable. The number of blades is selected according to the intended application. By increasing the number of blades the gap between the blades is reduced. This restricts the size of solid which can pass and hence may be useful when dredging in areas which contain debris, or large stones.

Serrated-blade cutters

These are used primarily in consolidated materials such as firm to stiff clays, dense sands or occasionally in very weak or highly weathered rocks. The blade edges can be replaced when worn.

Rock cutters

Rock cutters are of heavy construction with a generous blade section. The shape of the blades is designed to maintain the maximum number of teeth in contact with the face regardless of dredging depth. Blades have integral adaptors to which may be fitted different patterns of replaceable teeth. These may range from a chisel form for stiff clay and very weak rocks, to a pick point form for the dredging of weak to moderately strong rocks. Teeth are formed from highly wear-resistant alloy steel and are attached by means of a single sprung pin which allows a tooth to be changed in a few minutes. Plate 7.10 shows a number of replacement cutterheads on site during a large rock dredging contract.

Plate 7.10 Large replacement rock cutterheads on site during a rock dredging contract

Wheel cutters and bucket wheel cutterheads

Bucket wheel cutterheads, usually referred to as 'bucket wheels', consist of a number of buckets equally spaced around a circular carrier structure which contains the suction intake (see Figure 7.27). Each bucket is bottomless. Teeth may be fitted to the bucket rim. Bed material is dislodged by the bucket teeth, or rim, and scooped up. The combined effect of rotation, gravity and water flow move the material through the bucket to the suction chamber.

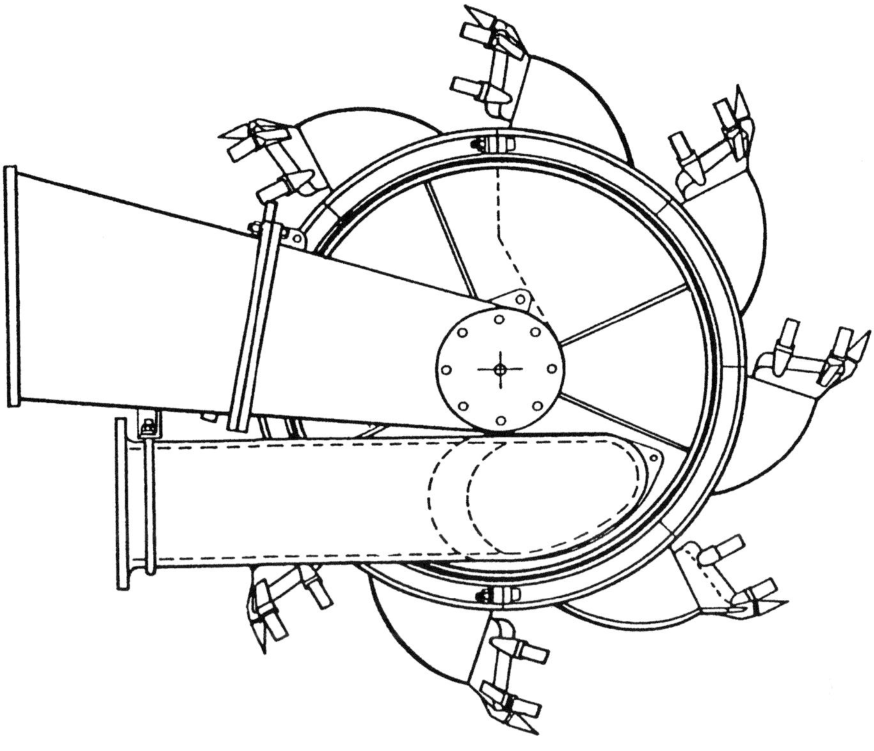

Figure 7.27 Example of bucketwheel cutterhead for cutter suction dredgers

Bucket wheels may be used in a wide variety of ground conditions, including soft rocks. Advantages include the following:

- equal cutting efficiency in both directions of swing;
- reduced spill relative to crown cutters;
- high solids concentration in suitable materials;
- improved break-out in certain ground conditions;
- reduced power requirements.

Disadvantages include the following:

- greater weight;
- greater sophistication;
- carriage mounted working spud necessary;
- unsuited to cutting by undermining or very shallow working (minimum 50 per cent wheel diameter).

On the alternative wheel-type cutter the individual buckets are replaced by a larger number of closely spaced cutters (see Figure 7.28). In side elevation the wheel cutter resembles the milling cutter used in machine-shop practice. The wheel cutter shares similar advantages and disadvantages to the bucket wheel, but is more suited to work in soft ground conditions. It is usually lighter than the bucket wheel.

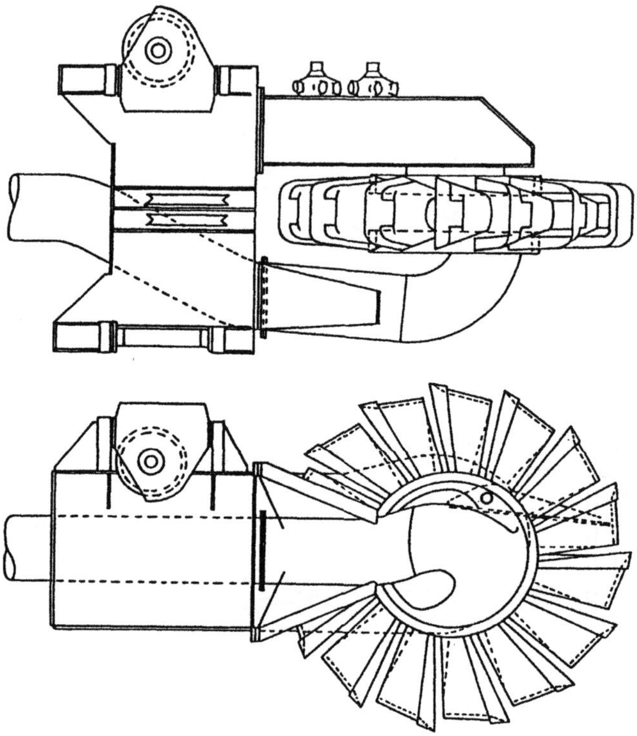

Figure 7.28 A wheel cutter cutterhead for cutter suction dredgers

Other suction heads

A wide variety of suction heads have been developed for specific applications, but few have been widely adopted.

The horizontal auger head has achieved quite wide acceptance (see Figure 7.29). This consists of two horizontally opposed continuous flight augers feeding inwards to a suction intake. It is particularly well suited to the dredging of low strength fine sediments. High solids concentrations are possible using this type of head. When fitted with an auger the dredger must be advanced directly into the face. A thin strip of material, usually slightly wider than the dredger hull, is removed on each successive pass of the dredger. The auger may be covered on the top side by an adjustable shield. This has the dual effect of controlling water intake and confining dislodged material. These characteristics may be beneficial when dredging contaminated sediments. The working method is similar to that of the dustpan dredger (see Section 7.7).

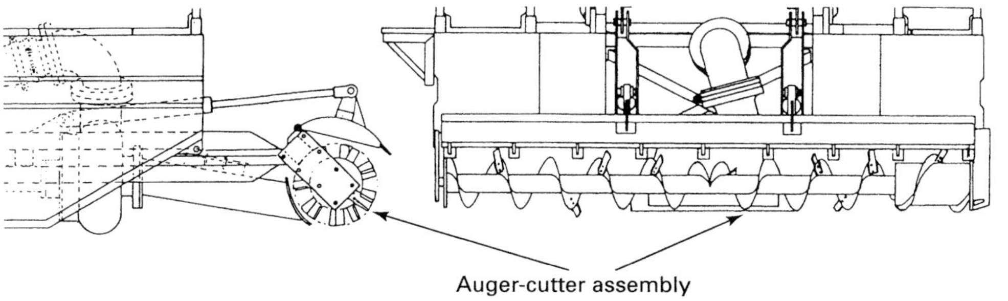

Figure 7.29 Example of horizontal auger for suction dredgers

The disc cutter, (see Figure 7.30), has a form rather similar to that of a simple waterwheel, but is operated in the horizontal plane. It is claimed to provide high solids concentration when used in fine-grained soft soils. It is best suited to the removal of relatively thin bottom layers. The pattern of working is similar to that of a conventional crown cutter suction dredger.

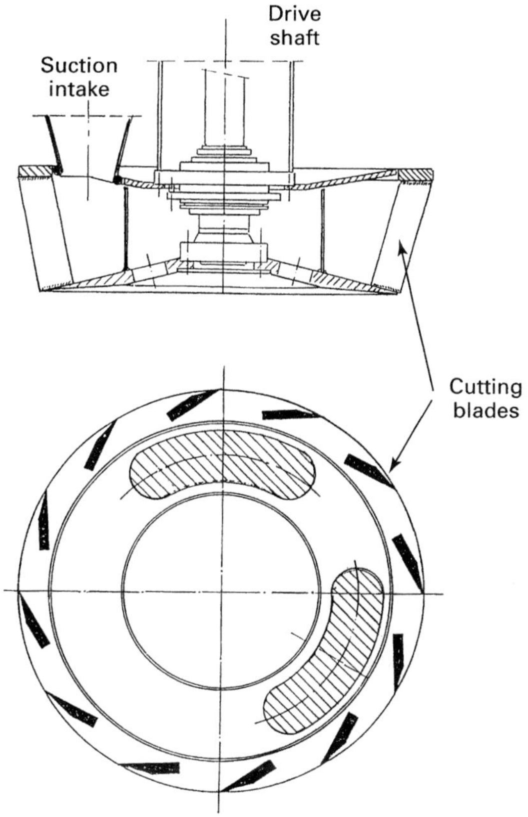

Figure 7.30 Example of disc cutterhead

7.7 Dustpan dredger

In its most common form, the dustpan dredger is a suction dredger which usually discharges through a short floating pipeline. In many respects it is similar to the cutter suction dredger (see Section 7.6). The most significant difference is in the design of the suction head, which is illustrated in Figure 7.31. Occasionally dustpan dredgers may be built for very specific purposes, such as accurate foundation preparation, but the basic principles remain similar.

The suction head has no mechanical cutting action, but relies instead on an array of water jets to loosen and fluidise the bed deposits.

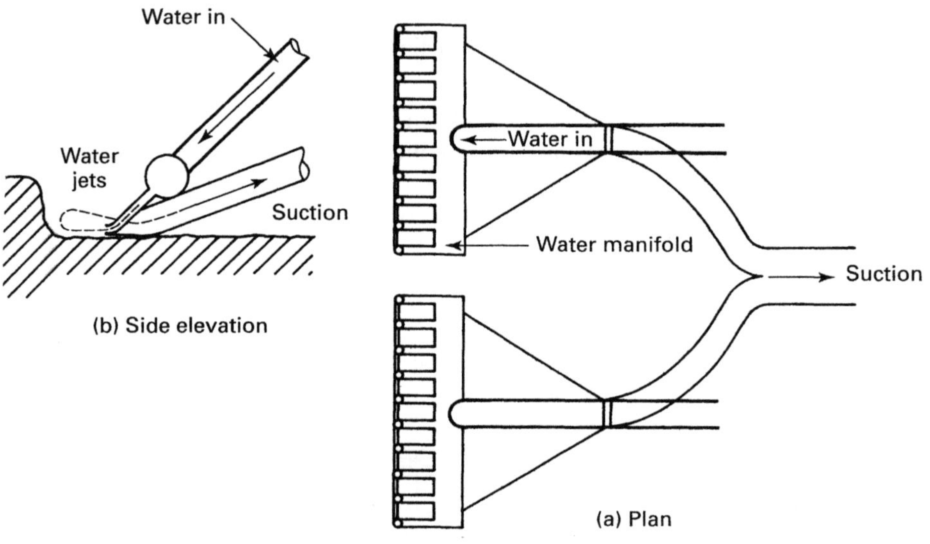

Figure 7.31 Suction head of dustpan dredger

Dustpan dredgers are usually positioned and moved by winches. This has the advantage that the dredger can move freely and quickly over the ground. In most situations, it is an advantage if the dustpan dredger is self-propelled. In normal operation the dredged material is discharged into deeper, or rapidly flowing water a short distance from the dredging area. The dustpan dredger is most effective in the removal of shoals in large rivers where mobile bed deposits may obstruct the navigation channel. If required, the dredger can pump discharge into onshore containment areas, but the requirement for a fixed onshore connection restricts the dredger's freedom of movement. The dustpan is also used for winning granular materials in sheltered sites.

The dustpan dredger is particularly useful where there is a need to remove thin deposits of loose material over relatively large areas.

The principal disadvantage of the dustpan dredger is the absence of any positive mechanical cutting mechanism. This limits efficient operation to the dredging of recent unconsolidated deposits.

The flexibility of the all-winch method of location means that accurate dredging in plan usually is not possible. However, specially designed dustpan dredgers may be used to produce a very accurate plane level surface to the seabed for the founding of pre-cast structures.

7.7.1 METHOD OF OPERATION

The dredger works on a pattern of anchors as shown in Figure 7.32.

Normally the dredger is employed in a river or channel with uni-directional flow. It arrives at the dredging area and proceeds to a point some distance upstream of the area to be dredged and drops the head anchor. The dredger then drops downstream paying wire off the head winch until the vessel arrives at the downstream limit of the area to be dredged, or at the limit of head wire reach. A stern anchor may then be dropped to increase stability. Two anchors may be sufficient, but in some cases additional side anchors may be necessary to allow coverage of a greater width of river bed, or a second head anchor, spaced apart from the first, may be employed.

The discharge end of the floating pipeline is moored in the area where the dredged material is to be placed, or dispersed.

With all necessary anchors positioned, the suction head is lowered to a depth just below the required finished level and the head winch is engaged. The dredger moves upstream into the face at a speed appropriate to the depth of material to be removed and continues as far as is required.

If the depth of deposit to be removed is large, then more than one pass may be made over the area in this way. Subject to the nature of the material to be dredged and to the water depth, dredging may take place in either direction.

If the width to be covered is greater than the width of the suction head, then side winches, or a second head anchor, may be employed to align the dredger along an adjacent cut. Alternatively, the head and stern anchors may be relocated. The particular method used will depend upon the dimensions of the overall area to be covered.

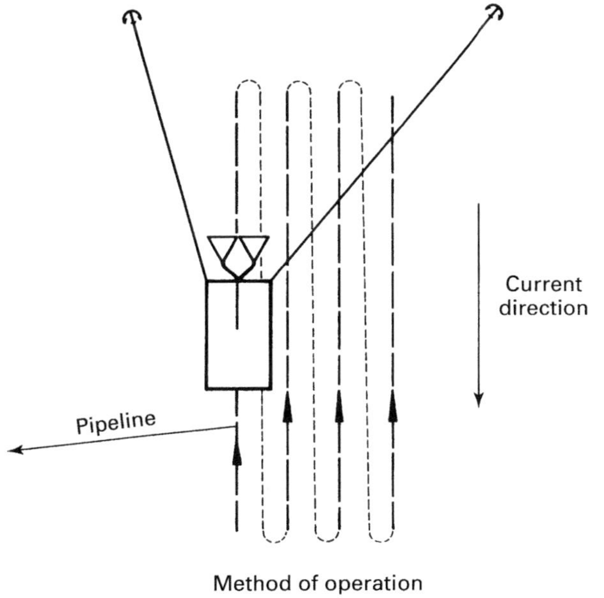

Figure 7.32 Typical pattern of anchor spread for dustpan dredger

7.7.2 PRODUCTION CYCLE

The dustpan dredger does not have a clearly defined cycle. The most important factor which affects production will be the area which can be covered from a given anchor pattern and the depth of material to be removed. The main elements of the limited production cycle are as follows.

Lay head anchor

Move the dredger to the upstream limit of the dredge area and drop anchor.

Position dredger

Move the dredger downstream on the head wire.

Commence dredging

Lower the suction head to the required depth and move ahead on the head winch adjusting the lateral position by use of the second head winch or side winches.

End of pass

Make return cut or drop back for a second cut or recover anchors and move to a new location.

Daily production is dependent on the extent of the area to be covered and, with shoal removal, the time to travel between shoals. Because of the normally free nature of the material to be dredged and the short discharge distance, unit production may be high. Daily production will be given by the product of effective working hours and the average hourly production of solids through the pump. The effective hours will be the total available hours for which the dredger is crewed and available to work, less any lost time.

7.7.3 LIMITING FACTORS

The limits on where the dustpan dredger can operate and what it can achieve will vary according to the size and characteristics of the particular dredger. An indication of the extreme limits which will apply is given as follows. These limits apply respectively to the smallest (minimum) and largest (maximum) in common usage.

Minimum water depth to operate	1.5 metres
Maximum water depth to dredge	20 metres
Maximum cross current	0.5 knots
Maximum cut width (single pass)	10 metres
Maximum discharge distance	500 metres

7.7.4 ANCILLARY EQUIPMENT

The dustpan dredger is a simple tool which works independently. Occasionally it may be reliant on barges for the disposal of dredged material (see Section 7.14.2). A variety of types of floating pipelines may be employed (see Section 7.6.4). A variety of anchor types may also be employed to suit particular ground conditions (see Section 7.6.4).

7.8 Grab pontoon dredger

The grab pontoon dredger consists of a lattice jib grabbing crane mounted on a simple pontoon (see Figure 7.33). It has no hopper. Instead the dredger loads into independent hopper

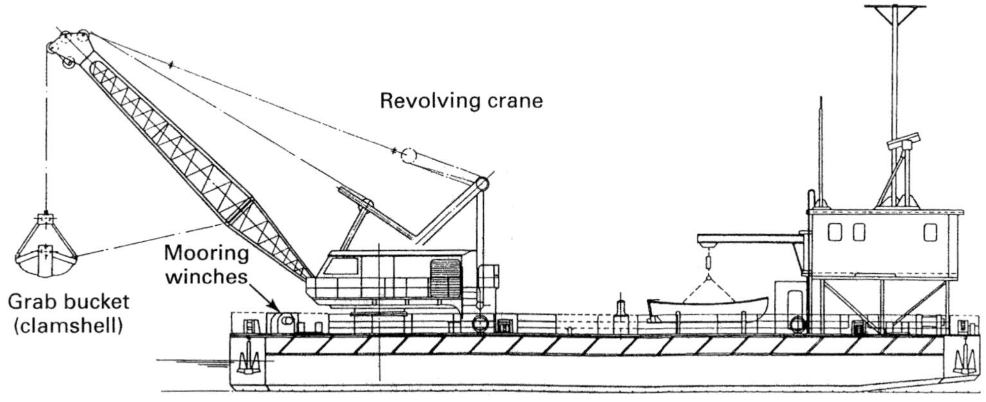

Figure 7.33 Main features of pontoon mounted grab dredger

barges (see Section 7.14.2). Subject to an adequate supply of hopper barges, uninterrupted dredging is possible, with the result that higher overall rates of production are possible than with the grab hopper dredger.

The grabbing crane is mounted towards one end of a pontoon which is usually approximately rectangular in plan, but may have a semi-circular or narrowed end projection on which the crane is mounted. The pontoon may be held in position by anchors and winches, or may combine the more positive location of spuds during dredging, with winches for pontoon relocation. The grab pontoon dredger is normally rated by its grab bucket capacity. The capacity of grab buckets may range from 0.75 to 200 cubic metres, although buckets over 20 cubic metres are rare; a statistical distribution is shown in Figure 7.34. A typical grab bucket is shown in Plate 7.11. Large units may incorporate living accommodation for the crew within the pontoon structure.

The grab pontoon dredger has several advantages:

- It loads dredged material with minimum disturbance and dilution of the soil compared to hydraulic methods, and the hopper can, therefore, be filled with a greater proportion of solids.
- Loading the hopper by grab, rather than by pump, greatly reduces the adverse effects of boulders, cobbles, rubbish, or debris. These can be handled with relative ease, although certain types of foreign matter, such as ropes, hawsers and chains, may cause problems during the discharge of dredged material by becoming entangled with the bottom door frame and seals of the hopper barge, failing to discharge and making it difficult to close the doors. These problems are avoided by the experienced crew, who will separate potentially difficult materials and stow them on deck, for later disposal onshore. This may be done with the aid of a moveable screen.
- The grab is well suited to the dredging of confined areas, such as alongside quays, in dock entrances, around jetties, and so on. Dredging in these situations may be very difficult with other types of dredger, unless aided by a bed leveller.
- The depth of operation of the grab is only limited by the rope capacity of the hoisting winch drum. It is, therefore, possible to dredge with a grab dredger to a depth which would not be possible with other dredgers of comparable size which require a fixed structure reaching to the level of dredging.

Plate 7.11 A typical grab bucket

- The pontoon draught is normally quite small and the grab pontoon dredger can thus work in shallow water, providing that the attendant hopper barges can gain access.
- A further and most important advantage is the ability to dredge narrow trenches.

The disadvantages of the grab pontoon dredger are as follows:

- Its relatively low rate of production when compared to most other types of dredger of comparable size. Unlike most other types of dredger, the excavating process (i.e. bucket filling) is only a small part of the production cycle.
- The difficulty of producing an accurate and level bottom finish. In order to ensure that no high spots remain upon completion of dredging, it may be necessary, particularly in cohesive soils, to overdredge by a significant amount. The actual amount will depend upon the skill of the operator, and the characteristics of the dredger and the soil.
- The combination of relatively low production and the necessity for significant overdredging, may result in high unit costs, particularly when it is required to remove thin layers of material from large areas.

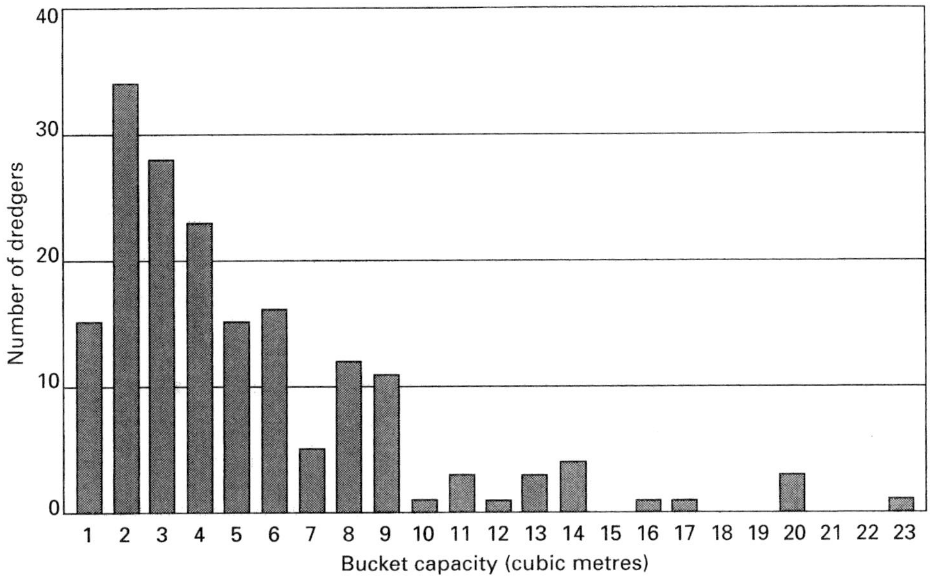

Figure 7.34 Statistical distribution of grab dredgers

Dredging to considerable depth may be possible, with the deep dredging capabilities generally being superior to that of other types of plant of comparable size. Construction and operation is relatively simple and hence inexpensive. The crane unit will generally be a standard production unit with all of the attendant advantages of series production.

The finished dredged level is very operator-dependent due to the difficulty of achieving a complete and overlapping coverage of the bottom, particularly in deep water, or where the strength of current is significant. Overdredging may be substantial, particularly in cohesive soils, and will tend to increase with increasing bucket capacities due to the bucket pulling into the bed during closure. No force can be applied to assist the penetration of stiff, or hard materials, other than the self-weight of the grab bucket. The range of materials which can be dredged economically without pre-treatment is, therefore, very limited.

7.8.1 METHOD OF OPERATION

The grab pontoon dredger is moored on a pattern of anchors, or with spuds, with the centreline of the dredger on the centreline of the proposed cut and with the grab crane closest to the working face.

All material within the working radius of the grab crane is methodically removed down to the required level and loaded into a hopper barge (see Section 7.14) moored alongside the pontoon. When no more material remains within reach, the pontoon is advanced towards the face by adjustment of the mooring winches, or spuds. This procedure continues until a barge is fully loaded, by which time an empty barge should be moored on the opposite side of the pontoon and the loading of this barge can commence, without any interruption to the dredging operation. On occasions, there may be insufficient room to accommodate barges simultaneously on each side of the pontoon, in which case a small delay will occur whilst barges are changed.

7.8.2 THE PRODUCTION CYCLE

The production cycle is illustrated by Figure 7.35.

Cycle

- Swing to mark – small variation, dependent on swing angle.
- Lower grab – directly related to water depth as a function of winch speed.
- Close grab – approximately constant, but dependent to some extent on material.
- Raise grab – directly related to water depth as a function of hoist winch speed.
- Swing to discharge – small variation, dependent on swing angle.
- Discharge – small variation normally, but may be extended in certain'sticky' cohesive materials.

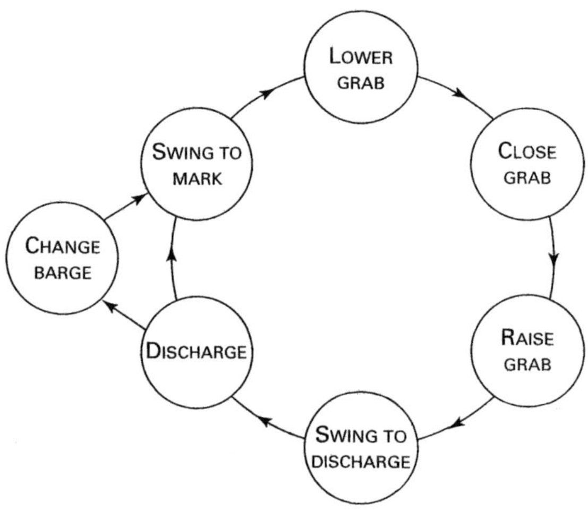

Figure 7.35 Production cycle for pontoon mounted grab dredger

The average cycle time of the grab crane may not vary very significantly for a given dredger working at a given depth, but the percentage bucket filling may be very variable. The time to fill the hopper is given by dividing the total hopper capacity by the product of the cycle rate, the grab capacity and the percentage filling of the grab (it may be greater than 100 per cent in soft clayey silts).

Daily production will be given by the product of (working hours divided by average cycle time) and (bucket capacity times the average percentage bucket filling). The total working time will be the total available hours that the dredger is crewed and available to work, less any lost time.

For a well-managed dredger under average site conditions, the lost production time may amount to between 20 and 35 per cent of the theoretically available working time. Special site conditions may, however, result in very much higher losses.

7.8.3 LIMITING FACTORS

The limits of operation of the grab pontoon dredger will vary according to its size and characteristics, but an indication of the extreme limits which will apply to economic operation is given below. These limits apply respectively to the smallest (minimum) and largest (maximum) grab pontoon dredger in common usage.

Minimum water depth to operate	1 metre
Maximum water depth to operate	50 metres (can be extended)
Wave height	2 metres
Maximum cross current	1.5 knots
Maximum shear strength (clays)	300 kPa
Maximum compressive strength (rocks)	1 MPa

7.8.4 ANCILLARY EQUIPMENT

Ancillary equipment is limited to the various types of grab bucket and hopper barges (see Section 7.14).

Grab buckets

The main types of grab bucket are illustrated in Figure 7.36.

Figure 7.36 Examples of various types of rope operated grab buckets

The bucket should be carefully selected according to the characteristics of the material to be dredged. When dredging in soft silts, muds and clays, a plain lightweight bucket, of the maximum size for which the crane is rated, can be used. For stiff clays or very weak rocks a heavy toothed bucket of reduced capacity should be employed. For most applications, a twin jaw configuration will be appropriate, but for special applications, such as the recovery of loose boulders or broken rock, a cactus, or orange peel grab may be more appropriate.

7.9 Grab hopper dredger

The grab hopper dredger is usually a ship with an integral hopper, which is loaded by one or more deck-mounted grabbing cranes (see Figure 7.37). Loading takes place with the dredger stationary at anchor.

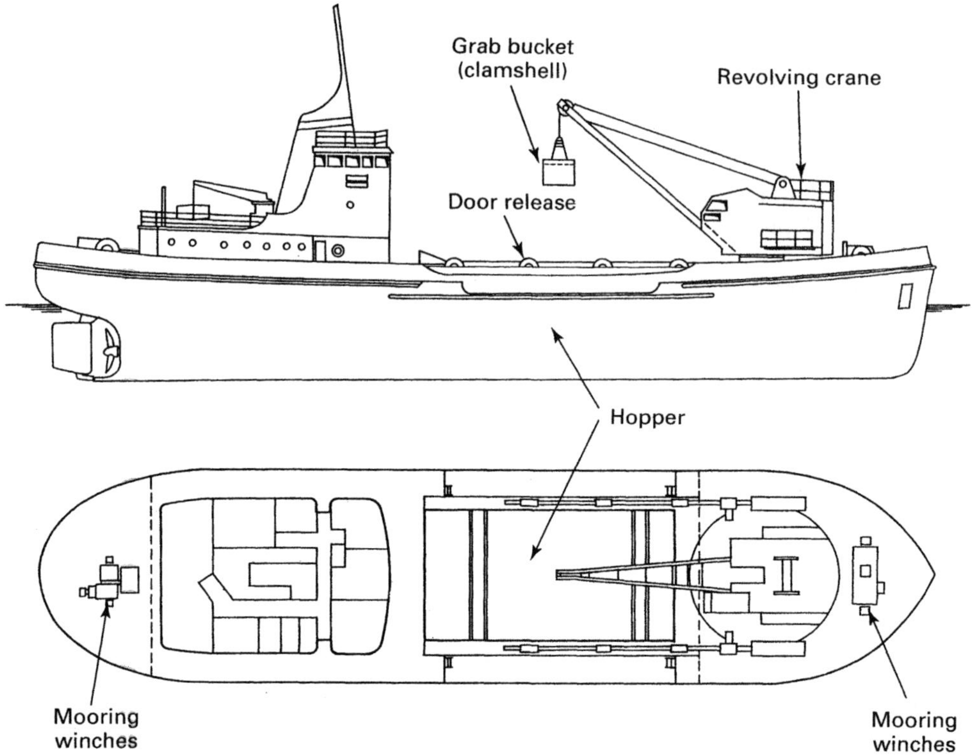

Figure 7.37 Main features of grab hopper dredger

The hopper capacity is rarely in excess of 1500 cubic metres. The hopper may be loaded by a single rope-operated grab crane, or by multiple cranes. The number of cranes rarely exceeds four. Occasionally, hydraulic grabs may be fitted in preference to rope operated grabs.

The method of discharge is usually through hinged bottom opening doors of design similar to those used by some trailing suction hopper dredgers.

The advantages of the grab hopper dredger are in many respects similar to those of the grab pontoon dredger (see Section 7.8). However, the integrated hopper and self-propulsion provides fully independent operation, thereby making the dredger well suited to small and dispersed dredging tasks. The self-propelled ship form of the dredger may permit continued operation in sea conditions which are more severe than those tolerated by the pontoon version and its attendant barges.

The disadvantages of the grab hopper dredger, additional to those described for the grab pontoon dredger (see Section 7.8), are its greater draught, reduced radius of operation and time used in the laying and recovery of anchors at the commencement and ending of each period of loading. This time may be considerable. Furthermore the pattern of anchors may interfere with other navigation.

The capital and operating costs of multiple grab hopper dredgers are high relative to a similar size of trailing suction hopper dredger.

The grab hopper dredger is normally rated according to its hopper volume, which may range from 50 to 2000 cubic metres.

7.9.1 METHOD OF OPERATION

Grab hopper dredgers dredge whilst moored on a pattern of anchors. The attitude of the dredger in relation to the working face is dependent on the number of grab cranes employed. Each crane should cover the maximum working area without encroaching upon the working area of other cranes (see Figure 7.38). Limited overlap is desirable to avoid leaving material behind, but if large overlap occurs and is unavoidable, then leading cranes may be required to dredge at a higher level than following cranes to equalise production.

When each crane has removed all material to the required depth within its radius of operation, the dredger is moved into an adjacent area. This procedure is continued until the hopper is fully loaded, whereupon the anchors are lifted and stowed, and the dredger sails to the disposal area.

7.9.2 THE PRODUCTION CYCLE

The production cycle is illustrated in Figure 7.39. The most significant variables in the production cycle are loading time and sailing time.

Moor

Mooring time depends on the situation and number of anchors deployed. It may range from a few minutes to half an hour in exceptional conditions.

Loading time

Loading time depends primarily on the characteristics of the material to be dredged and the effects that it may have on the percentage bucket filling achieved on each bucket closure and raising. If the soil has a high resistance to penetration of the bucket edge, or teeth, then the percentage filling will be reduced. If the material is very loose and very fine, some of the initial bucket contents may be lost as the bucket is raised through the water. Total loading time is dependent on the number of grab cranes.

Sub-cycle

See grab pontoon dredger (Section 7.8.2).

The time to fill the hopper is given by dividing the total hopper capacity by the product of the cycle rate, the grab capacity, the percentage filling (may be greater than 100 per cent in soft clayey silts) and the number of grab cranes.

Sailing time to disposal ground

Given by dividing the sailing distance to the disposal discharge point by average speed loaded.

Time to discharge

This is dependent on the characteristics of the dredged material. Normally for silts, or sands, it is only a few minutes, but for sticky clays the period may be extended significantly. Complete discharge may be impossible.

Sailing time to work site

This is given by dividing the total sailing distance to the working area by the average speed light.

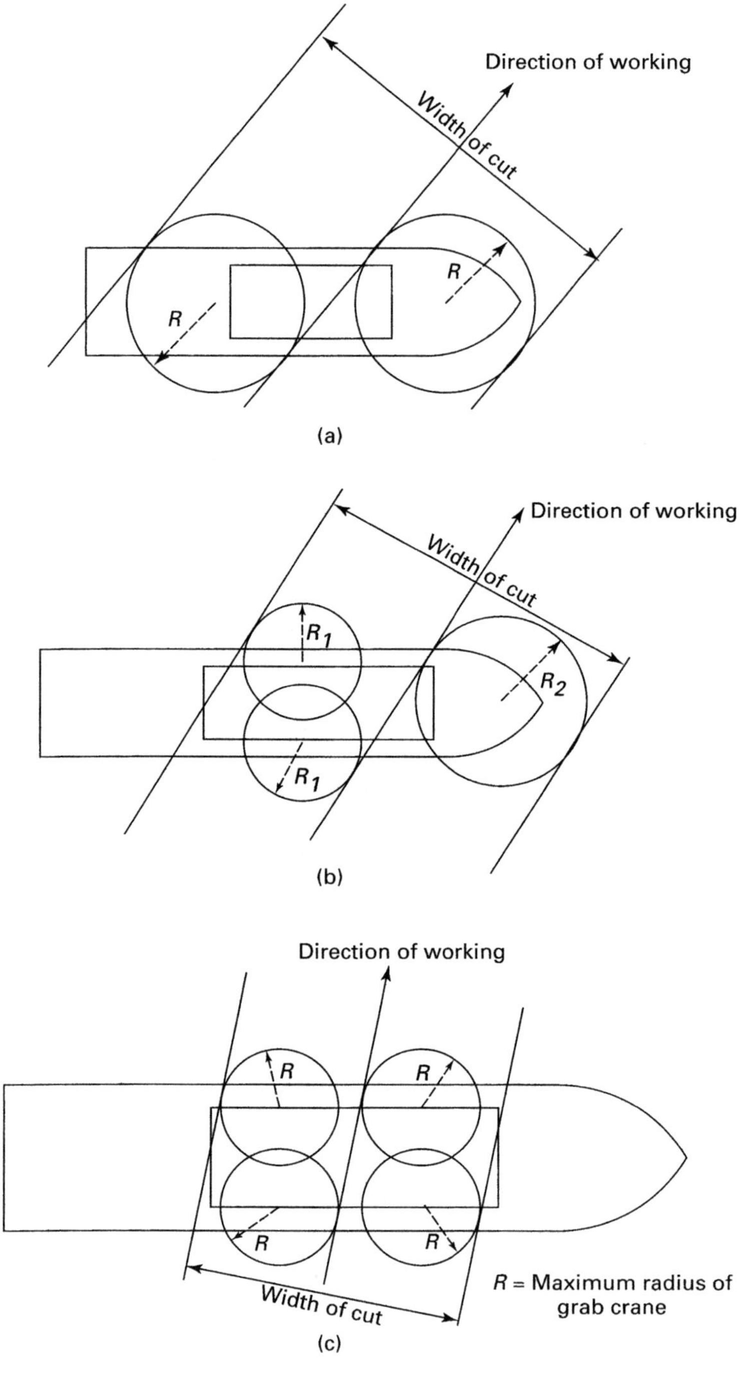

Figure 7.38 Working pattern for grab hopper dredger fitted with multiple grab cranes

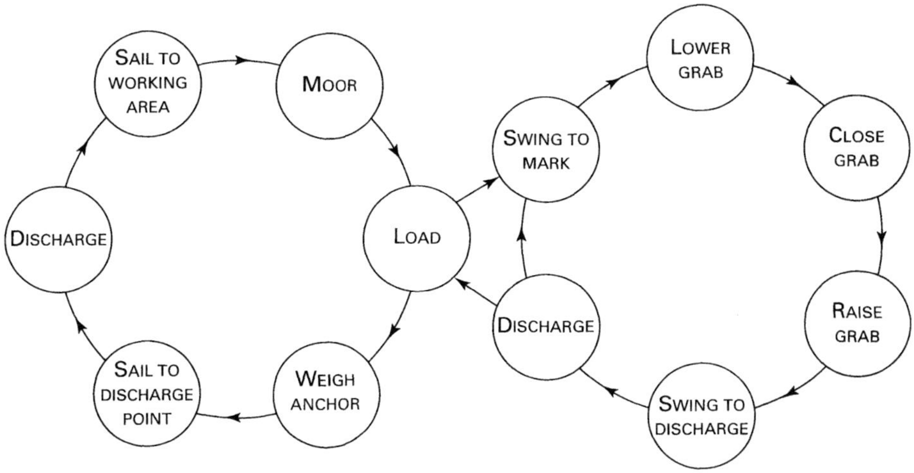

Figure 7.39 Production cycle for grab hopper dredger

The daily production is the sum of the hopper contents for each complete cycle achieved. The number of cycles will be given by the total working time divided by the average cycle time. The total working time will be the total hours that the dredger is crewed and available for work, less any lost time.

For a well managed dredger under average site conditions, the lost production time may amount to between 15 and 30 per cent of the theoretically available working time. Special site conditions may, however, result in very much higher losses.

7.9.3 LIMITING FACTORS

The limits affecting where the grab hopper dredger can operate and what it can achieve will vary according to the size and characteristics of the particular dredger, but an indication of the extreme limits which will apply to economic operation is given below. These limits apply respectively to the smallest (minimum) and largest (maximum) grab hopper dredgers in common usage.

Minimum water depth to operate	3 metres
Maximum water depth to operate	45 metres (can be extended)
Minimum turning circle	75 metres
Maximum wave height	2 metres
Maximum cross current	1.5 knots
Maximum ice thickness	200 millimetres
Maximum shear strength (clays)	100 kPa

7.9.4 ANCILLARY EQUIPMENT

Because of its independent operation the grab hopper dredger does not require a wide range of ancillary equipment. The main variations which may occur concern the type of grab bucket used (see Section 7.8.4).

7.10 Hydraulic backhoe dredger

The hydraulic backhoe dredger has evolved from the common land-based hydraulic backhoe excavator. Whereas the land-based machine is normally mounted on a tracked or wheeled undercarriage, the dedicated dredging machine usually is mounted on a fabricated pedestal at one extremity of a spud-rigged pontoon. However, tracked machines are occasionally employed with tracks secured to the pontoon deck. The main features of the hydraulic backhoe dredger are illustrated in Figure 7.40.

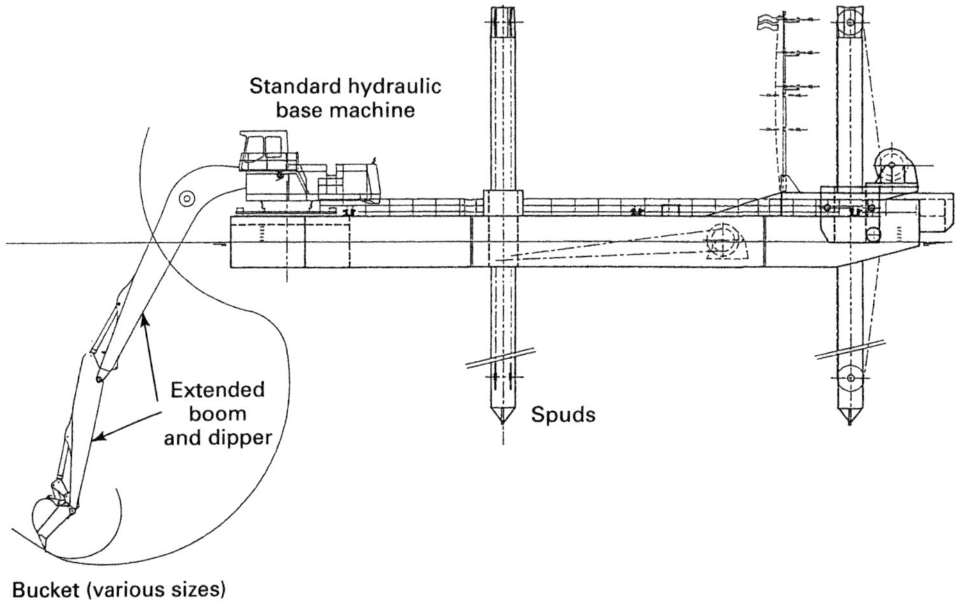

Figure 7.40 Main features of backhoe dredger

The backhoe dredger evolved with the introduction of larger backhoe excavators which provided the greater digging depth and power required for many dredging applications. A secure, shock absorbing pontoon mounting is important if the full digging potential of the machine is to be utilised.

Spud location of the pontoon is usually necessary to provide a positive reaction to the hydraulic digging action, particularly when dredging in difficult ground.

The backhoe dredger is normally rated according to the maximum size of digging bucket that the machine can handle. This may range from 1 to 20 cubic metres. A statistical distribution of sizes is shown in Figure 7.41.

The size of bucket employed will depend upon the nature of the material to be dredged and the maximum dredging depth. Bucket tear-out force decreases with increasing dredging depth, as does the total weight which can be handled. Thus, for dredging strong materials at maximum depth, a bucket smaller than the maximum should be used. Maximum dredging depth may vary from 4 to 24 metres.

The advantages of the backhoe dredger are:

- its ability to dredge a wide range of materials, including those which contain boulders,

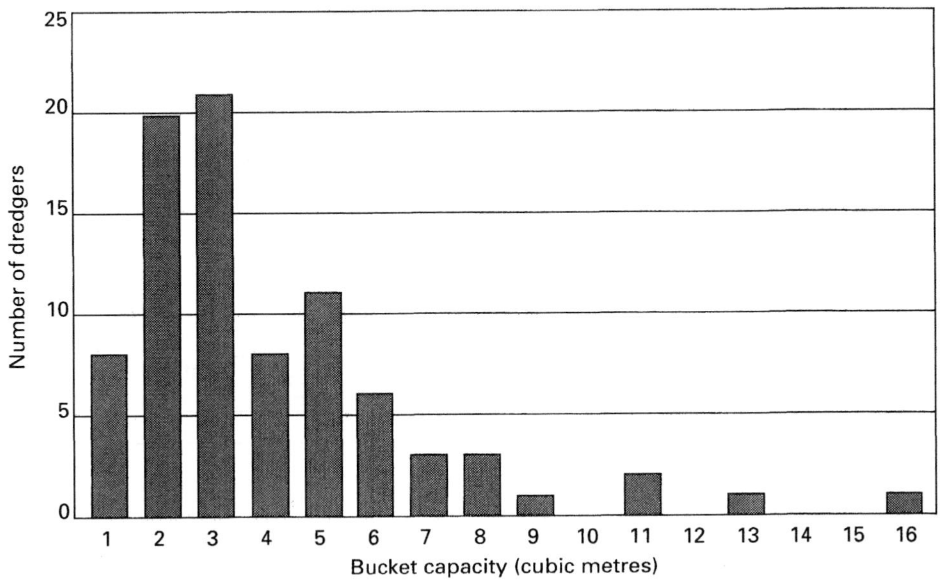

Figure 7.41 Statistical distribution of backhoe dredgers

or debris; difficult materials, such as stiff clays and weak rocks, can be dredged by the larger dredgers, providing that the maximum required dredging depth is not excessive;

- the ability to work in confined spaces;
- its accurate control of position and depth;
- the absence of anchors and associated wires;
- the minimum disturbance and dilution of the material being dredged;
- a faster cycle time than the equivalent size of grab dredger.

Because of the land-based origins of the backhoe dredger, the main machinery components are in series production, with a consequent reduction in capital cost and probably a more fully developed and tested product. The dredging operation requires only one man, although the total crew may number two, or more usually three, for safety and to assist in pontoon movement and maintenance.

Occasionally the pontoon may be self-propelled, but generally the self-propulsion is intended only for movement about the work site. A backhoe may be mounted on a self-propelled hopper in a similar manner to the grab hopper dredger. This provides independent operation and mobility.

The disadvantages of the backhoe dredger mainly relate to a low rate of production when compared with those dredgers in which the dredging process is relatively continuous (i.e. cutter suction and bucket dredgers). The finished dredged level may be irregular if operator control is poor or when dredging materials of variable strength or structure.

The backhoe is most efficient when working from behind the face, which means that the pontoon is located over the area to be dredged (see Figure 7.42). If water depth is at any time less than the maximum draught of the pontoon, this method may not be practical. Production is directly related to the depth of material to be removed. When this is small, as in some maintenance works, the resulting efficiency will be low.

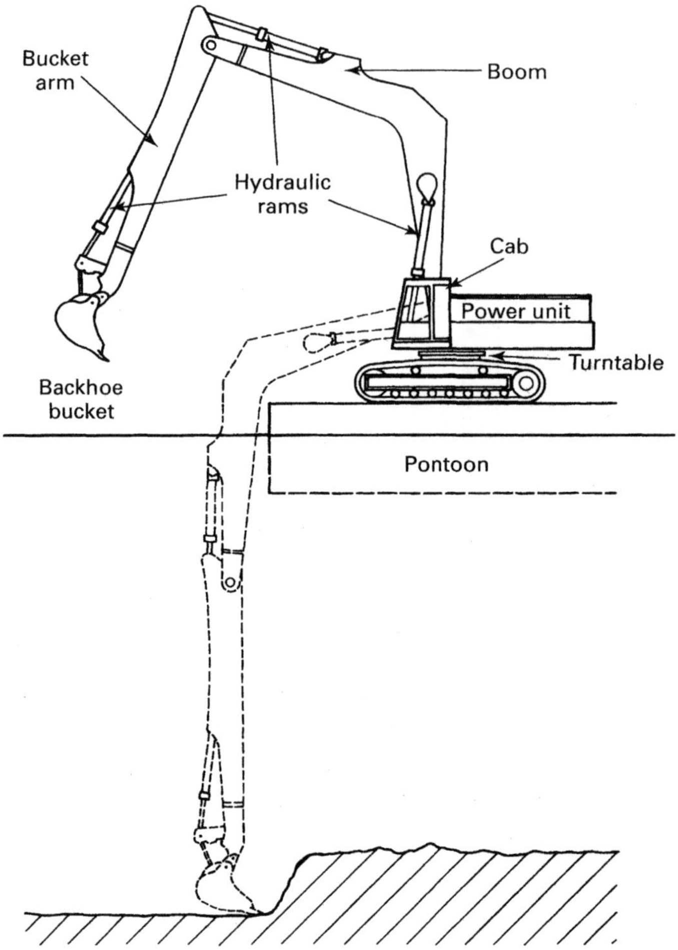

Figure 7.42 Example of dredging action of backhoe dredger

7.10.1 METHOD OF OPERATION

The dredging action of the backhoe can be likened to the action of the human hand and arm. The single digging bucket is located at the extremity of a pin-jointed arm. Between the bucket and the main structure of the machine there are three points of rotation, all in the same plane. The movement of each section, relative to the other, is controlled by high-pressure hydraulic cylinders.

The two components of the boom which carry the bucket are the dipper arm (or 'stick'), to which the bucket is attached and the monobloc, which is attached to the main machine structure. Boom and stick may be 'specials' to provide the additional downreach often required for dredging applications. In order to begin dredging, the bucket is lowered to present its open face, or in hard ground, its digging teeth, to the seabed. The bucket is then rotated, or 'crowded' and lowered, simultaneously, to achieve penetration and loading. The action is the same as that of the human hand when scooping up a handful of compact beach sand, or soil.

The backhoe is most commonly used to load a barge moored alongside the pontoon. Occasionally, such as in trench excavation, the dredged material may be sidecast, but outreach is limited and hence there is a strong possibility of the material re-entering the excavation.

The depth and configuration of the bucket and arm can be indicated in the operator's cabin by a simulating scale model. This device is an essential aid to accurate working. Without such aid the discontinuity of the dredging operation can easily result in an uneven dredged level. Alternatively, the position of the bucket and dredging arm may be represented on a VDU.

Upon completion of each cut across the face and before the pontoon is moved, a sweep across the face should be made with the bucket held at the target finish level to check for high spots and to level any minor peaks.

Movement of the pontoon is normally achieved with the digging arm. Prior to movement, the bucket is lowered into the seabed and the two forward spuds are raised clear of the seabed. The stern 'walking' spud, which may be tilting, or carriage mounted, remains in the seabed to maintain the pontoon on the centreline of the cut. The pontoon is then either pushed back from the face, or pulled into the face with the digging arm, aided by the spud carriage hydraulics, if fitted. The same method can be used to move the dredger over short distances about the work site.

7.10.2 THE PRODUCTION CYCLE

The production cycle (see Figure 7.43) for a given machine will be most affected by the material to be dredged and the dredging depth. There may be a significant variation in the minimum cycle time between machines of different manufacture due to variations in the capacity and configuration of the hydraulic systems. The rate of production is determined by the average percentage bucket filling, which will in turn be dependent on the characteristics of the material to be dredged.

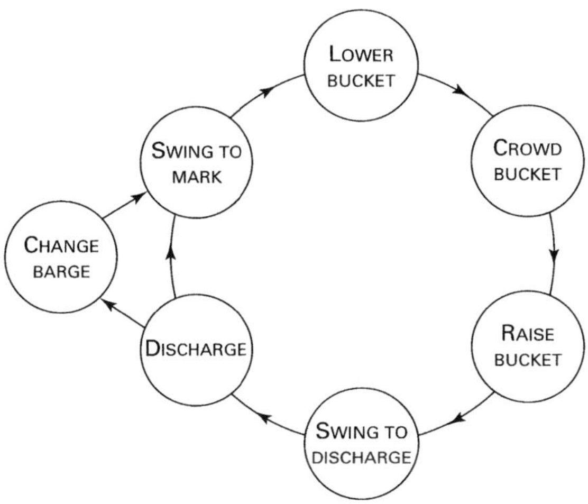

Figure 7.43 Production cycle of backhoe dredger

Loading

Swing to mark – small variation, dependent on the swing angle and slewing rate.
Lower bucket – directly related to water depth as a function of hydraulics speed.
Crowd bucket – dependent on the resistance to penetration of the material to be dredged.
Raise bucket – directly related to water depth as a function of speed of hydraulics.
Swing to discharge – small variation, dependent on swing angle and slewing rate.
Discharge – small variation normally, but may be extended in "sticky" cohesive materials.

Move pontoon

This is fairly constant, normally only a few minutes. The regularity of pontoon movement will depend on depth and width of cut face and dredging depth.

Change barges

This is dependent on sea conditions, current and working area. It is normally 5 to 15 minutes. Frequency of barge changing will depend on barge capacity.

Daily production will be given by the product of (working hours divided by average cycle time) and (bucket capacity times the average percentage bucket filling). The total working time will be the total hours that the dredger is crewed and available to work, less any lost time.

For a well-managed dredger under average site conditions, the lost production time may amount to between 20 and 40 per cent of the theoretically available working time. Special site conditions may, however, result in very much higher losses.

7.10.3 LIMITING FACTORS

The limits on where the hydraulic backhoe dredger can operate and what it can achieve will vary according to the size and characteristics of the particular dredger, but an indication of the extreme limits which will apply to economic operation is given below. These limits apply respectively to the smallest (minimum) and largest (maximum) backhoe dredger in common usage.

Minimum water depth to operate	2 metres
Maximum water depth to operate	24 metres
Maximum width of cut	25 metres
Minimum width of cut	Bucket width
Maximum wave height	1.5 metres
Maximum swell height	1.0 metres
Maximum cross current	2.0 knots
Maximum compressive strength (intact rocks)	10 MPa

7.10.4 ANCILLARY EQUIPMENT

Apart from the various types of hopper barge (see Section 7.14) which may be employed for dredged material disposal, the backhoe dredger requires little ancillary equipment. The main variations which occur are in the size of digging bucket employed and the size and geometry of the monobloc, or dipper arm. All of these should be selected to suit the particular operation required. Smaller capacity buckets should be used for rock excavation. Where the strength of rock is too great for direct dredging to be carried out economically, the bucket may be replaced with a single ripper tooth or with a percussion chisel.

7.10.5 PERCUSSION CHISEL ATTACHMENT

The 'blind' operation, which is a feature of all underwater excavation, or pre-treatment, detracts from the efficient deployment of percussion tools for rock destruction. However, where the area requiring treatment is small and the costs of mobilising additional large items of plant are uneconomic, the use of a hydraulic, or air-driven chisel may be attractive (see Chapter 14). Hydraulic hammers are undergoing considerable development which may lead to them becoming more applicable to dredging purposes.

7.11 Dipper dredger

The traditional dipper dredger is a heavily constructed, rope-operated machine. This has been largely replaced by the hydraulic face shovel, which is a variation on the hydraulic backhoe dredger (see Section 7.10).

Both the rope-operated and hydraulic dipper operate by digging forwards and upwards into the face. Both are mounted on spud rigged pontoons to provide the necessary reaction to the digging force. Both slew to discharge the bucket into a barge alongside. The basic features of the rope-operated type are shown in Figure 7.44. Hydraulic types are essentially similar to the hydraulic backhoe (see Figure 7.40), but with forward facing bucket and different arm geometry.

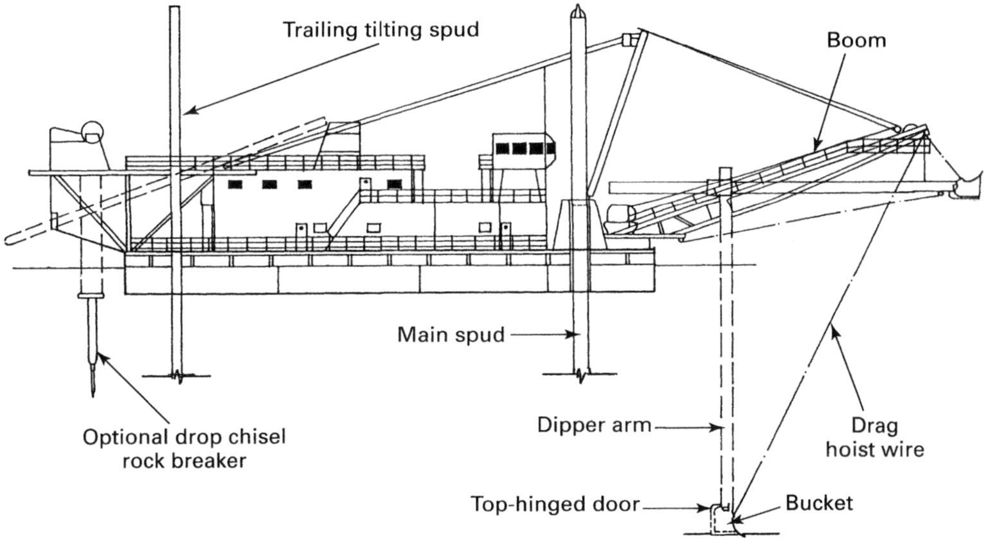

Figure 7.44 Main features of rope-operated dipper dredger

The main advantage of the dipper dredger is its ability to handle a wide range of difficult materials, including weak rocks, stiff clays and boulder clays. The heavy construction and spudded pontoon of the rope-operated type will permit operation in sea areas which are subject to moderate wave action, or swell. Because the dredger works by advancing into the face, it creates sufficient water depth in which to operate, regardless of pre-dredged level.

Disadvantages include a low rate of production relative to dredgers with a continuous dredging action, limited dredging depth, especially for the hydraulic machines and, for the rope-operated machines, high capital costs in relation to productive capacity.

The rope-operated machines can, size for size, dredge to greater depth than hydraulic machines. The construction is rugged and unsophisticated. Cycle times are, however, slower than for hydraulic machines. Designs and construction tend to be 'one-off', whereas the hydraulic machines are all derived from standard production models, only the pontoon being purpose built. The rope-operated machine can only bring into play the single leverage afforded by the dipper arm, but the hydraulic machine can also 'crowd' the bucket, which may help to prise loose boulders, or exploit discontinuities in rock.

7.11.1 METHOD OF OPERATION

The bucket is lowered to the required dredging depth. The bucket is then forced forwards and upwards breaking into and up through the material. Very considerable force can be applied which is resisted by heavy spuds which locate the main pontoon. The bucket is raised through the water and simultaneously swung to port or starboard in the direction of the barge to be loaded. When over the barge, the bucket is discharged, either by tipping in the case of a hydraulic machine, or by the release of a door in the back of the bucket on rope-operated machines.

The pontoon is located by heavy spuds, two towards the front end of the pontoon and one positioned centrally at the stern. Forward movement is normally achieved by tilting the stern spud, but alternatively, this may be carriage mounted. The dipper arm is used to guide the dredger during forward movement.

7.11.2 THE PRODUCTION CYCLE

The rate of production is most affected by the material to be dredged and by dredging depth. The production cycle is illustrated in Figure 7.45.

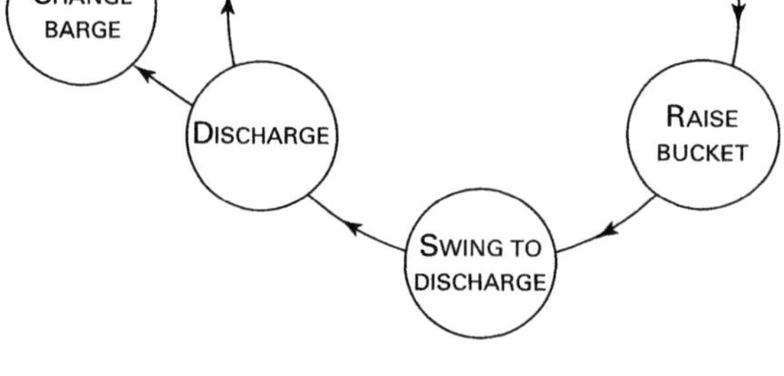

Figure 7.45 Production cycle of dipper dredger

Loading

Swing to mark – small variation, dependent on swing angle and slew speed. Average time consistent for a given machine.

Lower bucket – direct function of water depth.
Break out material – dependent on material type.
Raise bucket and swing to barge – direct function of water depth and dump height.
Discharge bucket – small variation, cohesive materials of high plasticity or large rocks may cause problems.

Move pontoon ahead

Fairly consistent, normally only a few minutes. The frequency of pontoon movement depends on the depth of face and width of cut.

Change barges

Dependent on sea conditions, current and working area. It is normally 5 to 15 minutes. The frequency of barge change is dependent on the barge capacity.

The daily production is given by the product of (working hours divided by average cycle time) and (bucket capacity times the average percentage bucket filling). The total working time will be the total hours that the dredger is crewed and available to work, less any lost time.

For a well-managed dredger under average site conditions, the lost production time may amount to between 20 and 40 per cent of the theoretically available working time. Special site conditions may, however, result in very much higher losses.

7.11.3 LIMITING FACTORS

The limits affecting where the dipper dredger can operate and what it can achieve will vary according to the size and characteristics of the particular dredger. An indication of the extreme limits which will apply to economic operation is given below. These limits apply respectively to the smallest (minimum) and largest (maximum) dredger in common usage.

Minimum water depth to operate	3.5 metres
Maximum water depth to operate	20 metres
Maximum width of cut	30 metres
Minimum width of cut	Bucket width
Maximum wave height	1.5 metres
Maximum swell height	1.0 metres
Maximum cross current	2.5 knots
Maximum compressive strength (intact rock)	12 MPa

7.11.4 ANCILLARY EQUIPMENT

Other than the various types of hopper barge (see Section 7.14) which may be employed for dredged material disposal, there is little ancillary equipment involved with the dipper dredger. The main variations which occur are in the type and size of digging bucket employed and the size of dipper arm, which should be selected to suit the particular operation required. Smaller capacity buckets are used for rock excavation.

7.12 Bucket dredger

The basic design of the bucket dredger has remained unchanged for many years (see Figure 7.46). The dredging action is achieved by a continuous chain of buckets which

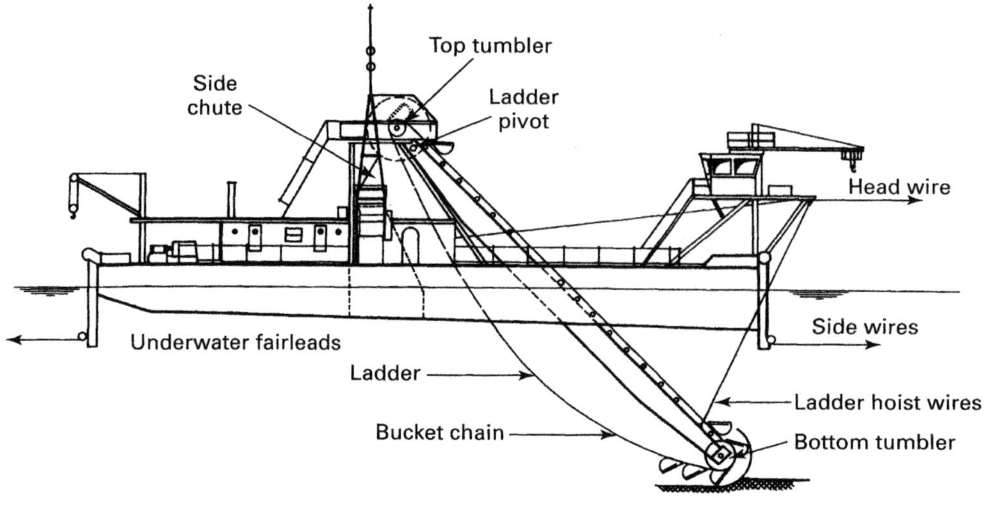

Figure 7.46 Main features of bucket dredger

scoop material from the seabed and raise it above water (see Figure 7.47 and Plate 7.12). The buckets are inverted as they pass over the top tumbler and discharge under gravity onto chutes which convey the dredged material to barges alongside.

The heavy bucket chain is supported by a fabricated steel ladder and driven electrically, or hydraulically, via the top tumbler. On older models the top tumbler may be belt driven. The ladder is mounted on the centreline of a long rectangular pontoon. Positioning and movement of the pontoon is by means of a pattern of five or six winches (see Figure 7.48). The head winch, which is usually the most powerful, provides the reaction to the digging force and is used to advance the dredger into the face. The four side winches are used to traverse the dredger across the face. The stern winch maintains an overall equilibrium in balancing tidal forces, or variations in the digging reaction. Subject to local conditions, the use of the stern winch may be eliminated by laying the side anchors aft of their usual positions.

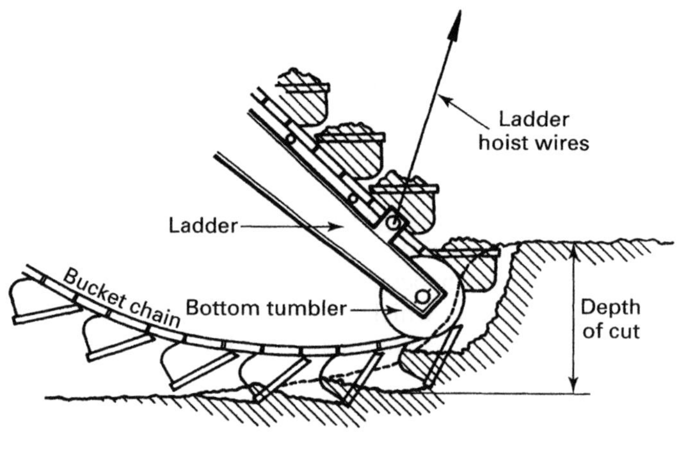

Figure 7.47 The bucket chain lower end

Plate 7.12 Heaped buckets move up the ladder of a bucket dredger

The bucket dredger is usually used only to load barges, but occasionally, in canal works, may discharge to a short conveyor system, or very rarely, to a mixing chamber for pump-discharge. The breakout force applied via the bucket cutting edge, or teeth, may be quite substantial. Breakout force is dependent on the size and mass of the dredger, the bucket shape, construction and capacity, and the power and characteristics of the drive system to the top tumbler. When dredging in strong materials, such as stiff clays or weak rocks, the dredger is fitted with special buckets of reduced capacity and heavier construction and the chain speed (normally expressed in buckets per minute) is reduced.

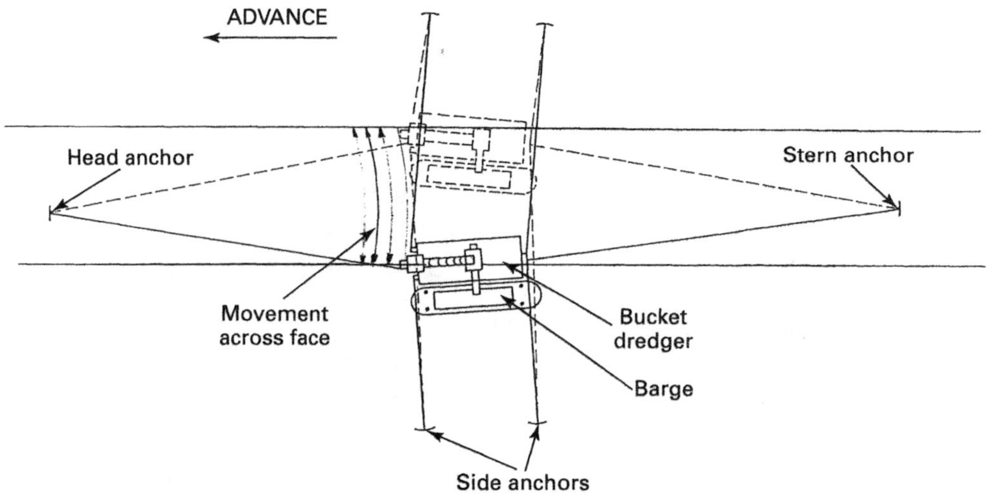

Figure 7.48 Working method of bucket dredger

When dredging rock, a ripper tine may be substituted for every second or third bucket. Bucket capacity may range from 150 to 1200 litres, and bucket speeds up to 30 buckets per minute. The distribution of bucket dredgers by size is shown in Figure 7.49.

The bucket dredger has the advantage of a continuous dredging process, which is unique amongst dredgers which dredge by bucket action. As with all bucket systems, dilution of the dredged material by water is not very significant and, therefore, high load factors can be achieved in the barges without excessive overspilling. However, spillage from the buckets may be significant, especially when dredging in weak silts and particularly when dredging firm clays which underlie weak silts. Spillage may also occur due to splashing as

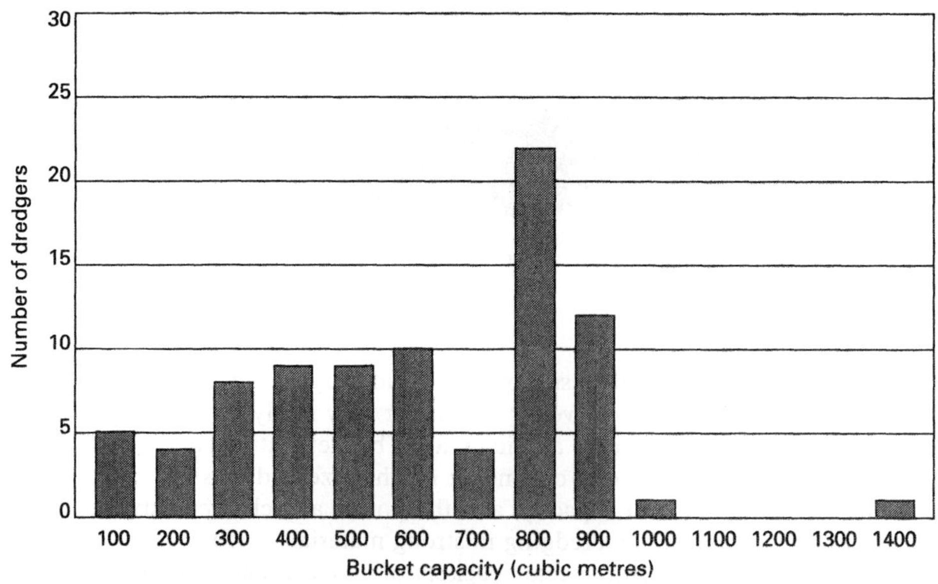

Figure 7.49 Statistical distribution of bucket dredgers

soft materials enter the barge from the discharge chute. This can be alleviated with careful operation, but with some loss of production.

The continuous dredging action allows a uniform dredged level to be achieved, with good control of dredging depth. In normal circumstances the amount of overdredging necessary to achieve the required level is not excessive. The dredging process is not particularly sensitive to debris, or small boulders, but lengths of cable or chain will be disruptive.

The heavy construction of the bucket dredger, coupled with its flexible all-winch mooring system, allows operation in moderate wave conditions when dredging soft materials.

The disadvantages of the bucket dredger include:

- a wide spread of anchors, which may disrupt navigation;
- dependence on barge loading, which reduces its suitability for land reclamation, or filling operations;
- low efficiency when required to remove only a small depth of material;
- noise levels which are higher than generated by other dredgers and may be a problem in sensitive areas;
- in sticky cohesive materials, the buckets may fail to discharge properly.

The bucket dredger is not able to work in very shallow water. When working in relatively shallow conditions the catenary of the bucket chain suspended beneath the ladder may become pronounced, with the result that the digging depth will increase. This may result in unintentional overdredging. Some bucket dredgers are constructed so that the height of the top tumbler may be altered to optimise the ladder angle in different water depths. However, the adjustment is usually laborious and time consuming.

Construction of the bucket dredger is heavy and the labour requirements are high compared to other dredgers of similar productive capacity. In consequence, both capital costs and operating costs may be higher. The bucket dredger is normally towed by tug to each new work site. If the distance is long, or includes areas where rough seas may be encountered, the bucket chain must be dismantled to improve the stability. Mobilisation costs may be high.

The bucket dredger is normally rated in accordance with the maximum capacity of each individual bucket. An indication of the characteristics of dredgers of different capacities is shown in Figure 7.50.

7.12.1 METHOD OF OPERATION

A pattern of six anchors is laid out, as shown in Figure 7.48. The heavy head wire anchor is laid well ahead, preferably several hundred metres away, to maximise holding power and minimise the radius of the cut face and the frequency of movement of the anchor.

In order that the digging forces are transferred to the head wire and not taken by the side wires the dredger is angled during the swing, as shown in Figure 7.48. The angling of the dredger causes it to move outside of the plan area of the cut. This can cause problems when working in shallow water and when cutting side slopes, because for such work the dredger must be angled into the cut, which will cause some of the load to be taken by the side wires.

The angle between the side wires and cut centreline is important. When the angle of wires exceeds about 40° it is necessary to move the anchors forward to new positions. If this is not done the effective force and control of the winches is diminished. Usually side winch force is approximately three to four tonnes per 100 litres of bucket capacity, whilst the head wire winch is eight to ten tonnes per 100 litres.

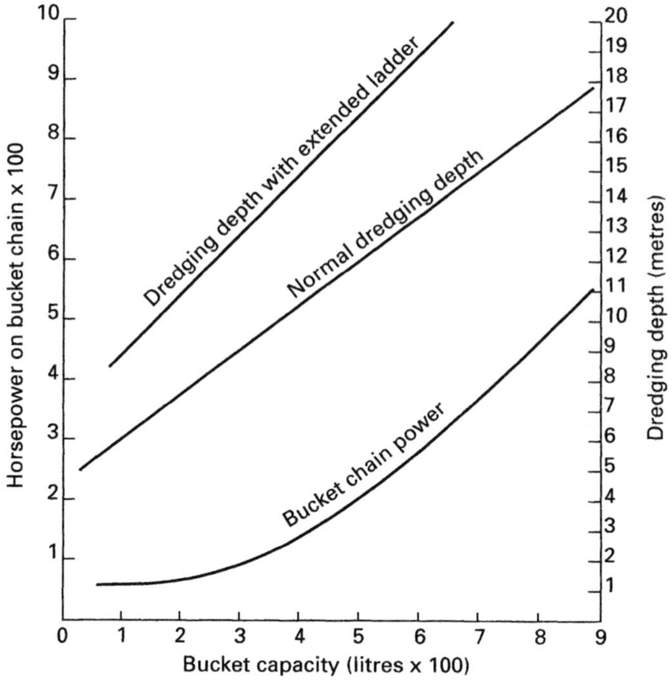

Figure 7.50 Characteristics of bucket dredgers

As is the case for all dredgers in which the cutting mechanism is supported by a ladder construction, the full dredging depth and face height must be established progressively. In soft materials, the bucket dredger may, depending upon its size, be able to remove several metres of material in a single pass, but the resulting spillage may be substantial. The maximum face height will be dependent on the stability of the face material. In more difficult materials, the depth of cut must be compatible with the power available on the chain drive.

When a face has been established, dredging is a continuous process of swinging the dredger across the face with the side winches. The speed of traverse across the face and the speed of the bucket chain are adjusted to suit the particular site conditions. Dredging efficiency is not affected by the direction of traverse. In difficult materials, speeds must be reduced.

When a cut across the face has been completed the dredger is advanced into the face by hauling in on the head winch. The increment of advance will depend on the size of the dredging buckets and the nature of the material. Normally the advance will be in the range of 0.3 to 2.0 metres. This process is repeated continuously until it is necessary to move anchors.

Barge loading is continuous. When a barge is fully loaded, a tilting hopper diverts the discharging material to an alternative discharge chute on the opposite side of the dredger and filling of another barge commences. Barges are secured alongside the dredger, and moved longitudinally to achieve an even distribution of dredged material, by means of warping winches on the dredger deck.

7.12.2 THE PRODUCTION CYCLE

The relative importance of the component parts of the production cycle will vary a little, according to the site conditions and requirements. The main effects, caused by difficult materials, will be to slow the whole process down and to reduce bucket filling. The main components of the cycle are as follows.

Traverse face whilst cutting

The speed of traverse is governed by the characteristics of the material to be dredged, the depth of cut, the amount moved ahead and the bucket chain speed. The time to traverse the face is given approximately by dividing the length of the face (in metres) by winch speed (metres per minute).

Number of cuts

The number of cuts across the face to produce the finished depth is dependent on the thickness of material to be removed and the depth of each cut. The depth of each cut is governed by the characteristics of material to be dredged. Sometimes, one final cleaning cut is required to remove spillage. The number of cuts is given by dividing the depth to be removed (inclusive of any necessary overdredge), by the depth of cut, and adding one, if a cleaning cut is necessary.

Advance into face for new cuts

This is dependent on the material to be dredged and the face height. The time occupied may be a few minutes if a ladder must be raised, but normally, is less than one minute. To establish a completely new face will require much more time.

Change barges

Normally there is no loss of production, but if restricted working allows access to only one side of the dredger, then between 5 and 15 minutes may be required.

Move anchors

This is very dependent on the experience of the crew. A good crew, under favourable conditions, may be able to limit disruption to those occasions when the head anchor must be moved. For this, between 15 and 30 minutes may be required. An additional 10 to 20 minutes may be required for each additional anchor move. When the thickness of material to be removed is small the time expended in anchor movement will be large and in extreme cases may exceed the productive time.

Daily production will be given by the product of working hours, average bucket speed in buckets per hour and average bucket filling expressed as a percentage. The effective working hours will be the total available hours that the dredger is crewed and available to work, less any lost time.

For a well-managed dredger, under average site conditions, the lost production time may be in the range of 20 to 40 per cent of the theoretically available working time. Special site conditions may, however, result in far higher losses.

7.12.3 LIMITING FACTORS

The limits on where the bucket dredger can operate and what it can achieve will vary according to the size and characteristics of the particular dredger, but an indication of the extreme limits which will apply to economic operation is given below. These limits apply respectively to the smallest (minimum) and largest (maximum) bucket dredger in common usage.

Minimum water depth to operate	5.0 metres
Maximum water depth to dredge	35 metres
Maximum cut width (single pass)	150 metres
Maximum wave height	1.5 metres
Maximum swell	1.0 metre
Maximum cross current	2.0 knots
Maximum ice thickness	100 millimetres
Maximum particle size	1500 millimetres
Maximum compressive strength (intact rock)	10 MPa

7.12.4 ANCILLARY EQUIPMENT

The bucket dredger may be used with a variety of different buckets for different materials, but the average dredger is unlikely to be equipped with more than two different bucket sizes, or types.

A special pontoon is often used to support the head wire above water level and prevent it from fouling the seabed.

Different anchors may be used in different ground conditions (see Section 7.6.4).

The number, capacity and type of hopper barge used for dredged material disposal will be selected to suit the particular site requirements (see Section 7.14).

7.13 Special dredgers

7.13.1 JET PUMP DREDGER

A jet pump may be incorporated into any dredger which works on the suction principle, but is most commonly used by plain suction types, such as the stationary suction dredger (see Section 7.5). The principal features of the jet pump are illustrated in Figure 7.51. Essentially it consists of a high pressure water system which injects a stream of water through a venturi section near the bottom of the suction pipe. The additional hydraulic head supplied to the system by this jet assists in lifting the solid/water mixture to the required height.

When employed in primary dredging applications the jet pump is positioned in the suction pipe, normally at the lower extremity. The advantages arising include:

- an ability to dredge to greater depth and to continue dredging with a buried suction intake without undue risk of cavitation (assuming a system designed to operate buried);
- a much reduced risk of pipeline blockage when discharging through long pipelines;
- the ability to work at any depth;
- reduced costs due to wear when handling abrasive materials.

The disadvantages of the jet pump include the following:

- a significant reduction in the overall system efficiency when compared to a solids handling centrifugal pump;
- a rather limited residual head available for the discharge of the dredged materials through pipelines.

The maximum discharge distance is therefore significantly less than with a centrifugal solids handling pump. For longer discharge distances, or high heads, it is necessary to

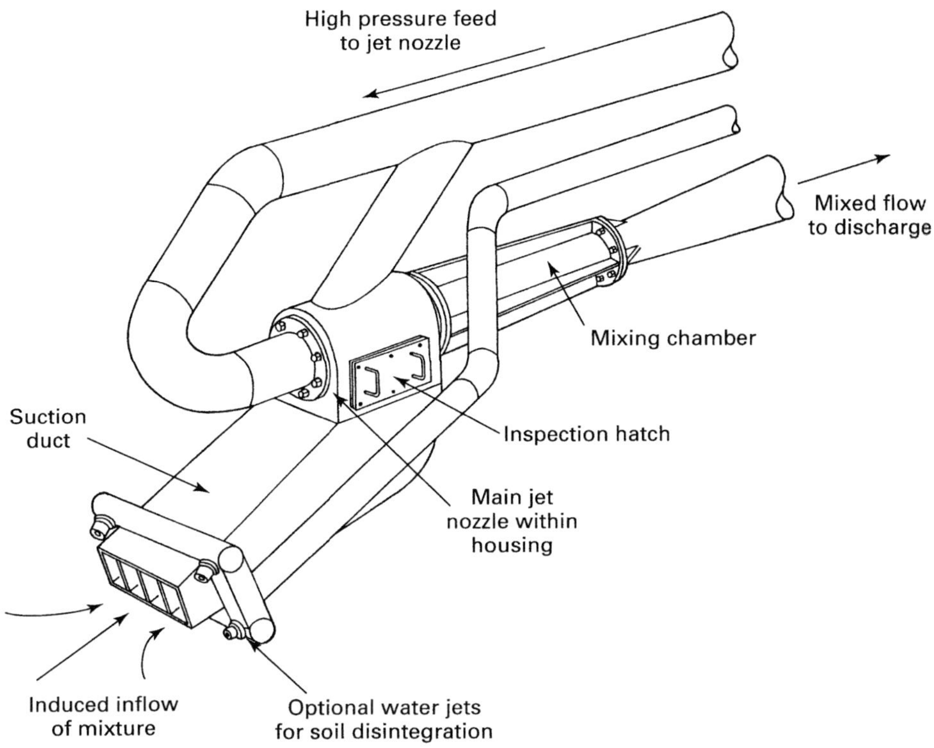

Figure 7.51 Main features of a proprietary jet pump (Genflo Ltd)

employ a conventional solids handling centrifugal pump in the discharge line to provide additional head.

The jet pump may be well suited to mining operations, such as aggregate extraction, where an abrasive, free-flowing material is to be dredged and pumped a relatively short distance to a stockpile or processing plant. It is also used in sand bypassing systems on coastlines where the littoral drift is interrupted by a river or inlet. In the latter case the unit is usually mounted on a jetty.

Jet pumps are also incorporated in small diver-operated devices for removing silt and sand from around wrecks and underwater structures.

7.13.2 AIR-LIFTS

The air-lift achieves flow in a pipe by the injection of compressed air at the submerged extremity of the pipe. The entrained air results in a reduction in the density of the air–water mixture within the pipe relative to the external water and an upward flow is induced. Fine loose soils may be eroded and carried in suspension by the flowing water. The system works best when removing fine materials in deep water. The basic principles are illustrated in Figure 7.52.

The air-lift system has the advantage of simplicity, with a complete absence of submerged moving parts. A complete small dredging assembly may be very light in weight,

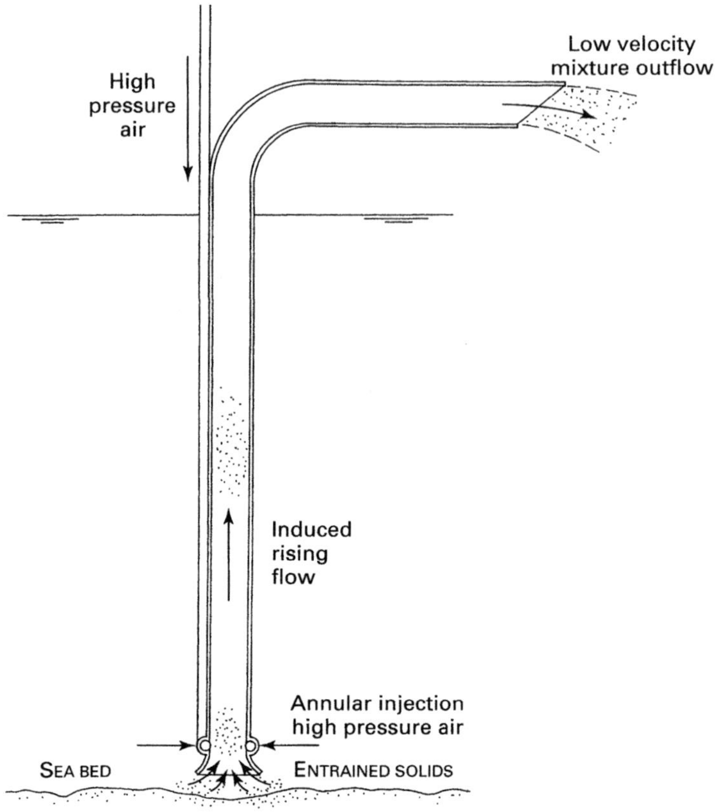

Figure 7.52 Principle of operation of the air lift

which makes it particularly suited to use by divers. A most important attribute of the air-lift is the virtual absence of any depth limitation.

Disadvantages include the following:

- very low head characteristics;
- a relatively low limit on the density of the entrained mixture;
- a restriction to loose and relatively fine materials, unless aided by a mechanical cutting head;
- because it is a low energy system it is prone to blockage which, in the case of hand held systems, results in handling difficulties;
- it has a low mechanical efficiency but this disadvantage may be outweighed by an absence of alternative dredging methods.

The air-lift most commonly serves as a simple underwater cleaner for the clearance of sediment from around wrecks, or structures and for the cleaning of submerged foundations.

Apart from the simple diver-operated types of air-lift there are few examples in existence which conform to any particular standard, or which share many common features. In consequence, maximum and minimum capabilities cannot conveniently be expressed.

7.13.3 POSITIVE DISPLACEMENT PUMPS

Whereas dredgers which pump-discharge conventionally use a centrifugal dredge pump, occasionally this may be replaced by a positive displacement piston pump, wherein the principles of operation are similar to those of a concrete pump. Material is moved mechanically into a hopper which feeds the piston pump. The exhaust stroke of the pump forces material via a non-return mechanism into a discharge pipeline. A proprietary version of a dredger incorporating a positive displacement pump is shown in Figure 7.53.

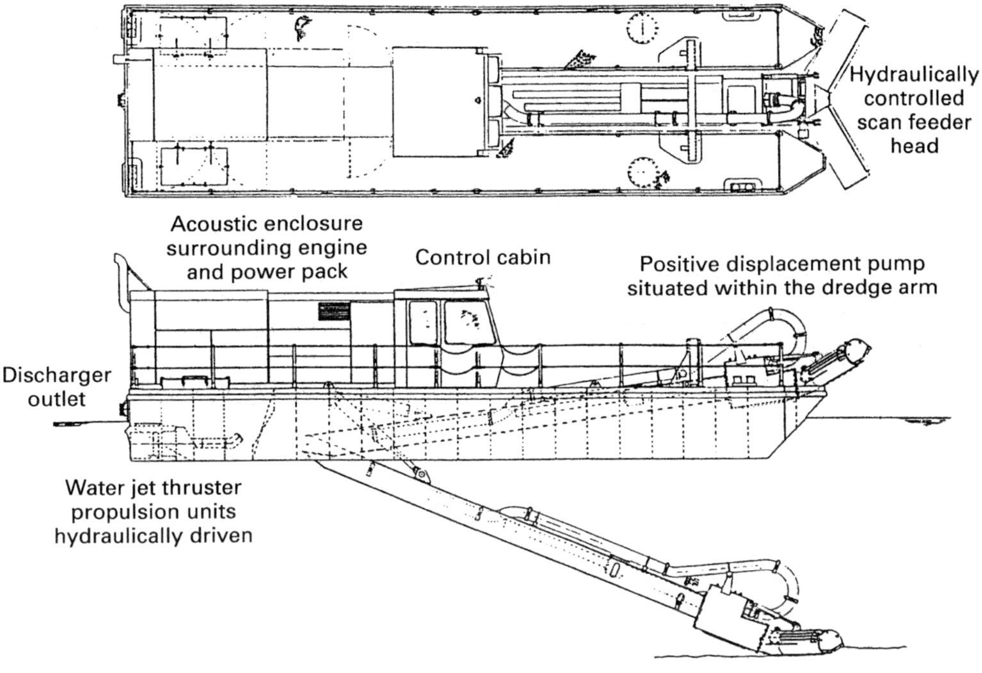

Figure 7.53 Proprietary version of dredger with a positive displacement pump (Conbar International)

The advantage of such systems is that the bed material may be moved without the addition of water. Disadvantages include the necessity to exclude all substantial solid material and a restricted discharge distance, which typically is limited to a few hundred metres, although this is dependent on the nature of the dredged material and the particular design of system. Some additional discharge distance may be gained by injecting compressed air into the discharge pipe to produce a 'plug flow' effect.

7.13.4 PNEUMATIC DREDGERS

Pneumatic dredgers are principally restricted to the dredging of very soft cohesive soils. They usually comprise one or more chambers which are evacuated using compressed air delivered from the surface. The valve in the bottom of the chamber is opened and the mud flows into the chamber under hydrostatic pressure. When full, compressed air is forced into the chamber, closing a non-return valve and ejecting the mud through a discharge pipe to the surface. When empty, the compressed air supply is turned off and mud again enters the

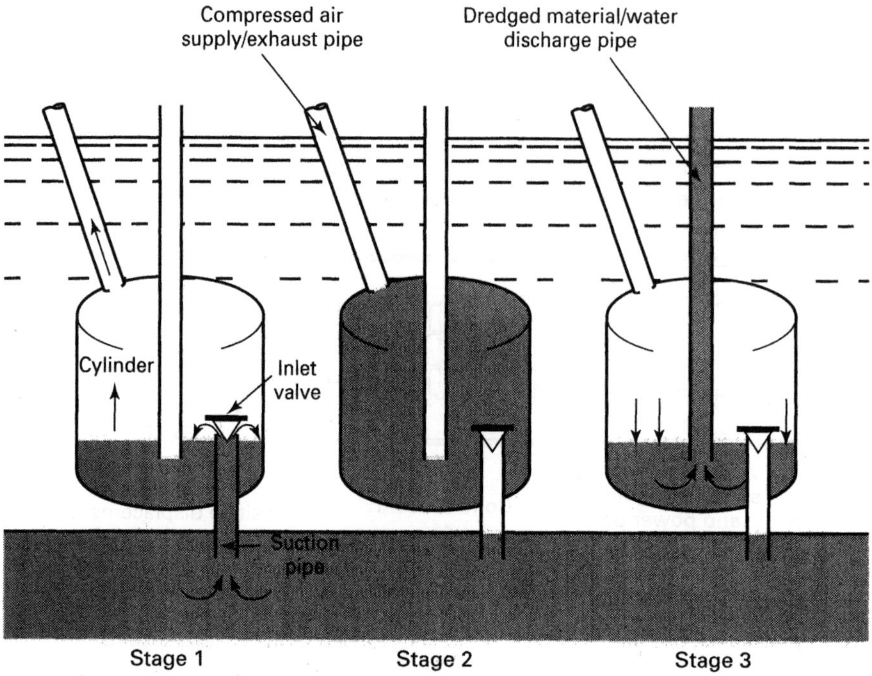

Figure 7.54 Main features of a proprietary pneumatic dredger (Pneuma)

chamber under hydrostatic pressure (see Figure 7.54). This cycle is repeated as the apparatus is moved through the mud to be dredged. The assembly may be fitted with shovels which guide the mud towards the suction inlet.

A number of proprietary systems based on this operating principle have been developed in Italy and Japan. They are most commonly used for the dredging of contaminated materials and at great depths, where conventional plant may be limited.

7.13.5 AMPHIBIOUS DREDGERS

Amphibious dredgers, which are usually small, are designed to work in shallow water to move out of the water and to travel on land. Usually the amphibious capability is confined to small grab, backhoe and cutter suction dredgers.

Most amphibious dredgers rely upon a track-laying mechanism for leaving the water to travel on land. However, a few employ hydraulically motivated legs which terminate in large pressure pads, or buoyancy pods.

There is no established standard method of providing dredgers with an amphibious capability and the function is not therefore dealt with in detail. A number of different arrangements are illustrated in Figure 7.55.

The most useful benefit of an amphibious dredger is the ability to bypass structures which restrict the waterway when employed in the maintenance of canals and channels. Occasionally, the transport mechanism may also permit continuous operation where water is too shallow to permit floating operation.

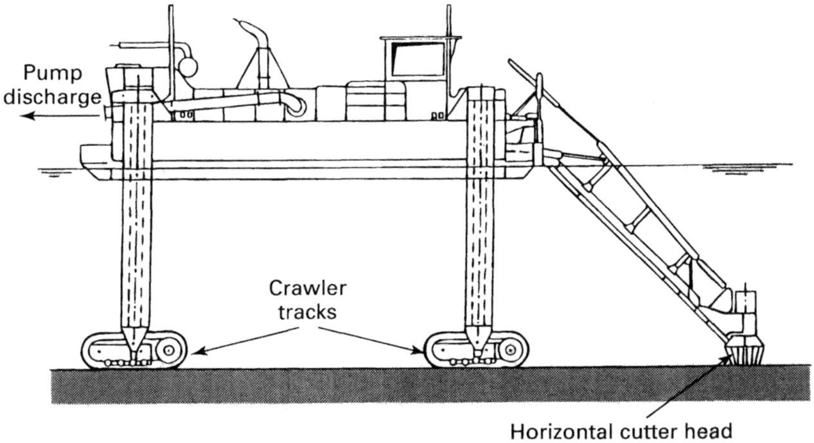

(a) The crawl-cat – with tracks (IHC Holland)

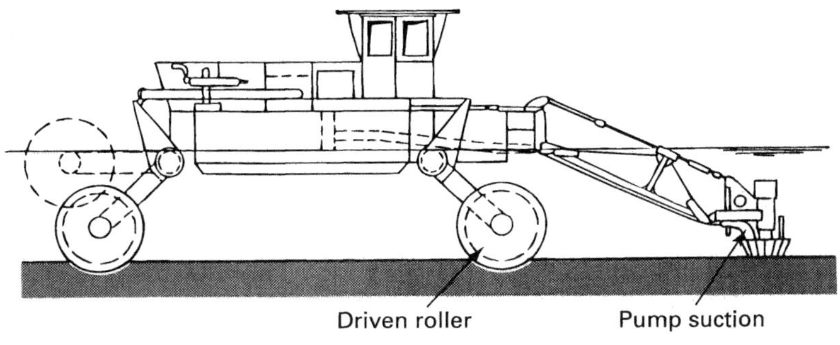

(b) The roll-cat – with rollers (IHC Holland)

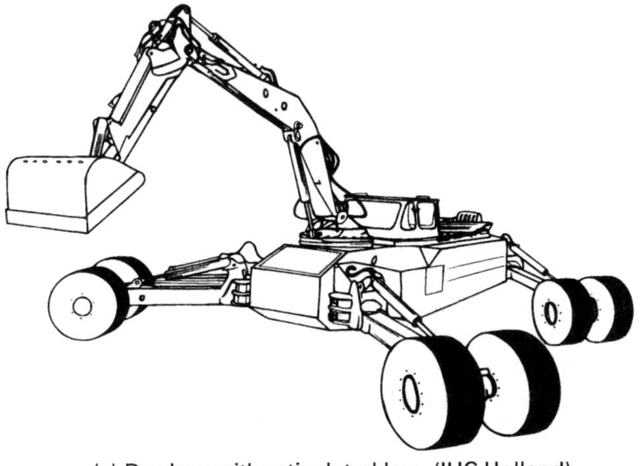

(c) Dredger with articulated legs (IHC Holland)

Figure 7.55 Examples of devices to allow amphibious operation for dredgers

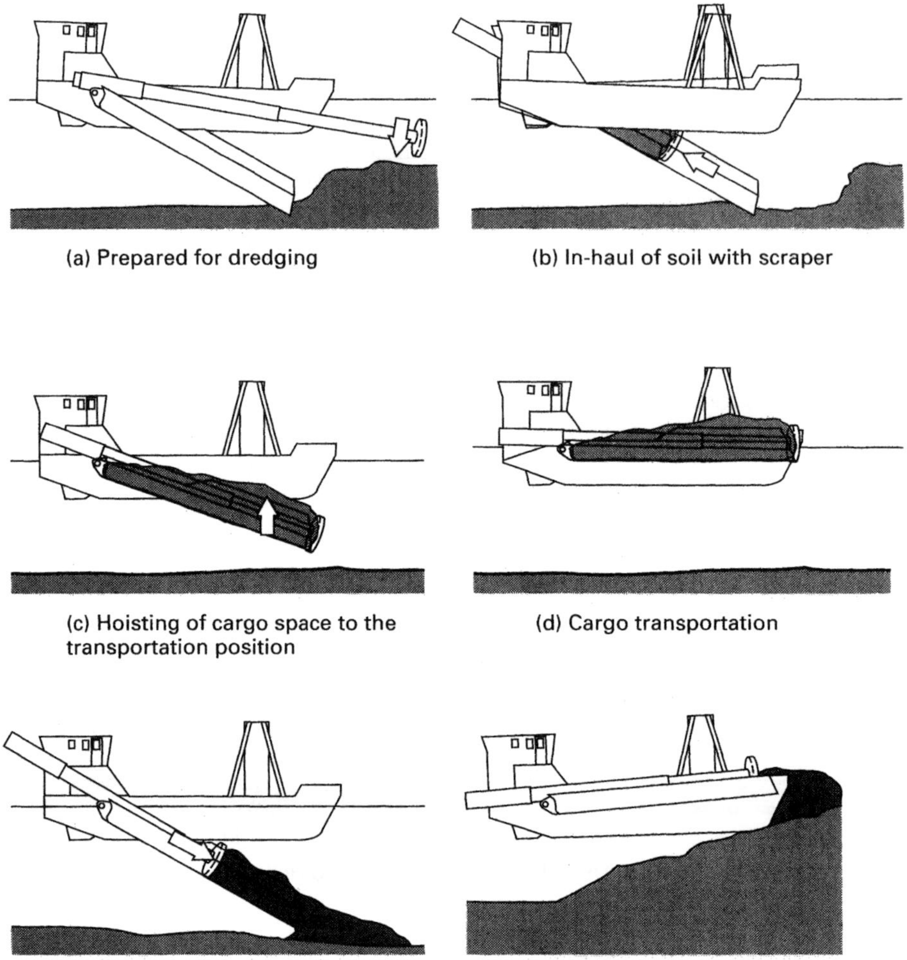

(a) Prepared for dredging

(b) In-haul of soil with scraper

(c) Hoisting of cargo space to the transportation position

(d) Cargo transportation

(e) Dumping of the soil in a discharge area

(f) Dumping of the soil by hauling it up on shore

Figure 7.56 Scraper dredger method of operation

7.13.6 THE SCRAPER DREDGER

The scraper dredger is a self-propelled independent hopper dredger which self-loads by a powered scraper action rather than the usual pumping method (see Figure 7.56).

The hopper floor is hinged at the aft end. The front, or bow end, may be lowered into contact with the sea bed by winches, or hydraulic cylinders. Seabed material is then scraped into the shallow tray formed by the hopper floor by a mechanical scraper. In early designs the scraper is activated by hydraulic cylinders, but winch operation is also feasible. The scraper is repeatedly operated until the hopper tray is full. The production cycle is shown in Figure 7.57.

A variation of this principle uses a fully enclosed scraper belt to feed soil into the hopper from whence it is pumped to the shore. This variant was developed to remove accurately thin layers of contaminated sediments with minimal loss to the water column.

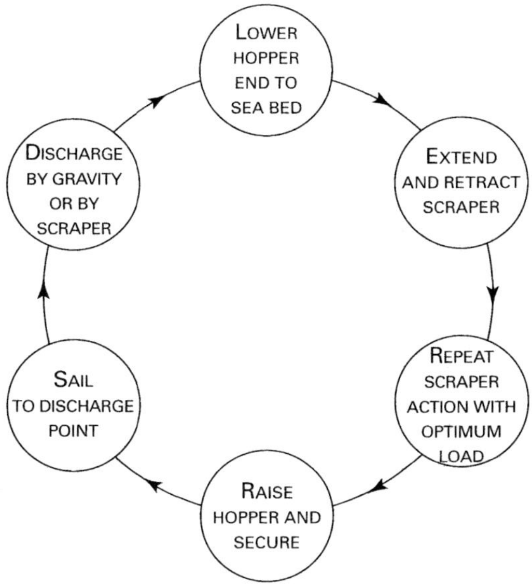

Figure 7.57 Production cycle of scraper dredger

The advantages of the system are those of independent operation, rapid loading, no dilution of the dredged material and a relative insensitivity to foreign matter, or small boulders. The disadvantages are shallow dredging depth, a high sensitivity to sea conditions and to the nature of the seabed materials. Strong materials or semi-fluids may be difficult, or impossible, to load.

7.14 Ancillary plant

7.14.1 BOOSTER PUMP STATIONS

Booster pump stations are used in long discharge pressure pipelines when additional pump head is required, to overcome either friction or static head.

Boosters are most commonly used in conjunction with cutter suction dredgers (see Section 7.6), suction dredgers (see Section 7.5) and pump discharging trailing suction hopper dredgers (see Section 7.2). Occasionally they may also be used with other types of hydraulic dredgers.

Booster pump stations consist of a centrifugal solids handling pump mounted on a floating pontoon, or on land a skid or trailer. The pump may be driven by a diesel engine or electric motor.

The additional pumping head provided by each pump will depend on the particular pump characteristics. The additional pumping distance which can be achieved will depend on the pipeline diameter, mixture velocity and concentration. For medium sand it will typically range from about 900 metres for a 300 millimetres diameter pipeline to about 3000 metres for a 800 millimetres diameter pipeline.

For maximum efficiency a booster pump should be positioned within the discharge pipeline at a distance from the dredger such that a positive pressure is maintained at the pump intake at all times. However, practical considerations may dictate that the booster be located closer to the dredger.

7.14.2 HOPPER BARGES

Hopper barges are usually referred to simply as barges, or hoppers. They are used to transport dredged material. The barge capacity may range from less than 50 cubic metres to 2,000 cubic metres, or more. The barge may be dumb (without a means of self-propulsion), or may be self-propelled. It may be suitable only for operation in sheltered water, or may be sea-going. The method of discharging the barge contents may be through bottom doors, a split hull, by pumping, or by grabbing from the shore.

Barges are most commonly employed with the various types of dredgers which excavate by bucket, but may also be used with the suction type dredgers, particularly in sand and gravel winning operations.

The construction detail may vary considerably depending upon the intended application. The most significant difference is between those designed for sheltered-water operation and those designed for open sea. The latter types are of similar construction to ships. Other design variations are influenced by the intended use. For example, if the barge is to be used for the transport of fine silty materials, then the door sealing arrangement will be important. If the barge is to be used to transport rock, then replaceable protective liners should be fitted.

Non-propelled (dumb) barges

The non-propelled barge is of simple construction. Travel between the dredging area and point of discharge is made under tow by a tug. The barge may have a small crew, or the dumping mechanism may be controlled by radio from the towing vessel.

Barge capacity may be as high as 2000 cubic metres, but more commonly will be in the range of 50 to 500 cubic metres for inland and small coastal works and 500 to 1000 cubic metres for sea transport. The laden draught is generally less than four metres (see Figure 7.58).

In general, non-propelled barges are not designed for operation in very exposed sea conditions and whilst commonly used for the disposal of dredged material at sea, this is usually in relatively sheltered coastal waters or during periods of calm weather.

The advantages of the non-propelled barge are as follows:

- its relatively low capital and operating costs – a fleet of non-propelled barges can be used to advantage where the relationship between the production of the dredger and the distance to the point of discharge is such that a number of barges can be handled by a single tug;
- the relatively shallow draught of most non-propelled barges may also be advantageous at times.

The disadvantages of the non-propelled barge are as follows:

- the relatively low speed of towing;
- the potential hazard to other navigation;
- the disruptive effect that rough seas may have on operations due to problems of safety.

Self-propelled barges

Self-propelled barges range from simple barges fitted with an outboard propulsion unit to those constructed to full sea-going ship building standards (see Figure 7.59).

The advantages are as follows:

- independent operation;
- in the case of those barges built to sea-going standards, the ability to continue working in more severe sea conditions and to travel at higher speeds.

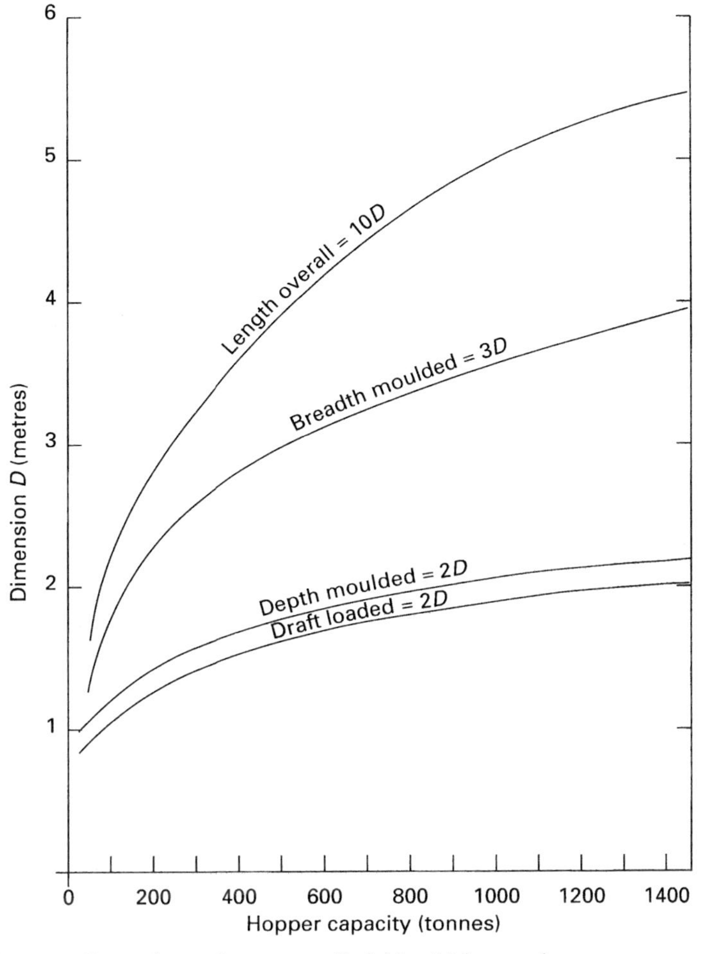

Figure 7.58 Average dimensions of non-propelled (dumb) hopper barges

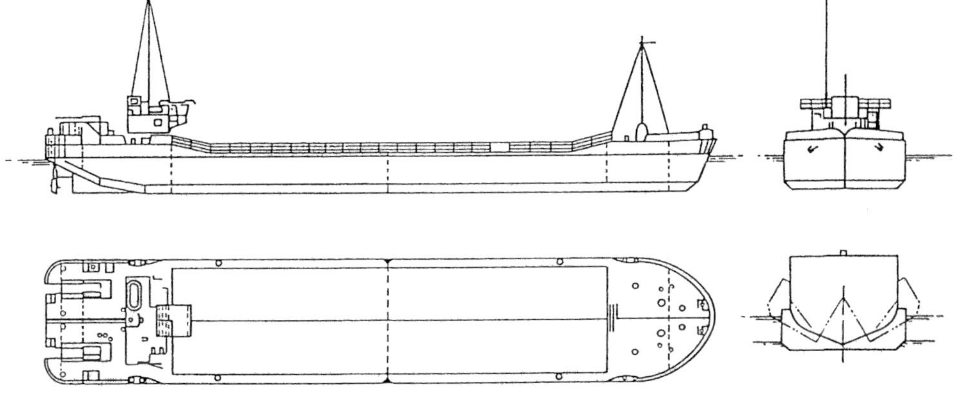

Figure 7.59 Main features of self-propelled split hopper barge

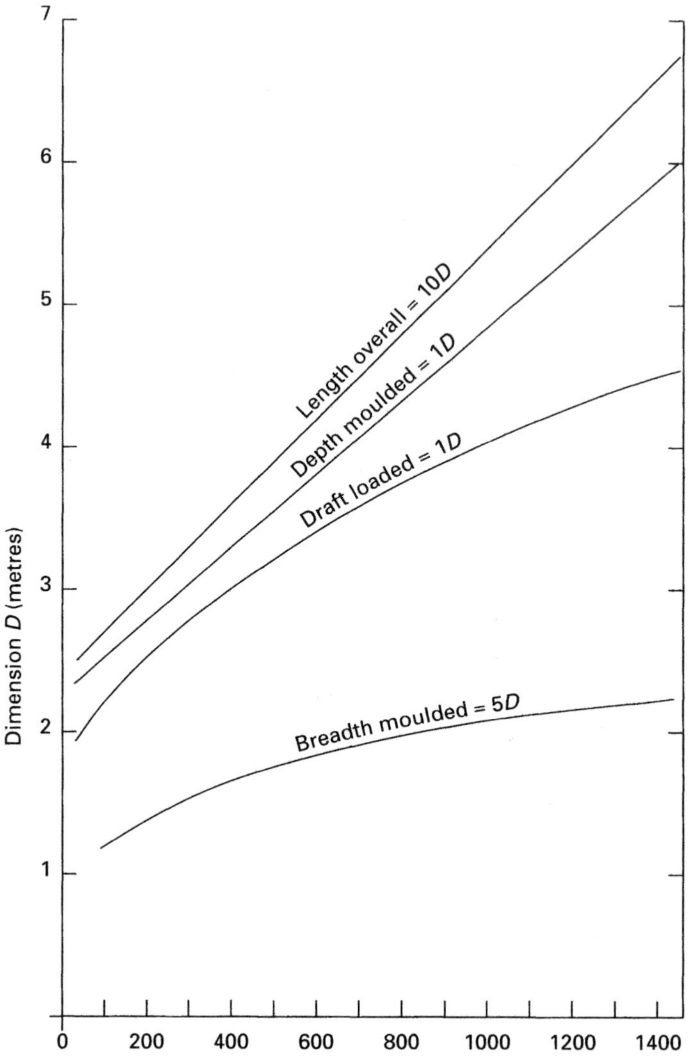

Figure 7.60 Average dimensions of self-propelled hopper barges

The disadvantages are as follows:

- higher capital and operating costs (costs for a sea-going barge may be almost as high as those for a small trailing suction hopper dredger);
- sometimes, a deeper draught (see Figure 7.60).

Bottom door discharge

A bottom door discharge mechanism may be used for both propelled or non-propelled barges. The doors are normally of the hinged type, opening and closing by means of winches or hydraulic cylinders which are attached to the doors by means of chains or links. Bottom door mechanisms are described and illustrated in Section 7.2.4.

Bottom doors have the following advantage:

- relatively simple construction and actuation. If the doors can be opened individually, or in a predefined pattern, better control of the dumping process may be possible which can be advantageous.

Two significant disadvantages of doors are as follows:

- the greater obstruction to discharge which occurs due to the door frames and control mechanism. This may cause problems when the dredged material includes large fragments of rock, or boulders, or debris such as old chains, ropes and hawsers;
- bottom doors are also susceptible to damage when discharging in shallow water, or in rough seas. A common problem is damage to the door sealing, either due to distortion of the door, or due to damage of the seal fabric.

Split-discharge

The split hopper barge discharges by splitting the entire hull. The hull is constructed in two halves which are joined by means of very heavy hinges at deck level. Opening and closing is usually by hydraulic cylinders.

The advantages of a split hopper arrangement are as follows:

- easier discharge of large particles of rock, or boulders which may be particularly advantageous when handling very cohesive clay materials;
- the absence of any protrusion below the keel permits dumping in shallow water.

The disadvantages of the split hopper arrangement are as follows:

- higher capital and maintenance costs;
- subject to design, less control of the dumping process.

Pump-discharge

When dredged material is to be disposed of on land, or used as fill, it may be necessary to discharge the barges by pumping. This is achieved with a barge unloader, sometimes called a reclaimer (see Plate 7.13). The barge unloader is simply a shore-based or floating pump set which discharges barges by suction. The material is initially fluidised with a separate water pump. Barges used permanently for pump-discharge do not have bottom doors, but consist only of hopper space, as in the case of any other bulk-cargo-carrying barge.

7.14.3 WORKBOATS

Many kinds of boat may be used as an attendant workboat in dredging operations, but particular characteristics are desirable to achieve the optimum operational efficiency. These include: sturdy construction, which normally means welded steel; adequate power; good manoeuvrability; adequate clear deck space; towing ability and shallow draught. Other characteristics which may be desirable include; lifting capability; survey capability; bunkering capacity; tolerance of silt laden water and personnel carrying capability.

Generally, it is not possible to satisfactorily combine all of the listed characteristics into a single craft. Two types of boat will, however, satisfy most needs:

Plate 7.13 A barge unloader in operation

Pontoon workboats

The pontoon-type of workboat is illustrated in Figure 7.61. Construction is of steel. An inboard diesel engine, or engines are fitted. The form is of a shallow rectangular box. This form provides the maximum clear deck space for the transport of dredging equipment and stores and the maximum stability for lifting duties. Lifting usually will be most conveniently achieved using a general purpose hydraulic loader, but a simple bow mounted A-frame and winch may sometimes be preferable. The draught of such boats is small. A keel cooling system can easily be incorporated. This will be an advantage if operating in shallow, or very silty waters. The pontoon form provides adequate carrying capacity, which may include fuel for other plant.

Pontoon workboats are usually restricted to operating in close contact with the dredging plant. Low speed and minimal covered accommodation render the boat unsuited to travel over long distances on a regular basis.

Launch-type workboats

A launch-type workboat is illustrated in Figure 7.62. The construction is in welded steel with in-board diesel engines. The hull form is shaped to provide a reasonable compromise between water resistance, carrying capacity and stability. Boats of this kind can be used for light towing duties, personnel transport, survey work and light anchor movement. Draught is reasonably shallow and if required, limited bunkering capacity can be incorporated.

The more streamlined hull form compared to the pontoon-types allows greater speed and hence the boat can range economically over greater distances.

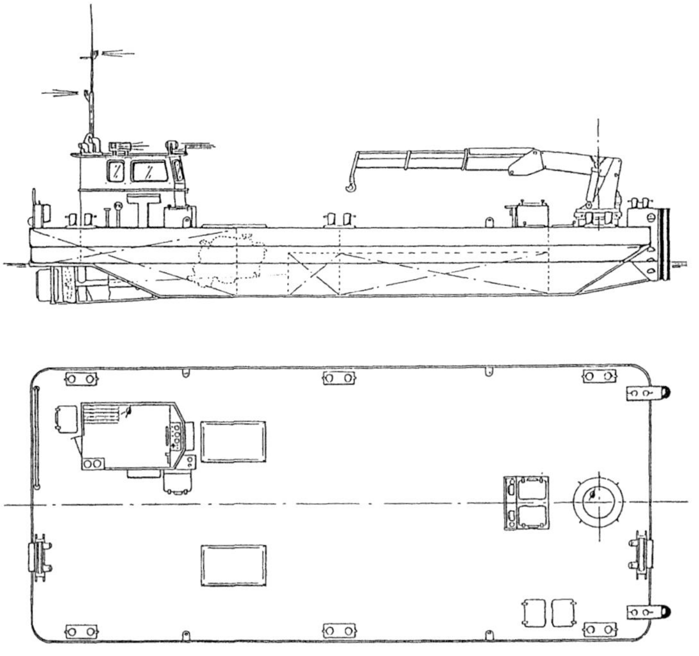

Figure 7.61 Example of pontoon type dredger tender, or workboat

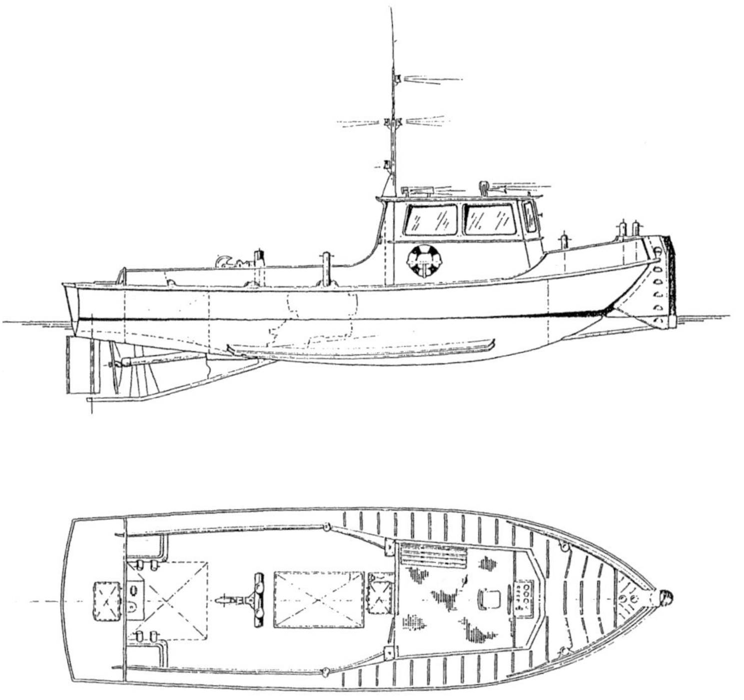

Figure 7.62 Example of launch type dredger tender, or workboat

8
Performance of Dredging Plant

8.1 Introduction

The performance of dredging plant depends on numerous factors. There are two measurements of performance:

- the instantaneous rate of production;
- the overall rate of production.

In any given situation, for a particular dredger the instantaneous rate of production is usually a function of the nature of the material being dredged. However, the overall production rate is strongly influenced by other factors such as the shape and size of the site, the weather conditions and the competence of the crew. In some cases, one or other of these factors may assume greater significance than the materials being dredged. In addition, dredging works almost invariably include the transport of dredged material, either to a point of use or to a disposal area. The choice and performance of the transport element of the works may also have a significant bearing on the overall performance and the cost of the work.

In this chapter, the factors which influence the performance of dredging and transport plant are reviewed. Basic methods of estimating the production rates of various types of plant are presented in Chapter 9.

Most of the factors discussed in this chapter need to be taken into consideration when undertaking pre-contract investigations (see Chapter 6).

8.2 Limit theory

The widely used concept of limit theory illustrates how various factors, or combinations of factors, can influence or limit the production of an item of dredging plant. Figure 8.1 illustrates the factors which contribute to the productivity of a cutter suction dredging operation. Broadly similar diagrams can be produced for all types of dredging plant.

The main factors which influence the performance of dredging plant can be divided into four broad groups:

- Site characteristics and conditions – these include constraints arising from the size and method of operation of the plant in relation to the shape and dimensions of the site, the location of the reclamation or disposal site, the use to which the site is put during dredging operations and the effects of weather, waves and currents.

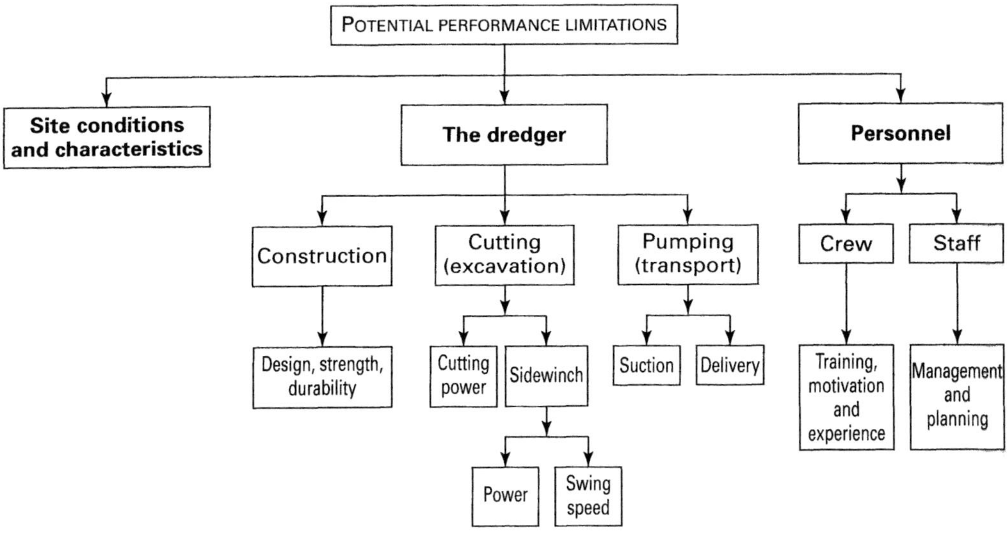

Figure 8.1 Potential limitations affecting the performance of a cutter suction dredger (simplified)

- Excavation – the nature of the soil or rock to be excavated.
- Transport – the nature of the dredged material to be transported, its influence on the suitability and effectiveness of the transport method and the distance over which the dredged material must be transported.
- Environmental factors – constraints which may be imposed to prevent or minimise the effects of dredging and transport operations on the environment.

Each group is considered in turn below. The influence of personnel is not discussed in detail here. It is evident, however, that the crew of a dredger, the site management and the overall planning of a dredging operation all contribute significantly to efficiency (see Chapter 9). In addition, the design and the construction strength, or durability, of the dredger plays an important role in determining overall efficiency.

In many cases, the dredging plant which is selected for a particular task is a compromise; whilst it may not be ideally suited to all of the site conditions and all of the materials to be dredged, it does offer the best overall performance. In particular, it is often the case that a dredging contractor will elect to use a vessel which, whilst being ideal for the majority of the soils to be dredged, performs relatively poorly in some of the soils. This may be the case, for example, when the cost of dredging some of the material at a low rate of production is less than the cost of mobilising an additional, more productive vessel.

It is stressed that the selection of plant and the estimation of its performance is a complex matter and the following pages can only give the reader some general guidance. In particular, absolute limits of, for example, strength of materials have not been defined when discussing the capabilities of individual types of dredger. This is because any such limits are generally economic rather than purely physical and economics can vary considerably from one job to another.

8.3 Site characteristics and conditions

Every dredging system (i.e. combination of excavation and transport methods) has optimum conditions in which its performance is maximised. The size, characteristics and method of operation of proposed dredging plant need therefore to be considered in relation to the general characteristics of the site.

Site characteristics may be divided into two groups: the basic characteristics of the site with respect to its location and dimensions, and those relating to the conditions obtaining on the site. The various factors falling within the groups are listed in Table 8.1 and are discussed individually below.

Table 8.1 Summary of main site characteristics and conditions affecting dredging performance

Site location and dimensions	Site conditions
Access	Wind
Location of reclamation and disposal sites	Rain
Depth of water	Fog
Length and width of dredging area	Temperature
Dredging profiles and accuracy requirements	Sea ice
Proximity to structures	Waves and swell
	Currents
	Anchorage
	Local traffic

8.3.1 SITE LOCATION AND DIMENSIONS

Access to the dredging area

Access to the dredging site may have a direct bearing on the size and type of dredger to be used. Some sites, such as inland lakes and the upper reaches of rivers, may not be accessible unless the dredger is either small enough to be transported by vehicle to the site or capable of being dismantled into a number of manageable pieces.

Constraints of access often arise in tidal situations and affect the time that the dredging site is accessible to the dredger or the time the disposal ground or reclamation area is accessible to the transport vessel. For example, locks to enclosed docks are usually opened for only a few hours during the high water period. This will be a hindrance to most cyclic dredging operations and particularly those dredgers whose working cycle, including disposal, is short. If transport of dredged material is by barge, it may be necessary to use more barges than would otherwise be the case so that dredging may continue whilst loaded barges are waiting to pass through the lock.

Similarly, intertidal zones are only accessible at high water but, in many cases, dredging in such areas involves the creation of access channels. If dredged first, these will provide access for the dredger or transport barges. Such dredging may have to be undertaken using plant which can excavate ahead of their hulls.

Location of reclamation and disposal sites

The distance between the dredging and the disposal sites will have a considerable bearing on the type of dredger to be used and its productivity. Over short distances, pipeline transportation (if applicable to the type of dredger being used) may prove to be the most economic method. As distances increase, transport by hopper dredger or by barges becomes more attractive on both economic and practical grounds. When considerable distances are involved it is the dredgers which remain on station and discharge their dredged material to separate hopper units for transport that will probably prove to be most economic (see Figure 4.1, page 61).

Pipeline dredgers and trailing suction hopper dredgers, particularly those with a pump-ashore capability, are the types most frequently used for reclamation. Mechanical dredgers and trailing suction hopper dredgers are equally suited to situations where disposal of the dredged material is to an underwater site. If mechanical dredgers are used for reclamation they are almost invariably combined with another dredger, or a rehandler, in a double-handling method. The only common exception to this is when water depths at the reclamation site are sufficiently deep to permit initial filling by dumping directly from barges or hopper dredgers.

Depth of water

The initial depth of water at the dredging site may restrict the selection of plant on the basis of both size and type. A distinction can be made here between those dredgers which are capable of excavating ahead of their hulls and those which are only able to excavate material from below their hulls. The former, which include cutter suction, grab and bucket dredgers, may be able to dredge in very shallow water and create sufficient depth for flotation as they advance. The latter group, which includes trailing and stationary suction dredgers require an initial depth sufficient for flotation. In both cases, the effects of tide need consideration as access and working time may be dependent upon the state of the tide.

On some shallow-water sites it may still be possible to use the relatively deep-draughted trailer dredger by commencing loading in the shallow areas and moving into deeper areas as the draught increases. Similarly, trailer dredgers can be provided with material by using a bed leveller (see Section 7.3, page 175) which moves the material from very shallow water into deeper water accessible to the trailer. This method is generally only economic when the distance over which the soil has to be moved is short and when volumes are small.

Some dredgers, whilst still being able to operate, are hindered by shallow water. Bucket dredgers and cutter suction dredgers, for example, can both be affected because their bucket chains or ladders may drag across the bottom in shallow water. The production rate of bucket dredgers may be markedly reduced because they are unable to fill the buckets completely. Dredging units which require cooling water for their engines must be allowed sufficient clearance between water intake and sea bottom to prevent soil being sucked in. Trailing suction hopper dredgers working in very shallow water are prone to fouling of propellers by debris, particularly when working in harbour areas. Spudded craft, whilst usually comprising shallow-draught pontoons, are often limited by the depth to which the spuds protrude below the hull.

The maximum depth of water after dredging is important; obviously the dredger must be capable of reaching to just below the required dredging depth. In some cases, the depth capability of a dredger can be maximised by working at low tide only but this has serious implications for productivity.

The maximum depth capability of most dredgers can be increased without altering the power requirement. In the case of a backhoe dredger this might be done by extending the dipper arm whilst decreasing the size of the bucket or for a bucket dredger by lowering the ladder hinge-point. However, these measures, whilst achieving the desired dredging depth, often reduce the capacity and may also increase the cycle time of the dredger, considerably reducing the overall output.

The rate of production of dredgers which incorporate pumps is strongly influenced by water depth. As water depths increase, so the ability of the pump to maintain a high concentration of material decreases. Pumps are usually mounted as low down as possible when installed in the hull of the dredger in order to minimise this problem. Modern dredgers, designed for working at considerable depths, have underwater pumps mounted on the cutter ladder or suction pipe so that high rates of production can be maintained at depth.

Water depths between the site and discharge point are also important since the hopper dredger or barge must be able to reach the discharge point.

Length of dredging area

The length of the dredging area is important for dredgers that pass over the dredging site during the dredging operation. Mobile dredgers, such as trailing suction hopper dredgers, are inhibited by short dredging sites. The shorter the passage over the site the greater the proportion of time is spent in manoeuvring rather than dredging. For trailing suction hopper dredgers any site less than 1000 metres long could be classed as restrictive. Most effective operations are achieved on sites which are considerably in excess of 1000 metres, whilst on sites which are less than 250 metres long the methods of operation may need to be modified to obtain the best results.

Width of the dredging area

The width of the dredging area is important for dredgers which swing during operation, such as cutter suction and bucket dredgers. Figure 8.2 illustrates a cutter suction dredger working at the side of a dredging area and shows how the width and draft of the pontoon limits the swing of the dredger. In the case of the bucket dredger, similar constraints apply and it is also necessary to angle the dredger to allow sufficient space for the barge to lie alongside whilst still being able to dredge at the bottom of the side slope. Trailing suction hopper dredgers require sufficient width to turn round at the end of a run. The minimum width needed will be reduced if the vessel is equipped with a bow thruster. Table 8.2 shows approximate minimum width requirements at water level for various dredgers.

Dredging profiles and accuracy

Most dredgers operate at peak efficiency when excavating thick horizontal layers of material. Thin layers and complex dredging profiles lead to a reduction of efficiency. In addition, the necessity to work to extreme accuracy can have a drastic effect on production and cost (Volbeda, 1983).

Dredgers which excavate by bucket or grab become inefficient when the thickness of material to be dredged is insufficient to allow the bucket to be totally filled during each cycle. Similarly, cutter suction dredgers may be inefficient when the cut thickness is significantly smaller than the cutterhead diameter. In both cases, very thin layers result in the dredger spending an increasingly large proportion of the time moving and re-laying anchors. In some ports, it is worth considering dredging to a greater depth than is strictly necessary for navigation requirements because the cost of the work might be only slightly greater but the period before maintenance dredging is required may be extended.

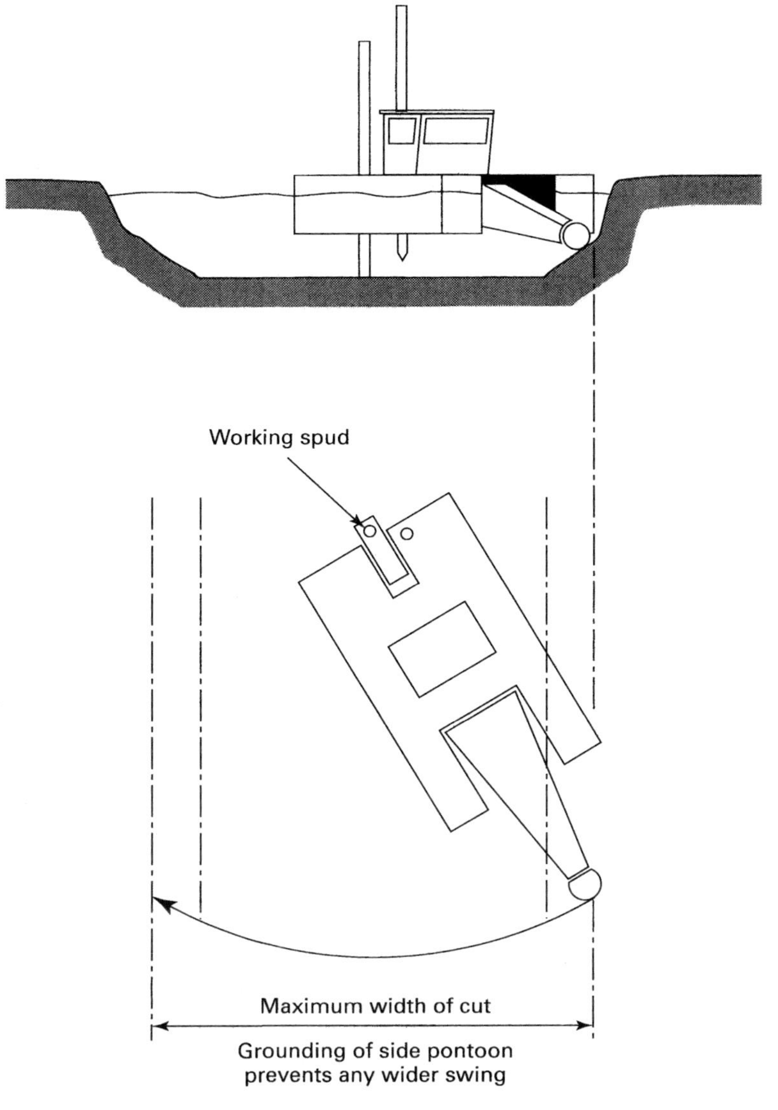

Figure 8.2 Swinging limit of a cutter suction dredger when working in a confined or shallow area

Dredgers which use suction are more efficient when dredging thin layers than those which use buckets. As a general rule, the thicker the cut the better, except when collapse of the dredging face is possible. When the cut is so high that a slip of the dredging face causes sufficient spillage into the area already dredged to inhibit movement of the dredging craft it is necessary to take additional time to cut stable side slopes. If the dredging face is liable to collapse onto the dredging unit, as may be the case with cutter suction or bucket dredgers, it may be necessary to dredge in several cuts to reduce the height of the face.

If there is a requirement for great accuracy, as for example in some pipeline or foundation trenches, it may be necessary to dredge with plant which is relatively slow compared with more productive but less accurate plant. Alternatively, working methods

Table 8.2 Minimum cutting (turning) widths at water level for various dredgers

Dredger type	Minimum cutting (turning) width at water level
Dipper, backhoe, dumb grab	Width of pontoon + 2 barge widths + 10 m
Bucket	1.5 lengths of dredger
Cutter suction	Length of dredger with ladder raised to water level
Trailer without bow thrust	4 lengths of dredger
Trailer with bow thrust	2.5 lengths of dredger

may require modification, almost certainly resulting in reduced production. Designers should be wary of setting unnecessarily tight tolerances as this almost invariably has the effect of raising costs.

Proximity to structures

In confined dredging areas, such as docks and in front of jetties, it may be necessary to dredge close to the sides of existing structures. In practice, regardless of the type of dredging plant employed, it is extremely difficult to remove the one or two metres of material immediately adjoining a vertical structure.

Most of the mechanical dredgers are suitable for working close to structures, particularly the grab and backhoe. Bucket dredgers are able to work close to a structure only if the dredging depth is relatively shallow and the ladder protrudes in front of the hull. In addition there must be suitable means of positioning the head and side wires.

The trailing suction hopper dredger is able to dredge as close as navigational conditions permit and the provision of bow thrusters is particularly helpful in this respect. Cutter suction dredgers are able to work reasonably close to structures when dredging at shallow depths when the cutter protrudes beyond the pontoon.

8.3.2 SITE CONDITIONS

Wind

Wind makes the manoeuvring of all vessels more difficult, particularly in confined areas. When hopper barges are being used they return to the dredger in a light condition with a large part of their superstructure exposed which may lead to difficulties when mooring against the dredger. Strong winds can also lead to dragging of anchors.

Rain

Rain does not usually directly affect dredging or transport, although it can affect the jacking operations on certain types of spudded craft. It does, however, cause flooding in rivers and this has the effect of increasing current velocities and water levels. Pipeline gangs may also be affected adversely on reclamation works.

Fog

Fog may be particularly bad for dredging operations in navigable waters and often causes them to be suspended. Even when the dredger's positioning system is operationally efficient in foggy conditions the dredger may still have to remain stationary to prevent collision with other vessels. Self-propelled vessels with radar and good positioning systems are sometimes able to operate.

Temperature

Temperature has an effect on personnel and mechanical working efficiency. Very high temperatures may require modification of the vessels in order to maintain efficiency and provide a good working environment. Very low temperature, if causing ice, may create adverse working conditions and restrict the movement of floating craft.

Sea ice

Sea ice hinders all marine operations and was a particular feature of dredging operations undertaken in the Beaufort Sea during the construction of artificial islands for oil exploration and production. Some trailing suction hopper dredgers have been constructed with strengthened hulls (e.g. Franje, 1983) and conventional vessels are often able to work in ice up to about 100 millimetres thick. Some development work has been undertaken on large ice breaking dredgers (Kitami *et al.*, 1983).

Waves and swell

The behaviour of a dredger subjected to waves and swell depends on many variables, such as orientation of vessel with respect to wave and swell direction, the method of mooring, the dynamic characteristics of the vessel and the periodicity of the waves and swell. These and many other factors contribute to the overall efficiency of dredging operation in open water.

Other related factors which may determine the limit of working, include the sea conditions in which the crew can be put onto or taken off the dredger, the warning period given before the onset of bad weather and the distance from the dredging site to sheltered water.

Dredgers which are located by means of spuds are particularly susceptible to waves which can lead to damage of the spuds themselves, the spud carriages and guides and even to the hull of the vessel. Anchored vessels are generally less susceptible except in the case of dredgers with rigid connections to the excavation face such as cutter suction and bucket dredgers. With these types, damage can occur when the ladder strikes the bottom. In the case of bucket dredgers, this problem is exacerbated by the high centre of gravity which makes it relatively unstable and prone to rolling.

Even before the point is reached when damage occurs, most dredgers suffer a reduction of efficiency due to lack of control of the excavation process and intermittent loss of contact between the ground and the grab, bucket, cutter or suction head. Similarly, operations can be hindered or even halted before damage to the dredger occurs in the case of vessels discharging into barges because of the relative motion between the dredger and the barge. In the case of cutter suction dredgers, damage to floating pipelines may be the limiting factor.

The trailing suction hopper dredger is specifically designed to work in open sea conditions and is the least susceptible of all types to waves and swell. Modern vessels are equipped with swell compensating devices which help maintain the draghead in contact with the bed but, inevitably, production tends to reduce as wave heights increase. The larger vessels are able to continue work in wave and swell heights of up to four metres.

An indication of the range of working conditions and limitations which have been recorded is provided in Table 8.3. It is based on waves or swells of periods between six and eight seconds and applies to dredging in soils or pre-treated rock. If virgin rock is to be dredged direct, the limitations will be more severe. Similarly, if wave or swell periods become much longer than eight seconds, their effect will be much greater and the limiting heights will be less than those shown.

Table 8.3 Limitations imposed on dredging equipment by wave and swell action

Dredging plant	Wave heights (m) – (period 6 to 8 seconds)	
	Limiting heights for efficient operations	Heights above which operations are dangerous and/or very inefficient
Drilling pontoon (floating)	1.0	1.5
Drilling pontoon (spudded) working	2.0	3.5
Drilling pontoon (spudded) moving	1.0	2.0
Dipper dredger	0.3	0.6
Backhoe dredger	0.4	0.8
Bucket dredger	0.4	1.0
Grab dredger (self-propelled)	2.0	3.0
Grab dredger (dumb)	0.4	1.0
Cutter suction dredger (small)	0.2	0.5
Cutter suction dredger (large)	1.0	2.0
Trailing suction hopper dredger (small)	1.5	2.5
Trailing suction hopper dredger (large)	3.0	4.0
Crew boat (small) under way	0.6	1.2
Crew boat (small) transferring crew	0.6	1.2
Crew boat (large) under way	2.5	4.5
Crew boat (large) transferring crew	1.0	1.5

Currents

Currents make manoeuvring of mobile dredgers more difficult. Manoeuvrability of the trailing suction hopper dredger will be affected by strong currents. This will be particularly important when dredging in confined sites, or cross currents.

Dipper, backhoe and bucket dredgers, given sufficient anchorage, can work in currents of up to three knots. It is recommended that the pontoons should be swim ended if currents exceed two knots. Fine dredged material may be washed out of the buckets in strong currents. Barges should be self-propelled to assist in mooring and manoeuvring when strong currents prevail.

Grab dredgers are very susceptible to currents which make positioning of the grab bucket difficult. Heavy-duty grab buckets are less affected by this problem. Fine dredged material may be washed out of the buckets in strong currents.

The cutter suction dredger suffers from the current in two respects; lateral pressure on the dredger and the floating pipeline. It is generally considered that a two knot cross current is the maximum allowable for large cutter suction dredgers.

The dustpan dredger is designed to operate in currents and uses the current as an aid to manoeuvring. The limitation, therefore, applies only in respect to the quality of the anchorage.

Anchorage

With the exception of trailing suction hopper and water injection dredgers all types of dredger require some form of anchorage, either using spuds or anchors. Generally, the stronger the material to be dredged the better the anchorage which is required. Anchorage problems are very time consuming and may result in considerable production losses.

Grab dredgers derive their digging forces from the vertical pull of the wires combined with the opposing action of the two halves of the grab. In this case the anchorage need only be sufficient to overcome currents. The other types of anchored dredger need to transmit their digging forces back into the ground and must therefore have good anchorage or be able to keep spuds well socketed into the sea or river bed. Anchorage becomes a greater problem for all dredgers as the current increases.

Cutter suction dredgers have side anchor wires leading from pulleys located near the bottom of the ladder. These are prone to wear and breakage due to rubbing on sharp edges when dredging rock.

Dredger anchors differ from conventional ship's anchors in that they are designed to maximise the holding power. A large number of special anchors have been developed in recent years in order to achieve this goal, using as light an anchor as possible, in order to minimise down time caused by lack of holding power (see Section 7.6.4, page 195). A detailed discussion of this is presented by Degenkamp and van den Haak (1992).

Local traffic

Much work, particularly maintenance dredging, is carried out in busy waterways. Many small craft are able to avoid interrupting the dredging operations by sailing down the side of the channel on the opposite side to the dredger. However, the larger vessels using the channel often require a considerable width of channel for safe navigation. In these circumstances it may be necessary for the dredging operations to be halted for a short period of time. Dredgers can be divided into three groups in terms of their vulnerability to this type of interruption:

- vessels which do not require anchorage during dredging operations, (e.g. the trailing suction hopper dredger);
- spudded and self-elevating plant, (e.g. some backhoe dredgers and drilling pontoons);
- plant which requires anchors and anchor wires for operation, (e.g. dustpan and cutter suction dredgers).

The trailing suction hopper dredger is the only true dredger which falls into the first group and is well suited to working in busy waterways. Occasional minor disruptions will inevitably occur from time to time to allow passage of large vessels but the loss of time will generally be slight. Ploughs, or bed levellers, which are not true dredgers, do not require anchors and can usually work without significant interruption.

Spudded dredgers and self-elevating platforms are generally of little hindrance to local traffic as they do not have long anchor wires. Vessels are therefore able to pass close by. In the case of drilling pontoons there may be a requirement for vessels to maintain a safe distance because of the explosives which are on board. If it is necessary to move the dredger or platform, several hours notice may be required. This applies particularly to drilling pontoons because of the necessity to blast before vessels pass.

Anchored vessels can be a considerable hindrance to shipping depending upon the number and length of the anchor wires. The complete removal of an anchored dredger can lead to substantial delays. In some cases it may be possible for the dredger to move to one side of the cut and to slack its wires. Bucket and dustpan dredgers, which have particularly long head wires, are very prone to such delays. To minimise this problem the side wires of many stationary dredgers are routed via underwater fairleads so that the wires leave the dredger at a depth of two or three metres below water level. Anchor wires may also be weighted with lengths of heavy chain.

In addition to the dredgers, the effects of ancillary equipment need to be considered. Barges transporting dredged material from the dredger may interfere with local traffic. Floating pipelines from cutter suction dredgers may be a serious hindrance. If submerged pipelines are used in a waterway, the reduction of navigation depth over the pipeline needs to be considered.

8.4 Excavation

The nature of the material to be excavated is usually a major factor in the choice of dredger. It is invariably a major factor in determining the instantaneous rate of production of the chosen dredger. From the point of view of dredgeability, materials fall naturally into two broad groups: soil and rock, although the distinction between hard soil and weak rock is somewhat blurred. This section considers the excavation process up to the point at which the material enters the suction mouth or the bucket or grab, as applicable.

8.4.1 EXCAVATION OF SOIL

Sand and non-cohesive silts

In the case of granular materials, a distinction must be made between three basic methods of excavation:

- allowing material to flow under gravitational forces towards a suction opening;
- excavation by erosion;
- excavation by mechanical cutting.

Although these processes are quite distinct, the soil properties which control the rates of production are generally similar.

Excavation by gravitational flow is the process used in stationary suction dredging and involves the creation of an unstable slope of soil. The production is determined by the rate at which the soil is able to flow towards the suction mouth. The processes governing the flow rate have been summarised by van Leussen and Nieuwenhuis (1984). When the suction mouth is inserted into the soil, excavation is initially achieved by the erosive forces of the water flowing towards the suction mouth. Within a short period, as the suction tube is further lowered, a circular pit is formed (see Figure 8.3), the walls of which propagate away from the suction mouth. The speed of propagation, or 'wall velocity' is solely dependent on the properties of the sand. Shearing and collapse of the slope induce negative pore water pressures within the sand which are compensated by water flowing into the sand body.

The slope of the wall is often significantly steeper than the natural angle of repose of the sand because of the negative pore pressures and the increased effective intergranular stresses. The speed with which pore water pressure equilibration occurs is therefore the controlling factor in the wall velocity and is a function of the porosity and permeability of the sand. Wall velocity is proportional to the permeability and inversely proportional to the porosity. The sand which falls from the wall is carried to the suction mouth as a density current which flows from the foot of the pit slope over the more gently sloping pit bottom.

Overall rates of production by this method are thus determined by the combination of wall velocity and the depth of the pit as long as the dredger is not limited by its pumping capacity.

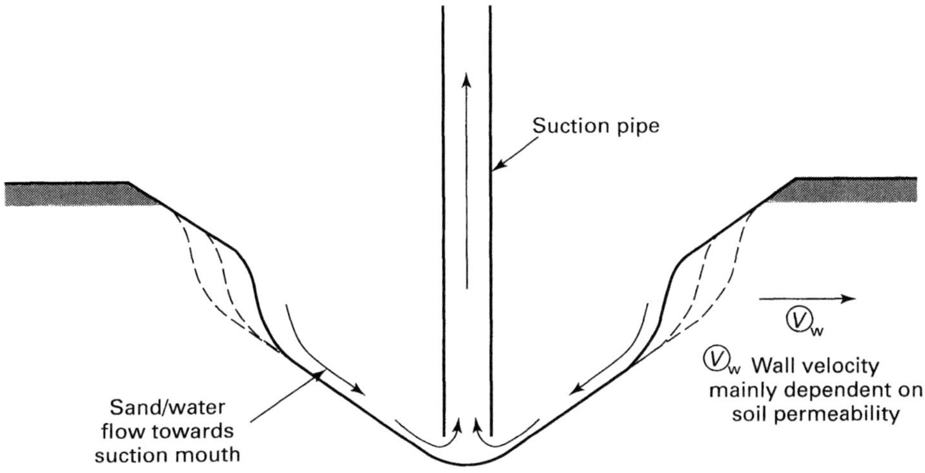

Figure 8.3 Development of pit when suction dredging in granular (non-cohesive) soils

Cutter suction dredgers can also be used in a broadly similar manner if large slope heights are required.

Stationary or deep suction methods are only appropriate in the case of sands with a relatively high permeability, where there are few cemented layers or intermediate layers of cohesive material and where the thickness of the sand layer to be dredged is sufficient to allow a reasonable pit depth. As such, they are generally only applicable in cases where the primary objective of the dredging is the winning of fill material rather than the creation of water depth. When dredging is for navigation or trenching, excavation must usually be undertaken using erosive or mechanical cutting methods.

Erosion is the method employed by trailing suction and dustpan dredgers. In both cases the erosion is caused by the flow of water towards the suction openings. The rate of production is primarily dependent upon the relative density, permeability and grain size of the material being dredged and the manner in which the dredger is operated. It is rare, however, that the dredging relies only on erosion and the dragheads of trailing suction dredgers often incorporate either linear cutting knives or water jets, or both, to assist the erosional process in non-cohesive soils. Water jets are also used on the heads of dustpan dredgers. These will also enhance production rates in sands containing thin clay layers and cemented bands when the water jets will undercut these layers and cause them to break up.

The cutting of non-cohesive soils is quite different from pure suction and erosion yet the soil properties which determine cutting forces and rates of production are similar. In the case of cutter suction and bucket wheel dredgers, the cutting of sand is essentially a high speed shearing process. Research undertaken in The Netherlands has shown that dilation in the shear zones ahead of the cutting blade creates negative pore pressures. At high velocities or when cutting dense fine sand the pore pressures may decrease to the point at which cavitation occurs. Overcoming these negative pore pressures is a significant component of the cutting force required. Miedema (1989) has stated that the cutting forces, in the case of no cavitation in the shear zone, can be summarised by the following general formula:

$$F_c = c.v_c.h_i^2.b.e/k_m \qquad (8.1)$$

where: c = coefficient of proportionality
v_c = cutting velocity
h_i = cut layer thickness
b = length of cutting blade
e = volumetric strain
k_m = weighted permeability

The coefficient includes the influences of under-pressure on the blade, under-pressure in the shear zone, the blade angle, the internal angle of friction of the sand, the angle of friction between the sand and the cutting blade, the ratio blade height:cut layer thickness and the rate of wear of the blade.

Thus, if the geometry and speed of cutting is excluded, it can be seen that the cutting forces are largely determined by the porosity, permeability and angle of internal friction of the sand; the same properties which influence the rate of production in pure suction. In order to cut with no cavitation, it is necessary to cut thin layers at slow speeds.

If full cavitation occurs in the shear zone, the above relationship simplifies to:

$$F_c = d.(z+10).h_i.b \qquad (8.2)$$

where: d = constant of proportionality
z = water depth

From these expressions, the basic rationale behind the design of cutterheads for working in sand can be seen. If cutting forces are to be kept as low as possible both cut layer thickness and cut velocity must be kept low. However, cut layer thickness is determined by the number of blades on the cutterhead and its r.p.m. (given a constant swing speed), and the number of blades is a function of the diameter. As a result we find that:

- the cutting speed must be low;
- the diameter of the cutterhead must be large;
- the number of blades must be high and;
- the r.p.m. must be high.

The slower mechanical cutting processes used by bucket dredgers and by grabs are influenced more by the ability of the bucket or grab to penetrate the sand and thus the shear strength of the material becomes the dominant controlling soil property (Becker *et al.*, 1992). In very dense materials with a high shear strength, small heavy duty grabs and buckets are used, sometimes with the addition of teeth to aid penetration.

The key role played by pore water in the excavation of granular soils has been described by several authors. In addition to the references cited above, readers are referred to van Kesteren *et al.* (1992).

The general characteristics of dredgers working in sands and non-cohesive silts are summarised as follows:

Dipper dredger, backhoe dredger and bucket dredger

Although these dredgers are able to dredge silt they are not commonly used since production is low compared to other methods. Fine sands and silts tend to wash out of the bucket during the dredging cycle. These dredgers have no difficulty dredging sand but the rate of production will generally decrease in dense and very dense materials.

Grab dredger

Although this dredger is able to dredge non-cohesive silts with relative ease, it is rarely used for this purpose in Europe, but may occasionally be used for this material in the USA and South East Asia. In coarse to medium sands the grab is able to operate in a satisfactory manner but in fine, well-graded sands it is necessary to have sharpened cutting lips on the grab. Due to the limited closing force of the grab the required shear force to dig is of great importance.

Cutter suction and bucket wheel dredgers

These dredgers have no difficulty in dredging silts and are able to pump the material a considerable distance. However, most silts are unsuitable for reclamation purposes and great difficulty is experienced in achieving a consolidated fill. All types of sand are suitable for this dredger but the rate of abrasion may be a significant cost element, especially if hydraulic transport is being used.

Dustpan dredger

This dredger is able to dredge silt. In its most common role, that of dispersing shoals in rivers, the high mobility of the silt is a positive advantage. There may be problems in discharging the dredged material if it is to be moved off site since the dredger is not capable of transporting it any distance and the silt is not ideally suited to hopper transport. In sands, the efficiency of the dredging operation depends on the ability of the dredger to fluidise the material with its water jets. Whilst it is not suited to dredging cemented sands, it can be used in materials containing thin cemented bands or thin layers of clay as long as the erosive power of the water jets is sufficient to undercut these layers.

Trailing suction hopper dredger

These dredgers are ideally suited to non-cohesive materials but rates of production diminish when cemented layers and clay bands are present. The rate of production can be greatly improved by the addition of high pressure water jets to the draghead. In fine materials there may be significant overflow losses from the hopper (see Section 8.5).

Gravels

In principle, the excavation of gravel is controlled by the same processes and soil properties as described above in the section dealing with sand and non-cohesive silts. The factors which influence the dredgeability of gravel are primarily grain size, grain shape and relative density. The relative density affects to a great extent the ease with which the gravel can be dug. The grain shape, or angularity, does the same since it determines the resistance to sliding during the shearing or erosive processes. The characteristics of dredgers with respect to their operation in gravel is summarised as follows:

Dipper dredger, backhoe dredger, bucket dredger

These dredgers are able to dredge gravel with relative ease. The relative density of the strata will affect the output.

Grab dredger

Grab dredgers are able to dredge gravel using general purpose grabs fitted with teeth. However, in dense gravels a heavy bucket is needed to achieve reasonable output. Bucket wear and damage may be high.

Cutter suction and bucket wheel dredgers

These dredgers are able to dredge gravel with relative ease but abrasion might be very high depending upon the mineralogy and angularity of the gravel.

Trailing suction hopper dredge

Some difficulty will be experienced in well graded dense gravels due to the difficulty in eroding such coarse material.

Dustpan dredger

This dredger is not usually used to dredge gravel.

Cobbles and boulders

Boulders and cobbles, which account for all naturally occurring particles in excess of 60 millimetres diameter, do not often occur in large quantities on their own. They are commonly found in glaciated or volcanic regions, often as a constituent of a mixed material such as glacial till or agglomerate, or at the sites of outflow of past and present fast-flowing rivers. They can also form an important constituent of the *in situ* weathering profile of some rocks, notably acidic igneous rocks. They are difficult to dredge and are widely recognised as being a particular hazard to dredging operations, either because they occur in unexpected quantities or are larger than expected.

Suction methods are not normally appropriate for dredging cobbles and cannot deal with boulders. All mechanical dredgers experience some difficulty depending on the size and frequency of the cobbles and boulders. There are two main problems. First, the effect of a boulder or cobble is to vastly increase the cutting or penetration forces required during excavation because the particle acts in the same manner as a shallow foundation, that is it spreads the load. The second problem is simply one of size. Cutter suction dredgers can, with difficulty, dredge particles up to about 300 millimetres, the limit being defined by the size of the apertures between the cutterhead blades and by the pump. They are rarely used, however, where deposits are predominantly cobbles because of the low rate of production and high wear and tear. The purely mechanical dredgers are more suited and their efficiency is a function of the size of bucket or grab and the available excavation power, characteristics which are usually complementary.

In some cases, particularly if the boulders are infrequent, it is necessary to consider the possibility of either dredging to uncover the boulders, which may then be removed in a separate operation, or to overdredge in the vicinity of the boulders so that when dredging is completed the boulders remain, but lie below the required navigation depth.

The characteristics of dredgers with respect to their operation in boulders and cobbles are summarised as follows:

Dipper dredger and backhoe dredger

These are probably the best suited to dredging boulders and cobbles when they are buried in other materials, particularly if large boulders are present. They can also cope with loose cobbles and boulders with reasonable ease. Both types are limited by the size of the bucket.

Bucket dredger

This is suitable for composite materials with small boulders and also loose cobbles. Much dredging of glacial moraine is carried out by bucket dredger for which the dredger should be equipped as for direct rock dredging (see Section 8.4.2). Large boulders may either get stuck in the well of the dredger or be pushed to one side.

Grab dredger

The heavier grabs are quite suitable for dredging cobbles and small boulders in composite materials. Large boulders, especially loose ones, can be dredged with special rock grapples, but these will not cope with smaller cobbles. Grab dredgers may, therefore, need to be equipped with at least two types of grab bucket.

Cutter suction dredger

These are generally unsuitable for bulk dredging of cobbles and are rarely used. Boulders greater than about 300 millimetres cannot be dredged.

Dustpan dredger and trailing suction hopper dredger

These are not suitable for boulders and cobbles.

Clay and cohesive silts

Cohesive soils fall into two categories with respect to excavation:

- those which are sufficiently soft to be excavated using suction methods only;
- stiffer soils which require mechanical assistance to suction methods.

Pure suction is generally only applied to soft cohesive silts and clays using the trailing suction hopper dredger. Such materials, typified by recent harbour sediments, behave as Bingham fluids with a defined yield stress. If the yield stress is exceeded, the soil will flow at a rate depending on the magnitude of the shear stress. When dredging with a trailing suction hopper dredger, the shear stress created by the pressure differential across the draghead causes the soil to flow towards the suction mouth.

In the case of extremely soft material, the shear stress may exceed the yield stress over a relatively large area of seabed and draw a considerable amount of material into the draghead. As the soil becomes stiffer, the shear stress required to maintain an adequate flow of soil becomes greater. The volume of material captured by the draghead progressively diminishes as the area of soil in which flow is induced reduces. As a consequence, the dredged mixture becomes progressively more dilute reducing the instantaneous rate of production. To counteract this, the trailing speed is increased.

As the strength of the soil further increases, it reaches the stage when it no longer behaves as a fluid and is drawn into the suction mouth as a mixture of clay lumps and water. A stage is reached when the suction is insufficient, on its own, to disrupt the clay and the mixture density falls to zero.

Thus, for very soft clays and cohesive silts, it is primarily the rheological properties of the soil which determine the efficiency of suction methods (Cox *et al*, 1986). When the strength of material has reached the point at which production levels are unacceptably low, it becomes necessary to introduce some form of mechanical excavation to assist the suction. In the case of trailing suction hopper dredgers, two methods are available; the draghead can be fitted with blades, which in a few cases may be rotating, rather like a cylinder lawn mower (an 'active' draghead) or, more commonly, ripper teeth can be fitted to the draghead. The latter are only effective in very stiff clays where the failure of the clay is essentially brittle and large lumps are broken out by the teeth. If teeth are used in firm clays, they tend merely to create narrow grooves in the surface of the clay with very little production advantage.

A characteristic feature of dredging of firm clays using trailing suction methods is that furrows are formed where the draghead has passed over the seabed and the soil which remains forms ridges which are difficult to dredge. It is common practice to employ bed levellers, dragged behind the dredger or, more commonly, behind a supporting tug, to smooth these ridges and increase the efficiency of later dredging passes.

Unless characteristics of the site such as wave conditions or particular requirements for the disposal of dredged material dictate the use of trailing suction hopper dredgers, they are not traditionally used to dredge large quantities of firm and stiff clays. This is because, despite the use of rippers, chisels and active dragheads, production rates are often uneconomically low. In the case of these firmer materials, mechanical dredgers will usually provide a better solution, but continued development of trailer dredgers has blurred the boundaries of application.

In all cases of mechanical excavation, the rate of production is most influenced by the shear strength of the soil which determines the force required to cut through the clay and to overcome the adhesion forces on the cutting tool. This latter factor can be very significant in the case of firm and stiff clays which can adhere strongly to metal surfaces. This can cause difficulties with respect to the emptying of grabs and buckets and the clogging of cutterheads.

Three different types of failure mechanism have been identified in the cutting of clay: shear, flow and tear. These failure mechanisms result from a combination of the shear and tensile strength of the clay and the adhesive and cohesive forces on the cutting blade. These are described by Joanknecht and Lobanov (1980). The force required to cut clay with a cutterhead is given by the general formula:

$$F_c = n.b.(c_1.V_c.h_i + c_2.h_i + c_3) \tag{8.3}$$

where: F_c = cutting force
n = number of blades on cutterhead
b = length of blade
V_c = cutting velocity
h_i = thickness of slice cut, and
c_1, c_2 and c_3 are constants.

An examination of this formula leads to the following conditions for minimum energy consumption per cubic metre of soil dredged (van Raalte and Zwartbol, 1986):

- the cutting velocity must be high;
- the diameter of the cutterhead must be small;
- the number of blades must be small;
- the r.p.m. of the cutterhead must be low.

Biogenic gas often occurs in very soft recent clays and silts and can have a marked detrimental effect on rates of production of dredgers which use suction. This, however, is related to dredging depth and is considered in more detail in Section 8.6, dealing with environmental constraints.

The main types of dredger with respect to their operation in clays and cohesive silts are summarised as follows:

Trailing suction hopper dredger

These are eminently suited to dredging soft and very soft cohesive soils but experience increasing difficulty as the shear strength rises. In firm and stiff clays, the material tends to be sucked up into the hold in lumps where it may form large masses and hinder dumping. The use of active dragheads with rotary cutters helps to improve the efficiency of trailer dredging in firm and stiff clayey materials. Split-hopper discharge arrangements simplify the discharge of cohesive soils.

Cutter suction dredger

This type of dredger can be used in clays of all strengths, depending upon the installed cutter and side winch power. The main problems which arise concern blockage of the cutterhead in firm and stiff clays which can be minimised by the use of a special clay cutterhead and by limiting the thickness of the cut slice.

Dipper dredger, backhoe dredger, bucket dredger

All these dredgers are capable of dredging clays with relative ease depending on the stiffness of the clay. However, highly adhesive clay will stick in the buckets and will slow the bucket emptying process. This is a particular problem for bucket dredgers where the dredged material may drop back into the water through the ladder well at normal bucket chain speeds, rather than emptying into the chute below the top tumbler. Problems may also occur when dumping from hopper barges.

Grab dredger

This dredger is frequently used to dredge clays of all strengths but, when working in stiff clay, requires a heavy grab bucket fitted with teeth. In some clays the adhesion between the bucket bottom and the virgin soil is sufficient to demand a line pull well in excess of that normally required for a full bucket load. Emptying of the grab can also be problematical in firm and stiff clays.

Dustpan dredger, stationary suction dredger

These dredgers are not suitable for dredging clay.

Peat and organic soil

This type of soil can be dug by all the bucket machines, including the grab dredger, with relative ease. Cutter suction dredgers are able to dredge both amorphous and fibrous peats but any gas which is present in the soil will reduce the efficiency of the system unless a degassing system is installed. Trailing suction hopper dredgers may also be used but are usually very inefficient in fibrous peats. These materials are not used for fill and hence disposal could be a problem.

Debris and demolition material

Large pieces of debris are most easily dredged by grab, dipper or backhoe dredger since by virtue of their bucket sizes they are able to pick up large lumps of masonry, rock, brickwork and so on. For wires and other awkward items a grab dredger is usually the best machine.

Cutter suction dredgers and trailer dredgers are not suited to dredging debris and may be subject to frequent delays when working, for example, in old ports where seabed debris is a common feature.

8.4.2 EXCAVATION OF ROCK

Modern, very powerful dredging plant can dredge rock directly but, as the strength of the material increases, rates of production fall off rapidly. A point is reached when direct dredging is no longer economic and the rock has first to be fragmented by some other means before it can be dredged. This section considers only direct dredging. Pre-treatment and dredging of pre-treated rock are discussed in Chapter 14.

Only four types of dredger are commonly used to dredge large volumes of rock directly without pre-treatment. These are the dipper, bucket, cutter suction and backhoe dredgers. Bucketwheel dredgers are used for rock dredging, particularly in the USA and in Australia but their use generally is restricted to relatively weak rocks and they are more commonly used in the mineral extraction industry than in capital dredging for ports and navigation channels. Trailing suction hopper dredgers are occasionally used to dredge weak and very weak rocks, but generally only in cases where these represent a very small proportion of the total volume to be dredged, or in exposed locations where the use of stationary types is impractical. In such cases the vessel, and particularly the trailing suction pipe assembly; must be of strong construction; the draghead must be heavy and fitted with ripper teeth to assist excavation.

Two factors control the ease with which rock can be dredged directly; strength and structure. In rock which is unstructured and free from joints, fractures and bedding planes, the excavation process is one of cutting and the ease with which the rock can be dredged will depend upon the ability of the dredger to penetrate the rock material and break out lumps. The strength of the rock, relative to the power and digging action of the dredger, is thus the primary controlling factor. As the number of discontinuities in the rock mass increases, the strength of the material becomes less dominant and, in closely fractured rock, the action of excavation is one of loosening rather than cutting, a process which is achieved with relative ease. However, it should be noted that rock which is described as of poor quality in engineering terms, may nevertheless be difficult to dredge directly.

Rock cutting, whichever type of dredger is being used, is achieved with picks which are fitted to the bucket or cutterhead. There are a number of complex rock cutting models which have been developed to describe and quantify the cutting action which involves the penetration of the pick into the rock followed by the formation of a loose rock chip. The theories are for the most part broadly similar, differing in detail principally with respect to the geometry of the cutting situation which determines whether the loose chip results from compressive, shear or tensile failure of the rock. The model described here applies to the cutting action of a cutter suction dredger, the type most frequently used to dredge large quantities of rock directly.

For rock dredging, the cutterhead is fitted with sharp picks. Penetration of the pick into the rock creates a small crushed zone around the tip of the pick. As the pick advances, the crushed zone grows, compacts and eventually dilates, resulting in the creation of tensile stresses in the zone which lead to tensile failure of the rock ahead of the pick and the formation of a loose chip of rock which is drawn into the suction mouth.

Chip formation, whilst being the result of a tensile failure of the rock, is preceded by crushing in compression and the cutting forces which are required are therefore dictated principally by the compressive strength of the material. Other properties of the rock do, however, contribute to the force required and the productivity of the cutting process, as in the following examples:

- The porosity of the rock has a controlling influence on the size of the crushed zone which is required to promote chip formation; more porous rocks require a larger crushed zone for chip formation and thus more cutting energy than less porous rocks of the same compressive strength.
- The elasticity will influence the amount of energy required to form the crushed zone; rocks with a high elastic modulus require less cutting energy than rocks with the same compressive strength but a lower elastic modulus.
- The tensile strength determines the magnitude of the tensile stress required to form the chip; the greater the tensile strength the greater are the cutting force and energy required.

In addition, the geometry of the pick itself and the angle at which it is presented to the rock will have a considerable bearing on the cutting forces required. Cutterhead and pick geometries are a compromise between cutting efficiency and the strength and durability of the pick itself and its fixing to the cutterhead.

In structured rock, the excavation forces will depend on the rock material strength and on the frequency of fractures, their condition (i.e. openness, smoothness, the presence of infilling materials, etc.) and their orientation with respect to the cutting tools. In very weathered, highly fractured rock, excavation may be achieved with a degree of difficulty similar to, or in some cases less than, that involved in the dredging of stiff clay or very dense sand and gravel.

Orientation of the dredger with respect to the rock structure can be very important and the authors are aware of one case where the rate of production of a heavy bucket dredger working in strong (75 to 100 MPa compressive strength) slate was more than doubled simply by changing the direction of the dredging.

The characteristics of the various dredgers which are commonly used in rock are summarised as follows:

Dipper dredger

The dipper dredger is used to dredge weak or decomposed rock, which it is able to do by virtue of the high point loads exerted by the bucket teeth. They normally only dredge untreated rock if it occurs as thin lenses in weaker materials. Some dipper dredgers can be converted to rock breakers.

Grab dredger

The grab dredger relies on the falling weight of its bucket to assist in the initial penetration into rock after which the line pull closing the grab assists the penetration. For this reason the grab dredger is better able to cope with rock dredging when using a heavy bucket. In very weak and weathered sedimentary rocks it is perhaps more efficient than a backhoe dredger when working at depth, but generally cannot compete in stronger materials.

Backhoe dredger

The backhoe dredger is able to dredge rock by virtue of its positive action and ability to prise up pieces of rock using the leverage of the bucket on the seabed. In this manner only indirect forces are transmitted back to the machine itself. It is generally more efficient than a grab in rock but can suffer from twisting of the boom when dredging at near its maximum depth. The biggest backhoes can dredge intact moderately weak to moderately strong sedimentary rocks.

Bucket dredger

The bucket dredger is capable of dredging weak to moderately strong sedimentary rock, particularly when the dip and strike of the rock are most advantageous and when the dredger is able to create a working face. The effectiveness of the dredger is dependent on it being able to apply a sufficiently high point load to the rock with a bucket tooth. This will depend on the bucket size relative to the machine power, tooth length, angle and position on the bucket rim and the ability of the machine to apply full power to the bucket chain at or near stalling speed.

Experience has shown that, for dredging in rock, for every 100 litres of bucket capacity there should be 100 to 150 horsepower on the bucket chain drive and 3 to 4 tonnes pull on the side winches. In addition to this, the angle of the teeth on the bucket should be not more than 15° off the tangent to the digging circle at the bucket's lip. The length of the teeth should be adjusted such that the precutting circle is only 25 to 50 millimetres in advance of the bucket's digging circle. The position of the teeth on the bucket is also important since it is necessary continually to have at least one tooth in a digging position whatever the angle of the dredger to the cut.

Cutter suction dredger

The cutter suction dredger is commonly used for direct rock dredging, particularly where large volumes of rock have to be dredged, and some dredgers are designed specifically for this purpose. Their capability to perform in rock is governed by the size and strength of the machine, particularly the weight of the ladder and the power supplied to the cutterhead and side winches. Rock cutter suction dredgers are as much as 2.5 times the weight of the equivalent dredger designed for soft soils. The most powerful rock cutter suction dredgers at the time of publication have 6000 horsepower cutter motors, a ladder weight approaching 1000 tonnes and side wire winches capable of exerting around 150 tonnes of pull.

Cutter suction dredgers can dredge rock directly only in relatively calm water, since severe impact of the cutterhead is likely to result in damage. When dredging abrasive rock, the wear on the cutter is high and much time will be required to replace the teeth and ultimately rebuild the cutterhead.

8.5 Transport

Transport of the dredged material is achieved through one of four methods:

- hydraulic transport, in which the material is pumped through a pipeline as a slurry;
- hopper transport, where the dredger transports the material in its own hold;
- barge transport, where the material is loaded into dumb or self-propelled barges;
- dispersion and density currents which carry material away from the dredging site in the case of agitation, sidecasting and water injection dredging.

8.5.1 HYDRAULIC TRANSPORT

Hydraulic transport is the pumping of the dredged material as a slurry through a pipeline. It is the most common method of transporting and placing material in the case of reclamation projects. It is also the means by which suction dredgers such as trailing suction hopper and cutter suction dredgers raise excavated material to the surface. The dredging unit's pumping capability is frequently a limiting factor in terms of the overall production rate and this can, on occasions, even apply to dredgers which are simply loading directly into barges or their own hopper.

In the vast majority of cases, centrifugal pumps are used in dredgers rather than displacement pumps. The basic principles of hydraulic transport using centrifugal pumps are best explained in stages, starting with the transport of clean water.

In order to move the water through the pipeline, pressure is required which is provided by the dredging pump. At the pipe outlet there is velocity, but no pressure and, therefore, there is a pressure loss along the length of the pipe. The magnitude of the pressure loss is dependent upon the velocity of the water being pumped, the frictional resistance offered by the pipeline wall and the total length of the pipeline. The pressure loss per unit length increases as the velocity increases and as the pipe diameter decreases (see Figure 8.4).

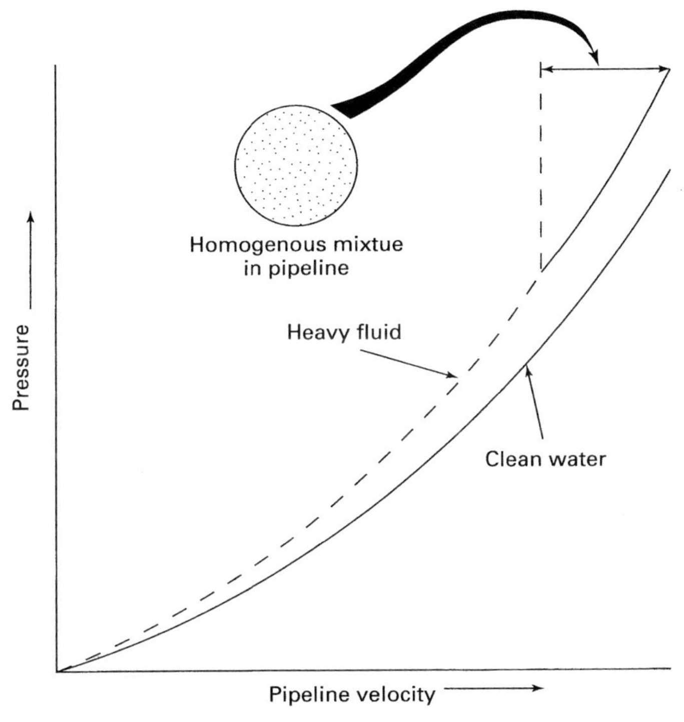

Figure 8.4 Pressure loss per unit length of pipeline increases as the velocity of water increases

When soil particles are introduced to produce a slurry there is a marked effect on the pressure loss curve. At high velocities, the soil is carried as a homogeneous mixture due to the very turbulent flow. Effectively, a more dense fluid is being pumped and the energy required is greater than that required for water. If the transport velocity is reduced, some of the solids concentrate near the bottom of the pipe and transport is partly in suspension, partly rolling along the bottom of the pipe. There is an overall decrease in the unit pressure drop due to the reduced velocity but the pressure required is much greater than would be required to transport clean water at the same velocity (see Figure 8.5).

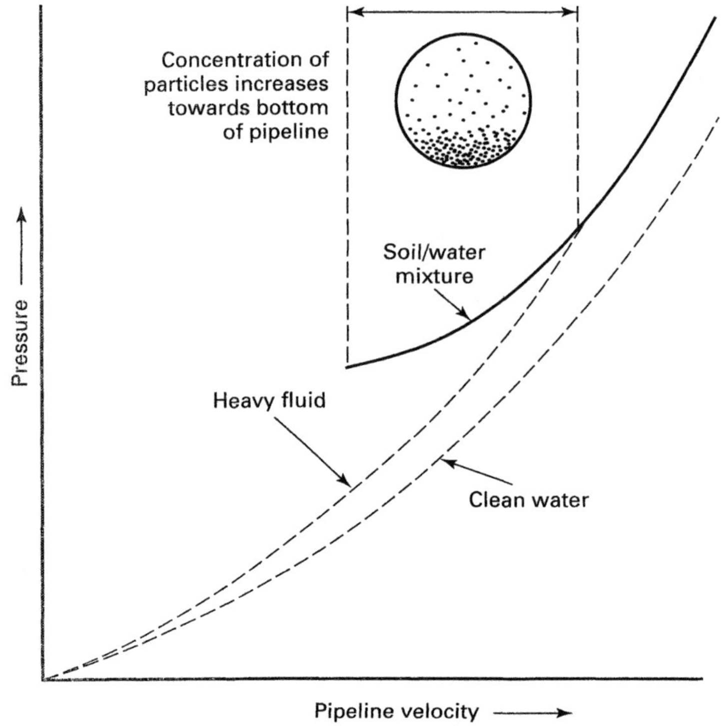

Figure 8.5 Pressure loss when pumping soil/water mixtures through pipelines

If the velocity is reduced even further, the solids settle at the bottom of the pipe, effectively reducing the diameter of the pipeline leading to increased resistance and, thus, pressure loss (see Figure 8.6). If allowed to continue, this would lead to a blocked pipeline, as the energy requirements rapidly build up and exceed the capacity of the pump. The velocity at which this occurs is termed the critical velocity, that is the minimum permissible velocity, below which the required pressure will build up beyond control. This velocity is matched by a critical pressure, that is the lowest pressure at which pumping can take place without leading to a blocked pipeline.

For a given pipeline length and diameter, the critical velocity is a function of the following:

- the size and shape of the particles being transported;
- the concentration of particles in the mixture;
- the specific gravity of the particles.

When designing hydraulic transport operations, the objective is clearly to obtain the highest possible output of solids, that is the combination of concentration and velocity, whilst remaining within the capability of the pump to provide the required energy. This is achieved by operating at a velocity slightly higher than the critical velocity and at a pressure slightly higher than the critical pressure. In cases where the length of the pipeline, or possibly the elevation of the reclamation area relative to the dredger, is so great that reasonably high outputs cannot be achieved, it is common practice to introduce a booster pump (see Section 7.14.1) into the pipeline between the dredger and the fill area in order to increase production.

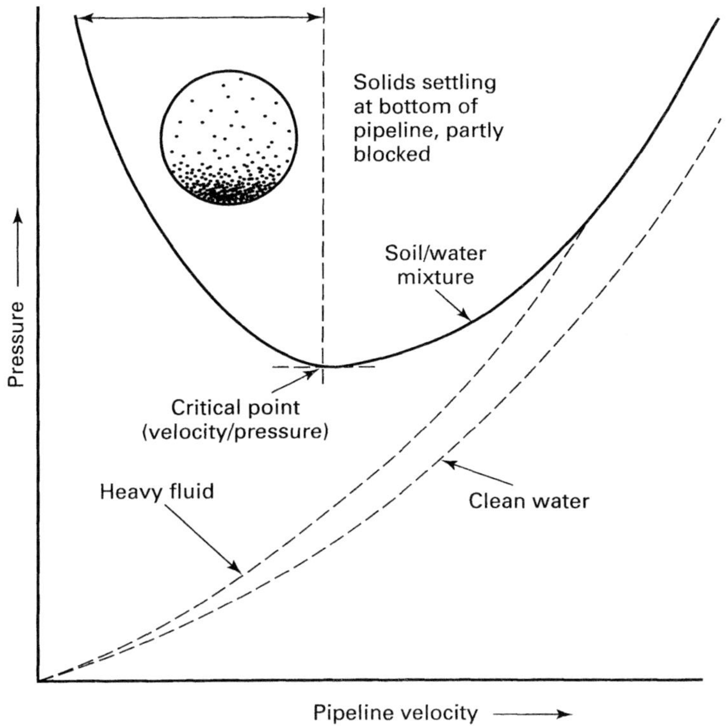

Figure 8.6 Pressure loss when pumping through pipelines with static bed load

Table 8.4 Transport velocities necessary for various materials

Material type	Velocity range (m/s)
Silt	2.0–3.0
Fine sand	3.0–4.0
Medium sand	3.5–4.5
Shell	3.5–4.5
Very soft clays	4.0–5.0
Coarse sand	4.0–5.0
Sand with fine gravel	4.5–5.0
Sand with medium gravel	4.5–5.5
Stiff clays	4.5–5.5
Sand with coarse gravel	5.0–5.5
Sand, gravel and cobbles	5.5–6.5

Note: Higher velocities than those listed above may be possible. This may result in higher production rates but at the expense of increased power demand and pipe wear.

When pumping cohesive soils, the material may form 'clay balls' which hinder transport. The balls comprise lumps of cut material. Generally, very soft clays disaggregate quickly during transport and do not constitute a problem. However, stiffer clays tend to degrade only slowly and the pumping effort required to transport them through a pipeline may be similar to that required for cobbles and small boulders.

Hydraulic transport is a complex issue and for a very detailed discussion of all of the numerous factors involved the reader is referred to Wilson (1992). Indicative values of hydraulic transport velocities for a range of materials are given in Table 8.4.

8.5.2 HOPPER TRANSPORT

The efficiency of hopper transport is primarily dependent upon the degree to which the hopper can be loaded. This is particularly important in the case of a trailing suction hopper dredger. The dredged material is diluted to some extent and enters the hopper as a mixture of soil and water. The heavier soil particles settle to the bottom of the hopper and the smaller, lighter particles remain in suspension due to turbulence. It is common practice, in non-cohesive granular materials, to continue dredging for a period after the hopper has been filled with mixture and to allow low-density mixture to overflow. This is done in order to increase the solids content in the hopper. The length of time that overflow is continued depends upon two factors: the nature of the material being dredged and the sailing distance to the disposal area.

With coarse materials long overflow times may result in a significant increase of load. The increase diminishes as the particle size becomes smaller because a greater proportion of the solids remain in suspension and, in a relatively short time, the density of the overflow mixture is very nearly the same as the density of the mixture being delivered into the hopper. Continued overflow in such cases is only justifiable when currents carry the solids away from the area being dredged, assuming, of course, that overflow is permitted and is environmentally acceptable.

It should be noted that there are three loading 'regimes' when working in clays and silts:

- In extremely soft materials which form a true slurry when pumped into the hopper, there will be little or no increase in hopper capacity if it is overflowed. However, these materials can often be pumped at or very close to their *in situ* density and the load in the hopper thus equates approximately with the volume which has been dredged at the point when overflow commences.
- In soft materials, the soil will be pumped as a dilute mixture of slurry and softened clay balls. The properties of the mixture are such that settlement in the hopper will be hindered and there may be only a small advantage in overflowing for a short period. However, because the soil will have been loaded as a mixture with added water, the hopper load will be significantly less dense that the *in situ* soil.
- In stiffer clays, although the soil is pumped as a dilute mixture of clay balls and added water, there is only a relatively small slurry component. This allows the clay balls to settle quickly in the hopper and the use of overflow can increase the hopper load in exactly the same way as is done in granular materials.

Figures 8.7 and 8.8 show, respectively, the progressive increase of the density of the mixture overflowed from a medium size trailer dredger when working in soft sandy mud and the density profiles in the hopper at the start and end of overflowing. It can be seen that there is an initial rapid rise in the density of the overflow over a period of about 10 minutes after which the density increases only slowly. After about 45 minutes, the overflow density is almost constant and it is evident that continued overflow will not result in any significant increase of the total hopper load. In this case, a very considerable amount of material will have been lost overboard during the 45 minutes of dredging.

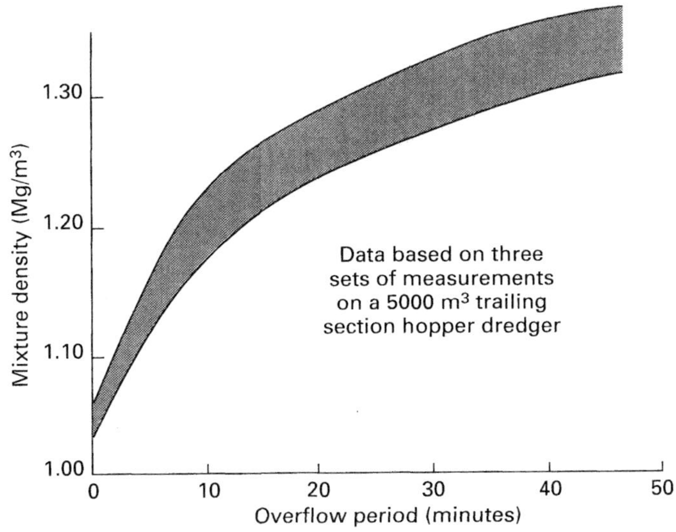

Figure 8.7 Density of the overflowed mixture during dredging of a soft sandy mud

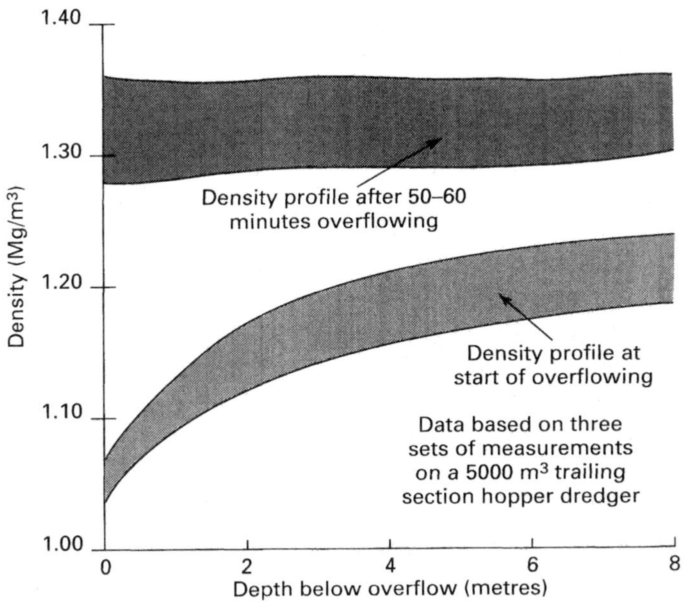

Figure 8.8 Hopper density profiles at start and end of overflowing during dredging of a soft sandy mud

Given a particular type of material and its loading characteristics, for predictive purposes the overflow time will be estimated on the basis of the time taken to sail to the disposal area, discharge the dredged material and return to the dredging site. The greater the overall sailing time, the more profitable it may be to extend the overflow time. The combined effects of overflow characteristics and sailing distance are illustrated in Figure 8.9. Point A is the time the dredger discharges at the disposal site or reclamation area. Point B is the time at which it commences dredging and B–C represents the duration of loading up to the point at which overflow commences. In terms of overall productivity, and hence cost, the ideal duration of overflow is determined by drawing a straight line from the start of the cycle (point A in Figure 8.9), tangential to the loading curve at Point E. In the example shown, the most economic overflow duration is seen to be C–D. At time D, the hopper is about 85% filled. If the distance to the disposal area were greater, and hence the sailing time were longer, a longer period of overflow would be indicated.

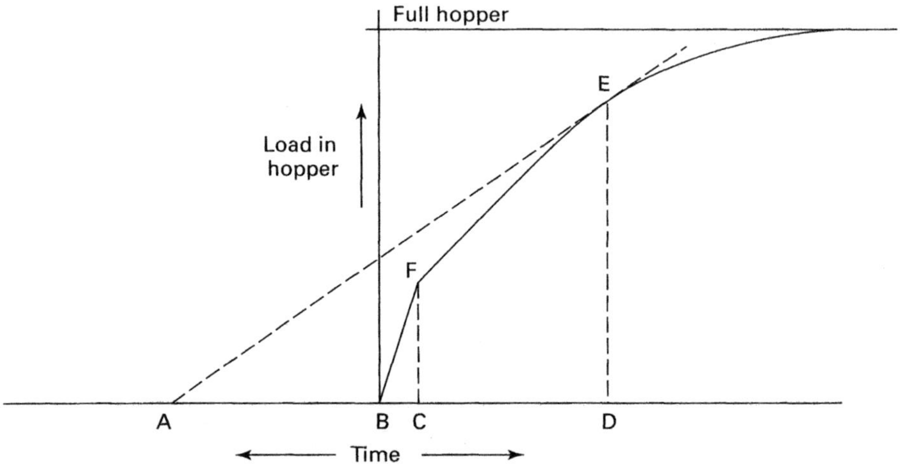

Figure 8.9 Graphic method of estimating optimum overflow duration

An alternative, non-graphical approach, which can be used during operations, is to record hopper load and total cycle time for a number of initial trial cycles and to determine the load time which gives the highest overall production per working hour.

In the case of a mechanical hopper dredger such as a grab hopper dredger, the material is excavated at approximately its *in situ* density and considerations of overflow do not generally arise.

The manner in which the hopper is discharged at the disposal site or the reclamation area will also influence productivity. Most modern trailer dredgers can discharge in any one of three ways:

- by dumping through valves or doors set in the bottom of the hull or, in some cases, by splitting the entire hull;
- by pumping the hopper contents through a connection to a floating pipeline which may extend, for example, to a reclamation area which cannot be accessed directly by the trailer because of water depth restrictions;
- by spraying the hopper contents through a bow-mounted nozzle. Sand can be placed in this manner up to a maximum of about 100 metres from the bow by some of the larger vessels.

Simple bottom-dumping is the fastest method of unloading and can usually be achieved within a few minutes but time must be allowed for manoeuvring the vessel into the right position. Discharging through a pipeline usually takes about 40 to 60 minutes depending on the material characteristics, including the time required to connect and disconnect the pipeline. For coarse granular materials the overall time may be nearer two hours, subject to pipeline length. Discharging by spraying through the bow nozzle will generally take about 45 minutes.

8.5.3 BARGE TRANSPORT

The efficiency and productivity of barge transport is determined by the same factors as those which apply to hopper transport.

Discharging from barges is usually achieved by bottom dumping, either through doors or by splitting the barge. This is usually achieved very quickly. Only on very rare occasions is it necessary to empty a barge by means of a mechanical, shore-mounted grab.

A significant disadvantage of bottom doors is the greater obstruction to discharge which occurs due to the door frames and control mechanism. This may cause problems when the dredged material includes large fragments of rock, or boulders, or debris such as old chains, ropes and hawsers. Bottom doors are also susceptible to damage when discharging in shallow water, or in rough seas. A common problem is damage to door sealing, either due to distortion of the door, or due to damage of the seal fabric.

The split hopper arrangement allows for the easier discharge of large particles of rock, or boulders and may be particularly advantageous when handling very cohesive clay materials. The absence of any protrusion below the keel permits dumping in shallow water.

When dredged material is to be disposed of on land, or used as fill, it may be necessary to discharge the barges using a barge unloader, or 'reclaimer' (see Chapter 7). Barge unloaders need to be able to form a soil-water slurry in order to remove the material from the barge and are thus most suited to non-cohesive sands or extremely soft silts and clays which can easily be slurrified. Significant difficulties will be encountered if the barges contain cohesive materials, gravel, cobbles and boulders, or debris.

8.5.4 DISPERSION AND DENSITY CURRENTS

In the case of agitation dredging and water-injection dredging, transport of material away from the dredging site is achieved by dispersion and density currents. The efficiency is measured in terms of the percentage of dredged material which is removed from the site, but the potential effects of sedimentation in adjacent areas must be examined carefully when any such operations are planned.

Dispersion during agitation dredging is most efficient when dredging fine-grained sediments in strong currents. However, in order not to return the soil directly back into the dredging area, it may, in some cases, be necessary to work only a part of the tidal cycle.

The density currents created during water-injection dredging generally travel downslope towards deeper water and even very shallow slopes are capable of maintaining a strong density current flow. In some cases, material can be carried up shallow slopes and removed from the dredging area but only if the slope is a 'threshold', that is it must lead to a slope break from which another slope leads away from the dredging area. The density current moves up the initial slope by virtue of the potential energy imparted by its thickness.

8.5.5 ABRASION OF DREDGING AND TRANSPORT PLANT

The effects of abrasion in pumps and pipelines are often an important cost item on a dredging project, particularly those involving cutter suction dredgers pumping over long distances. These effects need also to be considered when controlling the pumping operation. Abrasion in pipelines is a function of the following parameters:

- the hardness of the mineral grains being transported, relative to the hardness of the pump and pipeline materials;
- the shape and angularity of the particles;
- the size of the particles;
- the velocity at which they are transported;
- the pressure in the pipeline.

Surprisingly, perhaps, the concentration of particles in the mixture has little effect on the degree of abrasion, except in so far as it affects the total quantity of material being passed through the pipe. The strength and hardness of the material being dredged are the main factors which control abrasion. Quartz, one of the most common rock and soil components, is usually the only mineral which is of real concern with respect to abrasion in most situations but some dredge mining operations may be concerned with other, less common, minerals which have a hardness greater than that of steel. However, it should be noted that dredging plant can and will be abraded by minerals with a hardness less than that of the dredge components because some abrasion is the result of impact, rather than sliding, which results in small chips of steel being removed from pumps, pipelines and teeth. Thus even limestones, which have little or no quartz content, may cause significant abrasion if they are sufficiently strong.

In the case of floating rubber discharge pipelines and suction hoses, impact cutting and tearing may be more important than abrasive wear. The same applies to plastic pipelines which are sometimes used with small cutter suction dredgers. Such materials are particularly vulnerable to damage by large sharp solids such as broken flint.

Abrasion increases disproportionately as the particle size and the transport speed increase. Dredger operators will attempt to strike a balance between the requirements to achieve a good rate of production and the need to minimise abrasion but, in hydraulic transport, there will be limitations to what they can achieve in this respect because the mixture velocity in the pipeline must not be allowed to fall below the critical velocity with due allowance for a safety margin.

When very coarse particles are being transported by pipeline, abrasion is often found to be exceptionally severe over a small area of the bottom of the pipes. This leads to a requirement to rotate pipes frequently to maximise pipe life. Very aggressive abrasion such as this can be caused by clays which contain abrasive gravel. The gravel tends to protrude from the clay balls, making them abrasive, and the overall mass of the clay balls exacerbates the problem.

8.6 Environmental constraints

The increasing awareness of the environment and the effects which our activities have on the environment have become a prominent issue with respect to dredging operations. It is common, particularly in the case of large dredging projects or those which involve disposal rather than use of dredged material, to undertake environmental impact studies.

The studies are designed to establish the probable effects of the dredging works on the environment and can have a major impact on the choice of working method and its efficiency.

It is possible that, having undertaken an environmental impact assessment, it is decided that dredging operations would have a sufficiently adverse effect on the environment to warrant the imposition of constraints or mitigating measures. Many such measures can be implemented relatively easily but most have an adverse effect on the productivity of the dredging operation and, hence, on the costs of the work. Environmental aspects of dredging are considered in detail in Chapter 13 but the main restrictions on productivity which might occur in certain situations are summarised here.

8.6.1 RESTRICTION OF OVERFLOW

Trailer dredgers may be prohibited from overflowing, or the period of overflow may be restricted, in order to minimise the discharge of sediment into the water column. The effect on productivity and hence cost will be greatest in the case of granular soils and when transport distances are long. The effect will be least in the case of very soft muds which can be dredged at or close to their *in situ* density. Similar restrictions might be applied to barges loading sand with a high fines content.

8.6.2 RESTRICTION OF OPERATING METHODS

Other restrictions associated with reducing the discharge of sediment to the water also affect the trailer dredger. These are the prohibition of the use of degassing and ALMOB systems (see Section 7.2, page 163) and the pumping out of residual water ('restwater') in the hopper prior to loading. All of these restrictions may affect productivity.

If degassing is not allowed when dredging soils with a significant gas content, there will be a need to reduce suction to prevent cavitation occurring and this may produce a reduction in the maximum density of material in the dredged mixture. The full hopper will contain less solids than in the unrestricted case, and the dredger will have to sail to the disposal area with a reduced load, thereby lowering the productivity.

Similarly, prohibition of the use of the ALMOB system will mean that low density mixtures will enter the hopper during the dredging process and the dredger will sail to disposal with less material. The productivity will be adversely affected.

Restwater in the hopper which remains after the dumping of dredged material is normally pumped out prior to loading cohesive soils, particularly if overflowing of hoppers is prohibited. If the restwater may not be removed, the capacity of the dredger is reduced by the volume of the restwater.

8.6.3 RESTRICTION OF OPERATING RATES

Many dredgers can be operated at a relatively slow rate in order to minimise release of sediment. However, it should be noted that whilst turbidity levels may be reduced in this manner, the total amount of sediment lost, per cubic metre of material dredged, may actually increase. The rate of production will generally decrease in direct proportion to the reduction of operating rate.

8.6.4 USE OF SILT CURTAINS

Grab and backhoe dredgers can be surrounded by silt curtains in an attempt to contain some of the sediment put into suspension during dredging. These do not greatly influence the instantaneous rate of production but the overall rate is reduced due to the additional time which is required for installation and maintenance of the curtain and the increased frequency with which the dredger has to be moved, because the silt curtain restricts the area which can be dredged at any one anchoring location. Silt curtains are unlikely to be effective unless current velocities are low (significantly less than 0.5 metres per second).

8.6.5 NOISE RESTRICTIONS

The majority of dredging plant is relatively quiet but noise restrictions might be applied, either in the form of prohibiting the use of certain methods or by restricting working hours. The potential consequences for rates of production and costs can be very considerable and such restrictions should not be applied without full awareness of the consequences.

References

Becker, S., Miedema, S.A., de Jong, P.S. & Wittekoek, S. 1992. On the closing process of clamshell dredges in water saturated sand, *Proceedings of the 13th World Dredging Congress*, Bombay, India; World Organisation of Dredging Associations, Delft, The Netherlands.

Cox, C.M., van Deursen, P.P. & Verhoeven, F.A.1986. The relevant soil properties for dredging silt, *Proceedings of the 11th World Dredging Conference*, Brighton, UK; World Organisation of Dredging Associations, Delft, The Netherlands.

Degenkamp, G. & van den Haak, R. 1992. Increased holding of anchors for dredging practice, *Proceedings of the 13th World Dredging Conference*, Bombay, India; World Organisation of Dredging Associations, Delft, The Netherlands.

Franje, J. 1983. Construction of artificial islands in the Beaufort Sea – contractor's view, *Proceedings of the World Dredging Congress*, Singapore, BHRA Fluid Engineering.

Joanknecht, L.W.F. & Lobanov, V.A. 1980. Linear cutting test in clay, *Proceedings of the 3rd International Symposium on Dredging Technology*, Bordeaux, France, BHRA Fluid Engineering.

van Kesteren, W.M.G. , Steeghs, H.J.M.G. & Mastebergen, D.R. 1992. Pore water behaviour in dredging processes, *Proceedings of the 13th World Dredging Conference*, Bombay, India; World Organisation of Dredging Associations, Delft, The Netherlands.

Kitami, E., Kawamoto, T., Noble, P. & Semeniuk, A. 1983. An experimental investigation of ice impact load on drag arm of an arctic super trailing suction hopper dredger, *Proceedings of the World Dredging Congress*, Singapore; BHRA Fluid Engineering.

Leussen, W. van, & Nieuwenhuis, J.D. 1984. Soil mechanics aspects of dredging, *Geotechnique*, Vol. 34, No. 3, pp 359–381, Institution of Civil Engineers, London.

Miedema, S.A. 1989. On the cutting forces in saturated sand of a sea-going cutter suction dredger, *Proceedings of the 12th World Dredging Congress,* Orlando, Florida, USA.

van Raalte,G.H. & Zwartbol, A. 1986. The disc-bottom cutterhead: A report on laboratory and field tests, *Proceedings of the 11th World Dredging Congress*, Brighton, UK.

Volbeda, J.H. 1983. The price of accuracy in trench dredging, *Terra et Aqua No. 25*, March, International Association of Dredging Companies (IADC).

Wilson, K.C., Addie, G.R. & Clift, R. 1992. *Slurry Transport using Centrifugal Pumps*. Elsevier Applied Science, London and New York, 1992.

9

Estimating Output

Symbols used in Chapter 9

A	The area covered from one dredging position
a	Advance distance between side anchor movements
B	Bulking factor
b	Width of cut
C	Bucket capacity
D_s	Suction pipe diameter
D_{50}	Grain size
d	Water depth
d_{max}	Maximum dredging depth
d_n	Normal dredging depth
f_a	Advancing delay factor or anchor delay factor
f_b	Mechanical breakdown factor
f_c	Cycle factor
f_d	Delay factor
f_e	Proportion of hopper filled
f_f	Face height and dredging depth delay factor
f_h	Hopper changing delay factor
f_m	Modification factor
f_o	Operational factor
f_p	Spud moving delay factor
f_t	Traffic delay factor
f_w	Weather delay factor
f_θ	Tilt factor
g	Distance to disposal ground
H	Hopper capacity
hp_d	Horsepower driving the dredging pump(s)
hp_c	Horsepower driving the cutter
h_{min}	Minimum face height
L	Length of discharge pipeline
l	Length of dredging area
N	Number of grab cranes

n	Number of buckets per minute
P	Output
P_{nom}	Nominal uninterrupted output
P_{max}	Maximum potential output
P_t	Instantaneous theoretical output
p	Advance distance for each spud movement
T	Shift length
t_a	Time to advance dredger or move side wires forward
t_d	Time for soil disposal
t_h	Time to change hoppers
t_l	Time to load hopper
t_p	Time to advance on spuds
t_s	Duration of sub-cycle
t_t	Time taken to turn dredger
U_b	Base productive unit
U_m	Modified productive unit
V_g	Fully laden sailing speed
z	Thickness of dredged layer

9.1 Introduction

The estimation of the performance of a dredger in any given set of site and soil conditions is no easy task. However, it is not only the very experienced or technically specialised engineer who can attempt it. Naturally, those organisations which own and operate dredgers, and have built up detailed records of each dredger's performance, will be best able to assess the performance of a particular unit, but given sufficient information, any engineer should be able to estimate dredger output with sufficient accuracy to form the basis of programming and budget job costing. Having said this, it must be emphasised that important pre-tender cost estimates and detailed plant programmes should only be attempted by engineers who have specialised in this work, as there are many pitfalls for the uninitiated.

In the first edition of this book it was stated that the estimating methods put forward were of an experimental nature. Since that time most of these methods have been tried and tested in a variety of situations and it has become apparent that, although those estimating methods relating to mechanical dredgers are fairly reliable, the methods proposed for hydraulic dredgers needed improvement and these have therefore been revised.

Performance, production and output are all terms used to describe the rate at which a dredger moves soil or rock. In this chapter the term 'output' will be used and defined as the *in situ* quantity of material dredged in a given period of time. A dredger will have a number of outputs for any given set of conditions, depending on the time period considered. The output must, therefore, be qualified as being one of the following:

(a) Hourly output	average quantity dredged in a working hour.
(b) Shift output	average quantity dredged during a complete shift.
(c) Weekly output	average quantity dredged in a complete week.
(d) Annual output	total quantity dredged in a calendar year.

9.2 The division of time

Calendar time can be broken down into four standard divisions (see Figure 9.1).

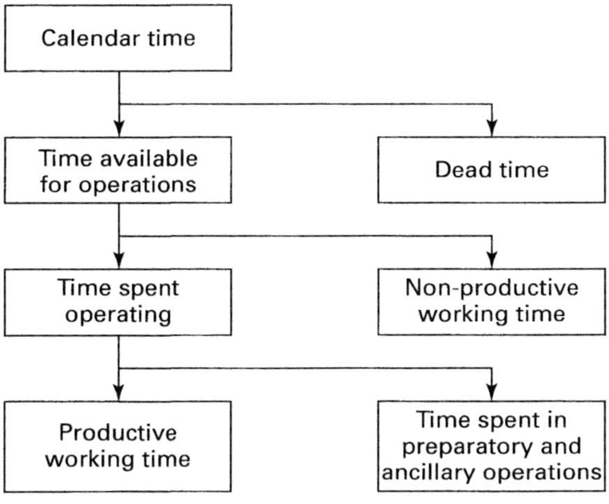

Figure 9.1 The division of time

1 Dead time

This is time during which it has been established that no productive work of any type will be carried out. It would normally include national holidays, annual overhaul of the dredger, moving the dredger between projects and the non-shift hours each day.

2 Time spent in preparatory and ancillary operations

This time is necessary to perform jobs that are essential to the dredger's operation but themselves do not allow it to operate. These could include conveying the crew to the dredger, bunkering, starting up and moving between locations on the same site, and carrying out routine maintenance when it must be done in operating time. Jobs which are part of the production cycle are not included.

3 Productive working time

This is time during which the dredger is performing its cyclic operation.

4 Non-productive working time

This time occurs due to weather delays, breakdowns and other unplanned interruptions to the dredging operation such as delays due to traffic, that is passing vessels, and management or crew inefficiencies. It is time when the dredger should be operating, but is not.

The output to be estimated is that which is achieved during the productive working time, modified to take account of non-productive working time. Both dead time and time spent in preparatory and ancillary operations are usually determined by the type and location of the site and its geographical context, as well as the type of the dredger.

9.3 Bulking factors

Output is defined here as the *in situ* volume of soil dredged in a given period of time. However, it should be noted that the volume of material will change during the dredging and transport processes, and placing in the reclamation area. In addition fill may be compacted after placement in order to improve its engineering properties, resulting in further volume changes.

During the dredging process the change in density is caused by the formation of additional voids in the soil or rock, reducing their dry density. Thus, when a dredger lifts material off the seabed, the volume which this material occupies in the hopper is usually larger than the volume it occupied in the ground. This increase can be expressed as a percentage of the *in situ* volume or as a ratio of the two volumes and is generally known as the bulking factor.

Bulking factors vary greatly for different types of soil, different particle size distributions, and for different methods of dredging. Mechanical dredgers usually cause the least bulking and silts with a high water content may actually increase in dry density during the dredging process. Hydraulic dredgers mix the soil into a slurry invariably reducing its bulk density and increasing its volume.

Most methods of dredging result in a loss of material, notably the overflow losses of a trailing suction hopper dredger. Such losses can be significant and need to be taken into account when estimating the volume of fill material which can be recovered from a given volume of soil *in situ*.

Similarly, material may be lost during placement in the reclamation area. Soil improvement techniques such as vibrocompaction and dynamic compaction will increase the dry density of the material leading to further volumetric reduction.

The volumetric changes which take place during placement in the reclamation area and during any subsequent soil improvement operations are not considered further here. However, when estimating *in situ* output when estimating output for mechanical dredgers, and hydraulic dredgers with hoppers, it is necessary to estimate the bulking factor for the particular soil in question in the barge or hopper. Table 9.1 gives an indication of the range of values likely to be encountered. For hydraulic dredgers the degree of bulking will vary according to the *in situ* density of the material to be dredged, the concentrations which can be achieved in the pipeline and, for hopper dredgers, the duration of overflow.

9.4 Basic principles

It is assumed at this point that the characteristics of the site have been identified and a possible dredger has been selected, after due consideration of the factors affecting performance (see Chapter 8). The method of estimating output is then as follows:

- identify the basic unit of production;
- modify this unit to account for soil and excavation/pumping conditions;
- identify the pertinent dredging cycle;
- apply the dredging cycle to the modified unit of production;
- apply suitable reduction factors.

9.4.1 THE BASIC UNIT OF PRODUCTION

The base productive unit, U_b

The base productive unit is a characteristic of the dredger alone. It is a function of the power available to carry out the work and the size of the excavating components. Thus, the unit is a figure which is known once the dredger has been selected.

The modification factor, f_m

The base productive unit must be modified to suit the site and soil conditions. The object of

Table 9.1 Bulking factor, *B*, for various soil types when excavated by mechanical dredger

Soil type	Bulking factor, *B*
Hard rock (blasted)	1.50–2.00
Medium rock (blasted)	1.40–1.80
Soft rock (blasted)	1.25–1.40
Gravel, hardpacked	1.35
Gravel, loose	1.10
Sand, hardpacked	1.25–1.35
Sand, medium soft to hard	1.15–1.25
Sand, soft	1.05–1.15
Silts, freshly deposited	1.00–1.10
Silts, consolidated	1.10–1.40
Clay, very hard	1.15–1.25
Clay, medium soft to hard	1.10–1.15
Clay, soft	1.00–1.10
Sand/gravel/clay mixtures	1.15–1.35

(NB B = dredged volume/*in situ* volume)

this modification is to take account of factors which affect the output of a single cycle, that is the volume of soil dredged during the cycle.

The basic units, and their modification, are described in detail in Sections 9.5 onwards.

9.4.2 THE DREDGING CYCLE

The cycle factor, f_c

The normal dredging methods and their associated cycles are described in Chapter 7. However, the duration of any particular cycle or sub-cycle will depend on a number of site and dredger characteristics which must be assessed for every new set of circumstances. These characteristics together effectively become the cycle factor which must be applied to the modified base productive unit to give an output figure. The methods of taking account of cycles and sub-cycles are described for each dredger in Sections 9.5 onwards.

9.4.3 THE OUTPUTS

It is necessary to consider various outputs during the stages of the estimating procedure and, depending on the dredger type, a number of the following are used:

Instantaneous theoretical output, P_t

This is an hourly output based on an optimum instantaneous working rate.

Nominal uninterrupted output, P_{nom}

This hourly output takes account of the basic dredging subcycle. It is smaller than P_t.

Maximum potential output, P_{max}

This output is obtained by reducing P_{nom} to take account of the main dredging cycle.It is, thus, the output which is achieved in the productive working time. It is a theoretical figure which cannot be achieved in sustained operations and represents the average hourly output in ideal circumstances with 100 per cent efficient crew and machinery in the given site and operating conditions.

Output, P

This is the final estimated output to be used for obtaining programme periods and budget cost estimates. It is obtained by applying reduction factors to P_{max}. It represents the output obtained during a period of productive and non-productive working time.

9.4.4 REDUCTION FACTORS

In practice, there are no ideal conditions, and machines and crews are not 100 per cent efficient. It is, therefore, necessary to reduce the maximum potential output by a number of reduction factors.

Delay factor, f_d

Delays due to bad weather and interruptions from passing maritime traffic can be combined to form a delay factor. It is necessary to express the time which will not be lost due to traffic delays as a fraction of the working time available. This figure should be multiplied by the fraction of time during which weather, and therefore sea conditions, are suitable for working.

$$\text{Delay factor, } f_d = f_t \times f_w \tag{9.1}$$

where: $$f_t = \frac{\text{Total working time available} - \text{time lost due to traffic during working hours}}{\text{Total working time available}}$$

and: $$f_w = \frac{\text{Total of days (hours) when weather is suitable for working}}{\text{Total number of days (hours)}}$$

Operational factor, f_o

Neither crew nor management in a dredging organisation can be 100 per cent efficient. A dredger operator cannot work his machine in the most efficient manner all the time, nor can the supervisory staff on site anticipate every conceivable contingency. An operational reduction factor is thus necessary to take account of inefficiencies. Table 9.2 shows some

Table 9.2 Operational factor, f_o, for given personnel ratings (valid for good climate)

	Crew rating				
Management rating	Poor	Mediocre	Average	Good	Very good
Very good	0.67	0.73	0.78	0.84	0.90
Good	0.65	0.71	0.77	0.82	0.88
Average	0.64	0.69	0.75	0.80	0.86
Mediocre	0.62	0.67	0.73	0.79	0.84
Poor	0.60	0.65	0.71	0.77	0.82

NB Climatic adjustments:
Poor climate. ×0.95; arduous climate. ×0.90

suggested factors based on the competence of the management and crew. The values reflect the fact that the skill and experience of the crew have a much greater effect than that of the site management.

The factors given in Table 9.2 are for work carried out in good climatic conditions. In poorer climates efficiency is inclined to drop still further and it is suggested that the efficiency should be adjusted as shown in the note below the table.

Some loss of efficiency may be offset by automation aboard the dredger, particularly for the more sophisticated dredgers, such as modern cutter suction and trailing suction hopper dredgers. However, it must be appreciated that some types of dredger, such as grabs, do not lend themselves to a high degree of automation.

Mechanical breakdown factor, f_b

In theory, machinery which is serviced and maintained should continue to work as if new. However, after a number of years it is inevitable that breakdowns will occur as parts become worn. Due to these breakdowns the output of the dredger will drop. On the basis that after 20 years a dredger will be completely overhauled, the following reduction factor is applied: for the first five years there is no reduction, and for every year thereafter a one per cent reduction, so that at year 20 the mechanical breakdown factor, f_b, = 0.85.

Mechanical breakdown can be substantially reduced by continually overhauling and replacing machinery before it reaches a stage when breakdown will occur (preventative maintenance). In this situation the mechanical breakdown factor will be almost unity. The percentage reduction given above, applies to equipment which is just given only normal preventive and running maintenance and not replacement.

9.4.5 SUMMARY

The basic principles of estimating output can now be restated as follows.

The base productive unit, U_b, is modified so that

$$U_m = f_m \times U_b \tag{9.2}$$

where: f_m = modification factor
U_m = modified productive unit

The maximum potential output, P_{max}, is obtained by applying a cycle factor, f_c, to the appropriately modified productive unit. It is often necessary to consider other theoretical outputs during this stage of the process.

The average output, P, is obtained from

$$P = f_d \times f_o \times f_b \times P_{max} \tag{9.3}$$

where: f_d = delay factor
f_o = operational factor
f_b = mechanical breakdown factor

(Note: Where appropriate the output is modified by the bulking factor, *B*.)

This simplified algebraic statement of the method cannot necessarily be applied to every dredger but it does indicate the logic of the estimating process. In the following sections the estimating method is applied to each type of dredger.

9.5 Trailing suction hopper dredger

Typical characteristics of trailing suction hopper dredgers are given in Chapter 7, Figures 7.6 and 7.7, pages 165 and 166. Although there is considerable divergence from these characteristics, particularly in the larger dredgers, the general trends remain fairly constant. The estimating method given below assumes that the standard characteristics apply and output should be adjusted accordingly if this is not the case.

The following input is required, or must be estimated:

H = the maximum hopper capacity of the dredger (cubic metres) for the soil type to be dredged;
V_g = the fully laden sailing speed of the dredger (knots);
g = the distance to the disposal site (kilometres);
l = the length of the dredging area (kilometres);
t_d = the time taken to discharge of the dredged material (hours);
t_t = the time taken to turn the dredger at each end of the dredging area (hours).

It is also necessary to know the soil type and, for non-cohesive soils, the grain size (D_{50}).

9.5.1 *THE PRODUCTIVE UNIT*

The productive unit for a trailer dredger is the hopper capacity, H. Sometimes two figures are quoted for hopper capacity, (e.g. 1100/1250 cubic metres). The difference is accounted for by the minimum and maximum overflow weir heights and the fact that dredgers are designed to carry a specific load, not volume, in the hopper. Generally the lower figure should be used for granular soils that settle well and the upper figure for fine soils which tend to remain in suspension in the hopper. A few dredgers, particularly aggregate dredgers, are designed with sufficient buoyancy to carry a full load of granular soil.

The productive unit has to be modified by the bulking factor, B, which takes account of the bulking of the dredged material, which in this case is the ratio of the volume of the hopper to the *in situ* volume. This bulking factor will vary widely depending on the soil type. For coarse granular material the figures given in Table 9.1 may apply, but for well consolidated fine cohesive sediments the factor could rise to 2.5 or more.

The modified productive unit thus becomes:

$$U_m = \frac{H}{B} \tag{9.4}$$

and is the total *in situ* volume of dredged material which, theoretically, can be contained in a full hopper load.

9.5.2 *THE CYCLE FACTOR*

The dredging cycle of a trailing suction hopper dredger consists primarily of four components:

- loading (dredging);
- turning the dredger at the end of each run across the dredging area;
- sailing to and from the disposal ground;
- discharging the dredged material.

Loading

Since trailers are fairly standard, the time taken to fill their hopper to the overflow point tends to be similar, irrespective of hopper capacity, that is the larger the hopper, the bigger the pumps and suction pipe(s). Variations in loading time therefore depend on soil type, overflow losses over the hopper weir and attainable concentrations in the suction pipe due to depth of dredging, obstructions and so on.

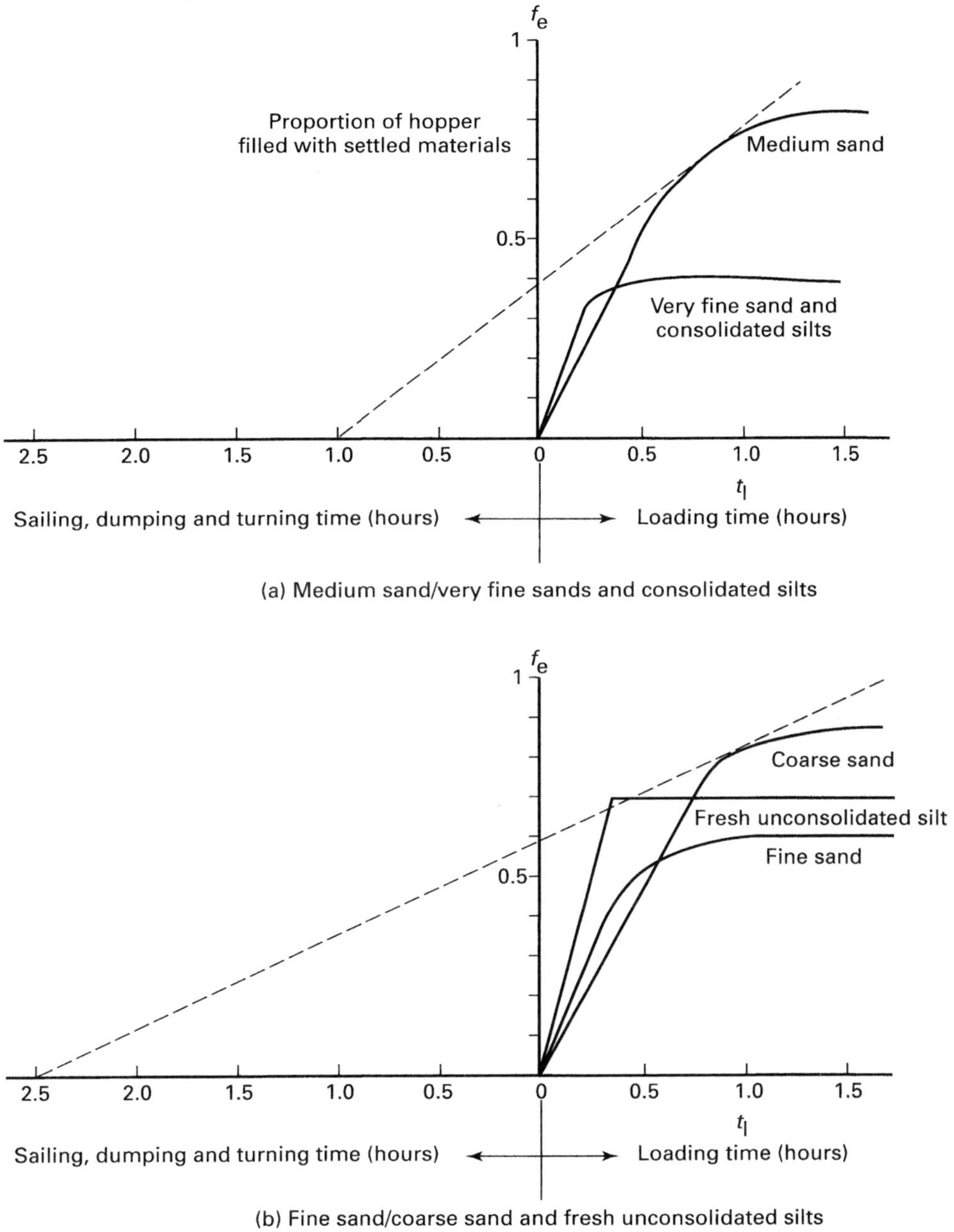

Figure 9.2 Trailing suction hopper dredger: loading graphs. a) medium sand/very fine sands and consolidated silt; b) fine sand/coarse sand/fresh unconsolidated silt

Typical loading curves for various soil types are shown in Figure 9.2. These relate to trailers working at their nominal design depth. Loading times will be greater when working at greater depths (unless an underwater pump is fitted in the suction pipe) or if the draghead is obstructed in any way. The graphs show the loading in terms of loading time, t_l, against proportion of hopper filled, f_e. The method of determining the optimum point at which to stop dredging and travel to the disposal site is explained later.

Turning

Each time the dredger turns at the end of a dredging run productive dredging time may be lost. Assuming that dredging is carried out at 2.0 knots, the number of turns is given by:

$$\text{Number of turns} = \frac{3.91 \times t_l}{l}$$

(Note: If the trailing speed is slower, as might be more normal in confined waters, the number of turns would be reduced in proportion.)

The time spent turning thus becomes:

$$\text{Turning time} = \frac{3.91 \times t_l \times t_t}{l} \qquad (9.5)$$

Sailing

The time taken to sail to the disposal ground and return to the dredging site is given, approximately, by:

$$\text{Sailing time} = \frac{1.02 \times g}{V_g} \qquad (9.6)$$

Although this expression does not take into account the slightly faster return journey in a light condition, this will be compensated by the fact that it also leaves out those periods of acceleration and deceleration at the beginning and end of each journey. If the distance to the discharge point is very short, the average speed achieved will be significantly less than the maximum.

Disposal

The time taken to dispose of dredged material, t_d, is known at the outset. The time will be short when material is bottom-dumped at sea, or much longer when material is pump-discharged. Pump-discharge periods of 60 minutes or more are common.

9.5.3 THE OUTPUT

Output can only be estimated when the loading time has been ascertained. This is determined from the loading graphs shown in Figure 9.2. First the total period of unproductive time is calculated from:

$$\text{Unproductive cycle time} = \frac{3.91 \times t_l \times t_t}{l} + \frac{1.02 \times g}{V_g} + t_d \qquad (9.7)$$

This time is then set off on the left hand scale of the graph in Figure 9.2. The tangent to the loading curve from this point will touch the loading curve at the time that loading should be terminated. Loading time, t_l, and hopper filling factor, f_e, can then be read off the graph. (Note: Since t_l is not known at the beginning of this exercise, a certain amount of iteration may be required to arrive at the final result.)

$$P_{max} \text{ is then given by } \quad \frac{\text{total load}}{\text{total cycle time}}$$

$$P_{max} = \frac{H \times f_e}{B(t_l + \frac{3.91 t_t t_l}{l} + \frac{1.02g}{V_g} + t_d)} \qquad (9.8)$$

Simplification

Standard dredging depths, sailing speeds and so on can be estimated from Chapter 7, Figures 7.6 and 7.7 pages 165 and 166. The following values may also be used if others are not known:

t_t = 4 minutes = 0.066 hours
t_d = 5 minutes (bottom-dumping) = 0.083 hours

9.6 Cutter suction dredger

Typical characteristics of cutter suction dredgers are given in Chapter 7, Figure 7.16 page 184. Some modern cutter suction dredgers are built for specific types of work, such as rock dredging or dredging at above average depths, and will not necessarily conform to the characteristics shown. The estimating method given below assumes normal characteristics apply and outputs should be adjusted accordingly if this is not the case.

The following information is required, or must be estimated:

hp_d = the horsepower driving the dredging pump;
hp_c = the horsepower driving the cutter;
L = the length of the discharge pipeline (metres);
d = the average dredging depth (metres);
d_{max} = the maximum dredging depth for the dredger (metres);
z = the average thickness of material to be dredged (metres);
a = the average distance advanced between side anchor movements (metres);
b = the width of cut (metres);
p = the average distance advanced with each spud movement (metres);
t_a = the time taken to move side anchors (hours);
t_p = the time taken to advance on spuds (hours).

Also the soil type (the D_{50} grain size for granular soils and the N-value) should be known.

9.6.1 THE PRODUCTIVE UNIT

The productive unit for the cutter suction dredger is the pump horsepower, hp_d. However, it is also necessary to ascertain that the cutter horsepower, hp_c, is sufficient to be able to cut the soil to be dredged. Figure 9.3 gives an indication of the cutter power required to dredge

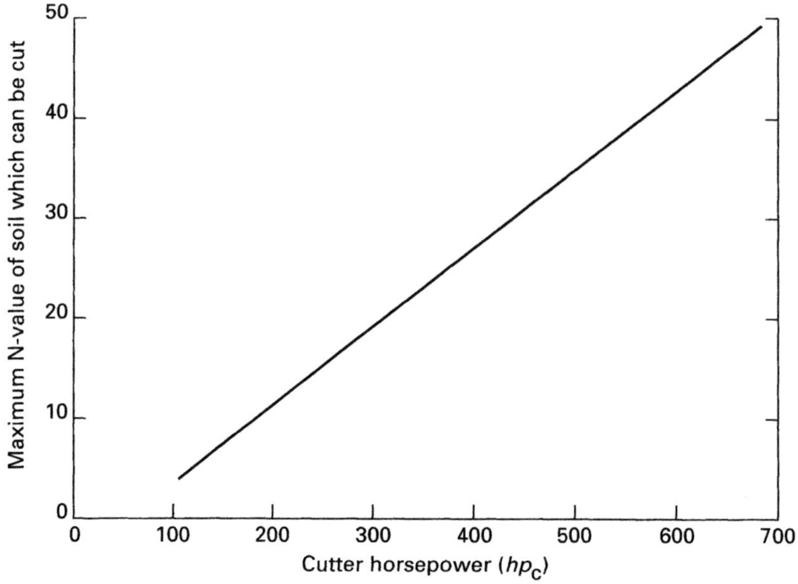

Figure 9.3 Cutter suction dredger: maximum N-value of soil which can be cut with cutter horsepower, hp_c

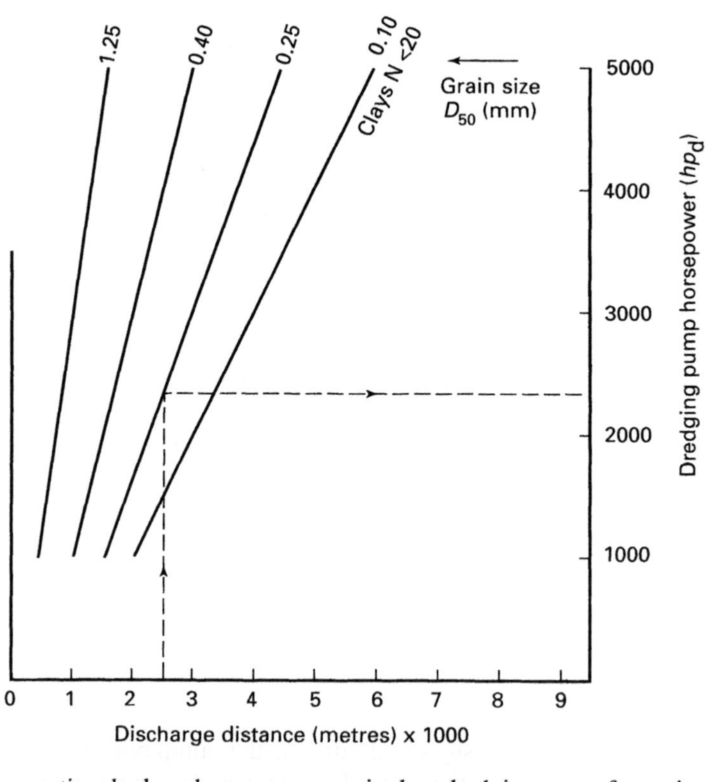

Figure 9.4 Cutter suction dredger: horsepower required on dredging pump for various discharge distances and soil characteristics

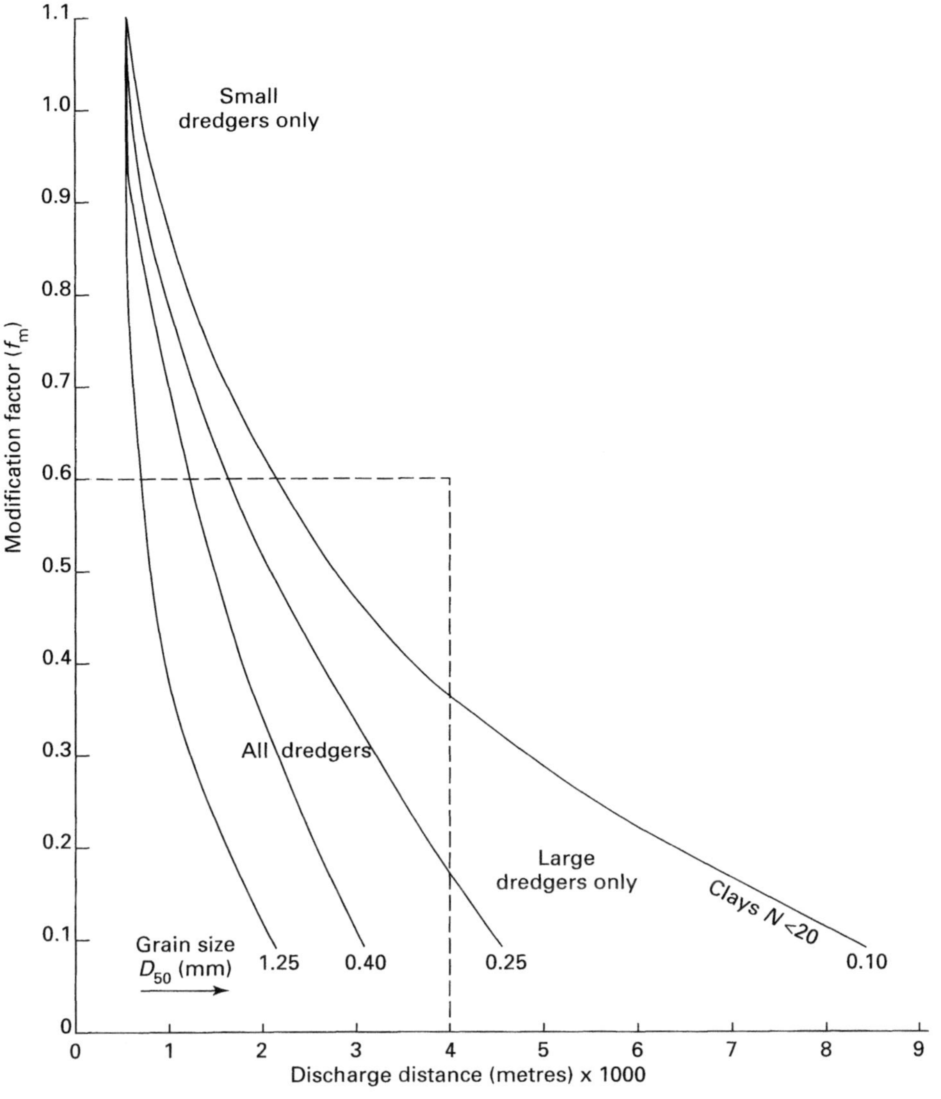

Figure 9.5 Cutter suction dredger: modification factor, f_m, for various discharge distances and soil characteristics

soil of a particular N-value. These values should be treated as limiting values and, for good production, the cutter horsepower should be well in excess of that indicated. In practice, production may be limited by the pump(s) or the cutterhead.

Having established that the cutter has sufficient power it is necessary to check that the pump power is sufficient to pump the soil the required distance. This may be done by reference to Figure 9.4 which gives indicative maximum discharge distances for varying pump power and soil characteristics. If there is insufficient pump power it is necessary to choose a larger dredger or add a booster pump to the system. When sufficient pump power is available, and a check has been made to ensure that the dredger is capable of reaching the required dredging depth, d, the theoretical output, P_t, can be established using the modification factor, f_m, obtained from Figure 9.5 as follows:

$$P_t = f_m \times hp_d \tag{9.9}$$

The theoretical output, P_t, is for a dredger limited by its pump. It is assumed that the cutter is able to supply the pump with the required amount of material. In the case of strong materials this will not necessarily be the case since, not only will the cutter not be able to cut material fast enough, but it may be necessary to cut only when swinging in one direction (see undercutting versus overcutting, Section 7.6.2, page 189). Production can also be limited by the height of the face to be cut and the relation between the dredging depth and the dredger's maximum depth. The maximum dredging depths for standard dredgers are shown in Figure 7.16, Chapter 7, page 184.

Optimum face heights for efficient dredging vary according to the type of material to be dredged and the size of the cutterhead. If the face height is so small that it limits production a reduction factor must be applied. The minimum face height may be estimated from the expression:

$$h_{min} = 2.1 \times D_s \tag{9.10}$$

where:

h_{min} = minimum face height (metres)
D_s = suction pipe diameter (metres)

(Note: The suction pipe diameter is usually just greater than or equal to the discharge pipe diameter.)

The reduction in theoretical output caused by the last two factors can be estimated from the graph in Figure 9.6. The factor, f_f, obtained from the graph is applied as follows:

$$P_{nom} = f_f \times P_t = f_f \times f_m \times hp_d \tag{9.11}$$

(see 9.9 above)

9.6.2 THE CYCLE FACTOR

To take account of the time lost due to the advancing of the dredger on its spuds a delay factor, f_p, is obtained as follows:

$$f_p = \frac{1}{1 + \dfrac{P_{nom} \times t_p}{z \times p \times b}} \tag{9.12}$$

There is an additional delay which occurs due to the time taken to move side anchors. This delay can be expressed as a factor, f_a, as follows:

$$f_a = \frac{1}{1 + \dfrac{P_{nom} \times f_p \times t_a}{z \times a \times b}} \tag{9.13}$$

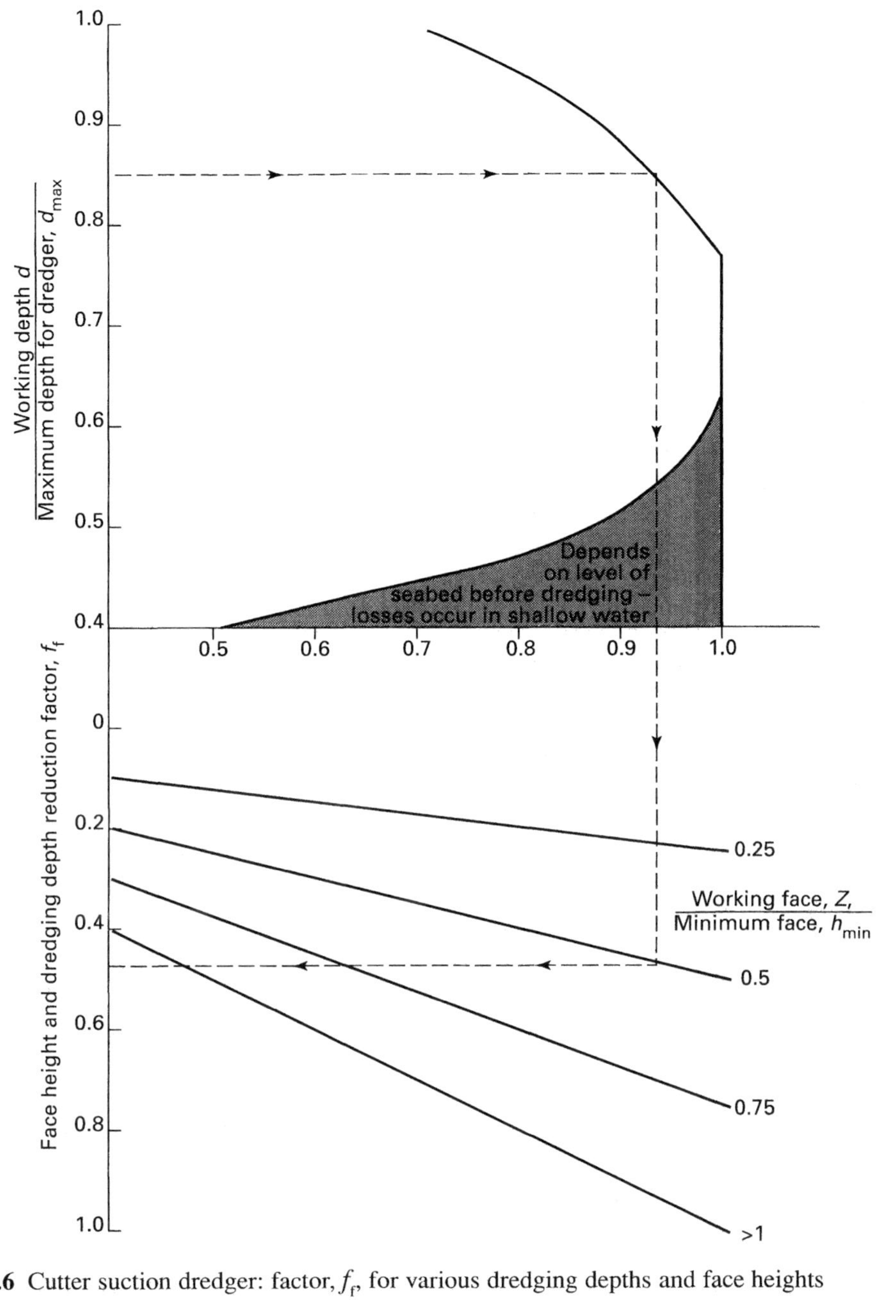

Figure 9.6 Cutter suction dredger: factor, f_f, for various dredging depths and face heights

The maximum potential output, P_{max}, is then given by:

$$P_{max} = f_a \times f_p \times P_{nom} \tag{9.14}$$

Note that no delay factor has been used to take account of moving the floating or land pipeline, or for replacing worn teeth and adapters at the cutterhead when dredging rock. It is assumed that this will be carried out during dead time. If this is not so, it will be necessary to reduce the output accordingly. In particular, when dredging strong or abrasive rock, the amount of time spent changing cutterheads or teeth may be considerable; 30 per cent or more in extreme cases.

Simplification

The following values can be assigned to some of the variables:

a = 50 to 100 metres depending on size of dredger;
b = 20 to 90 metres depending on size of dredger and depth of dredging;
p = 1.0 D_s to 6.0 D_s, where D_s = diameter of suction pipe, depending on strength of soil – smaller advances are taken in the stronger soils;
t_a = 0.33 hour;
t_p = 0.05 hour.

9.7 Grab dredger

Grab dredgers are either self-propelled or dumb. Self-propelled dredgers have their own hoppers. Dumb dredgers discharge their dredged material into a hopper alongside. For both types there is no particular relationship between grab and hopper size. The following information is, therefore, required to estimate output:

B = the bulking factor (see Section 9.3);
C = the bucket capacity (cubic metres) of the grab;
N = the number of grab cranes (grab hopper dredgers only);
H = the hopper capacity (cubic metres) (grab hopper dredgers or attendant barge) for the soil type;
d = the average dredging depth (metres);
z = the average thickness of material to be dredged in one cut (metres);
A = the average area dredged (square metres) by each crane in one dredging position;
g = the distance to the disposal site (kilometres) (grab hopper dredgers only);
t_a = the time required (hours) to advance to the next dredging position;
t_h = the time required (hours) to change hoppers (dumb grab dredgers only).

Also the soil type should be known.

9.7.1 THE PRODUCTIVE UNIT

For either type of grab the base productive unit, U_b, is the bucket capacity, C. Since most bucket capacities quoted are those for mud it is useful to note that the following relationship usually exists between buckets for the same grab crane:

Bucket type	Capacity
Mud	1.00 C
Sand/clay	0.72 C
Stones/rock	0.36 C

The productive unit of the grab must be modified according to the digability of the soil. However, unlike the other mechanical dredgers, the digging action is not powered by the dredger but by the weight of the grab bucket. For this reason the modification factor, f_m, is lower than for other dredgers and varies according to the size of grab. Table 9.3 gives values of f_m for various soil types and bucket capacities.

Table 9.3 Grab dredger; modification factor, f_m, for various soil types and bucket sizes

Soil type	Modification factor, f_m	
	2 m^3 bucket	4 m^3 bucket
Mud	0.75	0.80
Loose sand	0.70	0.75
Compact sand	0.60	0.70
Sand and clay	0.50	0.60
Stones	0.35	0.45
Broken rock (pre-treated)	0.20	0.30

NB: Factors will be higher for larger buckets and may exceed unity for large amounts of bulk dredging in some cohesive soils.

9.7.2 THE CYCLE FACTOR

The method of estimating the maximum potential output, P_{max}, for a grab dredger is different for dumb and self-propelled types.

Dumb dredger

The nominal uninterrupted output, P_{nom}, can be obtained from Figure 9.7, which gives values of P_{nom} based on different dredging depths, bucket capacities and modification factors. It should be noted that when the thickness of material to be dredged, z, is smaller than the usual digging penetration of the grab bucket there will be a reduction in efficiency. P_{nom} should be reduced to account for this.

Delay factors due to advancing the dredger and changing hoppers when they are full can be expressed as follows:

$$f_a = \frac{1}{1 + \frac{t_a P_{nom}}{Az}} \qquad (9.15)$$

where f_a is the delay factor for advancing, and

$$f_h = \frac{1}{1 + \frac{t_h f_a P_{nom} B}{H}} \qquad (9.16)$$

where f_h is the delay factor due to changing hoppers.

The maximum potential output can then be expressed as follows:

$$P_{max} = f_a f_h P_{nom} \qquad (9.17)$$

Self-propelled dredger

For the self-propelled dredger with N grabs, the total nominal output is obtained by multiplying the output of a single grab (obtained from Figure 9.7) by the number of grabs, N. The delay factor due to the advance of the dredger will then be given by expression (9.15).

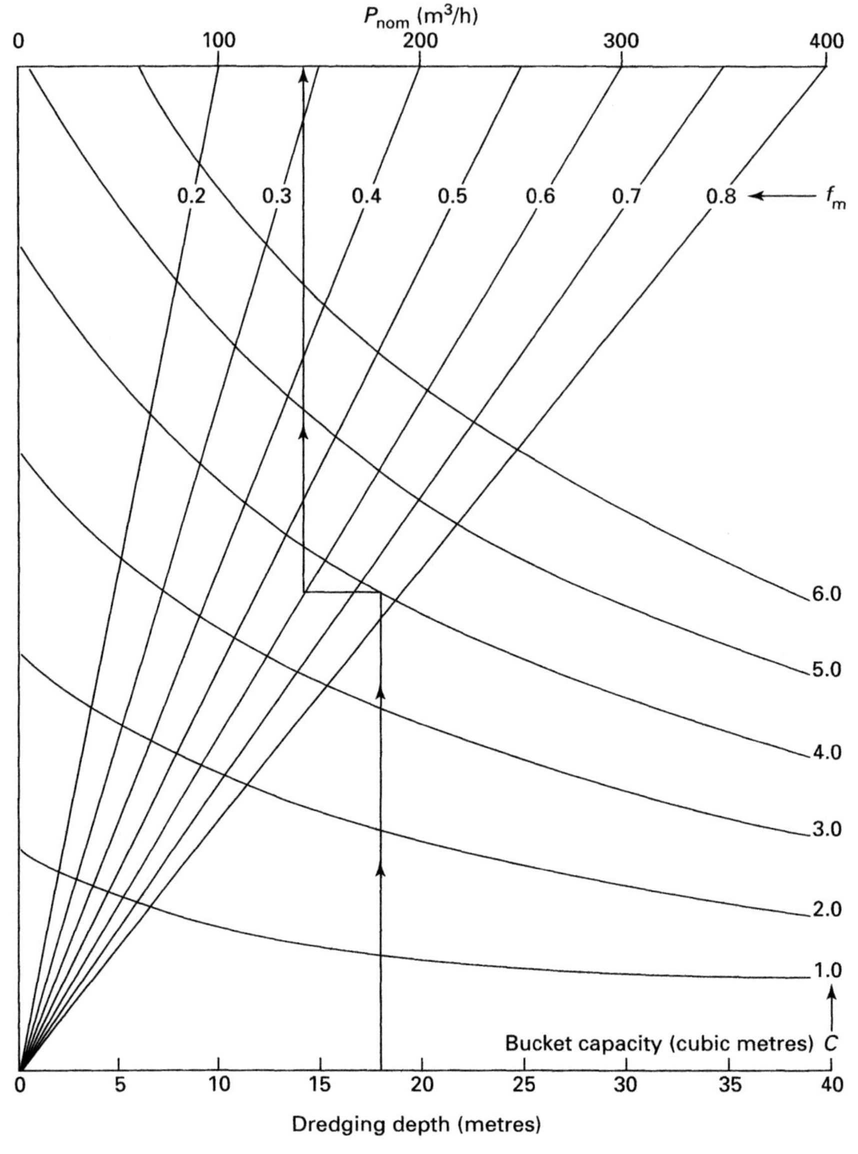

Figure 9.7 Grab dredger: nominal output, P_{nom}, for various bucket sizes and dredger characteristics

The delay factor, f_g, due to casting off, sailing to the disposal ground, sailing back and retrieving moorings is given by:

$$f_g = \frac{1}{1 + \frac{(0.66 + \frac{g}{3.25}) f_a P_{nom} B}{H}} \tag{9.18}$$

This allows the maximum potential output, P_{max}, to be calculated from:

$$P_{max} = f_a f_g P_{nom} \tag{9.19}$$

9.8 Backhoe dredger

The backhoe dredger is a versatile machine and there are many variations of power, bucket capacity and boom length. In the following estimating method it is assumed that the relationship between bucket capacity and power is reasonably standardised according to the normal range of off-the-shelf machines. If not, an equivalent bucket capacity should be used which equates to the horsepower of the machine being used. The following information is required, or must be estimated:

B = the bulking factor (see Section 9.3);
C = the bucket capacity (cubic metres) of the dredger;
z = the average thickness (metres) of material to be dredged in one cut;
d = the average dredging depth (metres) below water level;
H = the capacity (cubic metres) of the attendant hopper barge for the soil type;
A = the area (square metres) covered from one dredging position;
t_a = the time required (hours) to advance to the next dredging position;
t_h = the time required (hours) to change hoppers.

Also the soil type should be known.

9.8.1 THE PRODUCTIVE UNIT

The base productive unit, U_b, for the backhoe dredger is the bucket capacity, C. To take account of the soil conditions U_b must be modified according to the digability of the soil, as shown in Table 9.4.

Table 9.4 Backhoe dredger: modification factor, f_m, for various soil types

Soil type	Modification factor, f_m
Silts and soft clays	1.20
Sand and gravel	0.90
Sandy clay	0.80
Medium dense clay	0.75
Wet sticky clay	0.72
Broken rock (pre-treated)	0.55
Weak friable rock	0.30

NB: Factors will increase for large amounts of bulk dredging and may exceed unity for some cohesive soils.

9.8.2 THE CYCLE FACTOR

The sub-cycle of a backhoe dredger is influenced by dredging depth and, thus, the nominal output, P_{nom}, is a function of both dredging depth and soil type. In Figure 9.8 the lower half of the diagram can be used to assess P_{nom} for various combinations of d and f_m.

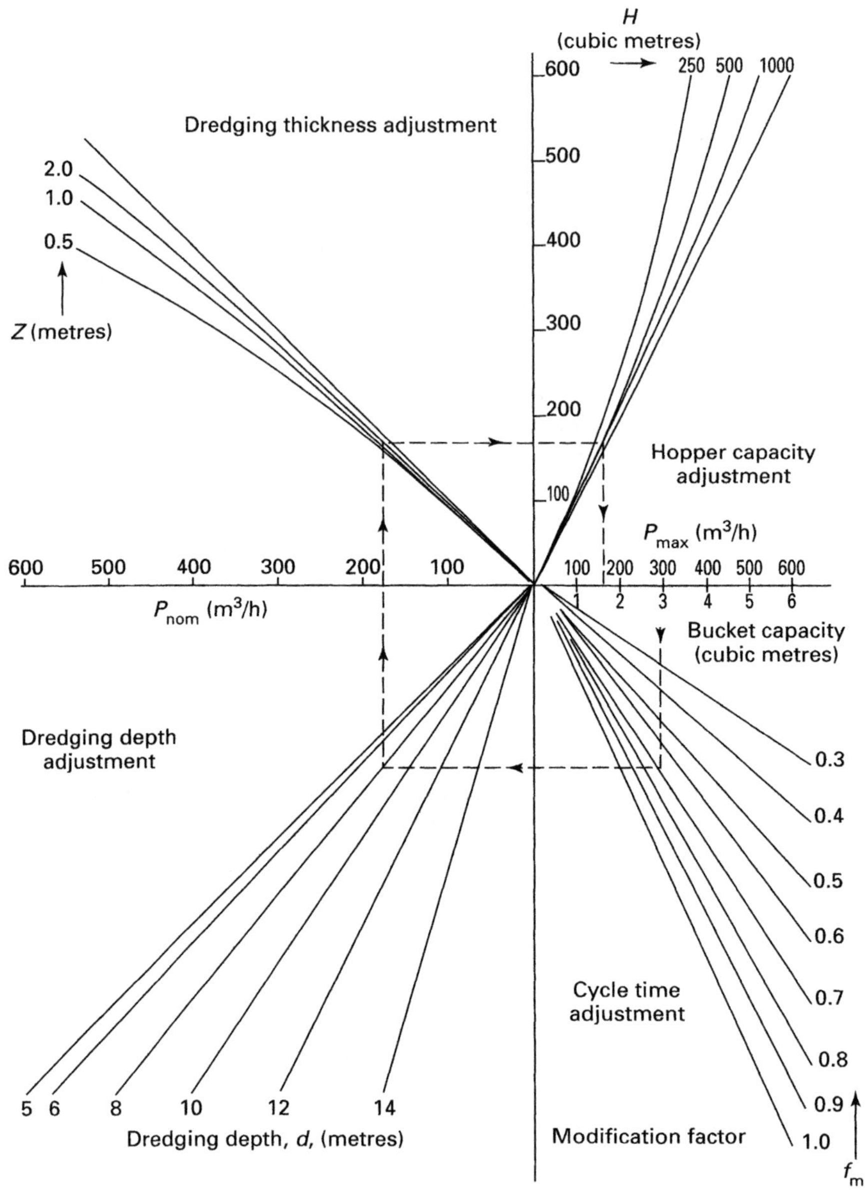

Figure 9.8 Backhoe dredger: outputs according to site and dredger characteristics

Delay factors due to advancing and changing hoppers can be expressed as follows:

$$f_a = \frac{1}{1 + \frac{t_a P_{nom}}{Az}} \tag{9.15}$$

where f_a is the delay factor for advancing, and

$$f_h = \frac{1}{1 + \frac{t_h f_a P_{nom} B}{H}} \tag{9.16}$$

where f_h is the delay factor due to changing hoppers.

The maximum potential output can then be expressed as follows:

$$P_{max} = f_a f_h P_{nom} \tag{9.17}$$

9.8.3 SIMPLIFICATION

The top half of Figure 9.8 makes the following simple assumptions.

- The area covered in one position = $(0.67.d_{max})^2$
- Advance time, t_a = 2.5 minutes
- Barge changeover time, t_h = 15 minutes.

If the actual dimensions or times are different to these, an appropriate pro rata adjustment must be made.

9.9 Dipper dredger

Dipper dredgers are uncommon. They are often built for a specific task so that their main characteristics of installed power, bucket size and maximum dredging depth are non-standard. The following information is required, or must be estimated:

B = the bulking factor (see Section 9.3);
C = the bucket capacity (cubic metres) of the dredger;
z = the average thickness (metres) of material to be dredged;
H = the capacity (cubic metres) of the attendant hopper barge for the soil type;
A = the area (square metres) covered in one dredging position;
t_a = the time required (hours) to advance to the next dredging position;
t_h = the time required (hours) to change hoppers.

The soil to be dredged must also be known.

9.9.1 THE PRODUCTIVE UNIT

The base productive unit, U_b, for the dipper dredger is the bucket capacity, C. To take account of various soil conditions U_b must be modified according to the digability of the soil, as shown in Table 9.5.

9.9.2 THE CYCLE FACTOR

Hydraulic dipper dredgers, which are now the most common type encountered, are somewhat different to the older, wire operated, machines, having a shorter digging period and being capable of dredging in much shallower water depths. In these circumstances dredging depth will affect output, much as for hydraulic backhoe dredgers (see Section 9.8).

Table 9.5 Dipper dredger: modification factor, f_m, for various soil types

Soil type	Modification factor, f_m
Sand	0.90
Medium clay	0.72
Gravel	0.60
Stiff clay	0.40
Pre-treated/highly fractured rock	0.33
Weak rock	0.30

NB: Factors will increase for large amounts of bulk dredging and may exceed unity for some cohesive soils.

A typical sub-cycle of a wire operated, non-hydraulic, dipper dredger can be broken down into the following components:

Operation	% of sub-cycle
Digging	40
Hoisting and lowering	26
Swinging	30
Discharging	4

Dippers are usually designed to work in water depths of between 10 and 20 metres and to raise the dredged material five metres above the water for discharging into a hopper barge. Thus, on the basis of the above sub-cycle breakdown and on the assumption that hoisting and lowering time is related to distance moved, an alteration of dredging depth of one metre is likely to alter the sub-cycle time by about one per cent.

It is found, however, that for the average dipper dredger there is a broad relationship between bucket capacity and nominal uninterrupted output. This is shown for various values of f_m in the lower right hand quadrant of Figure 9.9.

Dipper dredgers advance forward two to five metres after dredging the area ahead. Each movement forward takes about two to three minutes. The delay factor, f_a, caused by these movements can be expressed as follows:

$$f_a = \frac{1}{1 + \frac{t_a P_{nom}}{Az}} \tag{9.15}$$

In addition to the above delay there is another which is related to the time taken to change hopper barges and can be expressed as follows:

$$f_h = \frac{1}{1 + \frac{t_h f_a P_{nom} B}{H}} \tag{9.16}$$

where f_h is the delay factor due to hopper changing.

(Note: It should be checked that the hopper is capable of carrying a volumetrically full load without exceeding safe loading draft.)

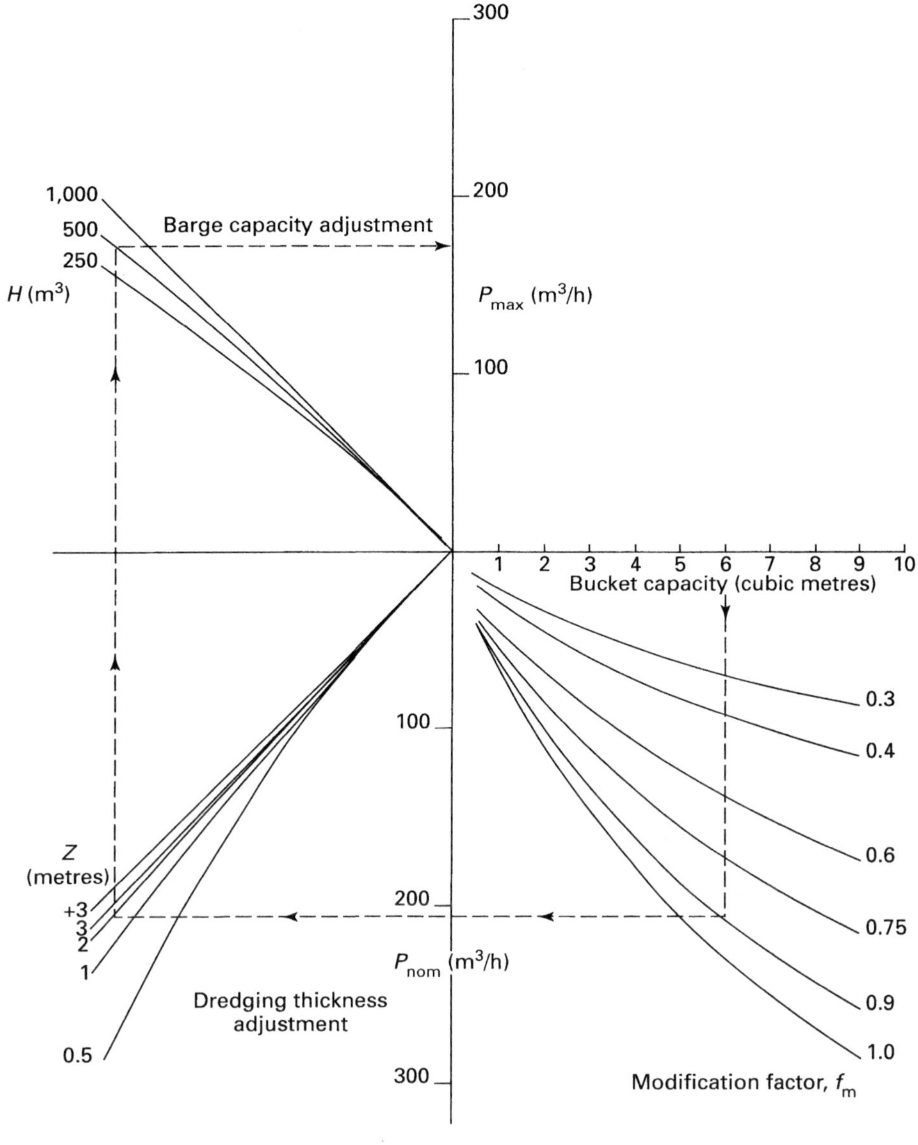

Figure 9.9 Dipper dredger: outputs according to site and dredger characteristics

This allows the maximum potential output to be expressed in the form:

$$P_{max} = f_a f_h P_{nom} \tag{9.17}$$

Simplification

On the assumption that the area, *A*, dredged by the dipper is around 50 square metres, and that the time taken to advance, t_a, is 2.5 minutes and that for changing hoppers, t_h, is 15 minutes, the maximum potential output will only depend on *z* and *H*. The effects of *z* and *H* on production are illustrated in the left-hand quadrants of Figure 9.9. Significant reductions occur only when *z* and *H* are relatively small.

9.10 Bucket dredger

The general characteristics of bucket dredgers are given in Chapter 7, Figure 7.50, page 224. In order to estimate the output of this type of dredger the following information is required or must be estimated:

B = the bulking factor (see Section 9.3);
C = the bucket capacity (cubic metres);
d_n = the normal dredging depth for the proposed dredger (metres) or the normal dredging depth with ladder extended;
d = the average dredging depth (metres);
b = the width of cut (metres);
z = the average thickness of cut (metres);
a = the advance distance (metres) between side anchor movements;
H = the capacity of the attendant hopper barge (cubic metres) for the soil type;
t_a = the time taken to move side anchors (hours);
t_h = the time taken to change hopper barges (hours).

Also the soil type should be known.

9.10.1 THE PRODUCTIVE UNIT

The base productive unit, U_b, for a bucket dredger is the capacity, C, of an individual bucket in the chain. This has to be modified to take account of the digability of the soil to be dredged. Modification factors for various soil types are shown in Table 9.6.

Table 9.6 Bucket dredger: modification factor, f_m, for various soil types

Soil type	Modification factor, f_m
Stiff clay	0.80
Soft clay	0.90
Sticky clay	0.60
Coarse sand	0.80
Medium sand	0.70
Fine sand	0.60
Broken rock (pre-treated)	0.40
Weak rock	0.20

The nominal uninterrupted output of the dredger, P_{nom}, can then be estimated from the following expression:

$$P_{nom} = 60Cf_m nf_\theta \qquad (9.20)$$

where n = bucket chain speed (buckets per minute)
f_θ = tilt factor

The bucket chain speed, n, is related to the type of soil being dredged as well as a number of other site factors. When the dredger is being operated with its normal ladder length and the horsepower of the machine is related to bucket capacity, as indicated in Chapter 7, Figure 7.50, page 224, then the bucket chain speed may be estimated from Table 9.7.

Table 9.7 Bucket dredger: bucket chain speed, n, for various soil types and site conditions

Soil type	Bucket chain speed, n (buckets per minute)
Very weak material	25 to 28
Weak material	18 to 22
Stiff material	15 to 18
Very stiff material	12 to 15
Broken rock (pre-treated)	8 to 12
Weak rock	3 to 5

NB

- if extended ladder is used, n is reduced;
- if horsepower of dredger is above or below average, n may be increased or decreased;
- if soil is sticky, n is reduced.

The tilt factor, f_θ, takes account of the reduction in effective bucket capacity when the bucket ladder is being used at an angle which causes soil to spill from the back of the buckets. The optimum angle is usually near 45°. For convenience, the tilt factor has been given in relation to the normal dredging depth, or extended ladder dredging depth, d_n, and the site dredging depth, d. Values of f_θ can be obtained from Figure 9.10.

9.10.2 THE CYCLE FACTOR

Time is lost during the operations of a bucket dredger when hopper barges are changed, unless loading is carried out on both sides of the dredger, in which case the delay is minimal.

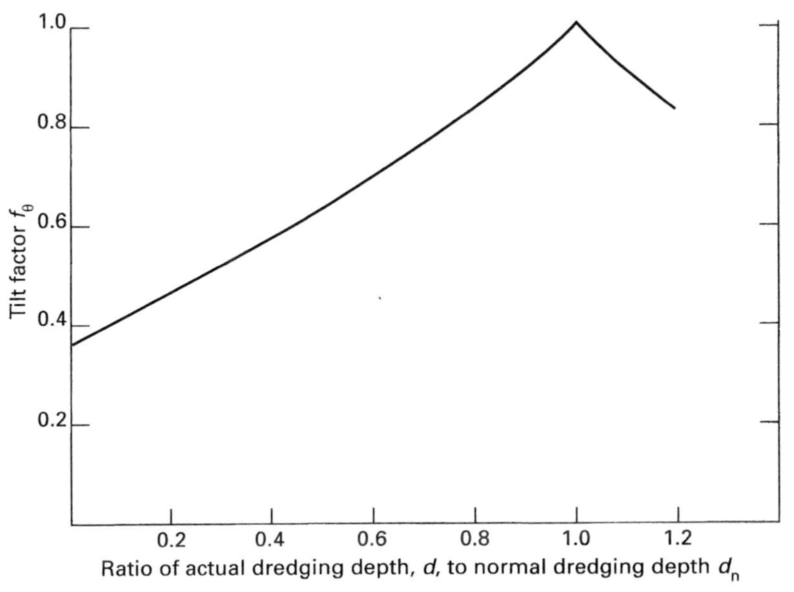

Figure 9.10 Bucket dredger: tilt factor, f_θ, for ratios of dredging depth, d, to normal dredging depth, d_n

The delay factor, f_h, caused by this is given by:

$$f_h = \frac{1}{1 + \frac{t_h P_{nom} B}{H}} \tag{9.21}$$

In addition to this, side anchors must be moved when the dredger has advanced a sufficient distance. The delay factor caused by anchor moving, f_a, is given by:

$$f_a = \frac{1}{1 + \frac{t_a f_h P_{nom}}{abz}} \tag{9.22}$$

(Note: the order of calculating f_a and f_h has reversed compared with previous dredgers. This is because of the relative frequencies of the two types of delays occurring.)
The maximum potential output can then be expressed in the form:

$$P_{max} = f_a f_h P_{nom} \tag{9.17}$$

Simplification

The following values could be assigned to some of the variables.

a = 75 m
b = 75 m
t_a = 0.33 hours
t_h = 0.25 hours (assumes barge on one side only)

Expressions (9.21) and (9.22) then become:

$$f_h = \frac{1}{1 + \frac{0.25 P_{nom} B}{H}} \tag{9.23}$$

$$f_a = \frac{1}{1 + \frac{5.86 \times 10^{-5} f_h P_{nom}}{z}} \tag{9.24}$$

It should be noted that expression (9.24) is for dredging to full depth in one cut, or in two cuts from the same anchor position. If the whole site is dredged in one cut to an intermediate level and subsequently in another cut to a final level, the value of *z* must be taken as the cut thickness and not the total thickness of dredged material.

10
Dredging Costs and Prices

10.1 Introduction

A clear distinction must be made between the cost to the contractor of executing a particular project and the price which he charges the client for that work. Costs are relatively easy to establish and the basic principles of cost estimation are discussed in this chapter. The pricing strategy which is adopted by contractors will vary significantly from time to time and does not have a fixed relationship with the costs of execution. For this reason it is often difficult to estimate with confidence the price of a particular dredging project without inside knowledge of the current state of the dredging market and the contractors' pricing strategies. Therefore, advice on this matter should be sought, either from specialist consultants or contractors.

The cost to operate dredging plant and the cost to execute dredging works are 'live' costs. Hence they are subject to change with time and with circumstances. However, given the necessary knowledge and experience, the cost to operate a particular item of plant or execute a particular work can be estimated with reasonable accuracy if the appropriate procedure is followed.

For most works, the total cost will include two important elements: the cost to mobilise and demobilise all necessary plant, labour and equipment, and the cost of actually carrying out the work. The cost of mobilisation is very heavily dependent on the time that the work is to be executed and its location. The price may be strongly influenced by market forces. It is, therefore, difficult, even for contractors, to estimate mobilisation costs long in advance of the work, particularly if the works are in a location which is remote from centres of regular dredging activity. Because of this, it is not normally possible to estimate mobilisation costs by any standard procedure.

In this chapter, a procedure for the estimation of the operating costs of dredging plant will be explained. When the operating cost per unit of time is known, this can be divided by the estimated production rate per unit of time to provide a cost per unit volume of dredging.

10.2 Mobilisation and demobilisation

Unless the distances vary significantly, usually there will be only small differences between the cost of mobilisation and demobilisation. The difference is mainly accounted for by the costs of site establishment, that is the erection of offices, connection of services and time and expenses of supervisory staff. The fact that when tendering, contractors will often enter significantly higher sums for mobilisation than for demobilisation usually owes more to cash-flow considerations than to actual costs.

For some types of dredging plant, mobilisation is a simple procedure and the costs can be estimated with reasonable confidence. For example, if a trailing suction hopper dredger is required for normal port maintenance dredging, with disposal of the dredged material at sea, then it is possible that no other plant will be required. The dredger has only to sail to the work site from its current location and begin work on arrival.

The cost per unit of time whilst travelling can be estimated, as will be shown later. The sailing time can be estimated when the sailing distance and sailing speed are known. Some allowance for adverse weather may be necessary, depending on the length of voyage, size of dredger and time of year.

From these estimates, the major part of the cost of mobilisation can be found. Further, but usually minor, costs may include such items as flying in additional crew and supervisory staff on arrival of the dredger and if necessary, setting up positional control and site offices. If the voyage is long, or if heavy sea conditions are expected, it may also be necessary to secure the ship for transit.

In contrast to the relatively simple mobilisation procedure for trailing suction dredgers, stationary dredgers (such as bucket dredgers, backhoe dredgers and most cutter suction and grab dredgers) require a more elaborate and usually more expensive mobilisation procedure. Except over very short distances and in calm seas, the mobilisation costs for stationary dredgers will usually include some of the following elements:

- prepare for transport;
- transport over land (perhaps at both ends);
- craneage (perhaps at both ends);
- transport over water;
- re-assemble/prepare for work;
- insurance;
- supervision;
- plant costs while idle (crew, etc.);
- overheads.

The most important of these costs can only be estimated given detailed knowledge of the particular plant to be mobilised and the journey.

10.3 Plant capital costs

The capital cost of most marine dredging plant is high. A small trailing suction hopper dredger may cost several million US dollars. Even a small cutter suction dredger for inland operation, if complete with all of the necessary pipelines and ancillary equipment, is likely to cost about a million US dollars (1993).

The capital cost of individual dredgers, even though of similar capacity, may vary widely. This is because the method of construction and sophistication of design and equipment is highly variable.

To provide an indication of the average relative costs, the capital cost for a variety of different types and size of dredging plant, as at January 1995, is provided in Figure 10.1. The cost indications provided by these figures should be used with caution for the reasons given above. A cutter suction dredger which is designed to dredge rock may cost several times more than one of similar size which is designed to dredge only loose superficial deposits.

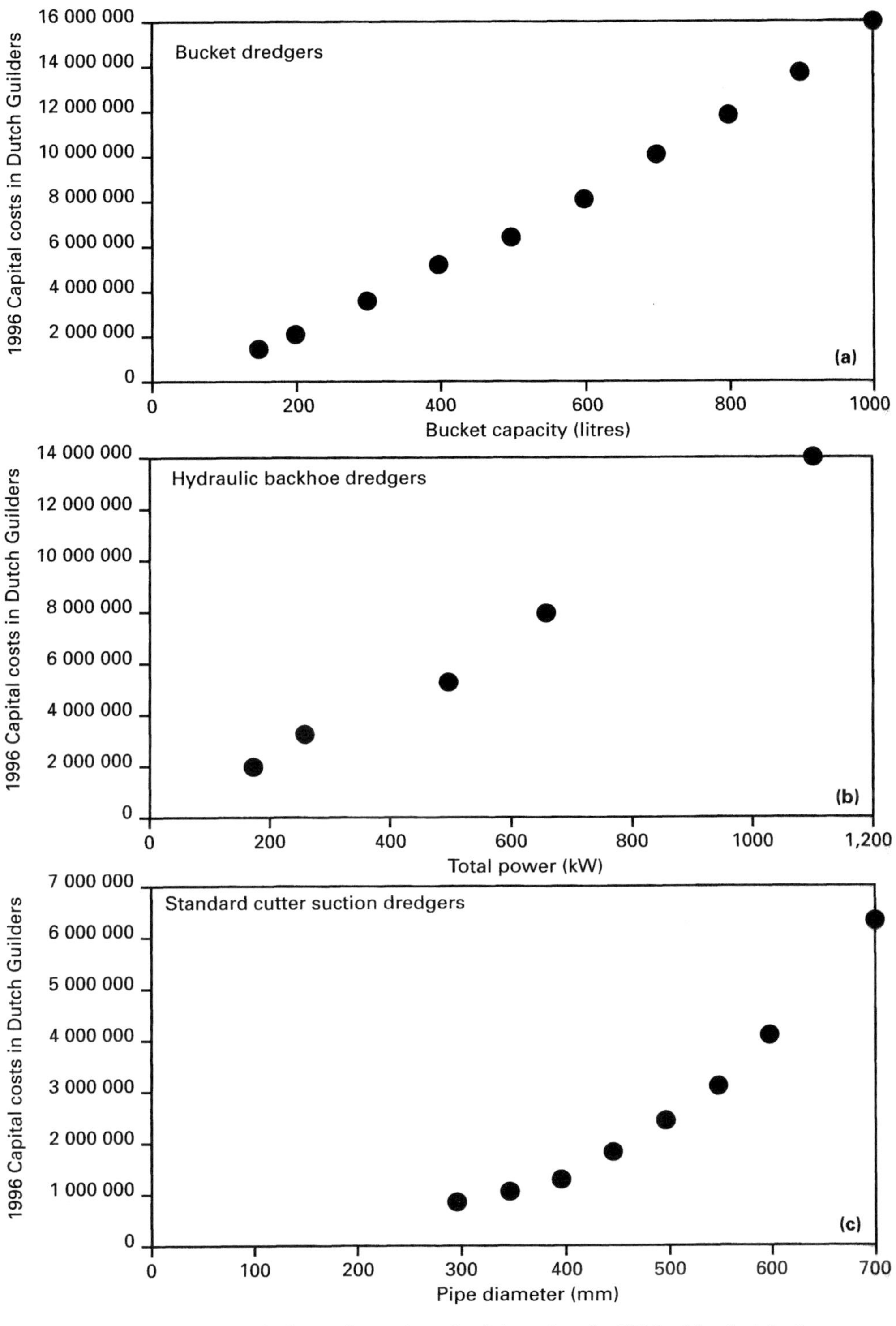

Figure 10.1 Approximate capital cost for various dredging plant in 1996: a) bucket dredgers, b) hydraulic backhoe dredgers, c) standard cutter suction dredgers

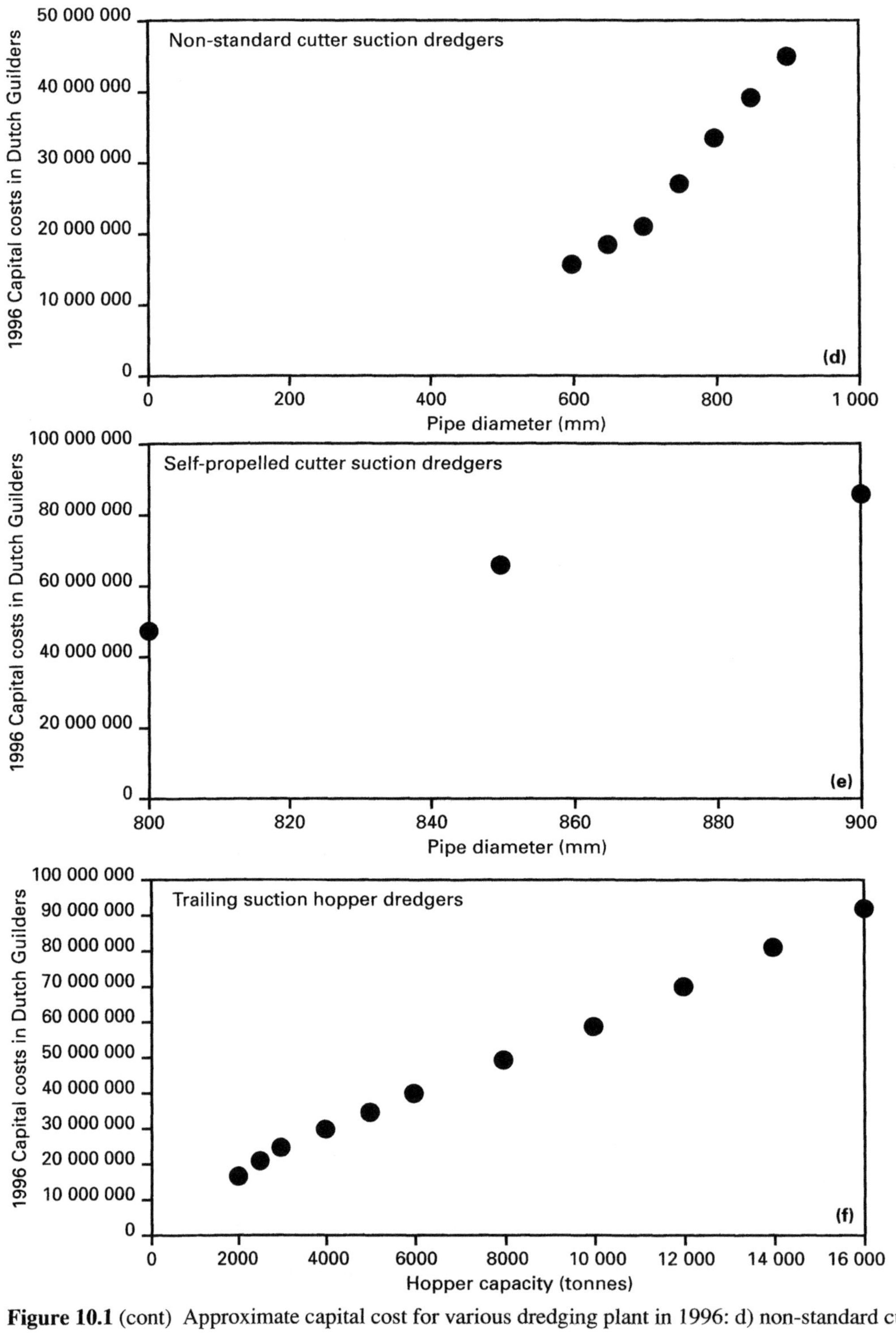

Figure 10.1 (cont) Approximate capital cost for various dredging plant in 1996: d) non-standard cutter suction dredgers, e) self-propelled cutter suction dredgers and f) trailer suction hopper dredgers

10.4 Plant running costs

Plant running costs can be estimated by following a simple procedure which is explained below. However, the running cost will be strongly influenced by the particular operating conditions and the quality of crew and management. Significant variation may therefore occur between different organisations and different sites.

Costs will usually occur under some, or all of the following headings:

- fuel and lubricants;
- other consumables;
- crew;
- survey;
- supervision;
- routine maintenance and running repairs;
- major repairs and overhauling;
- wear;
- insurance;
- overheads;
- financial charges (depreciation, amortisation and interest on capital employed).

Each of the above cost headings is a variable which can be estimated.

10.4.1 FUEL COST

The cost of fuel will depend upon the following factors:

- type of fuel used;
- local unit cost of fuel;
- power requirements of dredging plant;
- working hours of dredging plant;
- idle hours for dredging plant;
- efficiency of dredging plant;
- local losses (spillage and pilfering).

Power requirements are dependent on the work to be performed. For example, pumping through long pipelines at high velocity will require more power than pumping through a short pipeline of the same diameter at low velocity. Similarly, backhoe dredgers, working at consistently high hydraulic circuit pressure, will demand more power than if working at consistently lower pressures. To estimate power requirements, it is necessary to examine each element of the working cycle and working day, and to determine approximately what percentage of the available power will be used and for how long. It should be remembered that large items of plant may continue to use power, even when idle, due to the need to maintain vital functions.

Working hours and idle hours are dependent on the manning hours of the crews, site characteristics, nature of the dredging task and efficiency of operation.

10.4.2 LUBRICANT COSTS

Lubricants include oils and greases. Some equipment may demand special lubricants, for instance waterproof or high temperature. Lubricant costs vary from one type of plant to another, but it is usual to assume that costs are equal to approximately 10 per cent of fuel costs. However, there may be significant variation for particular plant, or circumstances. Old plant which is poorly maintained may consume more lubricants than modern, well maintained plant.

10.4.3 OTHER CONSUMABLES

Each type of dredging plant will consume a range of different products. These will include some, or all of the following:

- paint;
- packings;
- gaskets;
- teeth;
- ropes, wire;
- ropes, soft;
- fresh water;
- cleaning materials;
- food and cooking consumables.

10.4.4 CREW

There are wide variations between crew levels in different countries and cultures. Crew numbers are usually lower in developed countries where labour costs are relatively high. For example, a small trailing suction hopper dredger in Northern Europe may have a crew of only 3 men, if not required to work continuously. It is not unusual for a similar dredger to carry a crew of 10 men or more in countries where labour is cheap. Sometimes crew numbers are influenced by other considerations. For example, it may be possible to operate a modern hydraulic backhoe dredger with only one man, but safety considerations and occasional heavy work will dictate a crew of at least two.

Generally, the larger the dredger, the greater the number of crew required. The total number of crew is dependent also on the working hours per day and number of shifts. A complete crew will include all, or some of the following.

Afloat:

- captains;
- officers;
- engineers mechanical;
- engineers electrical;
- engineers electronic;
- operators;
- winchmen;
- welders;
- deckhands;
- cooks;
- boatmen.

Ashore:

- fill foreman;
- pipeline gang;
- machine drivers.

10.4.5 SURVEY

Regular survey is essential to the efficient management and execution of all dredging work. For stationary types of dredger the requirement for detailed surveys may be infrequent, perhaps at intervals of two weeks, or longer. Usually it will be possible for the dredgemaster to use simple hand line sounding methods from the dredger pontoon to check that the required depth is being achieved. For trailing suction dredgers regular survey may be required at shorter intervals, perhaps every few days, or even every day, in order that the dredgemaster knows which areas are finished and which require further dredging. Often, on small and simple projects, one surveyor with assistance from the crew of the survey boat can carry out the required work. Survey requirements and methods are discussed in detail in Chapter 6.

10.4.6 SUPERVISION

The level of supervision necessary is directly dependent on the scale and complexity of the work and must be a matter of judgement. However, for each working shift, one would expect a contractor to provide a land-based supervisory staff equivalent to:

- For a small project – One supervisor.
- For a medium sized project – One agent. One supervisor per major plant type. Basic office support.
- For a large project – One agent. Sub-agent per shift or section. One supervisor per major plant type. Major office support, including administration, financial control and payroll staff.

10.4.7 ROUTINE MAINTENANCE AND RUNNING REPAIRS

It is usual for all items of day to day maintenance and running repairs to be grouped under this heading. It will usually include some, or all of the items listed below:

- oiling and greasing;
- routine cleaning and painting;
- replacement of wearing parts;
- replacement of filters, packings (etc.);
- replacement of engine injectors (etc.);
- replacement of worn or damaged pipes and hoses;
- replacement of electrical consumables;
- replacement of small parts, motors, valves.

10.4.8 MAJOR REPAIRS AND OVERHAULING

It is usual for all major items of repair or replacement to be grouped under this heading. These will usually include some or all of the following items:

- replacement of major components;
- regular overhauling of engines and motors;
- complete repainting and protection;
- stripping and overhauling winches (etc.);
- substantial structural repairs;
- dry docking for repair or classification.

10.4.9 WEAR

The cost of wear and tear on machinery and other parts of a dredger or craft are usually included either under consumables, or maintenance and repair. However, for items with which wear is the primary cost, such as pipelines, it is usual to estimate the rate of wear and to make an appropriate allowance. For example, if for a specified task it is estimated that a pipeline will be completely worn out after passing 2 000 000 cubic metres, then the unit cost of pipeline wear will be found by dividing the capital value by 2 000 000. The daily cost of pipeline wear will be given by multiplying the unit cost by the estimated daily production.

Pipeline wear is difficult to estimate. Normally operators will allow their pipes to wear until a specified minimum wall thickness is reached, often between six to eight millimetres depending on the pressure in the line. In the period between changing from a new to a completely worn pipe the operator may have changed from one project to another. Even when wear rates are such that a pipe will be used totally on one project, the operator will have to turn the pipes to ensure an even wear rate. Pipe wear rates may vary from 100 000 cubic metres of dredged material per one millimetre of wear to well over one million cubic metres for the same wear.

10.4.10 INSURANCE

Insurance costs may vary greatly between different owners. An owner with a good record of few claims can expect to pay a lower premium than one who has suffered frequent, or substantial plant damage, or loss. A public owner, such as a port authority, which might have all port plant, including dredgers, cranes and so on covered under a single policy, may pay less than a dredging contractor.

Insurance premiums are usually based upon a percentage of the capital, or insured value of the plant. The percentage will be higher when working in a hostile environment, or politically unstable region. Annual premiums, based on the value of the plant, may range from a fraction of a per cent per annum for a public owner to 5 per cent or more for a contractor with a record of significant claims.

10.4.11 OVERHEADS

Overhead costs are also very variable. A figure of around nine per cent of the working costs might be taken as a starting point.

Note: Contractors will usually apply a fixed percentage overhead cost to each contract, but the actual percentage will vary between contractors and may be reduced during a period of recession, or in an attempt to secure a particular contract. Public owners may not charge realistic overheads, perhaps preferring to charge only those costs which are directly related, such as supervision.

10.4.12 FINANCIAL CHARGES

Depreciation

It is most common to depreciate dredging plant on a straight line basis. Small dredgers, such as portable cutter suction dredgers, may be completely written off over a ten year period. Large trailing suction hopper dredgers will usually be written off over much longer periods, with 30 years being a common choice. Annual depreciation costs are found by dividing the capital value by the assumed life. For plant owned by contractors, the daily cost of depreciation when working is found by dividing the annual cost by the estimated number of working days per year.

Amortisation

Amortisation may be combined with depreciation. A sinking fund may be required to provide for the eventual replacement of dredging plant at the end of its useful life. The amount set aside annually should allow for inflation which may occur over the term of the fund. Unless the rate of inflation is constant, which is unusual, it will be necessary at regular intervals, usually annually, to review the estimated cost to replace the equipment. The use of a sinking fund may have adverse taxation effects.

Interest on capital

Capital which is invested in equipment could instead be invested to earn interest, or in an appreciating stock or asset. Alternatively, if capital is borrowed to finance the acquisition of equipment, interest will be payable on the loan. If the cost of ownership is to be accounted for realistically, the cost of lost interest, or interest paid, must be shown.

Public owners may not make any provision for replacement and sometimes may neglect depreciation and interest on capital, particularly when equipment has been acquired with partial, or total grant aid. It can be argued that this is bad practice. Certainly, if a proper comparison is to be made between investment in plant for direct labour operation and contract dredging, then all financial charges should be taken into account, otherwise the comparison is invalid.

10.5 Building up a cost estimate

Cost estimates divide into two basic parts: the cost per unit of time to operate each item of plant and the cost to execute a particular task. The first must be known before the second can be estimated.

10.5.1 COST TO OPERATE DREDGING PLANT AND EQUIPMENT

It is usual to estimate the cost of operation per unit of time. The most convenient units of time are as follows:

- hours;
- days;
- months;
- years.

(See Chapter 9 for the normal production estimating periods and what they include.)

The particular unit of time used is not important. In the procedures which follow the units used will be days. The procedure used is similar for all types of dredging plant and for most ancillary plant. However, for some equipment, such as pipelines, a different procedure is necessary. The cost estimate procedure can be conveniently described within a standard cost schedule. If it is required to make repeated estimates, this can be most conveniently achieved by construction of a simple spreadsheet on a personal computer.

In the text which follows, it is assumed that a spreadsheet will be used for estimating purposes. The sheet should be constructed using the fixed elements and factors as proposed and the variables entered as appropriate.

Cost estimate schedule – for dredging plant
• Capital value of plant • Working days per year • Theoretical hours per day (hours crewed) • Operational efficiency (0 – 100 per cent)

1 Fuel

Assumptions: Rate of fuel consumption is equal to 0.182 litres per horsepower per hour.

Fuel consumption = 0.182 × horsepower hours

Enter: Total horsepower (hp) and theoretical hours

Estimate horsepower hours as follows:

Hours		percentage power		
( )	×	( )	=	( )
( )	×	( )	=	( )
( )	×	( )	=	( )
( )	×	( )	=	( )
Total			=	( ) × total hp = hp hours

For example:
Fuel consumption per day for plant with 1000 hp and crewed for 24 hours per day, working 4 hours at full power, 14 hours at 75% and 6 hours at 10% will be as follows:-

Litres/day = $1000 \times 0.182 \times ((4 \times 1.0)+(14 \times 0.75)+(6 \times 0.10))$

= etc.

2 Lubricants

Assume 10 per cent of (1)

3 Other consumables

Entirely dependent on type of plant and task to be performed.

Estimate:

- number of teeth;
- weld rods;
- wire ropes;

(etc.)

4 Crew

Enter the following information:

Type or rank	Number		Day rate		Cost per day
Captains	( )	×	( )	=	( )
Officers	( )	×	( )	=	( )
Eng. mechanical	( )	×	( )	=	( )
Eng. electrical	( )	×	( )	=	( )
Eng. electronic	( )	×	( )	=	( )
Operators	( )	×	( )	=	( )
Winchmen	( )	×	( )	=	( )
Welders	( )	×	( )	=	( )
Deckhands	( )	×	( )	=	( )
Cooks	( )	×	( )	=	( )
Total					( )

(Note: Cost per day is all-inclusive cost, not simply salaries.)

5 Survey

Entirely dependent on nature of task to be performed.

6 Supervision

Entirely dependent on nature of task to be performed.

7 Routine maintenance and running repairs

Usually expressed as a percentage of replacement cost, but account must be taken of theoretical hours and nature of the task to be performed. Average figures for continuous (168 hour) operation are as follows.

	Multiplier	×	capital value	=	cost per day
Trailing suction hopper dredger	0.000135	×	( )	=	( )
Cutter suction dredger	0.000140	×	( )	=	( )
Backhoe dredger	0.000140	×	( )	=	( )
Grab dredger (pontoon mounted)	0.000130	×	( )	=	( )
Self-propelled hopper	0.000130	×	( )	=	( )
Dumb hopper	0.000025	×	( )	=	( )
Workboat	0.000145	×	( )	=	( )

8 Major repairs and overhauling

Usually expressed as a percentage of replacement cost, but account must be taken of theoretical hours and nature of the task to be performed. Average figures for continuous (168 hour) operation are as follows:

	Multiplier	×	capital value	=	cost per day
Trailing suction hopper dredger	0.000275	×	( )	=	( )
Cutter suction dredger	0.000300	×	( )	=	( )
Backhoe dredger	0.000300	×	( )	=	( )
Grab dredger (pontoon mounted)	0.000250	×	( )	=	( )
Self-propelled hopper	0.000260	×	( )	=	( )
Dumb hopper	0.000050	×	( )	=	( )
Workboat	0.000300	×	( )	=	( )

9 Insurance

Very variable. Dependent on type of owner and insurance record, excess, location and nature of task to be performed and so on. For general contract work, assume an annual premium 2.5 per cent of insured value of plant.

$$\text{Daily cost} = \frac{\text{Insured value} \times 0.025}{(\text{working days per year})}$$

10 Overheads

Very variable. Dependent on type of owner, type of plant and so on. For general contract work, assume 9 per cent on cost on items 1 to 9.

$$\text{Overhead cost} = \text{Sum}(1\text{–}9) \times 0.09 = (\quad)$$

11 Depreciation

Dependent on owner's fiscal policy, nature of task and market conditions. Usual to assume straight line depreciation to zero value over between 10 and 30 years. For general contract work, assume as follows.

	Period (years)	Cost per day
Small trailer	20	
Medium or large trailer	30	
Small cutter suction	10	
Medium cutter suction	20	
Large cutter suction	25	
Backhoe	20	$\frac{(\text{Capital value})}{(\text{period}) \times (\text{working day/yr})}$
Grab hopper	25	
Grab, pontoon mounted	20	
Self-propelled hopper	25	
Dumb hopper	25	
Small workboat	10	
Large workboat, tug	20	

12 Amortisation

Dependent on owner policy, rate of inflation, local tax laws and so on. No standard quantification possible.

13 Interest on capital employed

Dependent on owner's fiscal policy and whether plant acquired by means of a loan, or cash purchase. If loan, apply loan interest rate; if cash, apply current rate of return on bank deposits, or other deposits or investments, according to policy.

$$\text{Cost per day} = \frac{(\text{Capital cost}) \times (\text{rate of interest})}{(\text{working day/yr})}$$

14 Miscellaneous

Enter any cost items not covered above.

10.5.2 COST OF EXECUTING A PARTICULAR TASK

The total cost of executing a dredging project is given by the sum of the following:

1 Mobilisation and demobilisation.
2 Ancillary work (containment areas, etc.).
3 Execution of the dredging and/or filling.

Items 1 and 2 are project specific and a standard approach is not easily defined.

The execution cost is directly related to the type, size and number of items of plant employed and the total duration of the work.

In order to estimate the dredging cost it is necessary to progress through the various stages listed below:

1 Decide the types of plant which are required.
2 Decide the size of plant required.
3 For large works, or works which must be completed quickly, decide the numbers of plant required.
4 Estimate the rate of production P for each item of plant and for each component of the work, or for each different type of material to be dredged (see Chapter 9).
5 Estimate the period of employment for each item of plant. For dredging plant this is found by dividing the volume V to be dredged by that plant (including any overdredging) by the estimated production rate P.
6 Estimate the cost per unit of time C for each item of plant to be employed on the work (see Section 10.5).

The cost of project execution is then given by:

$$\text{Execution cost} = \frac{(V1 \times C1)}{P1} + \frac{(V2 \times C2)}{P2} + \frac{(V3 \times C3)}{P3} \quad \ldots \text{etc.}$$

Where: $V1$ = volume in area (part 1)
$C1$ = cost per unit of time (may be approximately constant)
$P1$ = production per unit of time in area 1

If any ancillary plant, such as workboats, are involved, these must also be costed over the appropriate period of time. Also, in some programmes a period of prolonged idle time for a dredger may be unavoidable, and this time must be costed into the project at a stand-by rate. Alternatively, the cost of additional mobilisation and demobilisation must be added.

As is the case for the estimation of plant operating costs, the cost of project execution can be conveniently set up using a PC spreadsheet. The project execution cost spreadsheet can be linked to other spreadsheets, such as might be used to estimate production and plant costs. Thus, an arrangement of linked spreadsheets can be set up, as shown in Figure 10.2.

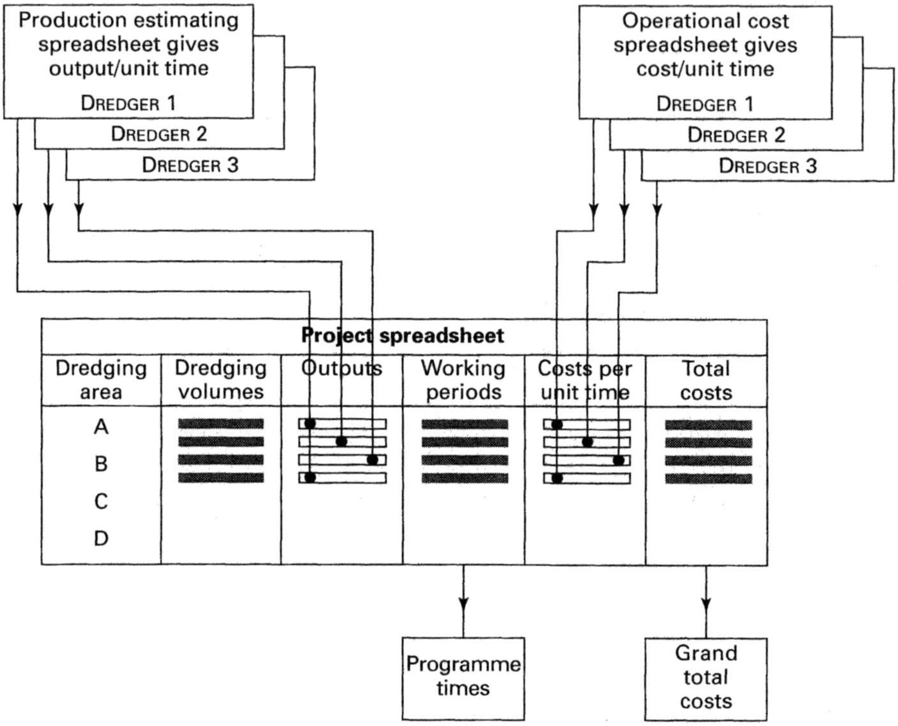

Figure 10.2 Illustration of linked spreadsheets for estimation of project costs

10.6 Contractor's pricing policy

It is important for those who intend to employ dredging contractors to achieve the lowest price compatible with the required quality and timely execution of the work. To achieve this objective it is necessary to understand at least the fundamentals of a contractor's pricing policy, although this may vary between different contractors and over a period of time. Pricing policy will be subject to a variety of influences, some obvious, such as market conditions and risk; others less obvious may include programme, strategy and even conditions which prevail during a site visit when tendering.

The contractor's main objectives are to achieve a satisfactory level of employment for his dredging plant and skilled labour and, so far as market conditions will allow, to recover his running costs, overhead costs, financial costs and a reasonable profit. In a depressed market a contractor may price work on the basis of recovering only his running and overhead costs with a small addition for profit. In very depressed conditions, he may seek only to recover running costs in order that he can at least retain his skilled labour and maintain his plant in reasonable working order. The latter may amount to a discount in excess of 40 per cent relative to a realistic economic rate.

Employers will understandably wish to have their work completed at the lowest possible cost. However, for those employers who rely regularly on dredging contractors, it is important that contractors generate sufficient income to provide good working conditions for their employees, maintain their plant in good order, keep abreast of new technology and, most important, remain in business. A reasonably diverse and healthy industry is an essential prerequisite for genuine competition and good quality work.

Programme

Higher levels of plant utilisation usually will be reflected in lower unit costs. However, it may be difficult, sometimes impossible, for the contractor to so arrange his contract work such that his major items of specialist dredging plant move from one contract to another without significant interruption.

Prospective employers have the potential to improve the situation by offering maximum flexibility in the programming of their required work. This may be difficult, particularly for capital work, but some flexibility is usually possible. In the case of maintenance dredging, particularly in areas of slow accretion, considerable flexibility, perhaps extending to several months, may be possible. The employer can reasonably expect that by offering a flexible programme he will receive lower tender prices.

Dredging work in exposed sea conditions is best executed during the summer. Hence demand for some types of dredging plant may be higher in summer. If work planned by the employer is at a sheltered location, offering a programme of winter execution may be advantageous. In such cases it may be important to consider conditions at both the dredging and disposal sites.

Market conditions

As for all commercial activity, a depressed market will result in lower prices and vice versa. Market conditions are not usually within the control of individual employers, or contractors, but it may be possible, particularly when planning capital works, to foresee future market influences.

For example, if it is known that large-scale dredging works are planned by others and that these works will create exceptional demand for a particular type and size of dredging plant, which is similar to that required by the employer, then it will be prudent, if possible, to reschedule the works.

Risk

It is to be expected that high risk will be reflected in high prices, but contractors are often inconsistent when assessing risk. When tendering, the adjustment of price to reflect risk usually is subjective and in a depressed market the risk may be ignored. The contractor then hopes that either the risk will not materialise, or if it does, he hopes to be successful in claiming additional payment. This is not satisfactory.

The prudent and responsible employer will take all measures which are reasonably within his power to minimise risk.

This implies programming the work to coincide with optimum working conditions at the site, providing sufficient and proper information such that the work and risks may be adequately assessed and employing fair and balanced conditions of contract and method of measurement.

In addition to variable site conditions, risk may arise from fluctuations in prices, especially fuel, and in currency exchange rates. The importance of such risks and extent to which they are shared, or provided for under the contract, is dependent on the interval between tendering and completion of the works and the policy of the employer and contractor.

Strategic pricing

A particular contractor may have his own reasons for offering an exceptionally competitive price. If he is well established and enjoys a monopoly in a particular region he may wish to prevent his competitors from gaining a local foothold. In order to achieve this he may decide to price new work at cost, or even below cost. The employer may gain advantage by this, either deliberately or inadvertently, but it should be recognised that the advantage may be short term. Eventually realistic, or even inflated, pricing will prevail.

A similar philosophy of low pricing may be adopted by a contractor who wishes to penetrate a new market, or region.

11

The Dredging Contract

11.1 Introduction

Almost without exception dredging work must be performed under conditions which cannot be defined accurately. Weather, waves, tides, currents and ground conditions are often very variable. Ground conditions for capital works, in particular, may be difficult to define.

As a consequence, the risks associated with dredging work are higher than in many other areas of engineering. When it is recognised that the value and daily costs of dredging plant may be very high it will be clear that variations in the physical conditions under which dredging work is performed may lead to large variations in the cost of executing the work.

Despite this uncertainty, and the fact that it is common in works of marine construction for the value of dredging to be a large part of the total, the attention given to dredging in the overall contract documentation is frequently inadequate.

The objective in the choice of contract and the preparation of contract documents should be:

- to describe accurately the work to be done and the conditions under which it is to be done;
- to apportion risk;
- to provide a fair and equitable method of payment for that work when it is satisfactorily completed.

Decisions concerning the apportionment of risk will influence the type of contract adopted. However, it is interesting to note that when the employer chooses to accept for himself a greater share of risk, he may not be rewarded by a proportional reduction in the tendered price, or final cost of the work. Indeed, experience has shown this to be unlikely. The reason is not clear, but probably stems from the rather arbitrary methods by which most contractors, large and small, assess and price risk when tendering. Usually estimates of production rates and costs will be made with considerable care, but risk may be priced by the addition of an arbitrary percentage on-cost. In a competitive situation, the percentage added may be unrealistically low, or indeed may be zero.

11.2 Types of contract

The choice of a suitable type of contract has been described in detail by Wallace (1974). Figure 11.1 shows some possible types of contract with decreasing contractor's risk towards

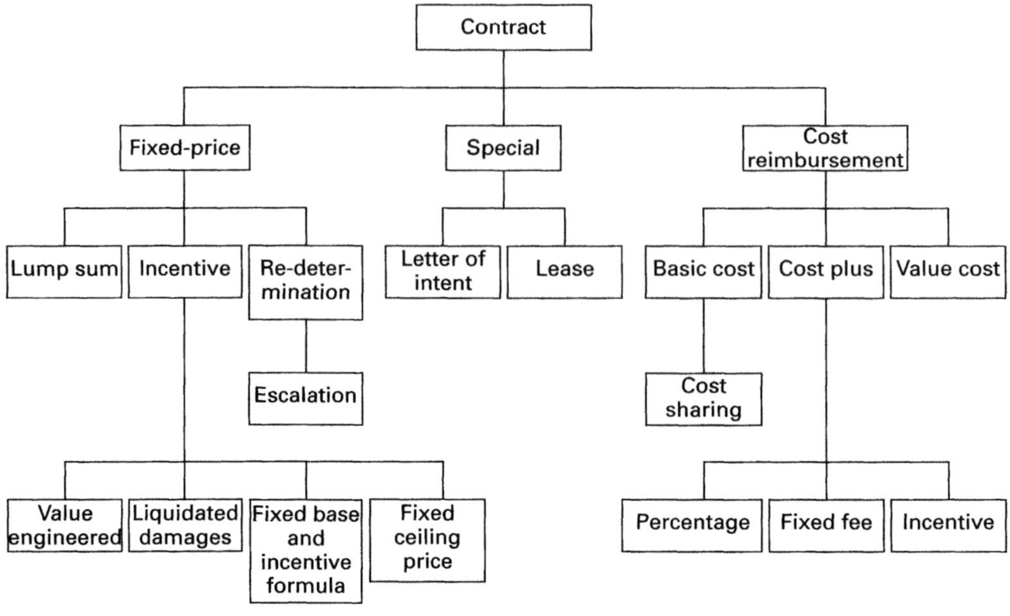

Figure 11.1 Basic contract forms (after Wallace 1974)

the right hand side of the diagram. The flexibility of contracts is further extended by the use of various mechanisms, one of which is the process of selection of the contractor. Figure 11.2 shows the several ways in which this selection may be made, again with decreasing contractor's risk towards the right. The diagram shows distinct types of contract, or mechanism, but it is better to think of a gradual merging of one type into another since in real situations it is often not clear which type pertains. Not all types of contract are well suited to dredging work and only those which are common, or potentially more appropriate, are discussed in detail below.

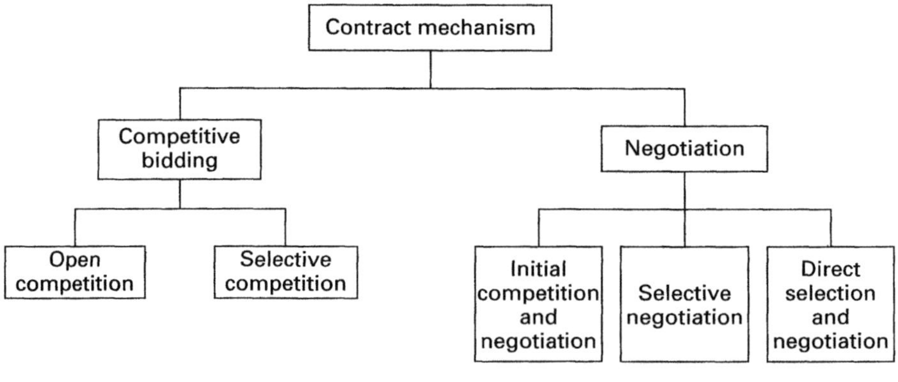

Figure 11.2 Selection of contractor (after Wallace 1974)

11.3 Fixed price contracts

In fixed price contracts a fixed sum is agreed between client and contractor for the work to be done, in advance of its execution. The contractor, therefore, accepts full responsibility for all costs, whether foreseen or not, and agrees to perform a total package for a total price.

Under the fixed price contract all risks are placed on the contractor. The only risks to the client are those of quality and the possible failure of the contractor through insolvency. This is only properly acceptable in circumstances where design and performance criteria and costs are well defined, well understood and not subject to large variation.

The fixed price contract has the following advantages to the client:

- there is a direct incentive to minimise costs on the part of the contractor;
- the contractor bears all risk and responsibility for management of the project.

The disadvantages are:

- that the contractor is predisposed to cut costs, perhaps in areas which may be inappropriate;
- if the contractor cannot bear the risk and goes into liquidation, the client suffers the consequences;
- there is less incentive for the client to operate his own control and inspection system to ensure compliance with the specification;
- the contractor will usually incorporate into his price additional costs to cover the risks and these costs are paid by the client even if the risk does not materialise.

Fixed price contracts may be modified in various ways to limit risk, or improve performance. Common modifications include:

- the introduction of a price fluctuation clause (Oosterbaan and Bean, 1977);
- the use of an incentive formula;
- the application of fixed unit rates to measured quantities.

Fixed price contracts with an incentive use a set formula which relates the final cost of the works to the level of profit which will be taken by the contractor (see Figure 11.3). The client and contractor share the actual cost savings or over-run on the target price. It is essential in this situation that all pricing is realistic and that the contractor does not load the contract with unnecessary or inappropriate plant or labour. The same applies to cost-plus contracts. A target price may be used, with the contractor taking the risk of excessive costs. This gives rise to a declining profit as the target is approached and a loss when it is exceeded.

Fixed unit rates is a common form of tendering for dredging work. A fixed unit rate is applied to all similar work. Sometimes the rate may be varied on a sliding scale related to quantity, but more commonly the approximate quantity is known in advance and rates are fixed. The sum paid to the contractor is the product of the unit rate and the final measured quantity. The fixed unit rate removes the risk of variations in quantities from the contractor, but is subject to the following difficulties in use:

- the accurate or timely measurement of quantities may be difficult;
- control and inspection are necessary to prevent inflation of the quantity of dredged material through overdredging or inaccurate measurement;
- the possibility of unbalanced tendering, front loading, cross-subsidisation and other practices used by contractors to improve their cash flow, or the apparent competitiveness of their tenders.

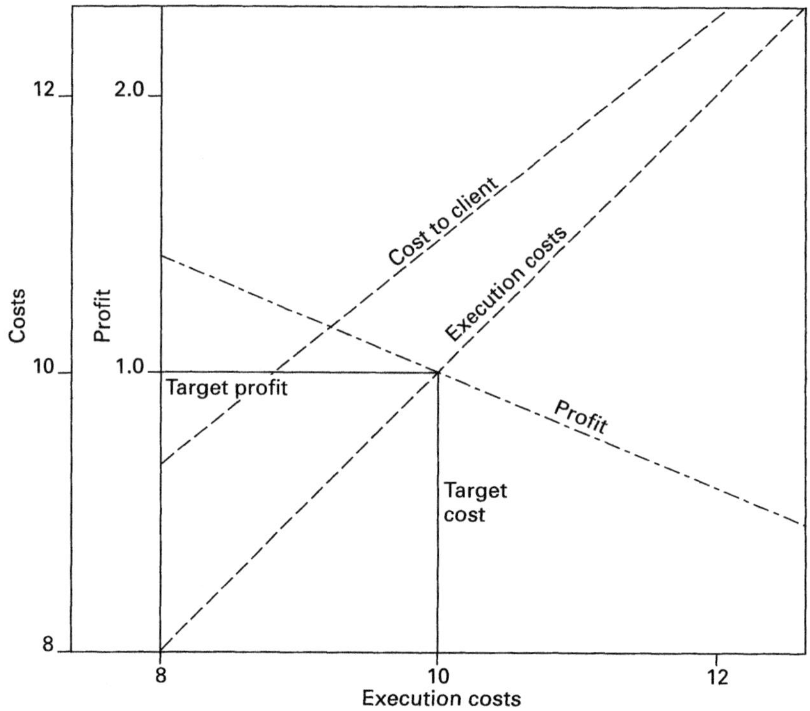

Figure 11.3 Incentive contracts (after Wallace 1974)

In the UK, government and local government procedures require that contracts be awarded on the basis of competitive confidential tenders. Usually these procedures may only be changed if the work to be undertaken is very urgent, or requires some unique skill, method or plant.

11.4 Charter contracts

11.4.1 INTRODUCTION

A charter contract for dredging plant can be likened to the plant hire contract which is a common feature of civil engineering works. However, in reality, the similarities are limited because of the long established tradition of vessel charter in the maritime world, and the high value of dredging plant, which exerts a special influence.

Under a charter contract the owner of the dredging plant will supply the plant to the charterer, with or without crew and consumables, for an agreed rate per unit of time, usually days, weeks, or months. Usually, the owner of the plant undertakes to maintain the plant in good working order, but may not assume any other risks.

The use of a charter contract may be attractive in a variety of situations:

- if work is very urgent, such that there is not sufficient time to prepare for a measured contract;
- if the work to be executed cannot be defined or measured easily;
- if there are uncertainties and risks associated with the work which cannot be assessed;

- if additional plant is required to supplement other plant engaged upon the work;
- if the charterer wishes to retain control of the method or rate of work;
- if it is a feature of the current market that the use of a charter contract will result in lower cost.

The control of the chartered plant is usually the responsibility of the charterer. Hence the quality of the work, the performance of the plant and the ultimate cost of executing the work are dependent on the quality of direction and management by the charterer.

It will be clear that it is most important that the charterer has a good understanding of the application of plant of the type chartered and has the resouces to manage the works properly. Thus, it is common for dredging contractors to charter plant from one another. Some ports with staff who have appropriate dredging experience may also charter plant, either to supplement the port dredging plant, or to carry out some particular task. For those organisations which do not have access to the necessary experience and staff resources, a charter contract is less likely to be attractive.

11.4.2 SCOPE OF CONTRACT

A charter contract may be 'bare boat', under which only the plant is provided, and all crew and consumables must be provided by the charterer. Such contracts are normally long term. The owners of dredging plant will not usually wish to place their plant, which may be sophisticated and very valuable, in the hands of a crew whose competence is unknown. A common compromise is for the plant to be supplied with a few key crew members, such as captain and chief engineer, with the balance of crew being provided locally by the charterer.

Because it takes a considerable time for a crew to achieve a satisfactory level of proficiency in the operation of a particular dredger, it is most common for dredging plant to be chartered with a full crew, or perhaps with only unskilled labour being supplied locally by the charterer. For works of medium duration this may be the most efficient arrangement. For works of short duration it will usually be preferable to charter with a full crew.

It is important that any plant supplied under charter be in good, safe working condition and, if appropriate, that any deterioration suffered during the charter period be assessed and valued. This may be achieved by means of 'on' and 'off' surveys by an independent marine surveyor agreed and appointed by the two parties. The importance of this will depend on the nature of the charter agreement, the length of the charter period and the nature of the work performed.

11.4.3 FORM OF CONTRACT

Because of the widely differing nature of dredging plant and applications it is most common for owners to draft, and hirers to agree, terms of charter which are specific to a particular job. The terms are not necessarily drafted by a professional in law, though usually some contribution by legal staff will be made. It is the view of many owners of dredging plant that their engineers understand better the risks and provisions which must be covered by a charter agreement than will a legal professional.

A standard form of charter which is in quite common usage is the 'Barecon Contract'. This applies to bare boat charter, that is the charter of a vessel without crew, or consumables.

Standard forms have some advantage in being well understood and tested, but may not meet the specific requirements of particular applications and, having their foundation in other marine activities, may be unduly onerous so far as the charterer is concerned.

If a charter contract is drafted for a particular purpose, there are a number of matters which should receive consideration. These include the following:

- the arrangements to be made for the on/off surveys;
- the arrangements to be made for any making good, valuation and payment;
- payments to be made for time during any making good, or modification;
- the period of charter;
- sub-letting (usually prohibited);
- non-lien – the owner may require indemnification against any lien, encumberance, charges, arrest and so on;
- the level and scope of insurance cover to be provided;
- payments to be made in transit during mobilisation and demobilisation;
- the payment of charges, such as pilotage, dues, royalties, taxes and so on;
- the rights and responsibilities of the respective parties;
- the rights and responsibilities of the master, supervisor and other persons;
- the issue of notices, acquiring of licences and so on;
- working hours;
- the start of normal payment;
- the finish of normal payment;
- payment during idle time – for instance during bad weather;
- payment and arrangements, if any, during standing periods due to maintenance and breakdown – it is usual to allow a period of some hours grace before non-payment begins;
- payments to be made during reduced working due to partial breakdown;
- the terms and currency for payment;
- price or currency fluctuation;
- the crew to be supplied, if any;
- the supervision to be supplied, if any;
- the surveyors to be supplied, if any;
- the survey equipment to be supplied, if any;
- the consumables to be supplied and by whom;
- the ancillary equipments to be supplied, if any – this may include special buckets, drag-heads, pipelines, position fixing, tide measurement, data logging and so on;
- the law of the stated country to apply;
- arbitration – the procedure in event of dispute;
- notice – the period of notice to be given of termination or change.

11.5 Measured contracts

11.5.1 INTRODUCTION

Generally, measured contracts are the most common in dredging works. The unit of measurement may be volume, weight, or time. Time contracts may be charter (see Section 11.4), or may be carried out on a daywork basis. However, the most common unit of measurement is volume. Details of particular methods of measurement are provided later in Section 11.6.

Under a measured contract the greater risk is taken by the contractor. In the preparation of his tender, or contract price, the contractor must estimate the productivity of his plant and hence the duration of the work or the capacity of plant required to complete the works within a specified period. He must also estimate the costs of his plant, labour and management for the estimated period of execution and thence, with the addition of any material costs, arrive at a contract price.

The greatest possibility for variation rests with productivity, which may vary enormously, subject to the nature of the materials to be dredged and handled, variations in weather, sea state and water depth. Unit costs may also be affected by conditions, including the material dredged (wear and tear) and by sea state (increased fuel consumption, damage, etc.)

11.5.2 INFORMATION TO BE SUPPLIED TO THE CONTRACTOR

In order to tender prices for any particular work a contractor must have detailed information concerning the work to be done and the site conditions. If adequate information is not supplied, the contractor may anticipate conditions which are more favourable than those actually encountered and, if this happens, he may make a claim against the employer for the additional costs involved. Conversely he may submit a price which is unnecessarily high. A variety of information is required, a discussion of which follows.

Ground information

The nature of the material to be dredged is one of the most important factors which will affect production of the dredging equipment. Soils and rocks must be properly characterised (PIANC, 1984). Some types of dredging plant are more sensitive to variations in ground conditions than others. For example, the trailing suction hopper dredger is sensitive to soil conditions, whereas a hydraulic backhoe is relatively insensitive.

It follows, therefore, that the nature of ground information provided will, to some extent, be determined by the type of plant which is to be employed. This does assume, however, that the employer, or his consultant, is able to foresee the type of plant which the contractor will use. This is by no means always the case. For small dredging works the convenience or availability of any dredging plant in the vicinity at the time that the work is required may be a more important factor than the absolute suitability of that plant for the task to be performed.

A detailed description of the effects of ground conditions on dredging plant is given in Chapter 8 and the methods of obtaining this information are described in Chapter 6.

Debris

Dredging in port areas, particularly alongside berths and within dock areas, will often involve the removal of accumulated debris. In some instances this may be relatively insignificant, perhaps in the form of general rubbish of small dimensions, but in other instances, such as at scrap rehandling berths, fitting-out quays and berths for fishing vessels, considerable quantities of difficult material may be encountered. When rivers which flow through developed areas discharge into the dredging site, high concentrations of debris may also be found.

The effects of debris are very much dependent upon the type of plant to be used for the dredging. Any type of suction dredger will be rather sensitive to significant concentrations of debris, whereas a grab or backhoe dredger will normally be relatively insensitive. When debris, particularly in the form of scrap metal, is loaded into hoppers, problems may arise due to the fouling and possible damage of the discharge mechanism. In the UK, it is not permitted to dispose of such foreign matter in licensed disposal areas at sea, and alternative handling and disposal is therefore necessary.

The concentration and character of debris are not easily quantified. The port operator should have a reasonable idea of the magnitude of the problem and should draw the contractor's attention to it by describing the use to which a berth or dock is normally put, the character of any debris pollution and whether it is considered to be merely an infrequent occurrence or if an area is severely polluted. In areas which are severely polluted, it is unlikely that suction dredging methods can be employed economically unless some prior cleaning of the area by raking or grabbing is undertaken.

Experience has shown that port staff may often underestimate the extent of pollution by debris, particularly where the use of suction dredging is unusual.

It will be reasonable in the contract to differentiate between debris, or solids, which pass through the dredger or which can be handled without separate equipment or undue disruption, and that which cannot be handled by the dredger, or which causes significant delay.

Bathymetry

Water depth and seabed levels will influence the size and type of dredging plant which can be employed. Where bed levels are high relative to tide levels continuous working may not be possible. In some instances the area to be dredged may dry out completely at low tide leading to very restricted working. It is important to supply adequate information concerning bed levels, not only within the area to be dredged, but within any areas over which dredging plant must travel in order to gain access to the dredging or disposal areas (ICE/RICS, 1984). Tidal information and the relationship between local datums should also be supplied.

Wind and wave data

All dredging plant will be affected to some degree by wave action. Waves are generally dependent (although not always) on local wind conditions. Ideally, a contractor will be supplied with wave records in the working areas affected, including the area where dredged material is to be disposed of and the route of navigation to reach the disposal area.

Wave information, to be meaningful, must be gathered over a period of years. Because of the expense and difficulty of obtaining wave records and the long lead time involved, these are not generally available. The next best thing are wind records relevant to the area. These can generally be obtained from the meteorological office or from any other suitable nearby recording station such as an airfield or large port complex.

The contractor will be most interested in the percentage of time, on a month by month basis, during which winds of particular strength and direction are exceeded. This information can be used to estimate the sea conditions which will normally arise in any particular month. It is helpful if the employer arranges for calculations to be made, based upon wind records, which provide an indication of the percentage time exceedence on a month by month basis of waves of a particular height for the critical working areas. Usually the heights of most interest to a dredging contractor will be 0.3 metre, 0.5 metre, 1 metre, and at one metre intervals thereafter, depending upon the nature of the work and size of plant to be employed.

Currents

Currents, if significant, may create difficulties, particularly for trailing suction hopper dredgers, if the current is across the direction of the channel to be dredged. Significant cross-currents will affect the positioning of the dredger and the attitude of the trailing suction pipes. Currents will also affect stationary dredgers such as grab, or bucket dredgers, due to the difficulty of mooring, and, in the case of grab dredgers, the flexibility of the grab suspension system.

Visibility

If it is known that a particular area where dredging is to take place may suffer from poor visibility, then the contractor should be supplied with information adequate to allow him to assess the impact of reduced visibility on the efficiency and safety of his operations.

Services

The line, depth or height of any services which traverse the dredging area should be clearly marked on the plans and any need for special precautions should be specified.

Shipping movements

The frequency of shipping movements in and around the works area may affect the contractor's operation. The average number of movements per day and any particular restrictions such as a prohibition on the passing in channels with certain types of vessels (e.g. liquid gas carriers, large bulk carriers, etc.) should be clearly stated and the contractor asked to so arrange his method and programme to minimise any disruption to himself and to others. Where the frequency and duration of delay to the contractor can not reasonably be foreseen, provision for the payment of delay time should be made.

Constraints to navigation

The dimensions of any structures, or other considerations, which may impose a restriction upon the beam, draught, height, or speed of plant to be employed should be provided.

General information

Recognising the relatively short time which contractors tendering for work will usually have to familiarise themselves with the areas of operation and facilities available, it will be helpful if the employer lists those important facilities such as fuel and water suppliers, towing companies, diving companies, local air services, accommodation, power, gas and telephone companies, and any other facility or organisation which is likely to be of interest to tenderers in assessing the logistics and costs of the works to be undertaken.

11.5.3 DESCRIPTION OF THE WORKS

The description of the works is important. It should convey in concise terms to the tendering contractor a clear understanding of the objectives of the work, the nature and scale of the work, and the conditions under which it must be performed. It provides an opportunity to draw attention to any matter which is unusual, or of special importance. By so doing, it may improve the tenderers early appreciation of the scope of the work and avoid misunderstanding.

The description will be different for every work, but some common matters should be described. These are:

- the objectives of the work;
- any dependence on other activities/contracts;
- the nature of the work;
- the work site environment;

- a brief description of the materials to be dredged/reclaimed;
- a brief description of the scale and extent of the work;
- unusual, or special factors, or restrictions;
- the overall progamme.

11.5.4 GENERAL CONDITIONS

The General Conditions of contract provide a legal framework which defines the responsibilities and obligations of the contracting parties. It is important that the General Conditions be clear, well understood, acceptable to all parties and well founded in law. These requirements will usually be best satisfied by standard conditions which are recognised and accepted at international or national level.

There is at this time no international set of general conditions drafted specifically for works of dredging and associated land reclamation. Such conditions have existed in the past and may appear again, but at present the majority of dredging contracts are performed under general conditions which have been drafted for works of civil engineering construction. The most common of these of international acceptance are the Federation Internationale des Ingenieurs-Conseils (FIDIC) Conditions (1987).

The FIDIC Conditions are now in their 4th edition, which was introduced in 1987. The 3rd edition contained a Part III of particular application to works of dredging and land reclamation. Part III has been omitted from the 4th edition, but some of its amendments have been incorporated into the revised Parts I and II. Part III of the 3rd edition contained in its introduction a summary of the important differences between works of dredging and of civil engineering construction. Although superseded, it will be informative to quote that introduction.

> 'In Dredging and Reclamation work the Contractor is not normally held responsible for the maintenance of the Works after takeover; the Works are usually taken over in sections as they are completed; the Contractor can only work economically if he is allowed to work continuously by day and by night; the incidence of Plant Costs (mobilisation, supply and demobilisation) forms a much higher proportion of total cost in the case of a dredging contract than is generally the case with construction contracts; as plant supplied by the Contractor almost invariably includes ships and at times includes ships taken on charter by the Contractor he cannot give to the Employer the unrestricted right to sell such plant. The Employer may find cover against the risks of non-completion by an increase of the amount of the performance bond. Quantities included in the tender documents must necessarily be estimates the accuracy of which is inherently less than normally experienced on construction contracts.'

A user's guide (Goudsmit, 1990) to the FIDIC 4th Edition has been published by the International Association of Dredging Companies (IADC). This is a trade organisation. However, the guide may be useful to those wishing to use the FIDIC Conditions for dredging work, providing that a balanced considered view is taken of the advice offered by the Guide.

By virtue of the almost inevitable uncertainty concerning ground conditions, dredging has much in common with tunnelling. Some employers have concluded that the FIDIC, and the similar British Institution of Civil Engineers 5th Edition Conditions (ICE, 1979), are not well suited to tunnelling work and have opted instead for the conditions produced in the UK by the Institution of Chemical Engineers. These adopt a more flexible approach to unforeseen changes in conditions. If difficult ground conditions are expected, some employers may wish to consider the use of the Institution of Chemical Engineers Conditions as an alternative to the FIDIC or the ICE 5th edition.

However, in the UK and some other countries, dredging works are most commonly carried out under the 5th edition (1979) of the ICE Conditions of Contract for Works of Civil Engineering Construction (and now the 6th edition, published in 1991). The Fifth Edition, which was the result of joint drafting by the Institution of Civil Engineers, the Federation of Civil Engineering Contractors and the Association of Consulting Engineers, is similar to the FIDIC Conditions – which borrowed heavily on the then current ICE Conditions.

As with FIDIC in the international arena, the ICE 5th edition Conditions are well understood in the UK by all those who regularly work under them and, importantly, they have been well tested and their meanings established in numerous court cases. It is, therefore, not advised that the conditions be altered, or at least no more than is absolutely essential to suit special or particular circumstances. Alterations made with good intent may alter the effect not only of the particular part of the document to which the alteration is made, but also, due to cross-relationships, of other clauses within the document to which no alteration is intended. Because dredging is, in most cases (inland waters being the exception), a marine activity which involves maritime craft and practice, some Special Conditions of Contract, or amendments to the Standard Conditions may be unavoidable.

The principal matters which may require special attention are discussed as follows:

Ownership of plant

Within the ICE 5th and 6th General Conditions, the ownership of plant is vested in the employer under Clause 53.

Dredging plant often has a very high value: a large trailing suction hopper dredger, for example, may have a value in excess of £10 000 000 sterling. It is often the case that the plant is not owned by the contractor, but is chartered. The contract draftsman must ask himself whether it is necessary, practicable, meaningful and proper to vest temporary ownership of such plant in the employer. It may be better to provide the necessary safeguards by means of a performance bond.

Insurance

The ICE 5th and 6th General Conditions address the matter of insurance under Clauses 21, 22 and 23.

Because these clauses have been drafted particularly with land-based civil engineering works in mind, they may prove inadequate or inappropriate to dredging works. It will usually be sufficient and more appropriate that marine plant be covered for the normal marine risks.

Insurance of the works, normally essential for civil engineering construction, may be quite inappropriate to dredging work, particularly to maintenance dredging where there are no 'works' as such to be insured. Maintenance dredging involves no construction, no materials of a permanent nature, and no finished product other than a hole in the seabed. The only conceivable damage that might occur to the works is the in-filling of any dredged area due to natural siltation. The contract draftsman must ask himself if this is a risk which should properly be covered by the insurance of the works. It may well be that such cover is not available.

Normally on a land-based project the insurance of construction plant which is employed upon the works is covered under the contract works insurance. This will be quite inappropriate for the insurance of marine plant of very high value and subject to the quite different laws which govern maritime activities. In consequence, it is advisable to seek expert advice from an experienced marine insurance broker in order to arrive at a properly drafted requirement in relation to insurance of dredging plant engaged upon any works.

Third Party and Public Liability insurance, addressed by Clause 23 in the ICE 5th and 6th General Conditions, is essential in dredging works, but the insured values which should be considered may be very much higher than is normal relative to the contract value. Even a modest dredging contract may be undertaken by a very large piece of dredging plant. Such plant is potentially capable of creating very serious damage if, for example, it is in collision with an oil loading jetty or some similar sensitive, expensive or highly utilised structure.

Maintenance period

The very nature of dredging, and more particularly maintenance dredging, will usually render a maintenance period inappropriate. Because of natural siltation the maintenance period may become in effect an unending extension of the contract works. It may be reasonable, depending upon circumstances, to require that a contractor remains responsible for the maintenance of any area in which he is working but, once work on the site or in a particular area is completed to the satisfaction of the engineer, there may be no justification for any further extension of the maintenance obligation upon the contractor nor can there be any justification for the retention of any money from the contractor in relation to such maintenance periods. An exception will arise if the contract includes actual construction, such as embanked lagoons for the storage of dredged sediment.

Price fluctuation

Unlike works of civil engineering construction, there may be no permanent materials employed in a dredging contract. The variable costs involved for the contractor are those of his labour and the consumables which the dredging plant requires during its operation, that is gas, oil, lubricants, steel, spare parts and so on.

Under some circumstances fuel costs may represent as much as 30 per cent of the contractor's variable costs. In consequence, although construction indices generally may be reasonably stable, a significant movement in fuel costs can quickly deprive the contractor of his profit margin. This is especially so when margins are eroded due to intense competition and depressed market conditions. For many years, the price of fuel oil has been particularly volatile and it is difficult for a contractor when tendering, particularly if he wishes to remain competitive, to make a reasonable allowance, even in the short term, which will cover him against some of the wilder fluctuations in the cost of fuel oil.

If it is accepted that the contractor should not be forced into the position of being a gambler, but should only be expected to make a reasonable allowance based upon reasonable foresight, there is a strong argument for including a price fluctuation clause which covers fuel and oils, even in contracts of modest duration.

Various formulae have been devised by which price fluctuation may be calculated. Some provide for full adjustment to compensate the contractor for increases in costs, usually after some initial period during which prices remain fixed. Others provide only partial adjustment. The actual formulae vary between countries and between industries. All rely upon recognised regularly published price statistics. The costs of most importance in dredging and their relative importance in percentage terms are dependent on the particular plant employed and the activity. Typical figures are as follows:

a Fuel	0.25
b Labour	0.30
c Construction plant	0.20
d Steel fabrication	0.15
Fixed	0.10
Total	1.00

The cost proportions are entered into the appropriate price fluctuation formula, together with the various base and current indices. In the UK, the increase or decrease in the amount paid is calculated by multiplying the effective value by the price fluctuation factor. This is the net sum of the products obtained by multiplying each of the proportions specified by a numerator, which is the relevant current index figure minus the base index figure, and the denominator which is the relevant base index figure. The resulting formula used in the UK is as follows:

$$P = p \times [1 + \text{Price fluctuation factor}]$$

$$P = p \times [1 + a\frac{(F-f)}{f} + b\frac{(L-l)}{l} + c\frac{(C-c)}{c} + d\frac{(S-s)}{s}]$$

where P = New price

p = Old price

a,b,c,d = Specified proportions for fuel, labour, construction plant and steel fabrication, or other as may be specified.

F,L,C,S = Current index for fuel, labour, construction plant and steel fabrication, or other as may be specified.

f,l,c,s = Base index for fuel, labour, construction plant and steel fabrication, or other as may be specified.

Currency exchange fluctuations

It is often the case that dredging contractors are of foreign origin and the dredging plant is either in foreign ownership or, in many cases, will carry a foreign crew (commonly Dutch, Belgian or American) although some labour may be recruited locally. The extent to which employers have an obligation to protect foreign contractors against the risk of currency fluctuations is inevitably the subject of debate. If, however, one accepts the basic premise that the contractor should not be placed in the position of being a gambler to any unreasonable degree, then it automatically follows that there is a sound argument for giving some consideration to sharing the risk which might arise through currency fluctuations. Against this, however, there are potential pitfalls which accompany this philosophy, particularly where major expense, such as repairs and overhauling, may be incurred outside the contract area.

If provision is to be made under the contract for price adjustments to reflect currency fluctuations it should be clearly stated. An initial base exchange rate between the currency of payment for the contract and another named currency must be mutually agreed. Thereafter full or partial price adjustment may be made in proportion to any increase or decrease, in the agreed exchange rate.

Weather delays

It is normal to place all risk arising from inclement weather conditions upon the contractor. This is certainly the safest policy for any employer to adopt. However, in some circumstances it may be necessary to decide the extent to which it is reasonable to expect the contractor to take all of the risk.

Where a contract is of significant duration, say in excess of three months, then, providing that adequate wave records, or at least wind records, are produced during the tender period, it will be reasonable to expect the contractor to make due allowance for weather when calculating his rates. It is less likely in a contract of more than three months duration that exceptional weather will occur for a significant proportion of the time.

However, for short contracts a week of bad weather which prevents the dredger operating, or which has a severe effect on production, may not only remove any possibility of the contractor making a profit but may result in a significant loss being inevitable. A reasonable employer may consider this to be unsatisfactory.

In this case, it may be argued that some provision should be made within the contract for the risk of exceptional weather to be shared between contractor and employer. This can be arranged in a variety of ways, but a method which is usually satisfactory is to require that the contractor take the risk for an initial period of time lost due to weather and the employer contribute to the cost of time lost which is in excess of the specified allowance at the rate of, say 50 per cent of the contract rate for demurrage.

Under this arrangement, the contract draughtsman makes an assessment of the probable time lost due to inclement weather over the whole period of the contract. The assessment is based on the use of the optimum dredging plant and a normal weather pattern. The contractor tenders demurrage rates for each item of plant to be employed and for site overheads.

If the lost time is greater than that which the contractor has been instructed to allow for, he is paid for each item of plant which is prevented from working at 50 per cent of the appropriate demurrage rate. Thus the contractor has a positive incentive to keep working for as long as is possible and to resume work as soon as possible, but does not carry the full cost of excessive lost time. The rate at which the employer contributes may be more or less than 50 per cent. This is a matter of judgement on the part of the contract draughtsman, made in relation to the particular work to be performed and the site conditions.

Ground variations

Some types of dredger, notably the trailing suction hopper dredger (which is today employed extensively for maintenance dredging works) are particularly sensitive to relatively small variations in ground conditions. Unexpected variations are only likely to occur, in most cases, if the programme of sampling and analysis has been inadequate at the tender stage. Providing, therefore, that a proper analysis of the material to be dredged has been carried out on the employer's behalf and assuming that the results of that analysis have been made available to the contractor when tendering, it should not be necessary to make any special provision for unforeseen variations in the character of the material to be dredged.

In the case of maintenance dredging unexpected variation in soils is unlikely because of the repetitive nature of the work. However, it is stressed that, because of the sensitivity of, for example, trailing suction hopper dredgers to even minor variations of soil density, investigations should still be undertaken.

Soil variations in capital dredging works are rather more difficult to deal with, as a wider variation in ground conditions may occur. If the presence of very difficult material, such as rock, is suspected, but for some reason can not be clearly defined, this should be made clear in the description of ground conditions and in the Preamble to the Bill of Quantities, where provision for a range of quantities should be made. See also Section 11.5.6.

Where unexpected ground conditions occur, it may be appropriate to provide under the contract a formula to be used to adjust the rates or sums to be paid for dredging. Any such formula should be based upon variations in productivity. This is a complex matter and it is not possible to provide a simple formula covering all situations.

Additional clauses

A number of matters may need to be covered by the provision of additional clauses. These could include:

- pilotage;
- the removal of sunken plant or materials;
- non-interference with navigation;
- non-interference with radio frequencies;
- environmental matters such as pollution (spillage or noise), non-interference with fisheries, etc.

11.5.5 SPECIFICATION

The specification should describe precisely the scope of work to be done and any restrictions upon the way in which it is to be done. It is usual to divide the specification into general items, concerning local conditions and the administration of the works, and particular items, which describe in detail the work required.

General specification

The location of the proposed work, the objectives of the work and the status of the Employer will all influence the range of general items to be specified. For example, works within a military port may be subject to restrictions or procedures which are unnecessary in a civil port. In most cases, however, there will be certain needs which will be common and which should be specified. The following list is a selection of these:

- location of the site;
- extent of the work required;
- list of drawings to be supplied which detail the work;
- units to be used, (i.e. metric or imperial);
- datums to be used, (i.e. ordnance (land) or chart);
- tidal conditions;
- conditions concerning pilotage, navigation, customs;
- security;
- provision of berths, offices, working areas or facilities;
- provision of facilities for the engineer;
- keeping of records;
- keeping of photographic records;
- safety and welfare;
- screening of lights which may confuse navigation;

- use and protection of services, power, telephone and so on;
- order of work;
- access and working areas.

The matters listed are generally uncontroversial and the views of the consultant or employer and those of the contractor are unlikely to differ greatly. However, the contractor might, with some justification, resist the divulgence of all records which relate to productivity. The contractor's performance and the performance of his plant are his main strengths or weaknesses. Any such information received, therefore, should so far as possible be treated in confidence.

The information may not be of great value to an employer, unless for some reason the contractor's performance falls short of that anticipated. If this happens and the reasons are, in the view of the contractor, beyond his control or reasonable foresight, then he may claim additional payment. In this situation, the records of production and productive time will become important. They may, however, be of limited value unless the consultant or employer has sufficient experience of work similar to that required under the contract to be able to judge what is normal and reasonable under the prevailing conditions. If, for example, the contractor suffers a 15 per cent loss of productive time due to blockages of the dredger draghead and pump, and claims this to be excessive, then the consultant or employer must be able to judge fairly whether this is indeed excessive, or in fact, normal.

Particular specification

The particular specification should define the work required in terms of dimensions, tolerances and any restrictions on the methods to be employed. It is in this area that some divergence between the views of the consultant or employer and those of the contractor may occur.

Matters which should normally be covered in the particular specification are included in the following list:

For dredging:

- description of the work;
- material to be dredged;
- restrictions on methods of dredging;
- method and place of dredged material disposal;
- sequence of dredging;
- dimensions of dredged area and sections;
- slopes;
- tolerances;
- overspilling of hoppers;
- natural siltation;
- siltation resulting from the contractor's operations;
- measurement of tide levels;
- setting out and positional control;
- method and density of surveys;
- movement of buoys or marks;
- dredging in the vicinity of structures;
- overdredging;
- shipping movements;

- records of dumping;
- speed limits;
- radio watch-keeping;
- noise levels;
- continuous or restricted working;
- environmental contraints.

For reclamation

- particle size distribution of fill material;
- permissable fines content, if any;
- conditions of placement (i.e. contained; free draining; submerged; layer thickness, etc.);
- drainage;
- quality of drainage water;
- method of compaction;
- degree of compaction;
- method of measuring degree of compaction.

Some of these matters deserve special comment.

Material to be dredged

As mentioned elsewhere, it is important that the material to be dredged be properly described.

Method of dredging

If practicable, the method of dredging should be unspecified. This will allow the contractor to choose the method best suited to his particular resources and maximises the chances of a competitive price being received. It is possible that a method may be proposed that has not been considered and which may require some change to the specification or programme.

Dredged material disposal

After the nature of the material to be dredged, the method and place of disposal are likely to have the greatest impact on the price for the work. The subject of dredged material use and disposal is covered in detail in Chapters 4 and 5.

Sequence of dredging

After the particular needs of the employer have been met, the sequence of dredging in each area of the works should permit for the optimum efficiency in dredging operations coupled with the minimum risk of resiltation of completed areas as a consequence of continuing work. Generally, it is preferable to allow the contractor to determine his own optimum sequencing.

Side slopes

It is likely that the natural stable slopes after dredging will vary considerably due to variations in the characteristics of the seabed material and the currents or waves acting upon them. It will be unrealistic to specify side slopes which are steeper than are natural, unless the rate of slope decay is slow.

Wide variations of slope which may occur make it difficult to satisfactorily specify a particular side slope. Different slopes may be specified for different areas, but this may be impracticable where there is a wide, natural variation over short distances. In this situation there are two main possibilities.

Maximum and minimum slopes may be specified. Ideally, all natural slopes within the work area will fall between these two. The disadvantage of this arrangement is that large quantities of material may be contained within the slope tolerance. Although the tolerance may be to protect the contractor, he may be tempted to exploit the situation and remove more material from the tolerance than is necessary. For most situations he will probably find this uneconomic, but the risk exists. It then becomes difficult to predict the final cost of the work. The risk to the employer can be minimised by placing a limit on the total paid volume removed from within the tolerances. Generally a limit of one third will be fair and reasonable, assuming that the initial tolerances are appropriate.

An alternative is to specify only the maximum slope, leaving the contractor to make his own assessment of any unpaid dredging and to allow for this in his rates.

Tolerances

Much of what has been said in relation to slope tolerances applies also to vertical and horizontal tolerances.

There are very significant differences between the accuracy which can be achieved with different types of dredger and under different operating conditions.

For navigation channels, vertical tolerances will normally be specified as plus nothing and minus any tolerance. The paid limits may be less than the overall accuracy of the dredging plant, indeed the paid tolerance may be zero. In some situations, where a wide variety of plant may be employed and where ground conditions are uncomplicated and well known, it might be preferable to omit any paid tolerances. This has the advantage, from the employer's point of view, that there is a positive and known limit on the maximum quantity measured for payment and hence on the total cost of the work; total cost being a product of the quantity lying above the design levels and profiles, and the unit rates for dredging. This is a matter in which contractors may hold opposing views for obvious reasons.

An alternative method of dealing with paid tolerances is to ask the contractor to specify his own, add the necessary quantity to the volume to be dredged and price his unit rates accordingly. This may be beneficial if the thickness of the tolerance is not of great importance to the client since it allows him to see what is the true unit rate for dredging. It also allows contractors to select their most competitive tolerance.

The employer may, on the other hand, also have good engineering reasons to wish to positively discourage overdredging, for example in the vicinity of structures.

Measurement

Measurement is of great importance but should be dealt with in detail in the Preamble to the Bill of Quantities. See Sections 11.6 and 11.5.6.

Overspilling of hoppers

The comments which follow apply mainly to the trailing suction hopper dredger, but may have some relevance to hopper barges loaded by stationary dredgers.

When dredging sandy materials with trailing suction hopper dredgers, continued pumping into the hopper beyond the time that overflow of water from the hopper commences will normally result in a gradual increase of load. When dredging in silts and soft clays, generally it will not.

It can be seen, therefore, that when dredging silt and soft clays in an enclosed system such as a dock, there will be no benefit in continued loading beyond the commencement of overflow. This will also be true when dredging in sensitive areas where increased suspensed sediment concentrations are undesirable.

When dredging in areas of reasonably strong water flow, there may be merit in permitting an extended period of overflow to gain some advantage through agitation dredging. Material will be drawn up by the suction pumps and loaded into the hopper but a large percentage will immediately pass out of the hopper with the overflowing water. A proportion of that sediment may remain in suspension long enough to be carried outside the dredging area and dispersed over such wide areas that the loss of depth in those areas of eventual deposition may be insignificant. The use of such agitation dredging methods may significantly increase production rates for the dredger in terms of the amount of material removed from a particular dredging area. As a consequence, unit costs for removing material from that area will be proportionally reduced.

The contract should clearly specify whether or not hopper overflow is permitted. Where overflow is to be permitted, any limitation on the period of overflow should be specified. It may not be necessary to impose any limits when the material is clean and granular, with a minimum particle diameter of 200 microns. Even for finer soils it may not be necessary to restrict overflow if there are specific water quality standards for the contractor to observe in or near the dredging area. In this situation, the overflow period will be limited by the necessity to comply with water quality criteria.

Natural siltation

When contract dredging takes place in areas with significant natural siltation, it is likely that the contractor will be obliged to remove some material which has not been measured for payment.

Where the dredging area is large, any problem of siltation during the work can be alleviated by splitting the area up into smaller zones and by measuring each zone immediately upon completion.

Usually, however, the contractor will be obliged to remove some material which is deposited during the contract period. This responsibility should be spelt out in the contract documents at the time of tender and sufficient information, such as recorded siltation rates, ideally on a month by month basis, should be provided to allow the contractor to assess the risks involved. The contractor can then make an appropriate allowance within his rates for any unpaid work which is likely to result due to natural siltation.

Positioning and survey

Accurate dredging will only be possible if accurate methods of positioning and survey are used. This may necessitate the use of electronic position fixing equipment and on-board displays. The accuracy of position, the ruling geometric configuration of signal stations, if used, and the method of recording and observing tide levels should be specified. If interim surveys are to be a contractual requirement, the frequency, density and accuracy of such surveys should be specified.

If it is required that the dredger be fitted with an on-board track-plotter, or display, which is usually desirable for trailing suction hopper dredgers and sometimes may be for other dredger types, then this should be specified.

11.5.6 BILL OF QUANTITIES

Usually, for straightforward dredging works, the Bills of Quantities will be quite simple. In common with other forms of civil engineering, three or four basic sections and bills will usually suffice. These are as follows:

- Preamble
- Bill 1 – general items
- Bill 2 – method-related charges
- Bill 3 – measured items
- Summary sheet
- Demurrage charges
- Day work rates.

Under these headings, some or all of the following matters should be described and items listed.

Preamble

Measurement may be in accordance with a standard method of measurement, such as CESMM in the UK, though these may not always be well suited to works of dredging or reclamation. Alternatively, the method of measurement may be tailored to the particular contract, in which case fuller item descriptions may be necessary.

The Preamble should clearly describe all aspects of what is to be measured for payment and any special conditions attached to payment. For example, it should be stated if measurement is to be *in situ*, if paid tolerances are provided and whether there is any limit on total volume within the tolerances which will be measured for payment.

The method of *in situ* measurement should be described. If by echo sounder, then the frequency to be used should be stated. The timing of survey, density of survey lines, form of data presentation, the method of computation of quantities, should all be described.

If measurement is to be in reclamation, the method of measurement of any allowance for sub-strata settlement should be described.

For demurrage, the circumstances under which payment will be made should be described, together with any restriction on payment. For example it may be stated that the first hour of each and every stoppage will not be paid.

General items

Usually the bill relating to general items will include the following:

- provision of bond;
- insurance – marine plant;
- insurance – persons and property;
- engineers, office, equipment and maintenance thereof;
- provision of sampling and testing equipment/facilities;
- collection and testing of soil samples;
- collection and testing of water samples;
- testing of water quality *in situ*.

Method-related charges

These will include:

- mobilisation of staff and site facilities;
- mobilisation/provision of survey/positioning equipment;
- mobilisation of plant by item;
- removal and demobilisation of same.

Measured items

These will include:

- dredging and disposal by area and/or type;
- reclamation by area and type.

Summary sheet

This will provide:

- a summary of preceeding bill totals.

Demurrage rates

There will be a list of rates per hour and per day for demurrage for each major item of plant, or equipment.

Day work rates

There will be a list of rates per hour and per day for day work for each major item of plant, or equipment, or operation.

11.6 Method of measurement

Measurement is clearly important when dredging is to be carried out by a contractor. It is, however, also important for a direct labour operation, as accurate *in situ* measurement will provide the means of assessing efficiency, productivity and a realistic comparison with contract prices. Experience has shown that more convenient methods of measurement, such as in the hopper, are rarely reliable, particularly in direct labour operations. Here, it is assumed that the work is to be carried out by contract.

Unless the contract is to be let on a lump sum basis, some method of measurement will be required to determine how much the contractor is entitled to be paid. In most cases, this will involve some form of volumetric or weight measure, but may also involve time charter. Time charter is discussed in Section 11.4.

The method of measurement which is to be used on a contract should be clearly and comprehensively described in the Preamble to the Bill of Quantities.

If volumetric measurement is to be used, then ideally, the volume removed *in situ* will be determined by means of pre- and post-dredge surveys (ICE/RICS, 1984)).

Alternatively, the measurement of volume in the hopper, or some form of load measuring instruments, might be employed. These methods are not recommended unless *in situ* measurement is, for some reason, not practicable.

11.6.1 MEASUREMENT IN SITU

In situ measurement is not without problems. Where the dredging area is fairly extensive, flat and sheltered, and the seabed firm, survey on a close grid or line spacing will allow quantities to be determined with acceptable accuracy. However, it is often the case, particulary for small to medium size ports, that dredging will be concentrated in a rather narrow, perhaps long, approach channel with fairly steep side slopes and a soft seabed.

In such situations, a grid method of survey and quantity computation is far from satisfactory. Closely spaced cross-sections, all sailed in one direction, with fixes at close centres, preferably using a narrow-beam echo sounder, will be more appropriate.

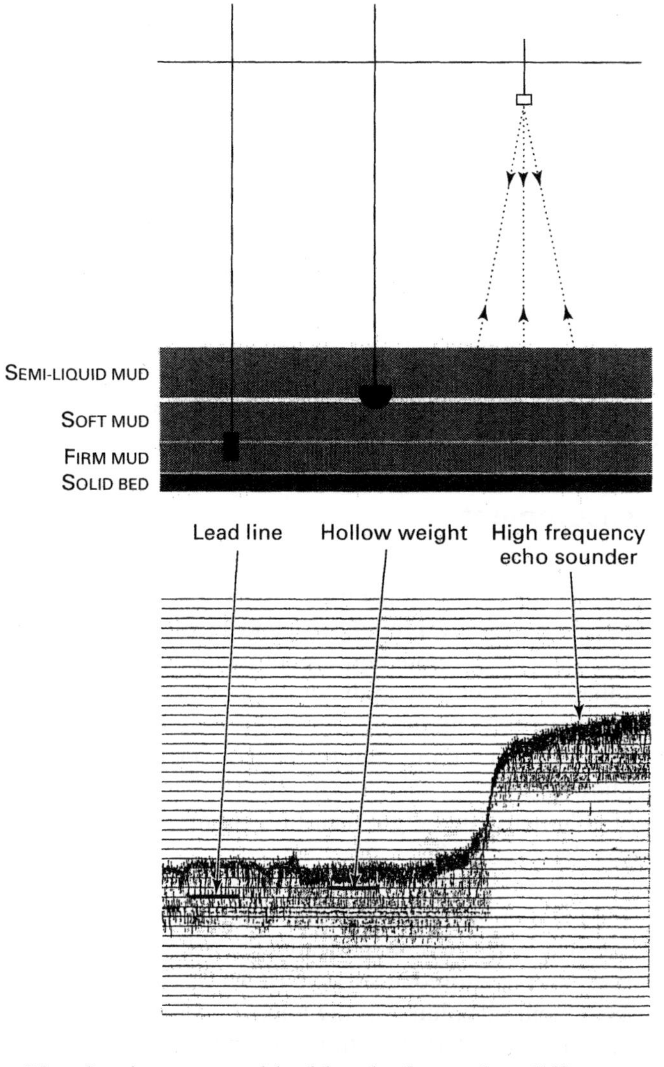

Figure 11.4 Varation in measured bed level when using different measurement methods in weak bed materials

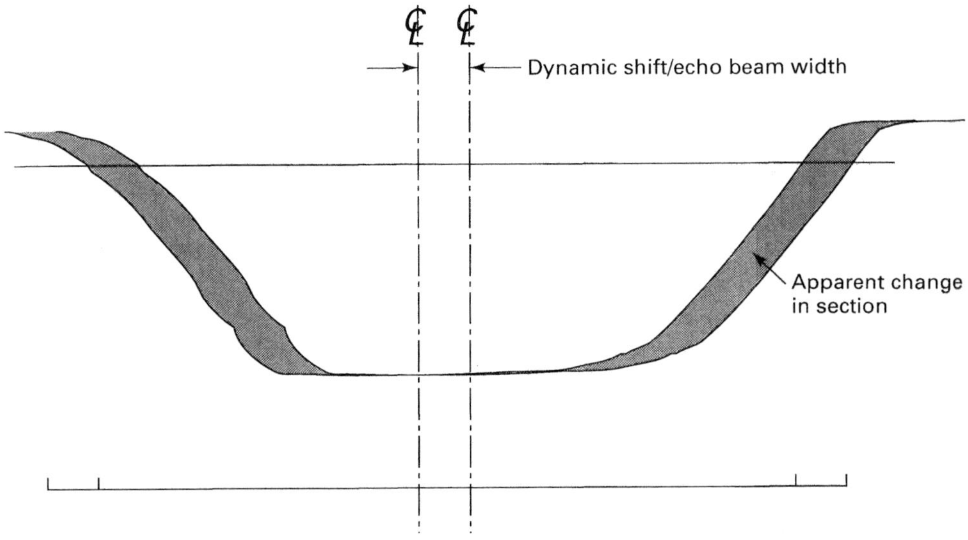

Figure 11.5 The effect of survey error on quality measurement

The choice of echo sounder used for survey may be important. If the seabed materials are very weak and contain a significant proportion of silt and clay size particles (i.e. less than 63 micrometre) it will usually be preferable to use an echo sounder with a relatively low frequency transducer (25 – 50 kHz). This is because high frequencies (150 – 250 kHz) may be reflected by low density deposits which do not represent the true dredged level (see Figure 11.4). The problem of accurate measurement may be particularly acute following dredging in fine sediments using a trailing suction hopper dredger, a commonplace occurrence in maintenance dredging work. The problem is usually reduced if an interval of several tides is allowed between the finish of dredging and the start of survey. This allows very low density deposits to be flushed away, or to consolidate. These problems usually are not significant when the seabed is firm, as in the case of sands, stiff clays and so on.

To allow true comparison, the post-dredge survey should mirror the pre-dredge survey. In practice, this is almost impossible to achieve and some compensation for survey error will be necessary. For example, a small shift in the position of pre- and post-dredge channel cross-sections relative to the channel centre line may seriously influence the apparent change in section and hence apparent volume removed (see Figure 11.5). Providing that each survey line extends well beyond the influence of dredging, the upper slope detail can be matched by overlaying the post-dredge section on the pre-dredge section and making an appropriate lateral adjustment before the change in section is measured.

It must also be decided who is to carry out the survey work. From a purely practical viewpoint, it is most logical for the dredging contractor to do the work as he will have a fully equipped survey team on site. The disadvantage with this arrangement is that it is the contractor who performs the measurement which will form the basis of his payment for the work and the evidence that the work has been satisfactorily completed. The contractor may be influenced by commercial considerations.

Hydrographic survey is imprecise and requires some degree of interpretation and judgment. It is possible that a contractor's interpretation of a survey, if not his execution of it, will suffer from some degree of bias, which is likely to be in his own favour. It is also possible that some bias will occur if the employer carries out the survey, but in this case it is unlikely to favour the contractor. It can be argued that bias will be eliminated if the work is carried out jointly. This may be true, providing that both parties are equally involved and equally skilled in the survey and plotting work. In practice, this is rarely so.

Equally involved means joint levelling of the tide gauge, reading of the tide level, co-ordination of all of the shore station positions, calibration of the position fixing equipment, reading of the position fix, calibration of the echo sounder, bar check, interpretation of the echo sounder chart record, plotting of all positions and levels, and the computation of all quantities. To achieve this, each party will, in most cases, require an experienced team of three people. In practice this rarely happens.

An alternative to a joint survey is for the employer to engage an independent survey contractor. A reputable independent surveyor has no motive to show any bias. The surveyor must, however, be chosen with care. Often there are only very few firms who fully understand, or can provide for, the special needs of dredging related survey. The introduction of a third party may also complicate matters in the event of any dispute.

Care must be taken to specify the objectives of the survey work carefully. An Admiralty trained surveyor may, for example, naturally plot levels of 'least depth'. This is fine for navigation purposes, but may be quite unsatisfactory for the measurement of quantities.

The employment of an independent surveyor may complicate the supervision of the work. Careful timing of surveys will be important to ensure proper measurement and to avoid an unreasonable delay in the issue of a certificate of completion.

Normally, the dredging contractor will be expected to arrange for his own progress surveys.

11.6.2 MEASUREMENT IN HOPPER

Due to the change in water content and bulk density of the dredged material, which occurs between the seabed and the hopper, it may be very difficult or impossible to relate hopper load to the actual volume of material removed from the seabed. Generally, therefore, it is only where regular dredging on a large scale allows a sound relationship to be developed that the method can be used with reasonable confidence.

Measurement in the hopper may be by volume or by weight.

Volume

If the material dredged is granular and of such particle size that it quickly settles from suspension it will form a firm deposit in the hopper. The density of the hopper load may not be the same as that of the material on the seabed. The volume of the hopper load can be estimated by the taking of soundings from some consistent feature of the structure, usually the hopper combing, to the surface of the hopper contents using a graduated hand line. Soundings should be taken at not less than eight points evenly spaced around the hopper. The average sounding depth will provide an average level for the surface of the hopper load. Reference to the official ships ullage tables will show the hopper volume at this level and hence the approximate volume of material in the hopper.

If the dredged material is not granular, or is very fine, or mixed, there may be a gradual transition in the hopper from surface water to solid deposit. The determination of the volume of solids in the hopper is then much more difficult. A common method is to sound in the hopper using a sounding line with a hollow hemispherical weight of a chosen specific displacement such that, when lowered into the hopper, the end weight will come to rest when reaching a level at which the density of the dredged material is equal to that of the weight. This is then assumed to be the surface of the hopper load. Material in suspension above this level may be ignored, or may be estimated by sampling at regular intervals of depth and analysing in a laboratory to determine the solids content. This is a relative method of measurement, with no fixed relationship between a cubic metre measured in the hopper and a cubic metre *in situ* on the sea bed.

Weight

The weight of solids in the hopper may be estimated by the recovery of representative samples and laboratory analysis, or by measuring the change in draught and hence displacement of the vessel.

Sampling of material in suspension may be carried out as described under volume measure, or by the recovery of cores from the whole depth of the hopper contents. Both methods are very laborious and expensive.

A method which is potentially more satisfactory is to measure change in displacement. For a dumb hopper in calm fresh water this is reasonably straightforward, providing that the relationship between draught and displacement is known and that the weight of water in the hopper is taken into account.

For large self-propelled vessels, such as trailing suction hopper dredgers, operating at sea, measurement is more complicated and more difficult. Factors other than cargo may affect displacement. These include changes in water density, the vessel's loads of such things as consumables, crew, spare parts, ancillary equipment and so on. For a large vessel, a small change in draught may represent an important change in displacement due to hopper contents. Small changes in draught are difficult to measure accurately. Not only is it necessary to measure the change in draught accurately, but it must also be measured in several different areas of the hull. This is because the vessel will usually be loaded unevenly with different draught forward and astern and port to starboard. There may be significant differences in displacement for different hull attitudes for the same average draught. The weight of water in the hopper must be taken into account, thus it remains necessary to measure the volume of the total hopper contents.

In spite of the difficulties of using the measurement-by-weight system described above, it has been introduced successfully by the Rijskswaterstat for dredging in the entrance to Rotterdam (van der Gouwe *et al.*,1993). The so-called 'tonnes dry solids system' (TDSS) has encouraged greater efficiency in maintenance dredging, thereby significantly reducing the total capacity of dredgers required annually to perform the specified tasks.

11.6.3 MEASUREMENT BY INSTRUMENT

Methods of measurement by instrument may include the measurement of changes in draught as discussed under weight in Section 11.6.2.

Another instrument method is the measurement of flow. Flow of mixture in pipes is determined by measuring velocity. The product of velocity and the cross-sectional area of the pipe gives flow. If the flowing mixture is perfectly homogeneous and if the specific gravity of the water carrier is known, then, by measuring the density of the mixture, the weight of solids in suspension can be calculated. If the particles in suspension have a fixed specific gravity, the volume of the solid mass can be calculated. However, this volume is quite different to that of the *in situ* material before dredging.

If the bulk density of the *in situ* material is known and is constant, then production of *in situ* cubic metres can be calculated. In reality, the bulk density of the bed material is likely to be highly variable. The specific gravity of the solid particles may also be variable, particularly if organic. The specific gravity of the carrier water may also vary, although the variation will be relatively small.

As if these complications were not enough, in reality most mixture flows are not homogeneous, but consist of part fully suspended, part saltating (bouncing) and part sliding bed load. Each element travels at a different velocity. Not only is the velocity variable, but so is the density, the mixture being most dense at the bottom of a pipe, or outside of a bend and least at the diametrically opposite point, with an infinite variation in between. Various methods may be employed in an attempt to overcome these problems and to measure a true average mixture density, but the resulting overall accuracy is variable.

Measurement by instrument cannot, at this time, be recommended as a satisfactory contractual basis for payment, unless experience of a particular system is sufficient for it to be well understood and meaningful. However, such methods may be very useful in relative terms as a guide to the efficiency of a particular dredging operation.

11.6.4 MEASUREMENT IN THE FILL

Measurement of fill, whether for land reclamation, beach nourishment, or other uses, should be made by means of before and after surveys of the site. Due to the effects of bulking and loss of fines, the volume delivered will rarely be the same as the volume measured in place.

If fill is to be placed upon weak deposits which are liable to significant consolidation under surcharge, some provision for the measurement of the resulting settlement is usual. This is most commonly by means of settlement beacons, or plates. Both consist of a square, circular plate or slab placed onto the foundation prior to filling. If beacons are used, a vertical rod, or pole is fixed to the plate, or slab. The pole is marked at a particular elevation and any change in elevation is determined by levelling. An objection to beacons is that they may interfere with the placing, levelling and consolidation of the fill material. They are often moved, damaged, or lost. It may be preferable, therefore, to place a large number of plates, or slabs and to carefully coordinate their position. On completion of filling the level of the plate can be determined by probing, or sonic methods.

If settlement occurs, it is usual to measure the resulting additional volume of fill necessary to achieve the specified final elevation for the fill surface and to make payment at the appropriate billed rate.

11.6.5 DELAYS

Dredging work may be delayed by natural events, such as weather and sea state, or by the employers activities, such as the movement of shipping. The contract should state clearly whether or not payment will be made for any delay time.

It is not usual to make payment for delays due to bad weather, or sea state, but if conditions are very unpredictable, or if delay may be long in relation to the anticipated contract execution time, then some risk sharing may be appropriate.

To avoid frivolous claims it may be desirable to require that the contractor bears the cost of an initial period of each and every stoppage. Thereafter the cost of delay may be shared, or divided in a specified way, and the employer's contribution calculated using rates for demurrage, or delay, entered into the Schedule of Rates at the time of tendering. It may be desirable to place a ceiling on the total payment for delay in any given period, for example, not more than 72 hours per item of plant per week.

It should be noted that the actual delay caused to an item of dredging plant may be more than the period of interruption. If a dredger has to move off station and return after the delay, the delay should be measured from cessation of operations to recommencement.

11.6.6 SILTATION

It is common for dredged areas to be subject to siltation, which may be slow, or rapid. This is clearly the case in all works of maintenance dredging. It is usual to require that the contractor be responsible for the removal of materials deposited due to siltation during the contract.

If the contractor is to be responsible for the removal of all materials deposited within the dredging areas during the period of the contract, this should be clearly stated. Furthermore, some reasonable indication of the rates of siltation within the areas of dredging and any seasonal effects, should be included in the tender information.

If the dredging area is extensive, or the work prolonged, or the rate of siltation high, then consideration should be given to the sub-division of the dredging areas with staged completion and acceptance.

If the dredging area is subject to severe siltation during periods of strong wind and heavy seas, then consideration should be given to re-measurement of the work following any such event. This will only be possible if surveys are made at regular intervals during the work. Such surveys may then form the basis of measurement up to the severe event and a new pre-dredge survey may be made following the event.

Usually the accurate measurement of siltation during the progress of the contract will be difficult, or impossible. It is preferred therefore that the contractor be required to make allowance for all siltation during the contract within his tendered rates, but in order that he may do so it is necessary that he be provided with information sufficient to properly assess the risks, or that such information be readily and freely available.

11.6.7 MEASUREMENT IN MATERIALS WITH WIDELY DIFFERING CHARACTERISTICS

It is commonly a requirement of capital dredging contracts that ground materials of wide diversity be dredged. For example, loose sands may overlay solid rock. The difficulty and unit cost of dredging the rock will be far greater than for the sand. The difference in cost may be as much as a factor of 50 times. This can cause significant difficulties in measurement.

If there is a clear vertical separation between materials of significantly different dredgeability, then usually, if water depth is adequate, it will be simplest first to dredge the upper strata, with separate measurement and payment, and then to dredge the remainder. Similarly, if there is a clear demarcation in plan between different materials, the same approach may be adopted. However, matters are rarely so simple.

Complications arise because often, not only does the rate of production of a given dredger vary greatly in different materials, but different materials may require dredgers of different types. For example, superficial deposits of loose sand may be efficiently removed by a trailing suction dredger, but underlying boulder clay may be impossible to dredge economically, except by using a different type of dredger.

If materials are highly variable and without clear boundaries, separate measurement may be impossible. In many areas of the world, such situations are common in coastal geology. There is then no practical alternative, but to measure all work for payment at a single unit rate. However, to arrive at a single composite rate, the contractor, or dredger operator, must estimate the average rate of production which will be possible. To do this the relative proportions of the component material types must be estimated. Hence, as has been said already and will be said again, an adequate and proper ground investigation is essential.

It is recommended that when a single composite rate is tendered for a mix of materials of widely different characteristics, the tenderer be instructed to show in his tender submission the make-up of his composite rate and his estimate of the proportion of each type of material. In the event that it is considered that the tenderer's assessment of the proportions of different materials is in error, he should be notified of the apparent error before formal acceptance of his tender.

If it is discovered during the progress of the work that there are important differences between the actual proportions of different materials relative to those which were foreseen, the component parts of the composite rate can be used as a basis for a negotiated variation in the composite unit rate. However, it should be noted that in practice it may be very difficult to determine the true proportions of the different materials dredged.

In the extreme case of undredgeable rock, the two stage approach described earlier may be preferable. In this case, all materials which can be dredged economically, and which are accessible, should be dredged first and the remaining rock mass be pre-treated and dredged separately at different unit rates. With this method it is important that the limit of dredgeability be defined in advance.

Ideally the boundary between superficial deposits and undredgeable rock will have been determined by prior ground investigation, but where this has not been done, or can not be done, a judgement is necessary as to the particular point at which the ground is deemed undredgeable. If practicable, this may be defined in terms of a minimum rate of production. Hourly and daily production rates would then be recorded. Even this approach may be difficult, because a point may be reached when production remains higher than the agreed minimum, but the stress imposed on the dredging machinery and operators is unacceptable. This may only be decided by very experienced judgement.

When significant uncertainty exists concerning the proportions of widely different materials that might be encountered, particularly when one material is rock, it is advisable to declare the uncertainty within the tender documents and to make provision for tenderers to tender a range of different lump sum charges and unit rates in relation to a range of possible quantities. The unit cost to remove ten cubic metres of undredgeable rock may be many times that to remove 1000 cubic metres, but such range of uncertainty is not unknown.

11.7 Evaluation of tenders

Tenders should be checked in three basic ways:

- for qualifications;
- for arithmetical accuracy;
- for sensitivity to variation in quantities, or items.

Any qualifications will usually be contained in a letter accompanying the tender, but may also less obviously be contained within the Method Statement, or Programme. Hence all of these should be checked carefully in each tender, but particularly for the lowest tenders.

Some qualifications may appear relatively unimportant, but each should be tested by asking, 'what if this happens?' For example, what if serious siltation, or serious overdredging occurs?

Arithmetical accuracy and sensitivity can be most easily checked by entering the appropriate Bill of Quantities and rates into a spreadsheet for analysis on a personal computer. The spreadsheet can be prepared during the tender period to allow rapid analysis on receipt of tenders.

When using a spreadsheet method, the sums and rates entered by each tenderer are entered into a spreadsheet which already contains the Bill items and quantities. It is then immediately apparent if the tenders are arithmetically correct, or not.

The various items and quantities can then be varied to determine whether, or not the relative position of the tenderers is affected. Quantities should be varied to reflect the most optimistic and pessimistic situations. Usually the original billed quantities should represent the situation which is considered to be most likely.

When analysis has provided the most competitive tenders, usually the lowest three are subjected to more detailed scrutiny to determine the following:

- financial stability;
- plant availability;
- management capability;
- experience;
- quality of recent performance on similar works.

If tenders have been invited on the basis of a pre-qualified select list, the above matters should have been investigated already.

Additional points relating to the evaluation of tenders are given in Section 2.7.4, page 27.

11.8 Commencement of work

Following the award of contract, the detailed timing and phasing of the work should be discussed and agreed between the Contractor and Engineer, or Employer's Representative. Usually this will take the form of a bar chart, or critical path programme prepared by the contractor following detailed discussion of the employer's requirements and contractor's planning.

The formal contractual date for the commencement of the contract should be agreed and recorded.

Before the work of dredging and/or reclamation commences on site there are a number of procedural matters which should be completed and actions which should be taken. These divide into administrative and practical matters as listed below.

Administrative matters

- the forms of contract agreement and bond should be completed;
- all necessary insurance cover should be in force and shown to be in force;
- names and positions of the principal staff representing the engineer and contractor should be notified;
- notices to mariners, if any, should be posted;
- all licences and consents should be formalised and displayed;
- procedures for the direction and copying of correspondence should be agreed;

- procedures for issue of instructions should be notified;
- the format for the recording of site activities and production should be agreed.

Practical matters

- accident and emergency procedures should be defined;
- setting out and control points should be established;
- pilotage arrangement should be made;
- standby and bunkering berths should be allocated if required;
- land should be allocated for temporary offices, workshops and storage;
- access routes to be defined;
- signs and notices to be agreed and erected;
- services should be arranged and provided;
- pre-dredge and/or pre-filling surveys should be completed;
- radio watch frequency and procedures should be agreed;
- traffic management procedures to be agreed (passing of vessels, making way, priority, etc.);
- arrangements should be made for the landing of debris and foreign matter which may not be disposed at sea;
- there should be a procedure for dealing with munitions, if necessary;
- placing and fixing of moorings should be arranged.

Further details of on-site arrangements and control are given in Chapter 12.

11.9 Common causes of claims

11.9.1 MAINTENANCE DREDGING

Generally, it is unusual for serious claims to arise on maintenance dredging work because the variation of physical conditions is usually less than on capital works.

Some matters do however give rise to claims or disagreement fairly regularly and, even though some of these may have been already referred to, it will be useful to highlight them here.

Soils to be dredged

By definition, maintenance dredging implies that only recent deposits of rather fine-grained materials are required to be removed. There are however exceptions to this.

Where the natural bed material is more difficult to dredge, such as firm or stiff clays, medium or dense sands and gravels, rock and so on, it is possible that some of this material may outcrop above the design level to which dredging is required under the contract. This may occur if the maintenance contract is the first to be undertaken since capital dredging was carried out. Of course, it may be presumed that during the capital work the required depth was achieved throughout, but in fact this is not always so. In particular, when capital dredging has been carried out by a 'non-sweeping' type of dredger, such as a grab or backhoe, and maintenance is by means of a trailer, cutter or bucket dredger, each of which will normally achieve close to 100 per cent bottom coverage, then it is to be expected that at least a few outcrops of original bed material will be encountered. Even when the maintenance work is not the first of its kind, it is still quite common, particularly when the method of position fixing has been improved, to discover areas of original material which outcrop above the specified level.

It is difficult to provide adequately for such an uncertain situation, but it is at least advisable to describe the recent history of dredging on the site and to warn those tendering that isolated outcrops of original material may be encountered. If very minor isolated outcrops, of a material which can be dredged by the plant employed, are encountered (not solid rock!), it will be reasonable to expect that the contractor will have made some allowance in his rates. If, however, significant quantities of original material must be dredged, then rates which have been tendered for maintenance dredging are unlikely to be appropriate.

Even assuming the more common situation in which only recent deposits are to be removed, the material may not always be fine sediment. This should be adequately covered by the ground investigation. Difficulty may arise when the bulk of the material is of fine sediment, but with isolated areas of other material. Such situations often occur when fast flowing rivers discharge into the dredging area, or when the area is flanked by slopes, or training works featuring loose stone protection. Stone may occasionally be moved by waves or currents from the slopes into the area of dredging. If this is known to occur, then the situation should be described and, so far as is possible, quantified, either in the ground description, or in the description of the works.

Refer to Chapters 6 and 8, and Section 11.5.2 for details of ground information which should be supplied if misunderstandings are to be avoided.

Debris

Some debris will almost certainly be encountered in maintenance dredging work. If the contractor has not allowed for the resulting disruption of his production process, a dispute may arise. There are particular situations in which heavy concentrations of debris must be expected.

If sufficient information can be provided at the time of tendering to allow the contractor to assess the effect on dredger production, then it will be reasonable to expect the contractor to make appropriate provision within his rates. If not, a separate arrangement for appropriate payment, based perhaps on delay time, should be made. The latter arrangement may be difficult to properly measure and administer.

Large items of debris, such as wrecks (vessels or vehicles), lost deck cargo, anchors and so on, may be removed only by special equipment or methods. If such items are known to exist, then clearly they should be described and the contractors obligations should be made clear. If not known, but suspected, then the contract should warn of the possibility and make some provision for removal by the contractor or, if preferred, by another party.

Stoppages

Stoppages, or delays which prevent a contractor from working, or cause reduced production may result in a claim, unless the contractor has been adequately informed in advance of the likely nature and extent of such events.

For some dredgers, particularly stationary dredgers, the passage of shipping may interfere with the dredging cycle. Normal shipping movements should therefore be described in reasonable detail, including number of movements, size of vessels, hazardous cargoes, passing restrictions and so on. If such movements are described, due allowance can be made within the rates for dredging. Alternatively, a rate can be entered for each stoppage, or each period of stoppage.

Delay may also arise due to industrial action by the employer's employees (dock, or lock operators, bridgemen, maintenance crews, pilots, etc.). Such events are outside the contractor's control or reasonable foresight and, if of significant duration, extra payment may be necessary.

If payment of a contractor is to be based upon pre- and post-dredge survey, a delayed pre-dredge survey may delay the start of dredging. A late post-dredge survey may delay the issue of a certificate of completion and hence the demobilisation of the dredging plant. Survey work is far more sensitive to sea conditions than dredging work, particularly if the latter is by trailing suction hopper dredger. The value of a survey contract (and perhaps the liquidated damages which apply under the contract) may be small compared to the cost of dredging plant. A medium-size trailer may well be charged at £15 000 to £20 000 per day (1995). In winter in the UK, it is not unusual to suffer several days duration of weather when accurate survey is impossible. Before contemplating inaccurate survey, it should be recognised that for a channel 200 metres wide and 5000 metres long, each 10 centimetres of error in sounding (if consistent) represents 100 000 cubic metres! Ten centimetre accuracy is not easily achieved, even in calm conditions!

Measurement

Whatever the method of measurement that is to be employed, it is most important that it is very clearly described and understood by both parties before any work commences. At each stage of the measurement, what has been measured should be agreed and that agreement should be recorded. The 'let's sort it out later' approach is tailor-made for progression down the path to dispute and eventual claims. If reasonable agreement cannot be reached at any stage of the measurement process, the measurement should be repeated until agreement is reached, even if work must be stopped or diverted to other areas in the meantime.

11.9.2 CAPITAL DREDGING WORKS

All of the potential heads of claim listed for maintenance dredging may also apply to capital dredging. However, in capital dredging by far the most common source of dispute arises from unforeseen ground conditions. These may include unexpectedly strong materials which are difficult to dredge, weak ground in which slopes are unstable, or on which reclamation is difficult, and unstable areas wherein rapid siltation of the dredged areas may occur. Of these, ground which is unexpectedly difficult to dredge is the most common cause of difficulty and dispute. The problem can and should be minimised by the prior execution of a comprehensive ground investigation (see Chapter 6).

References

Federation Internationale des Ingenieurs-Conseils, (FIDIC), 1987. *Conditions of Contract for Works of Civil Engineering Construction with Forms of Tender and Agreement*, The Federation of Civil Engineering Contractors, Cowdray House, 6 Portugal Street, London, WC2A 2HH.

Goudsmit, J.J. 1990. FIDIC conditions of contract, 4th edn – A user's guide, *Terra et Aqua* No. 42, April, International Association of Dredging Companies (IADC), The Netherlands.

ICE, 1979. *Conditions of Contract and Forms of Tender, Agreement and Bond, for use in Connection with Works of Civil Engineering Construction*, 5th edn, June 1973, revised January 1979. The Institution of Civil Engineers, 1–7 Great George Street, London, SW1P 3AA.

ICE/RICS, 1984. *Guidelines for the Preparation of Hydrographic Surveys for Dredging*, Thomas Telford Ltd, 1–7 Great George Street, London, SW1P 3AA.

Oosterbaan, N. & Bean, J.W. 1977. International dredging contract conditions, *Proceedings of the 2nd International Symposium on Dredging Technology*, November, Texas A & M University, 1977.

PIANC, 1984. Classification of soils and rocks to be dredged, *Report of the working group of the Permanent Technical Committee II*, Supplement to Bulletin No. 47, Permanent International Association of Navigation Congresses, Brussels.

van der Gouwe, J., 't Hart, M. & Blok, I. 1993. Results and effects of using the tonnes dry solids system on the efficiency of maintenance dredging, *Proceedings of the CEDA Dredging Days*, Amsterdam.

Wallace, H. 1974. Contract specifications in dredging operations, *Report of the Dredging Research Unit*, The University of Wales, DRU/MAST/008, April. Also published in Dock and Harbour Authority, Vols. LV 643 and 644, May and June, 1974.

12
Dredging Works and their Control

12.1 Introduction

Virtually all dredging works consist of the application of powerful, and often expensive, items of underwater soil-moving machinery to achieve relatively precise results in an inhospitable environment. Small changes in working parameters, job specification or the working environment, which might appear modest by land-based standards, can have dramatic effects on the ability of a dredger operator to complete his contract successfully. The progress of work may, all too easily, be delayed by interruption by third parties, inclement weather and sea conditions, unforeseen ground conditions and delays in decision-making. Every delay thus experienced increases the length of the contract period, increases the cost to the dredging contractor and the consequent risk of claims for extra payment from the client to cover these costs.

Both the promoter of the dredging works, and the contractor engaged to carry the work out, require a successful outcome of the work on site. In the case of the promoter there is a desire to see the work carried out according to programme and at the budgeted cost, whereas the dredger operator is looking to carry out the work to programme and to make a reasonable profit. Both should be committed to having the work carried out to the correct specification and with the minimum interference to the environment and to third parties. To achieve these objectives it is necessary to impose a high degree of control on the site operations and activities.

In the case of an autonomous port authority, or similar organisation, which owns and operates its own dredging fleet the role of the promoter and contractor becomes integrated. However there is still a need to control the works on site and to control the dredging operators. This is often achieved by giving the supervisory and operational roles to different departments of the same organisation. The guidance given in this chapter should apply to all dredging sites, whether the role of promoter and contractor is integrated or not.

In conventional contracts the promoter's supervisory role is often assumed by a supervising engineering consultant and embodied on site by the resident engineer and his staff. The contractor will normally be represented by a site agent or, on smaller sites, a dredging superintendent. The resident engineer and the site agent together have to interpret the requirements of the contract documents, control the site works to achieve the desired results and ensure that the manner of doing this is compatible with local and national legislation, and in harmony with the environment and other interested parties. A diagrammatic illustration of these relationships is given in Figure 12.1.

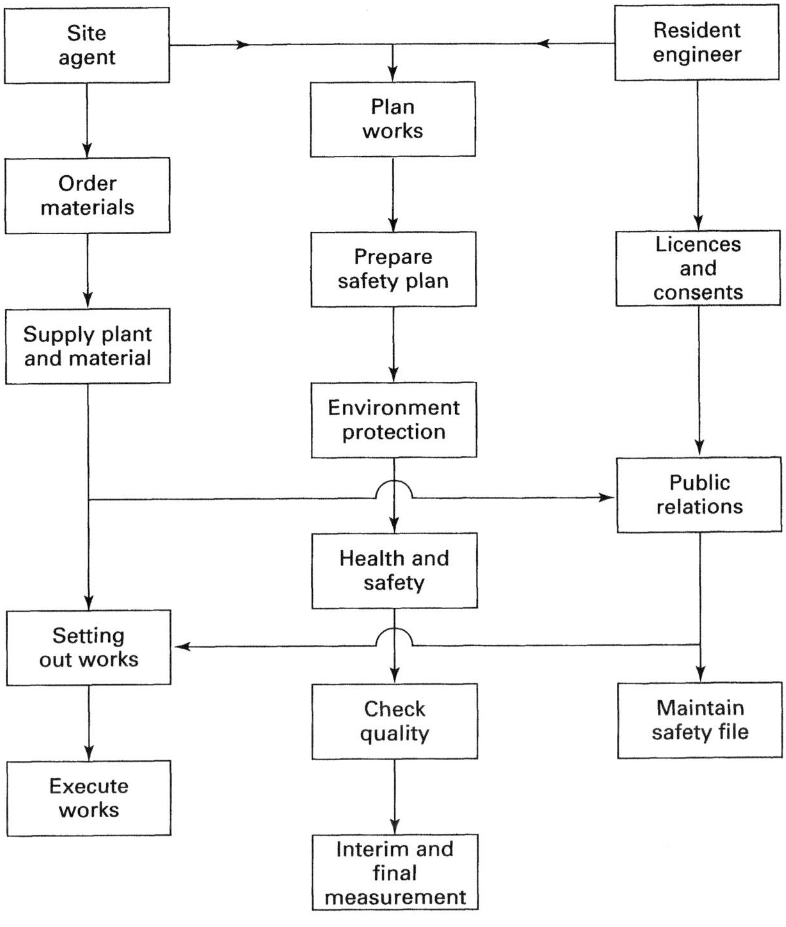

Figure 12.1 Relative responsibilities and relationships between site agent and resident engineer

The resident engineer's view of the control of the contract, and his detailed duties and responsibilities on site, will be different from those of the site agent. These duties and responsibilities are therefore described separately in the sections below. However, there are some broader aspects of the establishment of site operations, generally relating to public relations and safety, which are described first.

When the dredging is to be carried out by direct labour, the resident engineer/contractor situation does not exist. Although many of the points made in Sections 12.2, 12.3 and 12.4 are still applicable, there are a few additional aspects relating to the control of the works and these are described in Section 12.5.

Control of the works, in the technical sense, is also influenced by the ability of the dredger operator to use his equipment efficiently. To some extent this is affected by the control instrumentation available on the dredger. Section 12.6 in this chapter gives brief details of the instrumentation which may be installed on modern dredgers.

12.2 Public relations and safety

Generally, the degree of acceptance by the public of a particular contract is proportional to their awareness of the contract, what it entails and how it is to be carried out. This is especially true when such operations as pre-treatment by blasting and excavation of hard material by bucket dredging are to be carried out. There are few things more likely to cause irritation, bad feeling and, in some cases, legal action than the inhabitants of an area being shaken from their beds by an explosion of which they had no prior warning. It is, therefore, very important that every effort should be made to publicise widely the operations and details of the work to be carried out before they begin, because, otherwise, the initial publicity will be bad publicity.

Apart from general publicity there are a number of specific notices which should be considered, depending on the nature and location of the dredging site.

12.2.1 NOTICES TO MARINERS

In most nearshore waterways in the world there are specific authorities which have jurisdiction over navigation. It is their duty to ensure that any changes or obstructions to navigation in their areas are made known to mariners. This is usually done by issuing either a temporary or permanent notice to mariners, which informs them of:

- the charts which apply;
- the area in question;
- the nature of the change or obstruction;
- the method of marking the change or obstruction;
- the safe way past the change or obstruction and how it may be navigated.

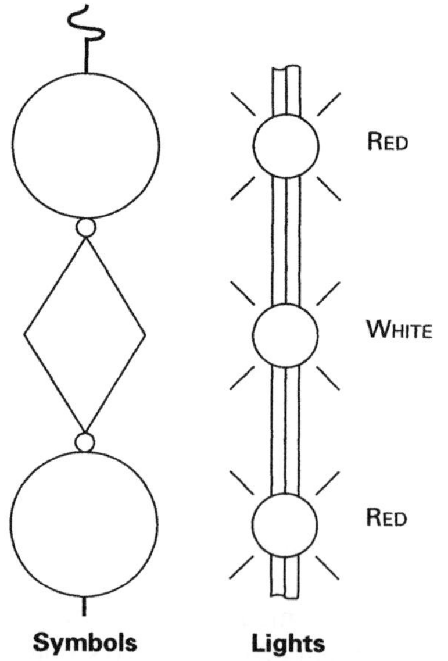

Figure 12.2 Dredging symbols displayed to alert mariners

In dredging operations, it is normal for the dredgers or pontoons to exhibit particular day marks and lights to indicate on which side vessels should pass them (see Figure 12.2). When blasting operations are to be carried out the appropriate warning and safety measures should be described (see Section 12.2.3 below).

12.2.2 NOTICES TO DIVING AND YACHTING CLUBS

Although general publicity will alert most of the local residents to the dangers of any particular dredging or pre-treatment operation being carried out it will not necessarily warn weekend visitors or holiday makers. In some cases, and particularly if blasting is to be carried out, it will be important that all relevant local clubs, especially those concerned with yachting and diving, are individually approached and informed of the nature of the work and how it is likely to affect their members. All agreed safety precautions and signals should be explained to each club in detail.

12.2.3 SAFETY MEASURES FOR BLASTING

When blasting is to take place it is normal to carry out a standard safety procedure, which has been previously made public. A typical procedure might be as follows.

Fifteen minutes before blasting
Safety vessels, displaying red flags, are sent up and down channel of the blasting area to warn off any approaching vessel. They remain on station until after the blast.

Immediately prior to blasting
Three hoots, each of three seconds duration, are made on the blasting pontoon's klaxon.

Immediately after blasting
An all-clear signal, ten seconds on the klaxon, is given and red flags are lowered.

12.3 Duties and responsibilities of the resident engineer

12.3.1 INTRODUCTION

The supervision of a dredging contract is an essential complementary function to the provision of good contract documentation. Without supervision good contract documentation is devalued, and without good contract documentation supervision becomes difficult, sometimes almost impossible. Supervision is an activity of many diverse aspects and the choice of suitable staff to carry it out should be given considerable attention. Some of the duties of the engineer and his/her representative on site are usually defined or implied in the contract conditions under which the work is controlled.

Apart from interpreting the contract and specifications and being an all-purpose co-ordinator, he/she will be expected to act as the go-between for client and contractor and as a link between the contractor and the general public, including other port and waterway users. He/she should also be an experienced engineer.

The engineer's main objectives can be defined as follows:

> *to ensure that the contract is completed to schedule, in accordance with the specification, at the agreed contract price, or other such varied price agreeable to both client and contractor, in accordance with the contract conditions, safely and with the minimum disturbance to the local populace and their environment.*

Supervision is likely, therefore, to include most, if not all, of the following activities:

- making the public, and, where appropriate, the port and waterway users, aware of the operations which are to be carried out;
- approving the contractor's working arrangements, the work programme and coordinating all site activities which are not the responsibility of the contractor;
- understanding and operating the method of contract control;
- understanding and operating the method of measurement;
- agreeing with the contractor methods of recording progress of the work on site;
- authorising and implementing variations to the contract and specification when desirable;
- checking on tolerances and specification, and ensuring that these are adhered to;
- ensuring that the contractor observes the contract conditions;
- checking interim measurements and agreeing interim valuations;
- checking on, and certifying, completion of work and authorising payment;
- assisting the contractor wherever possible, and particularly when there are problems.

12.3.2 SETTING UP

There are three main points to be established when setting up the site organisation:

- how the work is to be programmed;
- how the work is to be co-ordinated;
- how the progress is to be recorded.

Programming

Good programming is essential, particularly when the dredging work forms part of a larger contract. Since dredging machinery is expensive to mobilise, and to keep standing idle, it is important that the arrival and departure from the site should be well programmed. The completion of the work is as important as the commencement. Contractors will not normally demobilise their plant until the post-dredging survey shows that the work is acceptable, since re-mobilising is a costly operation.

Normally the dredging contractor submits a draft programme of operations at the tender stage. A revised programme should always be drawn up and agreed directly the contract is let. Any delay in letting a contract may alter plant availability and due allowance should be made for this.

Co-ordination on site

Site co-ordination is most important when works are being carried out in busy waterways or harbours. For normal operations of the waterway or port to continue without interruption, it is necessary for the central controlling organisation (harbour or waterway authority) to be fully informed of all dredging vessels' movements and intentions. It is also important that the dredging operations should only be interrupted when necessary, and then for the minimum of time.

The controlling organisation usually will be in VHF radio contact with all the relevant craft and the following procedures are recommended:

- that each dredging craft will report its position to the central control at regular intervals;
- that any movement from one position to another on the site will be carried out in full accordance with central control;

- that each dredging craft will have an agreed warning period before interruptions to dredging work occurs; it should be noted that this warning period may differ for different craft, different dredging areas, and so on;
- that the central control has ultimate responsibility for the coordination of all craft movement in the area and has the final decision in any case of dispute. (It should be noted, however, that the master of a vessel will always have ultimate responsibility for the safety of his vessel and crew.)

It is important to recognise that if dredging is to be carried out in berth areas which are heavily utilised, the contractor's access to the berth may be severly restricted and good co-ordination and forward planning are essential.

Progress reporting

To maintain the site records correctly and up-to-date, a system of progress reporting must be agreed and initiated from the outset. A detailed description of measuring progress on site is given in Section 12.3.3 below. The reports must include delays as well as progress, since delays are more likely to prove contentious and it is important that they should be agreed immediately. Any unresolved site conflict will almost certainly cause additional problems later on, when the memories of those involved may become unreliable. It will be helpful if the central control are encouraged to make and keep records of relevant vessel downtime and movements.

Progress reporting should include full details of the materials dredged, particularly if their characteristics are unexpected. Any incident which causes the contractor to suffer a lower production rate, stop his work, vary his working method or alter the equipment, men or material that he is using on the site, should be fully documented. In addition to this a detailed record should be kept of the plant, labour and materials on the site and the hours worked.

It is usual for sophisticated dredging contractors to keep detailed records of production, working times and conditions. These original records are usually reliable and can be copied to the engineer.

Records of progress should be kept in the form of bar charts, line graphs (volume against time) and also as areas marked up on the dredging plans. It is advisable to mark each shift's or day's progress in colour on the chart with the date, so that the work can be identified later if necessary. Areas pre-treated by blasting should also be recorded. This can be done by marking the position of each hole drilled, together with the depth of rock and the quantity of explosive used (see Chapter 14 for more detail).

12.3.3 PROGRESS MEASUREMENT

One of the main duties of the supervisory staff on a dredging site is to measure the quantity and quality of work carried out. The information gained from the measurement will be used to:

- record progress against programme, in order that any alteration due to slow or fast work may be identified;
- measure quantities for interim payments;
- measure quantities for final payment;
- check on the tolerances to which the work is being carried out;

- provide evidence in the case of disputes concerning such things as the dredging output, type of material dredged and delays.

There are various methods of measuring dredging progress and the choice of which to use will depend on the type of dredger, material being dredged and dredging site, as well as the type of work being carried out, that is capital or maintenance. The methods can be grouped into:

- operational;
- transport;
- difference methods.

Whichever system or systems are used it is essential that the measurements themselves are made by both the staff of the contractor and the client's representative, to an agreed and approved fashion. In many cases it will be necessary for there to be a hydrographic surveyor on site on the resident engineer's staff.

Operational

Operational methods are based mainly on the advance of the dredger across the dredging site and the assumption that all material in the path of the dredger has been removed. They are normally used for 'stationary' dredgers and are generally somewhat inaccurate. For this reason they are used only to establish quantities for progress measurement and interim payment. The success of measurement by advance of the dredger relies on the fact that the pre-dredging survey has been completed accurately and that the dredger is operated in a predetermined regular fashion and not at random on the dredging site. The method used is as follows:

- the position and attitude of the dredger at the beginning of the shift is determined and hence, from knowledge of the dimensions of the dredger and dredging depth, the position of the start of the dredging cut can be found and marked on a progress plan;
- the dredging level is set and the width of cut noted;

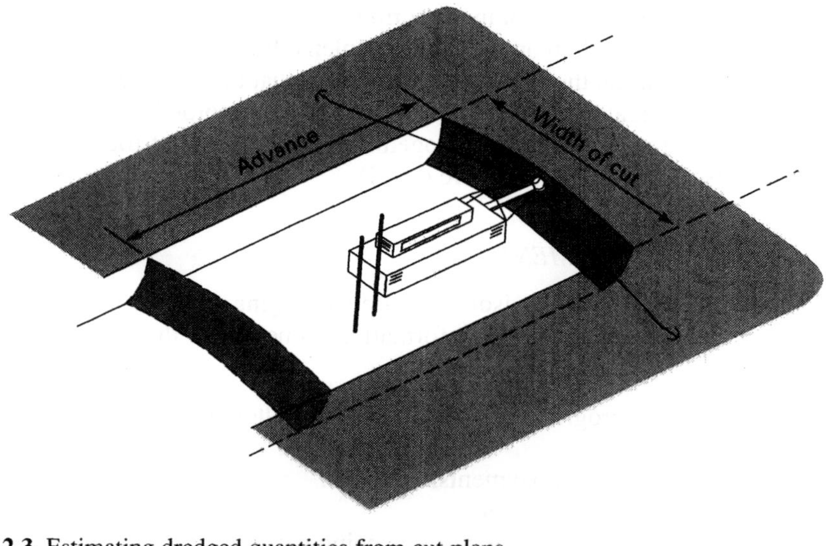

Figure 12.3 Estimating dredged quantities from cut plans

- during the dredging operations the position of the dredger is checked and sounding is carried out from the stern of the dredger to check that the set level is being attained;
- *in situ* quantities dredged are calculated from the dimensions of the shape cut by the advance of the dredger.

Figure 12.3 shows this method being used for a cutter suction dredger. However, it is also suited to the operations of bucket, grab, dipper, backhoe and dustpan dredgers. In each case the method of calculation of volume removed will depend on the dredger type and operating method. These methods should not be used to estimate dredged quantities for the final valuation. However, in rock blasting a similar method (see Chapter 14) may be used to estimate areas and volumes for final payment. This is because in the case of drill and blast methods, the process is usually very accurately controlled.

Transport

Measurement during transport involves measurement of quantities removed from the dredging site by means of hoppers or in pipelines. These methods are not very accurate and would normally be used only as checks on progress or as a basis for interim payments, except in cases when there is no alternative method, such as when dredging soft silts or fluid muds in areas of rapid siltation.

Measurement in the hopper

For most soils and rocks, measurement in the hopper consists of ascertaining the level of dredged material in the hopper. This method is explained in Chapter 11, Section 11.6.2, page 336.

Measurement in the pipeline

The measurement of quantities of dredged material transported through the pipeline may be carried out by measuring the velocity and density of flow through the line, and is described in Chapter 11, Section 11.6.3.

To achieve even a modest level of accuracy by this method requires sophisticated instrumentation. However, mixture velocity in the pipeline can be rudely determined by measuring the trajectory of the liquid being discharged from the end of the line. When the line is horizontal, the trajectory shape is given by the expression:

$$V = \frac{xg}{2y} \qquad (12.1)$$

where: V = velocity of emission
x = horizontal distance from pipe end to water jet
y = vertical distance from pipe end to water jet
g = acceleration due to gravity

If the measurement is made to the upper surface of the water jet it will give the maximum velocity in the pipe. The average velocity depends on the amount of solids in the discharge, but is probably about 85 per cent of this. Volume of discharge can be calculated by multiplying this average pipeline velocity by the internal pipe area.

The measurement of density in the pipeline, and in the seabed, is not so easy. A rough approximation of the solids content in the pipeline can be obtained by taking samples of the discharge water and allowing the solids to settle out, but the results will vary depending on the position from where the sample was taken. It will still be difficult to tell how much the settled solids have been bulked due to the dredging process. Hence the inaccuracy in the method.

Difference

Measurement by difference is the most accurate method of all and should be used wherever possible (see Chapter 11, Section 11.6.1). Since it involves making echo sounding surveys before and after the dredging operation, it has the added advantage that it acts as a check that the work is being carried out to the correct tolerances and indicates where re-dredging is needed.

Interim post-dredging surveys are usually carried out to allow estimation of quantities for payment purposes and to check on quality and progress. A final post-dredging survey is carried out at the end of operations to check that the whole area has been dredged to the correct level and that any re-dredging has been effective.

When the material to be dredged consists of soil and rock, and they are to be paid for separately, it may be necessary to carry out interim surveys after removal of soft material and before removal of hard. Although the difference method can be used for this purpose it should be noted that the conventional echo sounding methods do not differentiate between hard and soft material.

When rock without overburden is present, the difference method is ideal. However, quantities computed from pre-dredging surveys of jagged formations will tend to be slightly inflated due to the echo sounder's characteristic of recording the highest spots and masking potholes and small dips. The problem will be minimised by using a narrow beam echo sounder (see Chapter 6).

12.3.4 TIMING AND DELAYS

On most dredging projects it is inevitable that delays will occur. Some of these delays will be of the contractor's own making whilst others, such as weather delays, may be risks accepted by the contractor at the time of tender and will not warrant additional payment. However, there may be other delays which may be beyond the control of the contractor and for which he may seek reimbursement.

The tendered rates for demurrage are intended to cover the actual cost of that item of plant standing idle, together with its associated overheads, and should, therefore, be applied to the actual delay incurred. This is straightforward when the delay is ordered by the engineer's representative or when interruptions occur in such a way that the duration of the interruption is the same as the loss in dredging time. There are, however, often cases when the consequential loss of dredging time is greater than the actual duration of the interruption. This is particularly true of interruptions due to shipping when work is being carried out in a shipping channel.

The extent of the consequential loss in time will vary according to the type of plant working, the relation between the dredging site and the shipping channel, the point in the working cycle at which the interruption occurs and the weather, the sea state, strength of current and so on. The following examples give an indication of the situations where consequential loss of time may be incurred.

- A bucket dredger working in a shipping channel is ordered to move to the edge of the channel to allow a vessel to pass. The time of moving off and back onto station and re-establishing the dredging face must be added to the time spent waiting at the edge of the channel.

- A trailing suction hopper dredger is working in an estuarine site with a long haul distance out to sea. The contractor has programmed his cycles to take maximum advantage of the ebb and flood tides. The engineer's representative orders the dredger to stop work, thereby forcing the contractor to sail against the tide (or even wait for the next tide). The consequential loss in time due to reduced sailing speed, or lost tide, must be compensated.

From the above examples it can be seen that there are a number of rules which must be developed on the site to enable the contractor to operate in a rational manner whilst still allowing normal shipping to navigate through the site. The following points should be noted:

- The port or conservancy authority must have general responsibility for managing all shipping and dredging plant movements in the area. The officer on duty should keep a detailed record of all vessel movements and times of delay and so on.
- Captains of each dredger or pontoon must be ultimately responsible for the safety of their vessels and, therefore, must have the final decision concerning the action to be taken to move the dredger into the ordered position to allow a vessel to pass. The captain should also keep a detailed record of the movement of his vessel, delays and so on.
- Both navigation authority and dredger captain should exchange information on proposed movements in order that adequate warning may be given before interruptions to work occur.
- Where possible the client's representative and contractor should agree the consequential time losses to be added to actual delays in standard cases, prior to these delays occurring.

The analysis of consequential delays is often difficult as it involves the assessment of what would have happened had the delay not occurred. In situations where consequential delays are likely to occur the client's representative is well advised to become acquainted with the operational methods of each piece of dredging plant on the site and to establish what criteria affect their behaviour when the dredging cycle is interrupted.

12.3.5 PROBLEMS ON SITE

One of the main duties of the supervising engineer on the dredging site is to deal with the contractor's grievances and to help him overcome his problems. This does not necessarily mean committing the client to any further expenditure nor to accepting delay. It implies that wherever possible the client's representative should be endeavouring to encourage the contractor to succeed in completing the contract in the most efficient manner.

The contractor's problems may occur due to a number of causes. They may be self-made, accidental, or inherent in the contractual format, risk element and so on. Many problems can be avoided by improving the quality of the site investigation and the contract type chosen, but this is small consolation to the site staff who may be saddled with unravelling the difficulties of an ill-chosen contract format or a lack of site data.

The major causes of dispute (Dredging Research Unit, 1975) involve those factors which control the rate of production of the plant, such as:

- the material being dredged, its type, quantity and location;
- operational delays, interruptions due to traffic, wrecks, wartime debris and so on;
- wind, wave motion and strong currents.

In addition, there are other problems which may have been overlooked at the planning stage:

- siltation of the dredged area;
- disposal of material in the dredged area by other vessels;
- sharp increases in the contractor's costs;
- interfacing between different contractors working on the same site or hindrance by the client.

In recent years a concern for the environment has resulted in new restrictions being applied to dredging sites. Whilst such restrictions may be incorporated in the contract documentation and allowed for by the contractor, unforeseen consequences may arise during the contract relating to:

- the interpretation of monitoring data;
- the difficulty in accurately predicting the effects of dredging;
- the actions of pressure groups and other users of the sea.

These matters are discussed in Chapter 13.

It is of great importance that the supervising engineer keeps adequate records to enable him to identify and determine the cause, and quantify the consequences, of delays or disruptions to the work. This knowledge will be used to evaluate any claims submitted by the contractor.

Kyling (1972) gave the following reasons for poor operations when a contract is running badly, and it is found that these still hold true:

- poor control of dredging during the hours of darkness;
- inadequate operator training;
- lack of production standards and control by the contractor;
- delays in surveying, causing retention of certificates of completion;
- lack of direct cost control;
- poor pipeline control;
- lack of anticipation and longer range planning;
- poor shift arrangements and unduly long work periods;
- neglect of maintenance.

The client's representative should be aware of these potential problems and, if encountered, he should encourage the contractor to put things right. Successful completion of any major engineering works on schedule and within budget reflects with credit upon both contractor and employer alike.

12.4 Contractor's viewpoint

12.4.1 INTRODUCTION

It is the duty of the contractor's site management to carry out the contract works in accordance with the Conditions of Contract, Specification, good practice and, so far as is reasonably possible, within the agreed programme time and tender sum. In so doing, he should exercise proper care to ensure the welfare of his employees and others engaged about the works and to protect the general public and the interests of third parties.

Throughout the contract he/she should, by regular reports, provide to the engineer details of that work which is complete, in progress, or planned.

These responsibilities are common to most works of civil engineering construction, but in the case of dredging, usually matters are complicated by a tidal regime, the unpredictability of weather (and hence sea state) and, often, the unrelenting demands of continuous working 24 hours a day, seven days a week.

The process of continuous communication may at times be irksome to the contractor's staff, particularly when under pressure, but it is important that the contractor remembers that the engineer has a duty to follow closely the progress and quality of the work and to authorise payments to the contractor only when he is satisfied in all proper respects, that is:

- that the quantum of work has been carried out;
- that the quality of the work is to the specification.

Usually, the works will proceed more smoothly in an atmosphere of professional co-operation than if conflict exists.

There are also important matters for the employer to recognise. If a contractor has tendered rates, or sums, for the execution of specific tasks under specific conditions and those tasks or conditions change significantly, such that the work is made more difficult, the contractor may ask for additional payment. The response of employers often is surprise, or even indignation.

The request for additional payment is seen as a 'claim', which it is, but contrary to common perception, a claim is not necessarily unreasonable. It is unfortunate that the very word 'claim' often arouses an adversarial attitude.

It is important that the employer recognises that the contractor has come to the site for the purpose of providing a service to the employer, or to perform a specific task on behalf of the employer. In return he is entitled to expect fair payment and a reasonable profit for work which is performed satisfactorily.

If costs rise simply because conditions are more difficult than the contractor could have reasonably foreseen, it is not the fault of the contractor. Either the employer has failed to provide adequate information, or the adverse conditions were simply unforeseeable. If, as a result of the changed conditions, the reasonable cost of the work is greater than the employer has budgeted, the contractor should not be blamed, nor should he suffer financially.

Perhaps the attitude of employers to claims would be more favourable if, when conditions are found to be easier than foreseen, contractors offered a rebate!

12.4.2 THE WELL-BEING OF PERSONNEL

In many countries there exist stringent laws and regulations which control matters of health and safety in the work place. In dredging works matters may be complicated by the mixture of operations which are land and marine based. The laws which regulate each may be different. An additional complication arises from the fact that dredging contractors work internationally. There may be occasions when the regulations governing health and safety aboard the vessel may be those of where the vessel is registered and these may conflict with the regulations in the country where the work is being carried out.

In the UK, the main law is The Health and Safety at Work Act 1974, which is enforced by the Health and Safety Executive. This Act is intended to cover all types of work by virtue of the Statutory Instruments (SIs) issued subsequent to its original enactment and other Relevant Statutory Provisions (RSPs). These latter are Acts which were in existence prior to the 1974 Act.

RSPs of relevance to dredging work include:

- the Factories Act 1961;
- the Offices, Shops and Railway Premises Act 1963.

and for work involving the use of explosives:

- the Explosives Act 1875;
- the Explosives Act 1923.

Other matters relating to dredging works will also be found in:

- the construction industry regulations;
- the international maritime rules;
- the Department of Trade regulations (UK);
- the Construction Design and Management Regulations 1995.

12.4.3 MOBILISATION

Usually the first task facing the contractor is that of mobilising all necessary dredging plant, equipment and labour.

Dredging work will usually be executed by one, or more items of specialised dredging plant. Often the plant will move from one project to another. Hence, if work is running late on one project, the commencement of the next may be delayed. A large contractor may be able to supply alternative plant from within his fleet, but this will not always be possible. Such problems may sometimes be overcome by the contractor chartering suitable plant from a competitor, but this may not be economic.

It is important that these risks be recognised when inviting and assessing tenders, at the time of award of contract and when planning the works.

On all but the smallest of dredging works, major items of plant will usually travel by sea. The transport is therefore dependent on sea conditions. Adverse weather may delay arrival on site.

Within the mobilisation process, the contractor must arrange in advance for a number of different activities and facilities. These may include the following:

- installation of positioning systems;
- installation of tide gauges;
- site survey (if it is the contractor's responsibility);
- arrangement of pilotage for incoming vessels;
- arrangement of work permits for crew (where applicable);
- checking local hospital and first aid facilities;
- obtaining dredging and disposal licences (if it is the contractor's responsibility to do so);
- arrangement of temporary moorings;
- site offices and communications;
- supply of fuel and lubricants;
- supply of water;
- victualling;
- accommodation for staff and crew if required (crew may live aboard);
- transport arrangements for crew arrival and future changeovers;
- heavy craneage if necessary (to raise spuds, etc.);
- arranging local repair facilities;
- arranging local welding facilities;

- arranging local small boat hire;
- arranging for disposal of waste and/or harmful substances;
- recruitment of local labour/staff;
- assembling floating pipelines (where required);
- assembling and installing shore connection (where required);
- laying land pipelines (where required);
- hiring local earthmoving plant (bulldozers, loaders, etc.);
- constructing temporary bunds (where required).

Some of the above tasks may be arranged from head office, but the majority will usually be the responsibility of site staff and surveyors who will arrive on site in advance of the main dredging plant and crew. The last few items apply only in the case of work involving land reclamation, beach recharge, or onshore disposal of dredged material.

12.4.4 EXECUTING THE DREDGING AND/OR LAND RECLAMATION

When all necessary plant is assembled and manned, the work of dredging can begin.

During the initial period of work it is important for the operation of the dredging plant to be optimised. This is often the responsibility of the individual dredgemasters, but it is desirable that the project management maintain overall control, perhaps assisted by specialists from headquarters. Surprisingly, this is an area which is commonly neglected. This may be due to the very considerable pressure under which the project management may be working during the early stages of a contract.

It is important that the higher management of the company provide, on site, adequate staff resources, particularly during the initial phase of a contract. The additional staff necessary during this time can be provided by a 'flying squad' whose primary function within the company is to assist on new contracts in achieving the optimum contract performance in the shortest possible time.

Many dredging contracts are performed away from the contractor's home base, often in a different country, or even on a different continent. In this situation, the role of the project manager is not unlike that of the captain of a ship. The safety of personnel and vessels, the health, feeding, performance and morale of personnel, will all be influenced by the attitude and ability of the project manager and his supporting staff, if any.

The regular responsibilities of the site management and tasks which must be performed are many and will vary from project to project. The main tasks will usually include the following:

- maintaining good public and client relations;
- setting out of the work;
- recording progress;
- arranging for the replacement of all consumables, including food;
- arranging crew changes and transport;
- monitoring weather forecasts;
- planning work;
- advising employer's representative of work planned;
- reporting progress to the employer's representative;
- reporting progress to head office;
- arranging for repairs and renewals;

- quality control;
- agreeing measurement;
- preparing and submitting certificates.

Some of the above items deserve further explanation.

Maintaining good public and client relations

Usually the employer will make arrangements to inform the public of the timing, nature and implications of the project. However, the contractor often can assist by having a representative in attendance during public meetings to respond to technical questions, or to give reassurance concerning the nature of his planned operations.

In particular, he should provide advance notice of any temporary restrictions on road or marine traffic and any potential source of disturbance, such as explosions.

Complaints or enquiries by the public may often be directed to the contractor's staff who should deal with the matter politely. A procedure to be followed in such matters should be agreed in advance between the contractor and employer.

Setting out the work

It will usually be advantageous to set out the works relative to National Survey Grid using properly co-ordinated setting out points established in advance around the site. This is strongly recommended. The modest initial effort will be well worth while in the long term.

It is recommended that the same setting out points be used by all parties for all surveys and the positioning of all dredging craft. However, it is increasingly common for position to be established by DGPS (see Chapter 6). In such cases it is preferable that the same system be used by all parties.

Recording progress

Recording the overall progress of the works and of individual items of dredging plant has two important benefits. It provides a regular view of the performance of the contract and the individual progress of particular plant. In the event of difficulties or dispute, the existence of adequate and proper records will prove invaluable for the analysis of the circumstances giving rise to the dispute.

The progress of dredging plant should be routinely recorded by the senior crew members on a continuous basis in the form of daily log sheets. The detailed format of log sheets will vary depending on the type of plant and the nature of the work. Example log sheets suitable for trailing suction, cutter suction and backhoe dredgers are illustrated within this chapter (see Figures 12.4–12.6). Usually the daily and weekly logs provide the key element for providing the contractors head office with information concerning the progress of the work. The engineer may expect to be kept informed by the same means. It is usually the case that the detail of the records is directly related to the sophistication of the plant being utilised.

Quality control

It is usually the case in both dredging and reclamation works that work which is carried out wrongly is difficult and expensive to rectify.

Dredger: *Trailer suction*
Draghead: *Californian*

PROJECT
Navigation channel

Date:
Master:

Cycle No.	Start Loading	Finish Loading	Turn Time	Arrive Discharge	Leave Discharge	Load time	Cycle time	Stops from	Stops to	Time lost	Load m^3	Cumulative m^3	Production rate m^3/hour	Comment or Delay Code
1														
2														
3														
4														
5														
6														
7														
8														
9														
10														
11														
12														
13														
14														
15														
16														
17														
18														
19														
20														
21														
22														
23														
24														
25														
26														
27														
28														
29														
30														
Totals														
Average														

Figure 12.4 Typical record log sheet for trailing suction hopper dredger

Dredger: *Cutter suction*
Cutter: *Crown*

PROJECT
Land reclamation

Date:
Master:
Area:

Time	Bed	Dredge	Face	Cut	Carriage	Face	Cut	Cycle	Production	Stops		Time	Cumulative		Pipeline		Comment
	Level	level	Height	Width	advance	Area	Volume	time	rate	from	to	lost	prod.	Float	Land	Elevation	or Delay Code
	m	m	m	m	m	m^2	m^3	mins	m^3/hour			mins	m^3	m	m	m	
Totals																	
Average																	

Figure 12.5 Typical record log sheet for cutter suction dredger

Dredger: *Backhoe*
Bucket: *6.5m³*
Stick: *8m*

PROJECT
Sea outfall

Date:
Master:
Work Area:

Load	Start	Move pontoon		Finish	Cycle	Load	Production	Stops		Time	Cumulative	Comment
No.	Loading	from	to	Loading	time	m^3	rate m^3/hr	from	to	lost	prod m^3	or Delay Code
1												
2												
3												
4												
5												
6												
7												
8												
9												
10												
11												
12												
13												
14												
15												
16												
17												
18												
19												
20												
21												
22												
23												
24												
25												
26												
27												
28												
29												
30												
Totals												
Average												

Figure 12.6 Typical record log sheet for backhoe dredger

For example, if serious overdredging occurs alongside structures, it is possible that the engineer may, quite reasonably, instruct that the area be backfilled using materials and methods which secure the stability of the structure. Such an operation will be expensive, perhaps very expensive in relation to the unit cost of dredging. Conversely, if after dredging with a stationary dredger, areas are left above the specified level, it will be expensive to return to the area to carry out the re-dredging of thin layers or isolated high spots. Good site control is clearly essential.

It is for the dredgemaster to ensure that dredging is carried out to the specified levels. However, mistakes may happen, particularly during night shifts if men are tired, or when mental conversions between positive and negative datums are necessary. The site supervisor should routinely carry out simple checks by hand line astern of the dredged face to ensure that the correct levels are being achieved. More extensive checks by survey should also be made on a regular basis.

Agreeing measurement

It is usual to measure the work on a monthly basis for the purpose of interim payments. This may not require monthly *in situ* measurement by survey, though this is normally desirable. Occasionally, interim measurement will be based on hopper measurement, or an estimate of the percentage of work completed in a particular area, or areas.

Usually it will be beneficial if representatives of the contractor and employer can agree a mutually acceptable basis for interim measurement before formal submissions are made. Such an agreement should serve to reduce paper work and speed payment.

12.5 Direct labour operations

The narrative in the previous sections assumes that the dredging works are carried out by contract under a supervising engineer. However, in many cases works will be carried out by a direct labour organisation using dredging plant owned, or leased, by a port or public authority. In such cases many of the matters described earlier will apply equally, but a few points deserve special mention.

Operational efficiency

Whilst a public body may not be charged with making an annual profit, it is nevertheless a duty of such bodies that dredging or reclamation works be executed with a reasonable degree of efficiency and that appropriate methods of measuring and recording performance be set up and maintained.

Senior management should regularly examine weekly, or monthly synopses of performance to check that satisfactory standards are maintained.

It is important to ensure that dredging is carried out accurately to properly defined lines and levels. It is commonplace within direct labour organisations to instruct the dredgemaster to 'go and dredge over there'. The result of such lax instructions is that no proper record of the work exists and no accurate measurement of performance is possible. It may then happen that at a later date other work, perhaps by contractor, is ordered within the area, only to find that the existing conditions are not as expected, perhaps with resultant delays or claims.

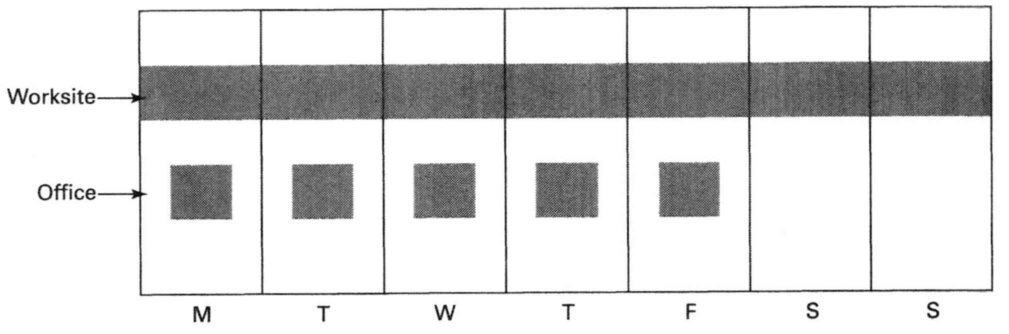

Figure 12.7 Relationship between normal office hours and dredging hours

Supervision

Efficiency will inevitably be affected by the level of supervision by management staff. If dredging is carried out on a continuous basis, which may be necessary when using large plant in order that the substantial investment be fully utilised, then supervision will be desirable outside of normal office hours. Figure 12.7 illustrates the inadequacy of supervision which is confined to office hours.

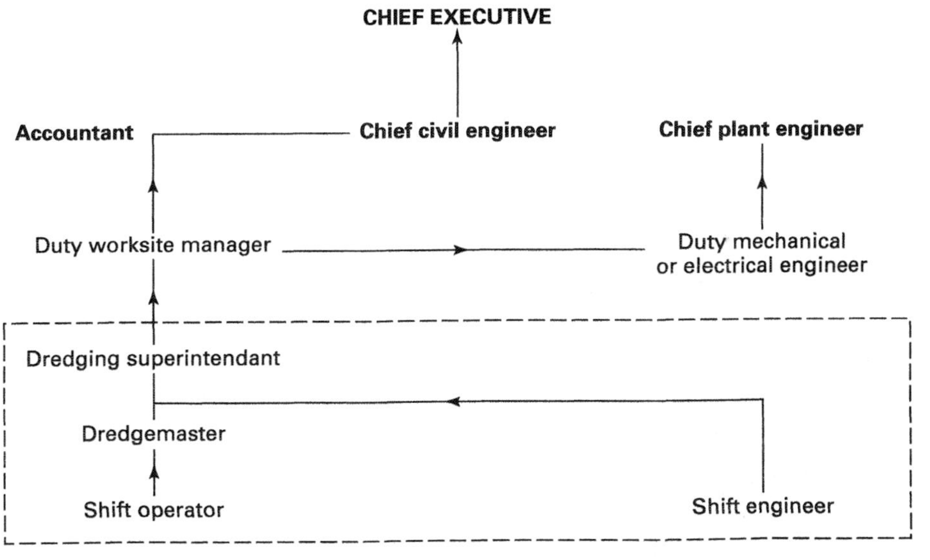

Figure 12.8 Typical chain of command for important decision making

To provide adequate supervision it may be necessary to extend the hours worked by management staff, or to provide for staggered working within a management team. It will also be necessary to make provision for major decision-making outside normal office hours. This is important to minimise delay and to ensure rapid response in the event of any threat to life or valuable plant, for instance as a consequence of a risk of vessel loss due to sinking. Decision-making should be via a well defined and clearly understood hierarchy (see Figure 12.8).

Training

Efficiency will be affected by the level of training given to staff and crew members. Training will usually comprise initial training at the time of new plant acquisition, or on setting up a dredging organisation and on-going training throughout the life of the project, or organisation. Failure to maintain an appropriate training programme will usually result in lower standards of efficiency than might otherwise be achieved. Investment in appropriate training should be self-financing through improved efficiency, but this will only be so if training is given to the right people. That is to say, those who are actively and routinely engaged in the dredging work. It is often the case that training, particularly attendance at conferences, is viewed as a perk for senior staff, but this is clearly counter-productive.

12.6 Control instrumentation

12.6.1 INTRODUCTION

Dredgers dating from the mid-20th century had the benefit only of simple instrumentation, usually comprising a depth gauge, perhaps supplemented on mechanical dredgers by gauges showing the electrical load on winches, and on hydraulic dredgers by vacuum and pressure gauges.

In recent years the scene has changed dramatically, thanks to developments in electronics. The benefits of these advances include improved control of production and accuracy, and a better understanding of the dredging processes. An understanding of the sophistication associated with a modern dredger may be gained from the dredging console of a large cutter suction dredger shown in Plate 12.1

Different types of dredger have been influenced in different ways.

Plate 12.1 The dredging console of a large cutter suction dredger

12.6.2 SUCTION DREDGERS

Modern suction dredgers, other than very small models, usually will have the benefit of flow, concentration, vacuum and pressure meters.

Flow meters

Flow meters measure the average velocity of liquid flow past a point in a pipeline. They are usually installed in the pressure pipeline a short distance downstream of the dredge pump, or pumps. An instrument mounted in the dredge operators control panel displays the instantaneous velocity.

Concentration meters

Concentration meters measure the average density of liquid flow past a point in a pipeline. As with velocity measurement, the measurement system is usually installed a short distance downstream of the dredge pump, or pumps. An instrument mounted in the operators control panel displays the instantaneous density.

The product of velocity and concentration gives the rate of production. By the continuous logging of these data the cumulative production over any period of time can be recorded.

It is common to display the instantaneous values of velocity and concentration in a variety of ways.

On both trailer suction and cutter suction dredgers, the velocity and concentration displays may be combined. The most common type of display takes the form of a square display with sides of approximately 300 millimetres. Within the square are two opposed quadrant scales, one each for velocity and concentration. The arms of the respective instruments rotate about the lower right and left corners of the display respectively. That on the right swings anti-clockwise and that on the left clockwise. Thus, each reads zero when horizontal. Maximum production is achieved when both meter arms cross in the middle of the display and at the highest attainable level.

Hence the operator strives to maintain levels of flow and concentration which, when combined, result in the optimum rate of production. The obvious advantage of such systems, compared to simpler systems, which often measured only vacuum and pressure, is that high vacuum is not mistaken for high production when the actual result may be low flow and hence low production.

Production may also be displayed in simple digital form in tonnes per second, or *in situ* cubic metres per second. However, conversion from tonnes to *in situ* cubic metres requires the manual input of a fixed conversion factor, whereas in fact in most situations the conversion factor is variable, perhaps over a wide range. Thus displays and records measured in cubic metres may not be accurate.

On trailer suction hopper dredgers it is common to duplicate the display of velocity and concentration at a location visible to the helmsman. This is because to achieve good and effective production with such dredgers requires combined action by the helmsman and the pipeman (suction pipe controller).

It is increasingly common on medium to large hydraulic types of dredger to supplement or replace fixed individual instruments with a VDU screen on which may be displayed a wide range of instantaneous and cumulative values. However, it is not common to find sophisticated measurement systems installed in small cutter suction dredgers. Usually it is not considered to be economic, or the lack of space and inhospitable environment render it impractical.

Vacuum gauges

Vacuum is measured on the suction side of the dredge pump a short distance from the pump inlet. The vacuum gauge is usually mounted in the operator's control panel and may be duplicated at the pump. The gauge measures suction head, which is of course limited to one atmosphere. Where more than one pump is installed, separate gauges may be provided for each pump.

In the past, the combination of a vacuum and delivery pressure gauge provided the only indication of pump performance to the operator.

A very experienced operator can learn much of what is happening in the pumping system from small variations in suction and pressure heads, but nevertheless, when relying only on simple gauges, operators often will prefer to be able to see the pump discharge in order to judge flow and concentration visually.

A relatively low vacuum, or suction head is usually accompanied by low solids concentration. Intermediate vacuum head usually indicates satisfactory solids concentration, but may be due to partial blockage of the suction inlet. A high vacuum head is usually indicative of a severe blockage.

Pressure gauges

Pressure is measured on the pressure side of the dredge pump, usually a few metres downstream of the pump. The pressure gauge is usually mounted in the operator's control panel and may be duplicated at the pump. Where more than one pump is installed, separate gauges may be provided for each pump.

The pressure gauge indicates the total pressure in the discharge pipeline. A relatively low pressure may indicate low solids concentration, or a leak in the pipeline. A progressive rise in pressure is usually indicative of an overloaded pipeline, that is too high a solids concentration. Consistently high, but steady pressure may be indicative of a partial blockage of the pipeline. A very sudden rise in pipeline pressure may precede total blockage of the pipeline.

12.6.3 TRAILING SUCTION HOPPER DREDGERS

There are a number of additional indicators on trailing suction hopper dredgers and these are described below.

Suction pipe display

The dredging depth, represented by the depth beneath water of the underside of the trailing end of the draghead, may be shown simply as a digital display of depth. Usually, however, this is supplemented by a moving scale representation of the complete trailing suction pipe and draghead. This shows the operator, at a glance, the level, relative to water level, of the point of connection of the suction pipe with the onboard suction pipework, the articulated centre pipe connection and the position and orientation of the draghead.

Draught and load indicators

These measure the draught of the dredger at various points, usually a minimum of two points, one well forward and the other well aft. Measurement is by means of pressure transducers. If the change in draught is known, then so too is the change in displacement and hence the total load in the hopper. Total load includes water. The changing trim of the vessel has traditionally been shown by means of two linked vertical slides set against a graduated scale background. The left-hand scale shows the draught forward and the right-hand scale the draught aft, usually in metres. The total displacement of the vessel is also shown.

The average displacement at any time may be plotted on a chart recorder where displacement is measured vertically against a time base. By regular reference to this 'time-displacement' graph, or 'loading graph', the dredgemaster may judge the optimum time to cease loading. The gradient of the load curve at any point represents the loading rate at that time. Alternatively, the information may be shown on a VDU.

More sophisticated systems of this type may measure displacement at more than two points, usually four, these being forward, aft, port and starboard. Water level within the hopper is also measured, as is the changing load of bunkers and so on. If all of these variables are measured accurately, the true weight of the hopper contents can be calculated. The more sophisticated systems of this type can provide an accurate method of measurement which in some circumstances may be an acceptable alternative to measurement *in situ* at the sea bed.

12.6.4 MECHANICAL DREDGERS

Excepting very large machines, the level of sophistication of instrumentation fitted to mechanical dredgers is low in comparison to hydraulic types. This is, in part, a reflection of the relatively simple mechanical dredging processes, but also reflects the more hostile working environment which is a feature of all mechanical dredgers. Space is often at a premium, shock loading and vibration are common and the risk of dirt and water intrusion is higher than on the larger ship-like vessels. However, despite the difficulties, the potential benefits of greater accuracy in dredging and improved recording of performance have encouraged the builders and owners of mechanical dredgers to overcome these problems and install measurement and data logging systems of increasing sophistication.

Grab dredgers

In grab-dredging work it is important to measure the depth, or level of dredging and the area covered. Whilst these are simple objectives, in fact they are quite difficult to achieve in a reliable way.

The usual method of measuring dredging depth is by means of a revolution counter on the grab-hoist wire-winch drum. The system is calibrated by lowering the grab bucket until it is just in contact with the water surface and, at that point, setting the system to zero. Each revolution, or part revolution, of the winch drum pays off a particular length of wire which can be remotely displayed on the operators control panel. The system may become inaccurate if more than one layer of rope is used from the drum or if the bucket is not suspended vertically. These problems are overcome in the modern systems by the use of pressure transducers to measure depth.

A simpler instrument, or indicator, may be used to indicate the angle of swing and the radius of operation. These have tended to be in the form of simple mechanical devices. In the simplest form these may consist only of a graduated scale and pointer which the operator moves manually, progressively, to indicate the area which has been covered. More sophisticated modern systems may display the areas dredged by colour-coding on a VDU display.

Backhoe dredgers

Due to the infinitely variable configuration of the bucket, dipper arm and boom, it is important to provide the operator with a visual display showing the current position of each component.

This may be achieved by means of a scaled-down representation of each component, shown in side elevation. Measurement devices are attached to each point of rotation within the hydraulic arm of the backhoe. Rotational movement is replicated by the scale model display in the operating cabin. Thus the operator can see at a glance the attitude of the bucket, dipper arm and boom and by means of a graduated scale fixed alongside can also see the depth of dredging. It is also important to measure and display the angular rotation of the backhoe in plan and, as for the grab dredger, this may be achieved by various means ranging from simple manually operated scales to more sophisticated systems which measure angular displacement and utilise VDU displays.

Bucket dredgers

Traditionally for bucket dredgers it has been usual to measure and display only the dredging depth, that is the depth of the bucket edge on the underside of the bottom tumbler, together with the loads on the various winches.

Modern dredgers may be fitted with systems which show the attitude of the bucket chain, in elevation, on a VDU, and with the plan area of completed dredging shown by colour coding on a separate VDU, or single split screen. Showing the bucket ladder and chain in elevation has the advantage that in shallow water the true depth of dredging can be seen. This is important as the deepest dredging will be achieved by the buckets at the lowest point in the suspended chain catenary, rather than the bucket beneath the bottom tumbler.

12.6.5 POSITIONING

The position of a dredger in real time may be indicated by means of a trackplotter, or shown on a VDU. All such systems require electronic position fixing systems.

With chart-type trackplotters, the position of the dredger is represented by a pen which moves over a fixed, or moving paper chart. Background shore detail, or survey grid, and the levels and limits of dredging, are usually drawn on the chart in advance.

VDU's offer greater versatility. The dredger can be represented to scale and with real orientation. The dredger, the dredging area, including sea bed levels, and shoreline details can all be shown simultaneously and at a variety of scales. Areas covered by recent dredging can be colour-coded automatically as dredging progresses.

These facilities are of great value in all methods of dredging, but are of particular value for trailing suction hopper dredging where the dredger may range freely over large areas.

References

Dredging Research Unit, 1975. *Contracting in Dredging Works,* Report No. DRU/Mast/014, University of Wales Institute of Science and Technology/Department of the Environment.

Kyling, I.C., 1972. Is a bad contract really necessary? *Proceedings of Wodcon IV*, Symcon Marine Publishing, pp 93–108.

13
Dredging and the Environment

13.1 Introduction

In recent years, dredging activities have increasingly been viewed in the light of their potential effects on the environment. In particular, the disposal of unwanted dredged material has been the focus of considerable attention.

The dredging industry has been particularly responsive to this growing awareness of the effects which it has on the environment. There are numerous examples of research and development undertaken by dredging contractors and organisations, such as the United States Army Corps of Engineers, the Rijkswaterstaat in the Netherlands and the Central Dredging Association, aimed at resolving conflict between the industry and the environment. A large proportion of this work has been directed towards the disposal of unwanted dredged material and particularly the contaminated sediments which are found in some ports and inland waterways.

There are other areas where the industry and the profession have been active and major contributions have been made in the identification of the beneficial uses of dredged materials, the development of plant and methods which can be used to improve the environment generally, prevent coastal erosion and provide coastal protection, amongst others.

The authors are firmly of the opinion that any discussion of dredging in relation to the environment should address not only the potentially adverse effects which could result from dredging but also the considerable benefits which can accrue from a well-planned dredging operation and the numerous ways in which dredging can be used positively to improve the environment (see also Chapter 4). They also stress that the most commonly perceived problem associated with dredging activities, that is the disposal of unwanted, sometimes polluted material, is one which has not been created by the dredging industry but by other industries. We all, as individuals and through our industries and representatives, have contributed to the massive pollution of our waterways and marine environment. The dredging industry can do much to alleviate the problems caused by this chronic neglect but, ultimately, the only sensible solution is to eliminate entirely the release of noxious substances into the environment.

In this Chapter a brief overview of the environmental aspects of dredging works and projects is presented. In spite of its importance, the subject has not been developed in great detail here, since it is known that a number of organisations are in the process of developing publications on the subject, notably the International Association of Dredging Companies (IADC), the Central Dredging Association (CEDA) and the Permanent International Association of Navigation Congresses (PIANC).

The environmental scene is changing fast and any reader who wishes to go into detail is advised to contact the organisations mentioned above to obtain information on the latest available literature. There is also a PC-based database of dredging-related environmental papers called 'DEBBY', which is kept up to date and distributed at modest cost by the Central Dredging Association.

13.2 The potential effects of dredging activities

The potential adverse and beneficial effects of dredging and related operations were summarised by Murden (1984). They can be categorised according to the nature of the activity:

- dredging (excavation);
- transport of dredged material;
- relocation of dredged materials on-shore;
- relocation of dredged materials off-shore.

They can be further divided into effects which are short-term, lasting only for the duration of the works, and those which are medium to long-term or permanent.

In order to study environmental impact it is necessary to examine the following aspects:

- areas of impact;
- types of impact;
- measurement of impact and;
- control of impact (mitigating measures).

In the following sections the types of impact in different areas are described, followed by impact measurement and control.

13.2.1 POTENTIAL ADVERSE EFFECTS OF DREDGING

Interference with marine and river traffic

Stationary dredgers can obstruct other marine traffic. Floating pipelines may also be obstructive and transport barges will increase marine traffic. This is an inevitable short-term effect but is one which can be minimised, if necessary by selection of suitable dredging and transport methods or, in some cases, by establishing some system of traffic regulation.

Noise generated by the dredging plant

All mechanised plant makes noise but the majority of dredging activities are relatively quiet and compare favourably with many other construction activities. However, bucket dredgers can be noisy particularly when working in stiff clays and weak rocks but a recently-constructed bucket dredger utilises caterpillar tracking rather than a bucket chain. This innovation is reported to reduce noise to almost negligible levels (Anon, 1995).

In addition, the pre-treatment of rock is often very noisy. In the case of the latter the noise of the rock drills and compressors may cause as much disturbance as blasting operations in areas close to dwellings. Noise problems are more significant at night when ambient noise levels are low.

Turbidity and suspended sediment

Dredgers release suspended solids into the water column during excavation and during overflow from hoppers and barges. This is a very important aspect and is considered in greater detail later.

Coastal erosion

Deepening of the nearshore area could cause the acceleration of coastal erosion resulting from changes of the hydrodynamic regime. In the case of regular maintenance dredging, the removal of sediment from the longshore transport system could deprive downstream coastal areas of sediment required to maintain coastal stability (see Section 2.2). Long-term marine mining activity may have significant impact on the local coastline.

Fishing

Dredging activities could disrupt fishing or damage spawning grounds. Oyster beds can be destroyed by influxes of silt due to nearby dredging activities. All marine life can survive a range of environmental conditions but extremes may be fatal or may result in increased stress and reduced rates of growth or reproduction.

Stagnation

Deepening, particularly if it results in the creation of closed seabed depressions such as those which are sometimes created during exploitation of offshore fill resources or the formation of silt traps, in severe cases may lead to anaerobic conditions.

Oxygen demand

Turbidity and the resuspension of sediment in the vicinity of dredgers working in contaminated or organic materials is often associated with a significant, but localised, increase in the oxygen demand. In most cases of open water dredging this has little damaging effect but may have an effect on fish and other organisms when dredging in closed areas such as small lakes and rivers.

Saline wedge intrusion

Lowering of seabed levels in stratified estuaries may result in the upstream migration of saline wedges (see Section 2.2).

13.2.2 POTENTIAL BENEFICIAL EFFECTS OF DREDGING

In addition to providing a means of creating, maintaining and improving ports and navigation channels, dredging has a number of less obvious possibilities with respect to environmental benefit (see also Chapter 4).

Contaminated sediments

Contaminated sediments can be removed to a safe disposal facility or can be capped *in situ* with clean materials.

Beach nourishment

Dredging is frequently the most economic method of beach nourishment and unwanted dredged material should always be considered for use as beach nourishment if a suitable site is located within reasonable distance of the dredging site (see Chapter 4). In addition to economic benefit, the use of marine aggregates reduces the scale of potentially damaging onshore extraction and transport operations.

Aggregates

Dredging is a major method of exploiting marine sand and gravel for the construction industry and diminishes adverse impact caused by land extraction.

Coastal protection

Just as an ill-planned dredging operation can result in coastal erosion, so too can dredging be used to improve coastal protection through beach nourishment and the construction of dykes and artificial dunes. The impact on communities of 'soft' sea defence by marine-based methods may be only a fraction of that resulting from 'hard' defences by land-based methods.

Water quality

Dredging activities may increase the oxygen content of the water or release nutrients trapped in bottom sediments. Eutrophication can be eliminated or controlled. Some dredging contractors have modified dredging plant to permit the fixation of phosphates in bottom sediments by the injection of lime, a particularly cheap and effective way of improving water quality in lakes subject to nitrate-rich run-off from farms using fertilisers.

13.2.3 POTENTIAL EFFECTS OF DISPOSAL OF DREDGED MATERIAL

The use and disposal of dredged material have been described fully in Chapters 4 and 5, together with some of their environmental effects. Because dredged material disposal has been studied extensively and is rapidly becoming controlled by the implementation of directives from international conventions, such as the London, Paris and Oslo Conventions, there is little point in examining the potential effects in detail here. For further information the reader is referred to Davis *et al.*(1990) and PIANC (1986).

13.3 Environmental investigations

It is evident that there will be few dredging projects where all of the potential effects of dredging described in the previous sections can be identified. Nevertheless, each project should be the subject of detailed investigations aimed at identifying all of the potential effects and, where necessary implementing measures to limit or eliminate any adverse effects, or to gain positive benefits.

Environmental investigations should be directed at:

- establishing the existing environmental conditions in and around the area of the works;
- identifying the potential impacts of the works on the environment both during construction and, in the longer term, after completion;

- providing baseline data which can be used to monitor the effects of the works;
- identifying methods by which impact may be reduced to acceptable levels.

The importance which is now widely attached to environmental protection, and indeed improvement, means that thorough environmental investigations and impact assessments are often vital to the success of project implementation.

Environmental investigations should be directed at all aspects of the works, including dredging and relocation of dredged materials, transport and placement of reclamation fill, and related construction activities such as piers, seawalls and wharves.

Within this framework, environmental investigations usually address the following subjects:

- water quality;
- sediment quality;
- air quality;
- noise;
- odour;
- marine traffic;
- employment and other social benefits;
- recreation and amenity;
- sediment transport (with respect to coastal and fluvial morphology);
- fishing;
- marine life.

13.3.1 WATER QUALITY

The investigation of existing water quality generally involves a programme of water sampling at locations in and around the works area (see Plate 13.1). Samples can be analysed in a number of ways, often depending on the nature of the surrounding area and the use to which it is put. In urban areas, for example, pollution of the water and of the sea or riverbed sediments from industrial and drainage discharges may be a significant feature and the detection and quantification of pollutants should be an important objective of the investigations. In other areas, the effects of suspended, but otherwise inert, solids, may have important consequences on mariculture and fishing activities. The effects on recreational use of the surrounding area need also to be considered. Water quality should be established in all areas where dredging and reclamation is to be carried out and along transport routes and at marine disposal sites.

Measurement of the following water parameters might be included in an investigation:

- salinity and temperature;
- dissolved oxygen;
- biological oxygen demand;
- turbidity;
- water chemistry.

13.3.2 SEDIMENT QUALITY

Investigation of the sediment quality is required to establish the potential effects of loss of sediment to adjacent areas during dredging and to establish the suitability of the dredged material for disposal to a particular site or for beneficial use. The following parameters

Plate 13.1 Water sampling in Hong Kong harbour with a multi-bottle sampler

might be measured:

- particle size distribution;
- percentage solids (dry matter);
- density;
- organic matter (loss on ignition);
- dissolved gases;
- P, PO_4, NO_3 and SO_2;
- chemical and biological oxygen demand;
- bacteria – *Clostridium perfringens;*
- hydrocarbons;
- toxins and pesticides;
- heavy metals (Cd, Cu, Hg, Zn, Cr, Pb, Ni);
- polychlorinated biphenyl (IUPAC Nos: 28,52,101,118,138,153,180) analysis on all material greater than two millimetres;
- arsenic.

In addition, analysis to determine the content of the following organohalogen compounds has on occasions been requested:

- tribromomethane;
- tetrachloroethylene;
- dichlorobenzine;
- bromotoluene;
- dichlorotoluene;
- trichloroluene;
- chloroprene.

Particularly around shipyards, accumulations of highly toxic TBT (tributlytin) may occur. TBT is a component of some anti-fouling compounds and is becoming recognised as one of the most widespread and dangerous pollutants in harbour areas and its presence should always be investigated in such areas.

Sampling methods are described in Chapter 6.

13.3.3 AIR QUALITY

Whilst dredging and reclamation works rarely have a serious adverse effect on air quality, potential effects may still need to be investigated, particularly if the works are located in a populated area.

Air quality is generally judged on the basis of medium to long-term monitoring of total suspended particulates (TSP), respirable suspended particulates (RSP), sulfur dioxide and nitrogen dioxide. In order to gain usable data, monitoring needs to be undertaken over a sufficiently long period to include seasonal climatological variations which have a significant effect on air quality generally and on the direction in which airborne pollutants are transported.

13.3.4 NOISE

Investigation of potential noise impacts involves the identification of sensitive receivers, monitoring of existing noise levels and estimating the noise levels generated by the plant which is to be used during the works. In some cases, there will be available noise data collected by the local authorities. If no data are available, it will be necessary to undertake monitoring to establish noise levels over a 24 hour period, preferably on both working days and public holidays so that the general range of noise levels can be established.

The basic unit of sound measurement is the sound pressure level, expressed in decibels (dB), as follows:

$$L_p = 20\log_{10}\left(\frac{P}{p_0}\right) \tag{13.1}$$

where L_p = the sound pressure level in dB
P = the RMS sound pressure in $\mathrm{N\,m^{-2}}$
p_0 = the reference sound pressure level, usually $2 \times 10^{-5}\ \mathrm{N\,m^{-2}}$

Sound level is the value measured with a sound level meter which is able to weight the various sound frequencies. An A-weighting network, which approximately corresponds to the frequency response of the human ear, is commonly used. Recorded sound levels are expressed in dB(A).

Sound levels fluctuate, so it is necessary to use a method of measurement to take account of this fluctuation. What is measured is the equivalent continuous sound level, L_{eq}, which is expressed in dB(A). The L_{eq} is a summation of the energy in the fluctuating sound which is expressed as a steady level having the same energy.

L_{eq} is given by:

$$L_{eq} = 10 \log_{10} \frac{1}{T} \int_{t=0}^{t=T} \left(\frac{pA^{(t)}}{P_0} \right)^2 dt \tag{13.2}$$

where L_{eq} = the equivalent continuous sound level in dB(A), over a time period T.
$pA^{(t)}$ = the instantaneous A-weighted sound pressure (in N m^{-2}) varying with time t
p_0 = the reference sound pressure, usually 10^{-5} N m^{-2}

L_{eq} can be determined over any time period, the period in hours being indicated in brackets, that is $L_{eq}(12)$.

National authorities do not usually set sound limits. Any noise limit would usually be set by a local authority, which would be more likely to be aware of local ambient noise levels and noise-sensitive areas. Beaman and Jones (1977) give tables of noise levels produced by construction equipment and methods of applying these to actual sites.

In order to understand the significance of the various noise levels which might be experienced near a site, the following examples of likely ambient levels, given by the Noise Advisory Council (1973), should be noted:

- The noise level in a bedroom with the windows open in a quiet urban area at night would be about 35 dB(A).
- The noise level in a large grocery store would be about 60 dB(A).
- The noise level in a busy workshop where raised voices have to be used for conversation would be about 85 dB(A).

Various methods of successfully reducing the noise level both on and near bucket dredgers are described by Dietze *et al.* (1977).

13.3.5 MARINE TRAFFIC

Dredging and reclamation activities result in increased marine traffic which, in some cases, may be disruptive to existing traffic or which may be hindered by existing traffic. In either case, it is necessary to review the existing marine traffic, preferably in consultation with the local maritime or river authority, and to estimate whether or not conflicts are likely to arise. If conflicts are identified, it may be necessary either to restrict or control the activities of the dredging contractor in some manner or to arrange the re-routing of existing local traffic.

Generally arrangements can be made whereby commercial traffic can interact with dredging activity without undue difficulty. Problems may arise with recreational sailors who lack the knowledge and experience of professional mariners. Hence traffic due to recreational craft, such as yachts and power boats, windsurfers and water skiers should be assessed carefully.

13.3.6 SOCIAL BENEFITS

It is often the case that dredging is to be carried out to develop a port or harbour, to provide a new industrial area or to enable some economic activity to proceed. The result of this action is to enhance economic activity in the locality. Enhanced economic activity usually signifies the availability of additional employment opportunities, a fact often ignored by those who object to the dredging being proposed.

It may not be understood by those carrying out environmental impact assessments that the availability of employment is an environmental criterion and should be taken into account in any assessment. The welfare of human beings is as important in their assessments as any other animal under consideration.

13.3.7 RECREATION AND AMENITY

Recreation and amenity may be enhanced by the results of dredging. Cleaner water, recharged beaches and the provision of lakes and navigable canals are all environmental enhancements having a positive amenity benefit to the community. In the United Kingdom, economic justification of coastal engineering works often includes economic analysis which ascribes a value to the amenity provided by the works. These techniques could equally well be applied to dredging projects.

13.3.8 SEDIMENT TRANSPORT (MORPHOLOGY)

The potential for dredging projects and activities to impact on the natural sedimentary regime has been discussed in Chapter 2 and the reader is referred to this chapter for further guidance.

13.3.9 FISHING

The needs of the fishing and dredging industries tend to come into conflict in two main areas; when there is a requirement to extract sand and gravel from the seabed and when dredged material has to be disposed of at sea. Other minor conflicts also occur, but tend to be less contentious.

Generally, the problems are:

- that the proposed dredging activities will result in the destruction, or covering, of habitat, and thus shellfish may be harmed and fish may move elsewhere;
- that fish stocks will be affected by the interrruption to fish breeding due to removal of bottom habitat or changes in water quality, or the impedence of migration during the dredging or disposal processes;
- that the actual fishing and dredging activities conflict.

To investigate these areas of concern in detail it is necessary to carry out a full study of marine life in the areas throughout the year. In many cases, particularly aspects relating to breeding, the problem is a seasonal one and may be mitigated by carrying the dredging work out at specific times of the year. This is particularly so in the case of migratory fish.

13.3.10 MARINE LIFE

The effect of dredging on marine life may extend well beyond the immediate area of work. Turbidity from dredging activities (see Section 13.4) may affect sensitive marine organisms and removal of the seabed may deplete the area of important forms of marine life which are an essential part of the food chain, or have a local, regional or international rarity value.

To evaluate the potential impact of dredging in such cases it is necessary to carry out a baseline survey of the area, identifying all items of significance, and then to assess the effect of the proposed works. Where the dredging is to be a temporary disruption to the environment it is essential to assess the potential for recolonisation of the area after the dredging activities have been terminated.

13.4 Dredging plant and sediment resuspension

13.4.1 GENERAL ASPECTS OF SEDIMENT RESUSPENSION

All methods of dredging result in some resuspension of sediment. The degree of resuspension depends on numerous factors, many of which are interactive and some of which are not directly related to the excavation process. These factors have been divided into four groups by Pennekamp and Quaak (1990):

- the soil being dredged;
- the method of dredging;
- the hydrodynamic regime in the dredging area;
- the existing water quality and characteristics.

Any assessment of the environmental effects of dredging operations (as distinct from disposal of dredged material) must recognise the complexity and the degree of synergistic interaction of these factors. It is evident from a study of the available data that it is virtually impossible to predict accurately the levels of suspended sediment around a working dredger. This is likely to remain true, irrespective of the amount of research effort applied to the problem in the future. It is possible, however, to describe in relative terms the sediment generation potential of various methods and to predict at an indicative level the probable range of suspended sediment concentrations.

Sediment resuspension is, in most cases, only likely to constitute a potential problem if it moves out of the immediate dredging area. The degree to which particular dredging methods cause turbidity may not, therefore, always be the prime consideration when selecting appropriate methodology. Dredging within a fully enclosed area such as a locked basin is unlikely to lead to significant loss of material to the wider marine environment, irrespective of the methods used. In addition, some types of stationary dredgers can be operated within silt curtains which may, in appropriate circumstances, reduce the quantity of suspended sediment which passes out of the immediate dredging area.

13.4.2 QUANTIFYING SEDIMENT RESUSPENSION

Sediment resuspension can be quantified in two ways, each of which may have important environmental implications. Resuspension can be measured in terms of sediment concentration in the water in and around the dredging area. Some dredging methods produce relatively high concentrations of suspended sediment, sufficient to have at least a short-term detrimental effect on the environment. Alternatively, resuspension can be viewed in terms of the total quantity of sediment put into suspension and lost from the immediate dredging area during the works, relative to the quantity of material which is dredged. This quantity may not be directly proportional to the suspended sediment concentration.

One difficulty which is often encountered, when comparing the amount of sediment generated by various dredging methods, is the inconsistency of observational techniques reported in the literature. Comparisons between the results of different studies must therefore be made with a degree of caution, recognising that an apparent difference between the sediment generated by two dredging methods could simply be a function of the manner in which the observations were made.

Blockland (1988) gives details of observational and interpretative procedures which could be standardised and which appear to have real potential as a means of comparing, or characterising, the sediment put into suspension by different dredgers. Measuring stations are established around the dredger on a grid spacing of between 35 and 50 metres extending up to about 150 metres from the dredger. At regular intervals during dredging, the suspended sediment concentration is measured at depth intervals of one to three metres at each station using a calibrated turbidity meter. In addition, observations are made before dredging to establish the background turbidity and after dredging to determine the rate at which the turbidity subsides.

For each measured vertical concentration profile, the depth averaged, maximum and minimum sediment concentrations (C_{ave}, C_{max} and C_{min} respectively) are established. These are plotted against time to establish the extreme values of each parameter during the dredging period. The extreme values can then be plotted as a contour diagram. 'Characteristic' concentrations are considered to be those values of C_{ave}, C_{max} and C_{min} which are equalled or exceeded at three or four stations near the dredger, that is within an area of about 50 by 50 metres. These values are corrected for background suspended sediment concentration.

Determining the total amount of material put into suspension inevitably involves a degree of speculation. Blockland starts by integrating and averaging the measured sediment concentrations to establish the mean amount in suspension during dredging. This is then multiplied by the dredging time and divided by the settling time. Thus:

$$\text{Amount resuspended} = \frac{\text{Dredging period}}{\text{Mean settling time}} \times \text{Mean amount in suspension}$$

Blockland suggests that settling times are of the order of about half an hour but this will obviously vary considerably from one situation to another. The expression can further be refined to compare the effects of different dredgers by introducing the amount of material dredged:

$$S = \frac{\text{Amount resuspended}}{\text{Amount of soil dredged}} \qquad (\text{kg/m}^3)$$

The 'S' factor can be used as a reasonable estimate of the amount of sediment which passes out of the immediate area of the dredging operations, that is the area defined by the nearest measurement points which should typically be about 50 metres from the dredger.

It should be noted that settling speeds will generally be much faster than the theoretical settling speed of the component parts of the soil. This is because cohesive soils are not completely disaggregated during dredging, even by comparatively aggressive dredging methods. In cases where non-cohesive soils are being discharged to the water column, particularly during overflow of hoppers or barges, the particles tend to sink to the seabed in a density current which also travels faster than the fall velocity of the individual particles.

13.4.3 DREDGER RESUSPENSION POTENTIAL

Because of the different working methods of each type of dredger, their potential for sediment resuspension varies according to dredger type and mode of operation. Brief details of how resuspension occurs are given below. More detailed information can be found in numerous publications on this subject (e.g. Hayes *et al.*, (1984); Pennekamp and Quaak (1990); Sosnowski (1984); Havis, (1988); Yagi *et al.* (1976); Volbeda (1983); Clarke and Imsand (1990); Okayama (1983); Kaneko and Watari (1983) and Koba (1982)).

The trailing suction hopper dredger

Causes of sediment resuspension are as follows:

- movement of the suction pipe and draghead at normal trailing speeds of 0.5–2 m/s causes turbidity and resuspension of sediment at the sea bed;
- action of high pressure water jets fitted to draghead;
- propeller wash and return flow under the keel when clearance is low;
- wash from bow thrusters when manoeuvring;
- discharge from automatic lean mixture overboard systems (ALMOBs);
- discharge from degassing systems;
- overflow from the hopper during loading and on completion of loading;
- escape of gas from the disturbed soil;
- blockages of the suction pipe which require the pipe to be emptied;
- discharge of 'rest water' (water remaining in the hopper) after disposal of the hopper load;
- leakage from the hopper.

It is generally acknowledged that the greatest single contribution to suspended sediment generation during trailer dredging arises from overflowing during loading (see Plate 13.2).

Meaningful measurements of the suspended sediment around trailer dredgers are difficult to make and interpret because, in contrast with most other types of dredger, trailers are very mobile, covering large areas before the hopper is filled. Nevertheless, a reasonable amount of data has been obtained, probably reflecting the importance of the trailer dredger in terms of its wide application and the large numbers of vessels in the world dredging fleet.

The main causes of suspended sediment generation are overflowing and the use of degassing and automatic lean mixture overboard systems. When overflowing whilst dredging sand, '*S*' factors can exceed 500 kilograms per cubic metre in fine or dirty sands but may be almost negligble when dredging clean clean, coarse sand. In fine, dirty materials, suspended sediment concentrations can be of the order of several thousand mg/l near the base of the water column but near surface concentrations are generally less than 100 mg/l except immediately behind the dredger. Overflowing whilst dredging mud effectively constitues agitation dredging in which all of the dredged material is released to the environment for disposersion by water currents.

Plate 13.2 A trailing suction hopper dredger dredging clayey sand in Australia, showing the overflow plume

Lean mixture overboard and degassing systems can give rise to '*S*' factors in excess of 10 kilograms per cubic metre.

If overflow, ALMOB and degassing discharge is prohibited, the trailer dredger is substantially 'cleaner'. '*S*' factors of the order of 5 kg/m^3 might be expected and sediment concentrations are likely to be significantly less than 100 mg/l and restricted to the lower part of the water column.

The grab dredger

Causes of sediment resuspension are as follows:

- material adhering to the grab is washed off as the grab descends;
- the impact of the grab on the bottom, exacerbated by flow of water around the grab;
- pulling the grab out of the soil, particularly if the soil contains biogenic gas;
- sediment washed from the grab as it ascends, particularly when overfilled;
- sediment lost from the grab due to poor closure (either a badly maintained grab or due to debris preventing closure);
- sediment-laden water spilling from grab when it surfaces and during slewing of the crane to the disposal barge;
- overflow of barges;
- splashing during barge loading.

Grab dredgers, although frequently used for the dredging of contaminated material, are not inherently 'clean'. '*S*' factors well in excess of 20 kilograms per cubic metre are likely, even for grabs of limited capacity (say three to five cubic metres), and suspended sediment concentrations may exceed 100 mg/l throughout the water column. The use of watertight grabs appears to reduce the sediment concentrations in the upper parts of the water column but there is some evidence to suggest that they may increase concentrations near the bottom. Dredging within silt curtains appears to significantly reduce the loss of material from the dredging area, perhaps by as much as 75 per cent, but subject to current velocities being very low. If the current velocity exceeds 0.5 metres per second silt curtains are likely to be ineffective.

The bucket dredger

Causes of sediment resuspension are as follows:

- sediment is washed from the buckets as they descend;
- air trapped in the descending upturned buckets is released at the bottom causing upwelling of sediment-laden water;
- sediment spills from overfull buckets;
- leakage from the discharge chutes, particularly that which is not in use for barge loading;
- release of biogenic gas during excavation;
- dragging the bucket ladder over the bottom when dredging in very shallow water;
- overflow of barges;
- splashing during barge loading;
- anchor movements.

The limited data available relating to bucket dredgers suggests that they are similar to grab dredgers in respect of '*S*' factors and suspended sediment concentrations. It is not practical to surround a bucket dredger with silt curtains. However, a bucket dredger was recently converted to overcome many of the drawbacks associated with type of dredger in terms of its effect on the environment (Anon, 1994). The modifications included enclosing part of the bucket ladder so that any spillage returns directly to the digging area and the installation of water jets to wash the buckets after discharging, in order to ensure that no material adheres to them as they descend the ladder.

The cutter suction dredger

Causes of sediment resuspension are as follows:

- disturbance of the ground around the cutterhead;
- biogenic gas release from the soil;
- impact and removal of spuds;
- cutter ladder dragging on the bottom when dredging in shallow water;
- anchor and side wire movements;
- leakage from discharge pipelines.

When loading barges using cutter suction dredgers, the following will also be important:

- overflow of barges;
- splashing during barge loading.

Usually it is disturbance of the ground around the cutterhead which contributes most to sediment resuspension.

It is evident that cutter suction dredgers can, if operated with care, dredge with minimal sediment resuspension and loss. The available data suggests that high suspended sediment concentrations are confined to a small area close to the bottom within a few metres of the cutterhead.

With the exception of the refresher system (see below), most of the variations on the cutter suction dredger (i.e. the bucket wheel and disc cutter) are likely to be broadly similar in terms of sediment resuspension and, therefore, are suitable for the dredging of contaminated material if operated in a careful manner. Cutter suction dredgers and their variants have the advantage that they are reasonably accurate both vertically and horizontally because they are able to work in fairly confined areas and in shallow water.

The refresher system (Kaneko and Watari, 1983), having been specifically designed for the clean removal of contaminated sediments, is reportedly associated with negligible levels of sediment resuspension.

All of the research conducted on cutter dredgers and their variants has clearly shown that sediment loss increases with increasing cutter rotation and swing speeds.

The dustpan dredger

Causes of sediment resuspension are as follows:

- disturbance of the ground due to the action of the water jets;
- gas released from the soil puts sediment into suspension;
- anchor and wire movement;
- pipeline leakage.

When used for shoal dispersion, which has been a common application of dustpan dredgers, particularly in the USA, sediment resuspension occurs on a massive scale.

The backhoe dredger

Causes of sediment resuspension are as follows:

- disturbance of the soil by the excavator bucket;
- soil washed from the bucket during descent and recovery;
- release of biogenic gas from the soil;
- sediment and dirty water spilling from the bucket during slewing to the disposal barge;
- overflow of barges;
- impact and removal of spuds;
- splashing during barge loading.

Although there is little published observational data available concerning sediment resuspension by backhoe dredgers, it is likely to be comparable with that of grab and bucket dredgers in terms of '*S*' factors but absolute levels of suspended solids might be lower. In mitigation, it is possible to operate backhoes inside silt curtains and they are suited to working in shallow water and confined spaces, in addition to being rather accurate and able to dredge the material at near-*in situ* density.

The auger dredger

The causes of sediment resuspension are broadly similar to those described for cutter suction dredgers. No published figures are available concerning the levels of suspended sediment created by conventional auger dredgers but they might reasonably be assumed to be fairly low except close to the auger, especially if the dredger is operated in a careful manner.

There are several examples of auger dredgers which have been specially constructed, or modified, for the specific purpose of dredging contaminated materials accurately and with minimal loss of sediment (e.g. Anon, 1994). These dredgers almost invariably have an enclosed auger, sometimes incorporating an upper rigid hood with a flexible screen enclosing the digging area.

Pneumatic dredging systems

Little data is available relating to sediment resuspension levels associated with pneumatic dredging systems. Herbich *et al.* (in press) report concentrations of 48 mg/l at one metre above the bottom and 4 mg/l at seven metres above the bottom and 4.9 metres in front of the pump. The Osaka Port Authority (1972) undertook trials with a Pneuma dredger and reported that polluted muds had been dredged at solids concentrations of 82–95% and that no change in water transparency had been observed at a distance greater than 0.2 to 0.3 metres from the bottom.

In the USA, the pneuma system was successfully used to dredge the polluted bottom sediments of Gibraltar Lake. In The Netherlands, in 1989 trials were undertaken by the Dutch Government in Berghaven during which 'no turbidity was observed during operation' (Faldi, 1989).

Other small dredgers for shallow water work

Very little measured data is available for machines of this type. However, it is expected that most could be operated in a relatively clean manner, either because they excavate by means of suction or enclosed mechanical methods or because they could be enclosed by silt curtains.

The IHC Dosco scraper cannot be enclosed and excavates by 'shaving' off a layer of mud as it advances. However, it is claimed to produce very little turbidity, mainly because of its slow rate of advance of about 0.3 metres per second.

The bed leveller

Sediment is put into suspension by disturbance of the soil as it is bulldozed by the leveller and by propeller wash from the towing vessel. Blockland (1988) reports that high sediment concentrations are confined to the lower parts of the water profile reaching as high as 900 mg/l. Concentrations in the middle and upper parts of the water profile are typically around 30 mg/l. He quotes C_{min}, C_{ave} and C_{max} values of 35, 60 and 500 mg/l respectively for a leveller moving 600 cubic metres per hour and 'S' factors of about five to six kilograms per cubic metre are reported. However, the distribution of solids through the water column is very much dependent on water depth during operation and the characteristics of the sediment being moved.

13.4.4 COMPARISON OF RESUSPENSION POTENTIAL

A number of broad conclusions can readily be drawn from the data presented in the preceding sections.

- The characteristics of the dredging site have a significant bearing on the type of dredger which can be used and on the extent to which precautions to minimise sediment resuspension are needed.
- When evaluating dredging methods it is important to distinguish between suspended sediment concentrations and the total amount of sediment lost to the wider environment.
- Subject to appropriate modification, most types of dredger can be operated in a manner which does not cause excessive loss of sediment to the surrounding environment.

An additional factor which must be taken into account is the possibility of enclosing the dredger with a silt curtain to reduce the escape of sediment. The use of silt curtains is not practical with most of the larger types of dredger or with mobile types such as trailers, but they may be used with grab or backhoe dredgers. In some areas, such as docks, the dredging operations will effectively be enclosed and escape of suspended sediment from the dredging area is likely to be very limited, even when silt curtains are not used.

Early drafts of this book included a table summarising indicative sediment release rates for a variety of dredgers. The authors have decided that the considerable number of variables which combine to determine release rates, and the varying performance of individual dredgers within each class, make the publication of any such table potentially very misleading. The prediction of release rates is a complex procedure involving a detailed evaluation of site and soil conditions, the type of dredger which is to be used (including any modifications designed to restrict sediment release) and the manner in which it is operated. Even with a specific type of dredger, release rates might vary by one to two orders of magnitude depending on these factors and each proposed project should be evaluated on an individual basis by experts.

References

Anon, 1994. Environmental concerns breath new life into the maintenance dredging industry, *Dredging and Port Construction*, Vol. XXI, No. 1, pp 17–30.

Anon, 1995. IHC design a noiseless bucket dredger ladder, *Dredging and Port Construction*, Vol. XXII, No. 6, pp 17–18.

Beaman, A.L. & Jones, R.D. 1977. Noise from construction and demolition sites – measured levels and their prediction, *Report No. 64, Construction Industry Research and Information Association, (CIRIA)*, London.

Blockland, T. 1988. Determination of dredging-induced turbidity, *Terra et Aqua*, No. 38, pp 3–12, International Association of Dredging Companies (IADC), The Netherlands.

Buchberger, C. 1993. Removal technologies for contaminated sediment, *Proceedings of the CEDA Dredging Days*, Amsterdam.

Clarke, D., and Imsand, D. 1990. An environmental assessment of overflow dredging in Mobile Bay, Alabama, *Information Exchange Bulletin*, Vol. D-90-1, USACE, Waterways Experiment Station, Vicksburg, Mississippi, USA.

Davis, J.D., MacKnight, S., Staff, I.M.O., *et al.* 1990. Environmental considerations for port and harbor developments, *World Bank Technical Paper Number 126*, Transport and the Environment series.

Dietz, W. *et al.* 1977. Untitled, *Paper of the 24th International Navigation Congress*, Section II, Subject 5, Permanent International Association of Navigation Congresses (PIANC), Leningrad.

Faldi, G. 1989. Italian dredging system eliminates secondary pollution, *Dredging and Port Construction*, Vol. XVI, No. 11.

Havis, R.N. 1988. Sediment resuspension by selected dredges. *Environmental Effects of Dredging: Technical Notes*, EEDP-09-2, USAEWES, Vicksburg, Mississippi, USA.

Hayes, D.F., Raymond, G.L. & McLellan, N.T. 1984. Sediment resuspension from dredging activities, *Dredging and Dredged Material Disposal*, Vol. 1, pp 72–82, ASCE.

Herbich, J.B. & Brahme, S.B. 1988. *A literature review and technical evaluation of sediment resuspension during dredging*, C.R.-88, USAEWES, Vicksburg, Mississippi, USA.

Kaneko, A. & Watari, Y. 1983. Special dredgers for fighting pollution, pp 527–546, *Proceedings of the 10th World Dredging Congress*, Singapore.

Koba, H. 1982. Dispersion of sediment resuspension caused by dredge operation, *Proceedings of the 9th US/Japan Experts Meeting on Toxic Bottom Sediments*, Tokyo, Japan. Army Corps of Engineers Waterways Experiment Station, Vicksburg, Mississippi, USA.

Murden, W. 1984. Dredging and the environment, *Dredging and Port Construction*, December 1984.

The Noise Advisory Council, 1973. *A Guide to Noise Units*, Department of the Environment, HMSO, London.

Okayama, Y. 1983. *Effective removal of soft mud layer by front-open type draghead*, pp 137–155, Wodcon.

Pennekamp, J.G.S. & Quaak, M.P. 1990. Impact on the environment of turbidity caused by dredging, *Terra et Aqua*, No. 42, pp 10–20, International Association of Dredging Companies (IADC), The Netherlands.

PIANC, 1986. Disposal of dredged material at sea, *Report of a working group of the Permanent Technical Committee II*, Supplement to Bulletin No. 52, Permanent International Association of Navigation Congresses, Brussels.

Sosnowski, R.A. 1984. Sediment resuspension due to dredging and storms: an analagous pair, *Dredging and Dredged Material Disposal*, Montgomery, R.L. & Leach, J.W. (Eds), Vol. 1, pp 609–618, ASCE.

Volbeda, J.H. 1983. Overflow effects when trailing very soft silts and clays, pp 547–556, *Proceedings of the 10th World Dredging Congress*, Singapore.

Yagi, T., Koiwa, T, & Miyazaki, S. 1976. Turbidity caused by dredging, *Proceedings of the 7th World Dredging Conference*, WODCON VII, San Fransisco, California, USA.

14
Pre-treatment of Rock

14.1 Introduction

In this chapter the term pre-treatment is used to denote treatment of rock prior to the dredging operation. It usually consists of a separate operation carried out independently of other dredging operations, but there may be occasions when a hybrid method incorporating both pre-treatment and dredging could be appropriate. There are two basic methods of pre-treatment; mechanical and chemical.

Chemical methods involve the use of explosives or expanding gas cartridges, use of the latter being uncommon. The most common method is to place explosives in boreholes drilled vertically into the material to be blasted. A grid of boreholes is drilled, covering the area to be pre-treated. Loading and firing of the charges is carried out in convenient groups of boreholes. The work is usually carried out from a floating or self-elevating pontoon but is occasionally carried out underwater by divers with submersible drills. Details of the drilling and blasting method are described in Section 14.3.

High explosives are also used underwater in the form of surface, or lay-on, charges. These charges are laid on the surface of the material to be broken up and are used in three forms; cartridges, slabs, or shaped charges. Use of these methods is described in Section 14.4.

Mechanical methods are less common than chemical methods, but are still sometimes used in cases where explosives are undesirable or inefficient for some reason. They are described in Section 14.2.

The various methods of pre-treatment are illustrated in Figure 14.1.

14.2 Mechanical methods

The standard method of breaking rock mechanically, and one which has been around for at least the last century, is the simple rock breaker, consisting of a pontoon on which is mounted one or more heavy needles or chisels. This device, which could be formed from a heavy pile with a suitably hardened tip, is hoisted and dropped vertically on the material to be broken. Fragmentation is caused by shearing or crushing the rock, depending on its formation and crystalline structure.

Modern versions of the rock breaker sometimes take the form of pneumatic or hydraulic hammers which strike the rock with a frequency of 1.5 to 2 blows per second. The advantage

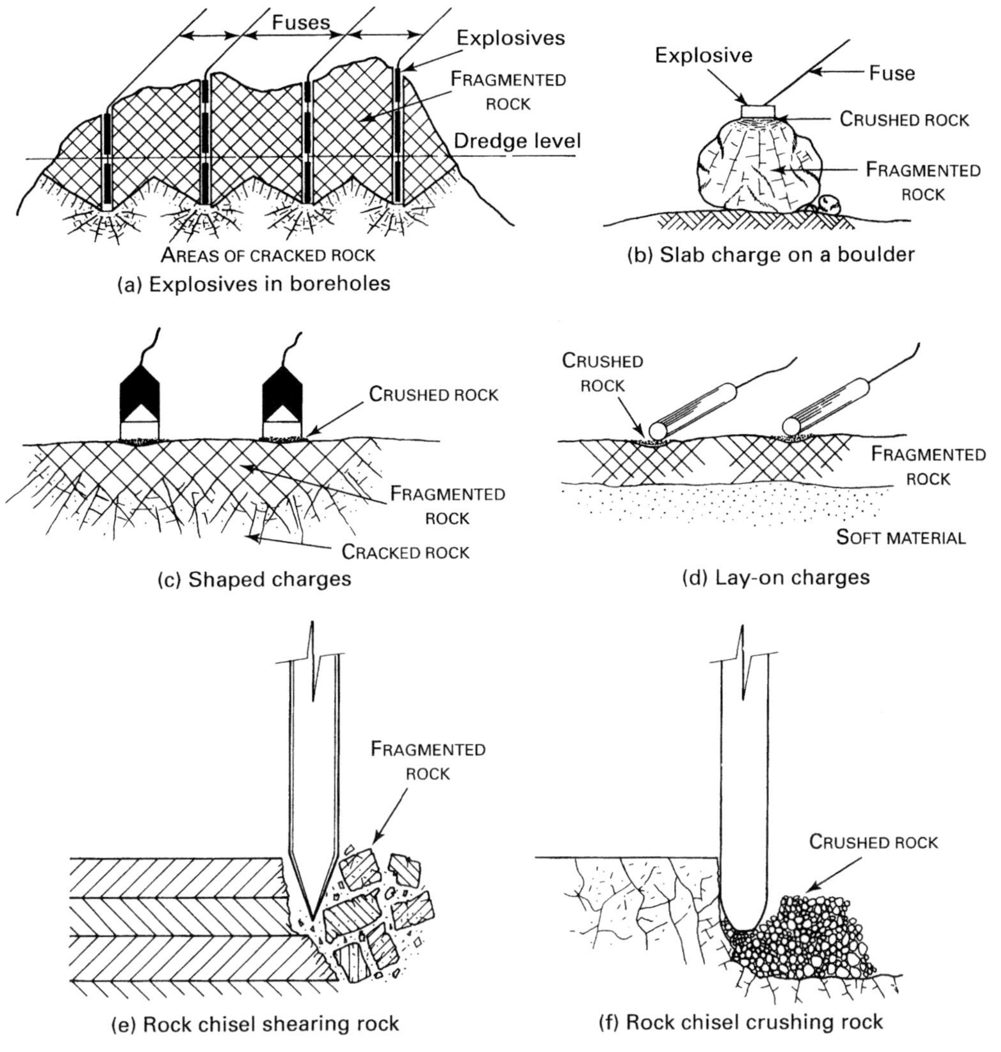

Figure 14.1 Methods of pretreatment and rock breakage

of these is that they may be mounted on the booms of hydraulic backacting machines, or similar, as an alternative to a dredging bucket.

14.3 Drilling and blasting

14.3.1 METHODS OF OPERATION

Drilling pontoons are of two types: floating or self-elevating (spudded). Spudded pontoons may have the capability to elevate above water, or may be limited to a partial transfer of weight to the spuds. Many drilling pontoons are constructed specifically for one project and are dismantled on completion of the work. There are many different possible configurations but most drilling pontoons have common characteristics which are as follows:

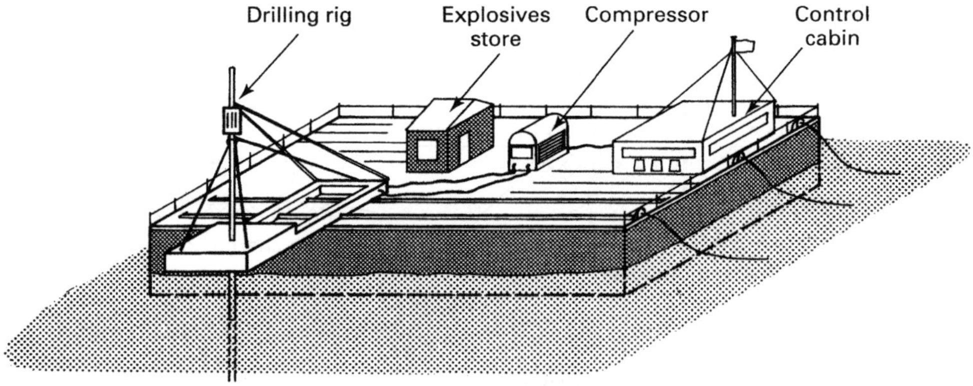

(a) With drills cantilevered over the side

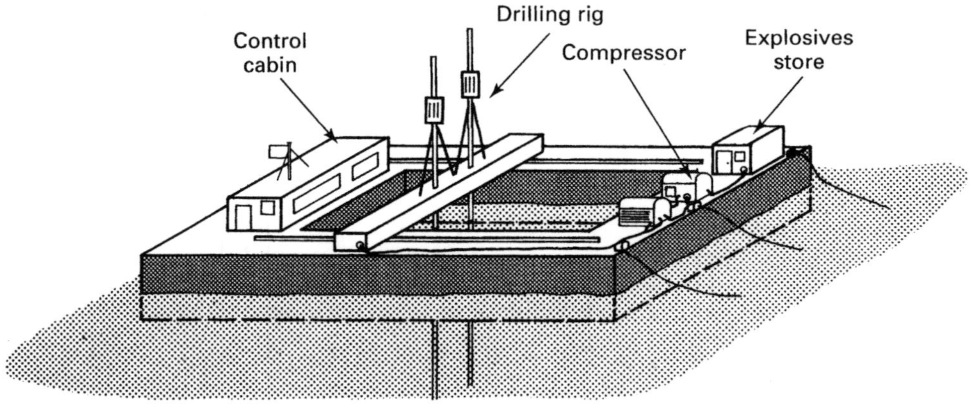

(b) With drills over a central well

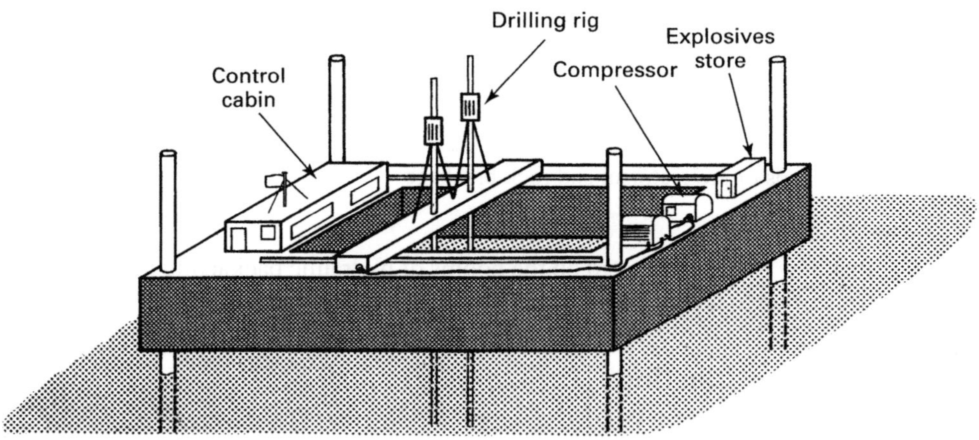

(c) As (b) but spudded

Figure 14.2 Drilling pontoons

- A hull of either bulkhead construction or formed from unit floats. The shape of the hull will depend on how the drills are to be worked, either over the side or through a central well.
- Drilling rigs mounted to drill over the side of the pontoon or through a central well. Up to 12 rigs have been used on a single pontoon. The drills are usually capable of lateral movement and, if working through a well, are mounted on a bridge unit which moves across the well.
- Winches and anchor wires capable of holding the pontoon on station and moving it regularly by small increments over the working area. Alternatively anchoring may be by means of spuds.

The three most common configurations of drilling pontoon are shown in Figure 14.2. These are floating pontoons with drills cantilevered over the side or operating through a well, and a self-elevating pontoon with a well.

The drilling rigs used for underwater blasting work take various forms. Most of them operate with rotary and rotary-percussive drills which are mounted on a drill mast in such a manner that the drills can be raised and lowered on the mast by winches or hydraulics. Tall drill masts are often constructed to eliminate frequent coupling and uncoupling of drill rods and casings, and some masts extend almost to the seabed in their lowered position to provide additional rigidity. A relatively tall drill tower mounted on a bridge unit is shown in Plate 14.1.

The basic objective of the drilling pontoon is to drill, charge and fire as many vertical shotholes on a pre-arranged grid as possible in a standard shift. The drilling cycle is shown diagramatically in Figure 14.3 and the production cycles are illustrated in Figures 14.4 and 14.5.

All drilling pontoons, whether they be self-elevating or floating, have a similar basic subcycle:

- lower the outer casing to the sea or river bed, drive through any soft material and collar into the rock;
- drill out the rock with inner drill rods to the desired depth;
- charge the holes with explosive;
- lift the outer casing and retrieve the initiating line with the capture ring;
- move the drill rig or pontoon.

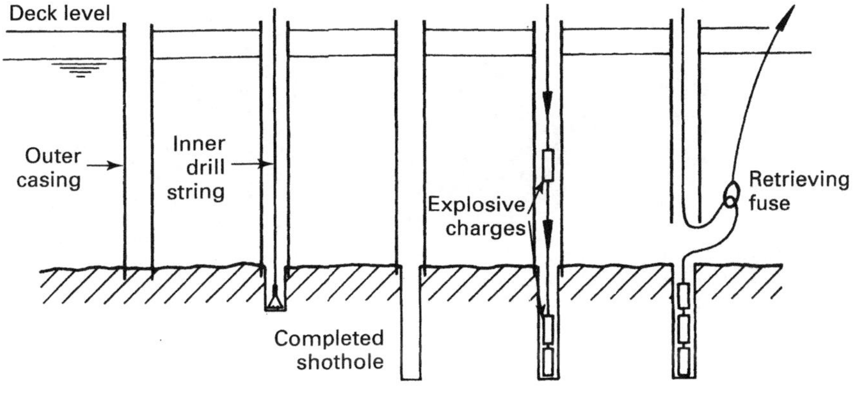

Figure 14.3 The basic drilling cycle: outer casings collared into rock; inner string drilling shothole; shothole drilled; loading explosive charges; retrieving fuse

Plate 14.1 A shothole drilling rig, mounted on a bridge unit, on a drilling and blasting pontoon

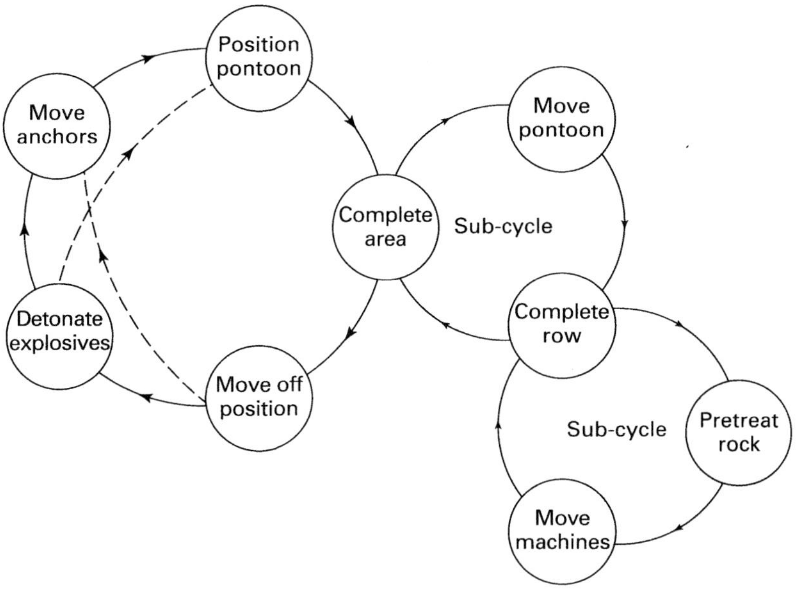

Figure 14.4 Production cycle for floating pontoon; drilling machines mounted over the side

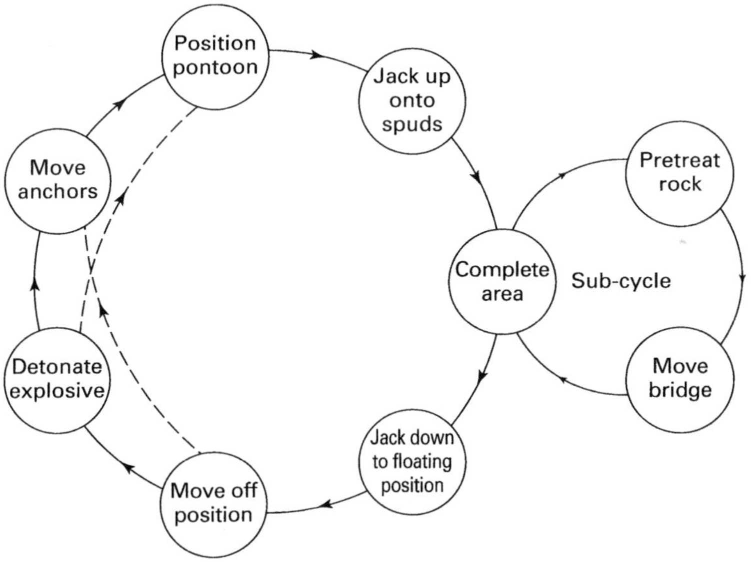

Figure 14.5 Production cycle for a spudded pontoon; drilling machines mounted on a bridge over central wall

Drilling methods have been developed for softer rocks which have one drill string and when using those methods the first two actions are combined.

The number of times that this sub-cycle will be repeated before blasting takes place will depend on the type of pontoon, the speed of the sub-cycle, the number of drills being used, the length of the working shift and so on, but in any event, the main cycle of operation follows a general pattern.

Self-elevating pontoon:

- move onto position;
- elevate platform to correct height;
- sub-cycle, repeated as many times as necessary;
- lower platform to water;
- move off position;
- blast.

Floating pontoon:

- move onto position;
- sub-cycle, repeated as many times as necessary;
- move off position;
- blast.

During the work it will be necessary to move the pontoon anchors to new positions. However, for floating pontoons this operation takes place when moving onto or off position for blasting. It is not normal, and could be dangerous, to move a floating pontoon's anchors when drill holes have been charged or when drills are in operation. The anchors of a self-elevating pontoon can be moved whenever the pontoon is elevated on its spuds and the operation does not interrupt the drilling cycle.

A typical sub-cycle for a drilling pontoon might be as follows.

• lower outer casings to bed and collar into rock	10 minutes
• drill out hole to desired depth	20 minutes
• charge hole with explosives and recover initiating line	8 minutes
• move to next hole	5 minutes
Total sub-cycle time	43 minutes

Naturally these times would depend to a great extent on water depth, tidal range, depth of drilling, type of drill mast used and nature of the rock.

When using pontoons which have drills cantilevered over the side it is normal to drill as many holes in a row as possible. The pontoon is then moved backwards until the drills are placed over the next parallel row (see Figure 14.6a). The number of rows drilled and charged before blasting will depend on the number of drills and the working method but blasting is often arranged to take place at the end of a shift.

Pontoons with drills mounted over a well will normally be kept stationary until the drills have covered most or all of the area under the well (see Figure 14.6b). Blasting will then occur before the pontoon is repositioned in the next area. When a self-elevating pontoon is used it is necessary to lower the pontoon to the water and float off position before blasting occurs.

The time taken to carry out blasting operations from a floating pontoon varies between 20 and 60 minutes depending on circumstances. Self-elevating pontoons may take considerably longer, perhaps as long as ten hours if the operation of the self-elevating system is slow. If sea conditions do not allow the pontoon to return to the water, the blasting period could be extended even further.

A good general description of underwater drilling and blasting operations is given by Abrahams (1974).

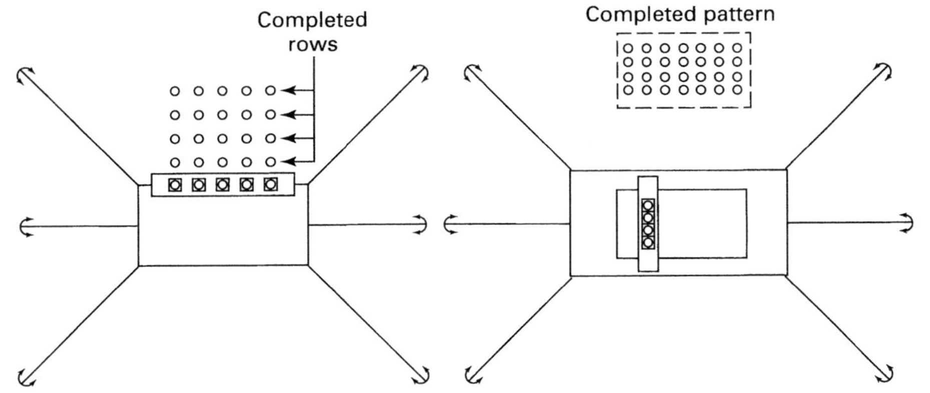

(a) Pattern produced by a cantilevered drill arrangement

(b) Pattern produced by drills over a well

Figure 14.6 Drilling operations

14.3.2 FACTORS AFFECTING PERFORMANCE

Dimensional constraints

Water depths will have a bearing on the ability of a drilling pontoon to operate satisfactorily. Drilling pontoons without spuds have a shallow draught, typically 0.6 to 1.0 metres, and are therefore able to operate in shallow waters. Maximum operating depth is around 20 to 30 metres, depending on drill type, rock formation and water conditions.

Pontoons with spuds will not be able to work in shallow water, unless they can gain access at high water, since their spuds cannot be lifted above a certain level. A minimum depth of five or six metres might be applied for larger pontoons. Maximum depth of water for operating will depend on the length of the spuds and the requirement for clear space between the bottom of the hull and the maximum wave height at high water when in the elevated position.

Sea conditions

Waves and swell will affect the operations of drilling pontoons. Pontoons constructed from a number of small floats are unsuitable for work in swell conditions unless specifically strengthened for the job. Rigid flat-topped barges are susceptible to swell, particularly when of long period. Limiting conditions are normally determined by the degree of movement of the craft which prevents shothole charging operations. Drilling operations can often be carried out in more severe conditions.

Spudded pontoons are susceptible to swell when floating due to the height of the spuds. In the elevated position they can cope with much more severe conditions than the floating pontoon. Limitations to work tend to be due to other factors, for instance excessive wind, inability to operate the jacking mechanism (due to spud friction loss), and so on. Delays often occur when the sea conditions are too severe to allow the pontoon to be lowered to the water surface. The critical point is when the pontoon is just in the water and when any lifting of the pontoon will result in the spuds striking the seabed with force.

Currents will also have an effect on performance. Floating pontoons have to be kept on station and re-positioned accurately after each row of holes has been completed. Positioning is more difficult in strong currents and will take longer. Any work by divers, usually to sort out problems with the initiating line, is hampered. A particularly serious problem is the bending of drill casing tubes due to the force of the current. Figure 14.7 shows the effect of various current velocities on a standard 100 millimetre diameter casing tube. Deflection may be reduced by the addition of special underwater casing guides or extension of the mast towards the sea bed. Elevating pontoons, which are obliged to work from well above the highest normal wave, may be badly affected by casing tube deflection.

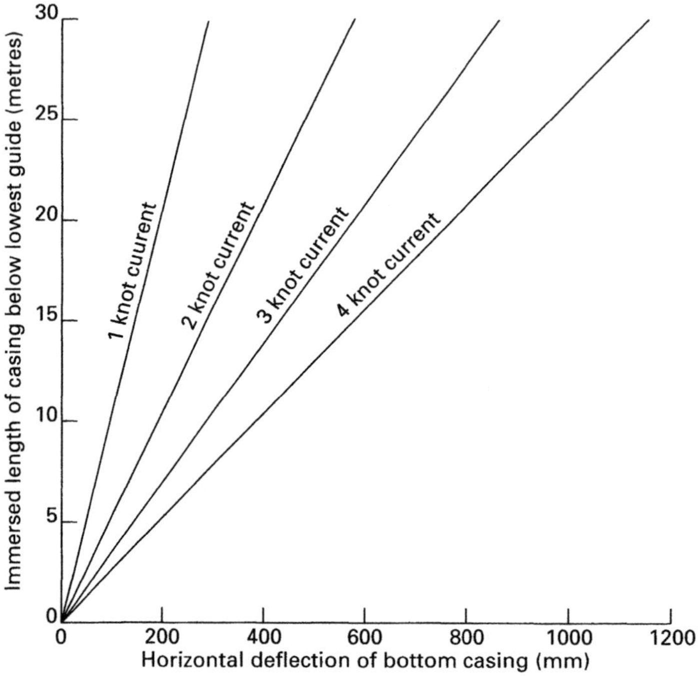

Figure 14.7 The deflection of 100 millimetre diameter casing tube due to currents

Rock type

The rock type will have a dramatic effect on both the drilling and the blasting process.

The ease with which a rock can be drilled depends on its resistance to drilling, as well as the relative ease with which the hole may be started and kept open. With respect to the latter, it should be noted that sharp, saw-toothed formations are difficult to drill when they are not covered by soil because the casing tubes tend to slide off location. Rock with internal discontinuities is also troublesome, due to difficulties in keeping the hole straight and the drill bit free.

The speed with which a rock can be drilled also varies considerably. Macgregor (1967) relates the ease of drilling different rocks to a drilling index which is based on the speed at which pink granite from Aberdeen, UK, can be drilled (see Table 14.1). In underwater drilling and blasting a rotary percussive drifter drill is normally used which would have a drilling speed of approximately 12 metres per hour in pink granite when drilled with a 70 millimetre diameter cross-drill bit.

Table 14.1 The ease of drilling rock: the drilling index for various rock types (MacGregor, 1967)

Rock type	Location	Drilling index
Calcareous sandstone	Bargate beds	52*
Fine grained sandstone	Culm measures	80
Granite	Penmaenmawr	99
Pink granite	Aberdeen	100
Brecciated and recemented limestone	Saudi Arabia	105†
Vein quartz		138
Basalt	India	149
Shale	Culm measures	166*
Dolomitic limestone	Lower magnesian limestone	272
Thin-bedded sandstone with carbonaceous partings	Yorkshire coal measures	313‡
Coarse friable grit	Millstone grit	463‡

* Index figure low due to air hardening of specimen
† Dense stone: specific gravity 2.66; compressive strength 1600 kg/cm^2
‡ Parallel bedding: normal-to-bedding figures were 275–403 respectively

The ease with which a rock can be blasted depends on its behaviour under shock and other dynamic loads and depends on the strength, structure and homogeneity of the rock. A rock with a high blastability is one which fragments and bulks to the required degree with a low blast ratio (ratio of weight of explosive to unit volume of rock treated). Bulking is the movement of the rock fragments which causes an increase in volume of the rock mass and is dependent on the size, shape and juxtaposition of the fragments. Blast ratios can vary from 0.5 to 3.0 kg/m^3, depending on rock type and fragmentation required. To achieve these blast ratios the drill hole spacings have to be varied from between one to three metres depending on the diameter of the holes. Smaller spacings are inadvisable for safety reasons and larger spacings tend to result in patches of unblasted rock.

Table 14.2 from Langefors and Kihlstrom, (1978), shows the variation in fragmentation which would be obtained from various blast ratios. It can be seen from this table that to double the blast ratio will reduce the average size of fragmentation to one-sixteenth of its previous size. The fragmentation required for a particular job will depend on the type and

Table 14.2 Underwater blasting: variation of fragmentation with blasting ratio (Langefors, 1978)

Blasting ratio (kg m^{-3})	Size of fragmentation – average side length, actual dimensions dependent on rock type
0.24	100
0.30	50
0.40	12.5
0.50	6.4
0.60	3.7
0.70	1.5
0.85	0.8
1.00	0.4

size of dredger to be used to excavate the rock (see Chapter 7) and whether or not the rock is to be put to some specified use. Care must be taken when assessing the blastability of coral since, in weak corals, much of the explosive's energy is absorbed in crushing the rock and crack propagation is therefore limited. Hole spacings should, thus, be reduced to ensure complete fragmentation.

Navigational constraints

When blasting work is carried out in waterways which carry other maritime traffic there will be a need to ensure that the operational use of the waterway is maintained as far as is reasonably possible. Large vessels will require a considerable width of channel for safe navigation and the minimum of depth restrictions. In these circumstances, it may be necessary to halt drilling operations for a period of time or occasionally to complete a cycle and move off position.

In the case of floating pontoons, if the wind and sea conditions are favourable, the pontoon may be able to slack its wires on the channel side and remain in position whilst the vessel passes by. If this is not possible the pontoon will most probably have to move off position and blast its charged shotholes before the vessel passes. This is a time-consuming operation and requires plenty of warning prior to the vessel's passage.

Spudded pontoons do not rely on wires to maintain position in the elevated mode and thus vessels may pass quite close to the pontoon. However, if the pontoon does have to move, it may require many hours notice since it will have to jack down to the water, move off position and blast before the vessel passes.

Proximity to structures

Drilling pontoons are able to work in close proximity to structures, particularly if the drills are cantilevered over the side. The effects of blasting, however, are severe at such short range. These effects are discussed in Section 14.7.

14.3.3 ESTIMATING OUTPUT

The basic principles of estimating production rates for dredging craft are outlined in Chapter 9. In that chapter the maximum potential output, P_{max}, is derived from the various site parameters. To estimate the average output, P_{max}, would need to be reduced by the delay, operational and breakdown factors described in Section 9.4 and applied to the relevant available time.

In order to estimate production the following data must be assembled.

- T – The average available working time (hours) in each shift (i.e. the total shift length minus breaks, travelling to the pontoon, etc.).
- t_b – The time taken to move off position, blast and return to a new position (hours). This also includes time taken to move anchors if necessary.
- t_o – The time taken to make and break inner and outer drill strings during one cycle (hours).
- t_c – The time taken to charge the shothole with explosives and retrieve the initiating line (hours).
- t_m – The time taken to move an individual drilling rig, or move the pontoon back to the next row of holes (hours).

N – The number of drilling rigs on the barge.
A_w – The total area of rock (square metres) which can be covered from one barge position. For pontoons with wells only.
A – The area (square metres) treated by each shothole.
D – The average drilling speed (metres per hour) in rock.
z – The average thickness (metres) of the whole rock volume to be treated. (Note: for large variations in rock thickness it is best to divide the area into zones of similar thickness and to separately estimate production for each area.)
d – The average water depth (metres) over the whole area to be blasted, measured from mean sea level.

The unit of production

The unit of production is the number of rigs operating, N. This is modified by multiplying by the volume treated by each rig, Az. The modified unit of production, or volume treated per cycle, is, thus, given by:

$$U_m = NAz \tag{14.1}$$

The production cycle

Both types of drilling pontoon have a basic sub-cycle whose duration, t_s, is given by:

$$t_s = \frac{t_m + t_o + t_c + (\surd A + z)}{D} \tag{14.2}$$

where $(\surd A + z)/D$ represents the time taken to drill the hole and $\surd A$ represents the drilling necessary below dredged level to ensure total treatment. In order to establish how many sub-cycles will be performed before blasting, it is necessary to make two assumptions; that a floating pontoon will blast at the end of each shift and that a spudded pontoon will blast on completion of a well area.
(Note: if this is not true, then it will be necessary to modify the estimating procedure to take account of the altered circumstances.)

For a floating pontoon

If n is the number of cycles per shift, the total shift time, T, can be expressed by:

$$T = nt_s + t_b \tag{14.3}$$

or

$$n = \frac{T - t_b}{t_s} \tag{14.4}$$

Since the shift output $= nU_m$, the hourly production is given by:

$$P_{max} = \frac{nU_m}{T} \tag{14.5}$$

where P_{max} is in cubic metres per hour. P_{max} can be expressed in terms of the basic variables by substitution to give:

$$P_{max} = \frac{NAz(T - t_b)}{T(t_m + t_o + t_c + \frac{\sqrt{A} + z}{D})} \tag{14.6}$$

For a spudded pontoon with well

It can be shown that, using a similar analysis to that given above, the hourly production is given by:

$$P_{max} = \frac{NAz}{t_m + t_o + t_c + \frac{C\sqrt{A} + z)}{D} + \frac{t_b AN}{A_w}} \tag{14.7}$$

Expressions (14.6) and (14.7) can be used to estimate hourly outputs if all the variables are known. Since the area of rock blasted by pontoons is usually more consistent than the volume treated, it may be more realistic to express the output in square metres per hour by dividing P_{max} by the average rock thickness, z.

When the relative merits of floating and spudded pontoons are assessed, it should be noted that, not only will a spudded pontoon suffer delays when elevating or descending in bad weather, but that any increase in the frequency of blasting will cause the spudded pontoon to lose production rapidly due to the long period required to carry out the blasting operation.

Simplification

Since many of the variables connected with the cycle times are often unknown at the estimating stage it may be necessary to make some assumptions, as follows:

T = 9 hours (for a nominal 10 hour shift);
t_b = 0.6 hours (floating pontoon);
= 10 hours (spudded pontoon in exposed location);
$t_o = \frac{d}{50}$ hours;
t_c = 0.12 hours;
t_m = 0.13 hours;

With these assumptions expression (14.6) becomes:

$$P_{max} = \frac{NAz}{0.93(0.25 + \frac{d}{50} + \frac{\sqrt{A} + z}{D})} \tag{14.8}$$

and (14.7) becomes:

$$P_{max} = \frac{NAz}{0.25 + \frac{d}{50} + \frac{\sqrt{A} + z}{D} + \frac{10AN}{A_w}} \tag{14.9}$$

Drilling speeds and patterns for various rock types are given in Section 14.3.2.

14.3.4 EXPLOSIVES

Explosives used for underwater work must have two very important characteristics; they must be water resistant for a suitable period of time, and they must detonate at a high velocity.

The water resistance of the explosive should be such that it remains reasonably unaffected by water for a period of about 24 hours, but that after two weeks to a month underwater the substance becomes inert. The reason for this is that the explosive must be active for the time required to complete a cycle of operations prior to detonation (with an allowance for delays) but if there is a misfire of any cartridge, due to some problem in the initiating system, the explosive must be rendered inert by the time that the excavating machine reaches it.

High detonation velocity is essential in that much of the fracturing process in an underwater blast is caused by the shock wave and its effect on the rock structure. A high detonation velocity will propagate a strong shock wave. Some explosives are affected by the hydrostatic head and will detonate at inadequate velocities unless initiated by a booster charge.

A third characteristic which is important is that the explosive must be of high power. Particularly when the explosive is being placed in drilled holes it is important, for reasons of economy, to get the maximum amount of power into each hole and this implies using explosive of high density.

Generally the more powerful types of explosives, either gelatinous nitroglycerine (NG) based (of around 30 to 40 per cent NG content) or high-strength slurries, are used for underwater work. These not only have the right properties but their strength is also high. They have densities of 1.4 to 1.6 and 1.3 to 1.4 Mg m^{-3} respectively, and detonation velocities of 5000 to 6000 and 4000 to 5000 m/s.

To ensure that the maximum amount of explosive is packed into a drilled hole a good design of cartridge is required. It must have a thin wall, a tapered nose cone and be completely filled. Some cartridges are designed to connect together so that they can be loaded into the hole as a long string.

14.3.5 INITIATING SYSTEMS

There are currently at least four distinctly different methods of initiating shotholes under water. These are:

- detonating cord;
- long lead electrical detonators;
- toroidal detonators;
- the Nonel system.

Detonating cord

Detonating cord consists of a thin core of high explosive, usually pentaerythritol tetranitrate (PETN), covered by layers of protective material to enable it to withstand strain, abrasion and the ingress of water. When used for underwater work, any necessary cuts in the cord are usually sealed at the ends with a cap or a sealing composition. The cord is threaded through the explosive charges and brought to the water surface, where it can be linked, with other cords, to a main detonating line. The main line is initiated by electrical detonators.

Detonating cord is relatively simple to use and has the advantage that any charges which have misfired will not contain detonators, a possible risk to excavating equipment. However, it also has a number of disadvantages. It is prone to 'cutoffs', where one line cuts another thereby preventing it from firing, it can only be used for instantaneous firing of all charges and the portion on the water surface is noisy during detonation.

Long lead electrical detonators

Long lead electrical detonators, which allow a detonator to be placed in every shothole, have a number of advantages over detonating cord. They enable delay detonation to be achieved (and thus assist in reducing vibration), they allow the explosive charge to be detonated from the bottom of the hole (thus improving fragmentation) and the firing circuit can be checked prior to detonation.

Toroidal detonators

These are electrical detonators whose lead wires are coupled to ferrite rings in such a way as to act as the secondary winding in a transformer. The primary winding is effected by passing a length of plastic-covered connecting wire once through the centre of each toroid. This assembly gives significant improvements in safety properties compared with conventional detonator systems, since it is immune to AC mains and stray direct currents.

The Nonel system

This proprietary system is based on a flexible plastic tube whose inside surface is coated with a reactive substance which maintains the propagation of a shock wave. This shock wave has sufficient energy to initiate the primary explosive or delay element in a detonator. The reaction within the tube leaves it intact and hence there is no lateral shock effect.

The Nonel system has a number of relays and delay devices which enable it to be used for instantaneous or delay detonation sequences.

14.4 Surface blasting

On land, surface (or contact) charges are seldom used, except for breaking boulders with very small charges. Underwater their use is more common. The reason for this is that, due to the confinement of the charge by a head of water, more of the explosive energy can be harnessed for rock breaking than on land. In addition, airblast and noise, which prevent most applications on land, are not present when the charge is placed underwater.

The main advantages of using surface charges in underwater rock blasting are that it eliminates the need for drilling. This means that small volumes of rock can be blasted with rapidity and at depths which might not be possible with conventional drilling equipment. It is, however, a relatively inefficient method of using explosive energy and large quantities of explosive are often required, with a commensurate effect on the local environment.

Surface charges may be formed either from cartridges (or boxed charges) or made up specially in the form of shaped charges. The latter are often formed of two-part liquid explosives. The use of these is not permitted in the UK, where cast-shaped charges are usually supplied. Whatever the method used, it is essential that surface charges are made from high velocity explosives.

Surface blasting is only feasible when the rock to be blasted is bare or it is possible to remove the overburden prior to placing the charges. Even with perfect contact, penetration is poor and it is unusual to find that blasting has been effective in fragmenting more than one metre of rock. Best results are usually obtained in the weaker sedimentary rocks and in coral (see Section 14.5). Fragmentation is often poor, with secondary blasting being required, and the heave of the rock, or bulking, is also poor.

Individual charges are normally in the range 10 to 25 kilograms and maintaining the correct spacing is important. For this reason the charges are often lowered to the seabed in a frame (see Plate 14.2) and each charge will be weighted with a concrete block, or collar, to ensure that it remains in position. Even then it is possible for charges, or their fuses, to become accidentally snagged and dragged out of position. It is essential that the position of all the charges and the initiating system is checked by divers prior to blasting.

Plate 14.2 A typical frame (top of picture) for lowering shaped charges (bottom of picture) to the sea bed on a pre-determined grid pattern

14.5 Blasting coral

Coral is one of the few rocks which may be strong enough to require blasting prior to dredging by small dredger, and yet is often more easily blasted by surface charges than by charges placed in boreholes.

Coral is a formation of calcium carbonate, of infinitely variable shape, which is normally sufficiently porous to be crushed by the shock wave and bubble pulse from a surface charge, rather than being shattered by the shock wave reflection from a free face. Thus, on detonation, a charge placed in a borehole tends to crush the area surrounding the borehole but does not adequately fracture the remainder of the rock mass.

In addition coral tends to be found in shallow waters, often in the surf zone, which do not lend themselves to the operation of a drilling pontoon.

Coral blasting is usually carried out in such a manner as to provide at least a metre of water cover above the charges and, where practicable, large numbers of charges are fired instantaneously. This reduces the number of times when the water is muddied by explosions, thereby improving operational efficiency and reducing environmental impact.

14.6 Contract documents for blasting work

Information concerning contract documentation for dredging works has been given in Chapter 11 and most of this is also applicable to rock pre-treatment. However, there are some additional points which should be covered for blasting work. These are described below.

14.6.1 CONTRACT CONDITIONS

It must be made clear in the conditions whether the contractor is permitted to use explosives and, if so, whose responsibility it is to obtain the required permissions for storage, handling, use and (where necessary) import of the explosives.

Any laws or regulations appertaining to the use of explosives should be noted (see Section 14.8).

14.6.2 DATA SUPPLIED TO CONTRACTOR

The contractor should be supplied with a survey of the dredging area which contains the rock, preferably to a scale of 1:500 if the rock is exposed since this will give an impression of the configuration of rock outcrops. In addition he should be provided with the following:

- full details of all ground investigations carried out;
- details of any diving surveys carried out;
- details of any trial drilling or blasting operations, together with results including vibration measurements taken.

14.6.3 DATA REQUIRED FROM THE CONTRACTOR

The contractor should be asked to provide with his tender:

- details of the plant he intends to use;
- a detailed method statement;
- a programme showing how the blasting work is to be integrated with the dredging operations.

14.6.4 SPECIFICATION

Record keeping

The contractor should be asked to provide a daily log showing the position of drill holes, weight of explosive charges placed, delay sequencing (where applicable) and blasting vibration records.

Where the site is in close proximity to inhabited buildings the contractor should be encouraged to make prior records of cracks and other imperfections in these. This will assist in the identification of spurious complaints and will facilitate the assessment of legitimate claims.

Warning signals

The contractor should be instructed as to the warning signals he should give prior to firing his charges and those precautions which are to be taken to prevent vessels and persons straying into the blasting area.

Noise

In many cases there will be a need to restrict blasting to daylight hours. Drilling equipment may also be very noisy and it should be made clear whether any operations are to be permitted at night.

Vibrations

Specific vibrations limits must be specified (see Section 14.7) and the manner in which they are to be applied. This will include the specification of the vibration measuring instrumentation and where it must be positioned at the time of the blasts.

Watershock

If watershock is expected to be a problem, limits should also be specified. However, this is very difficult to measure and it may be preferable to specify mitigating measures, such as a bubble curtain.

Overbreak

In some instances it is found necessary to restrict the amount of vertical or lateral overbreak from blasting. This is usually required when the rock under a quay wall or at its toe has to be protected, or when a foundation is to be laid in the area after dredging. It should be appreciated that severely restricting overbreak is a difficult, and normally very expensive, operation. It should be avoided wherever possible.

14.6.5 MEASUREMENT FOR PAYMENT

Drilling and blasting

Drilling and blasting involves setting up an expensive piece of equipment to drill holes vertically over an area of rock, no matter what the thickness, and so there is a substantial cost element which relates to area. Even if the rock is only a few millimetres thick above the dredged level it is still necessary to drill down to a depth well below this level to achieve adequate fragmentation all over the area.

Figure 14.8 illustrates how the costs of drilling and blasting vary according to the thickness of rock to be excavated. With accurate drilling, and careful placing of explosive charges and detonation systems, it is possible to pre-treat considerable depths of rock in one pass. However, when deviation of the drilled hole is likely to jeopardise accuracy and consequently fragmentation, it is preferable to blast in two passes. The effects of this on blasting costs are also illustrated in Figure 14.8.

It is clear from the above that method of operation, type and thickness of rock all affect the cost of pre-treatment. It is, therefore, recommended that payment for rock drilling and blasting should be based on the area of rock covered at a certain thickness, that is as follows:

item A	drilling and blasting rock above pay level, not exceeding one metre thickness;	(x) m^2
item B	drilling and blasting rock above pay level, exceeding one metre but not exceeding two metres thickness	(y) m^2

and so on.

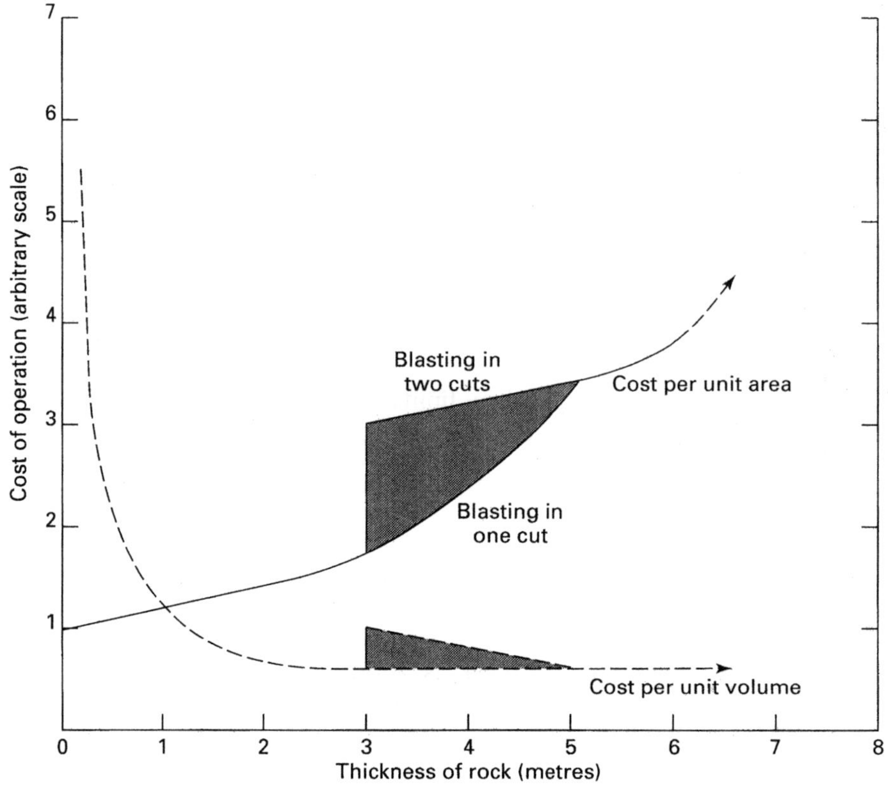

Figure 14.8 Relative costs of drilling and blasting with thickness of rock

The pay level mentioned in these items should preferably be the same as that for payment for excavation. Whatever level is chosen it should be very clearly defined. Where the extent of areas requiring pre-treatment are unknown at the time of tender or where unforeseen isolated outcrops may be found, it will also be prudent to invite rates for areas of different size, for instance 10 square metres, 50 square metres, 100 square metres and so on. Small areas may be very expensive.

It will also be necessary to measure the progress of drilling and blasting works during the project, for programming and payment purposes. This may be effected by using the drilling and blasting contractor's daily record sheets which will show the position of the holes drilled and the depth of rock above the pay level. From these the weekly or monthly area covered and volume blasted may be computed.

Contractors tend to use the level of the bottom of the casing tube to denote rock head level. This has been found by experience to correspond fairly well to the level at which rock becomes difficult to dredge by mechanical means. The advantage of using this method is that rock pre-treatment can be measured without the necessity to remove the overburden.

Surface blasting and non-explosive methods

Since these methods are not so tied to a method of work which relates to area, it is possible to treat them on a volume basis. However, it should be noted that wherever very thin layers of rock have to be removed it is preferable to measure them on a area basis since this allows the contractor to include realistic allowances for non-productive work on the periphery of the area.

Mobilisation and demobilisation

All pre-treatment requires special plant to be mobilised, and associated items, such as explosive stores, to be installed. Separate mobilisation and demobilisation items should, therefore, be included in the Bill of Quantities.

Interruptions to work

Most contracts will include a standing, or demurrage, rate to take account of interruptions caused by factors beyond the control of the contractor (other than weather delays, breakdowns, etc.). If a drilling and blasting pontoon is interrupted during its operations to make way for the passage of a vessel it is often necessary for the contractor to pull off position and blast his charged holes prior to the vessel passing. This causes considerable disruption to the normal flow of work and delays the contractor by a period of time far in excess of the time taken for the vessel to pass. For this reason standing rates are likely to be high to take account of this hidden loss of production, unless the full loss of production time is to be measured for payment.

14.7 Environmental factors

Blasting of rock underwater has a severe impact on the environment in terms of the following:

- noise;
- watershock;
- ground vibration.

14.7.1 NOISE

The general principles of noise measurement and regulation are described in Chapter 13. In underwater rock blasting operations noise arises from a number of sources, such as the hammering of rock drills, the whine of compressors, the crack of detonating cord and the detonation of the main explosive charge.

When work is carried out near residential areas it is common to find that blasting is not permitted between the hours of 7 p.m. and 7 a.m. In some cases the client, or local authority, may allow drilling to be carried out at night, but not blasting, except in emergencies. In such circumstances the noise from the drilling machinery may still prove to be an irritant to the local community unless steps are taken to mitigate its propagation.

14.7.2 WATERSHOCK

Physical process

When an explosive charge is detonated underwater the explosive shockwave and rapid expansion of the explosive gases give rise to a series of watershocks. The primary shock wave, which produces the maximum peak water pressure, travels outward from the source of the explosion, initially at around 5000 metres per sec but soon slowing to the velocity of sound in water (around 1500 metres per second). This is followed by a number of smaller, secondary shocks caused by the oscillating expansion and collapse of the explosive gas bubble.

Generally speaking the primary shock wave is the effect most likely to inflict damage on structures in the water, vessels and persons in the water, particularly at a distance. The secondary bubble pulses become relevant when charges are detonated in close proximity to an object in the water. This is because the explosive gas bubble behaves in a curious manner, being attracted to rigid boundaries, such as the sea floor, quay walls and the like.

The pressure in a shock front, as the wave reaches any given point, is called the peak shock pressure. It is usually measured in bars or kilobars, the latter being 100 MN m^{-2}. Another important parameter is the impulse of the shockwave. This is a measure of the pressure and time of the shockwave and is proportional to the area under the pressure-time curve graph for the positive duration of the wave. Impulse is measured in bar-milliseconds.

Actual peak pressures from the shockwave of an explosive charge, freely suspended in water, may be estimated from empirical formulae, the most widely accepted being:

$$p_{\text{dmax}} = \frac{k(W^{1/3})^A}{L} \tag{14.10}$$

where:

p_{dmax} = Peak pressure (bar)
W = Charge weight (kg)
L = Distance from the charge (m)
k = A constant which relates directly to the detonation pressure of the charge. For most underwater explosives it is taken as being 510.
A = An index which is assigned the value 1.16 for TNT and differs slightly for other explosives.

The value of impulse may also be estimated, from the following expression:

$$I = k_I W^{1/3} \frac{W^{\beta}}{L} \tag{14.11}$$

where:

I = Impulse in bar-milliseconds
k_I = A constant usually taken as being 0.057
β = An index which depends on the explosive and its density. For TNT of density 1.52 g/cm^3 it is taken as 0.89.

W and L are as in (14.10) above.

The values for peak pressure and impulse obtained from the expressions (14.10) and (14.11) above are valid for a charge suspended in water and apply to a situation where the shockwave travels directly to the point of measurement. In practice this is rarely the case, since charges are usually on, or in, the seabed and the shock wave is often reflected at the seabed or water surface as it travels towards a particular point. For this reason it is extremely difficult to estimate accurately the maximum pressure at a point some distance from an explosion.

The effect of placing charges in boreholes is normally to reduce watershock to some 10 to 20 per cent of that which would be obtained from an unconfined charge. In contrast the maximum pressure produced by a surface charge may increase by as much as 15 per cent due to the reflection of the wave off the seabed.

Effect on others

To assess the effect which an underwater explosion will have on a nearby vessel, structure, or on adjacent human and marine life, it is necessary to determine how these react to the peak pressures produced. Here, again, it is discovered that there is a wide variation in response which depends on type, condition and position.

For vessels it is usually recommended that the peak shock pressure should not exceed eight bar. This is a British Royal Navy limit which is considered to be very conservative in terms of machinery and hull damage. However, it would be unwise to exceed this limit and for commercial and pleasure craft the limit could well be lowered to reduce the nuisance value of the effect. Another measure to reduce the effect which can be taken is to point the bow of the vessel towards the point of the explosion, thereby preventing the pressure from acting normally to the vessel's plating.

Some vessels have delicate instruments or machinery on board and special precautions have to be agreed to take account of this. On a contract in Milford Haven (Swansbourne and Dudley, 1971), the detonation of large instantaneous charges near tankers with automated pumping systems caused the pumps to reverse, and charges had to be restricted when vessels were discharging. Maximum instantaneous charge weights were limited, as shown in Figure 14.9.

To determine the effect of the water shockwave on a structure it is necessary to know the characteristics of the structure in terms of mass and resonant frequency. From these parameters, and from considerations of blast design, range and shallow water correction

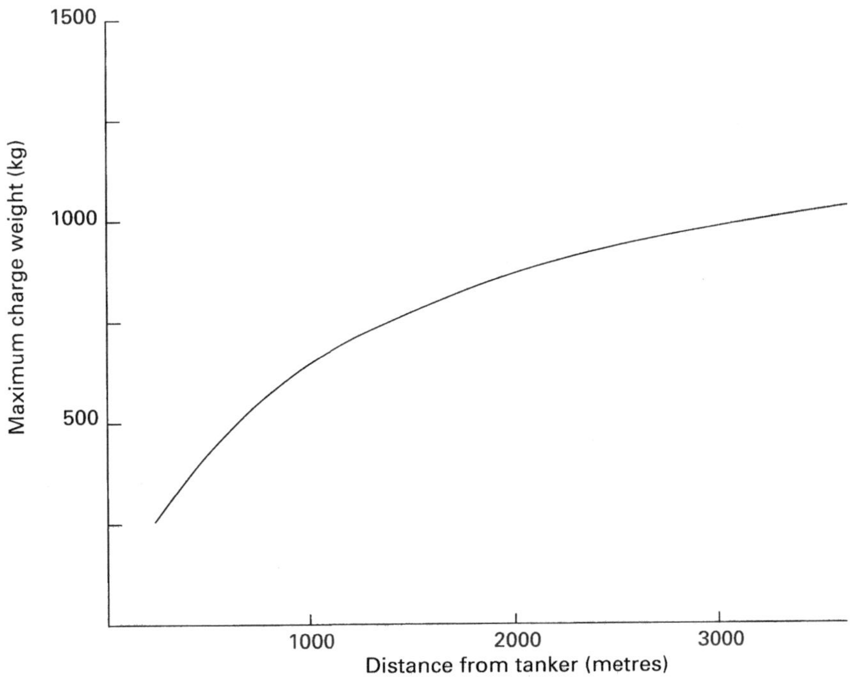

Figure 14.9 Maximum allowable instantaneous charge weights at given distances from automated tankers (as used in Milford Haven in 1969)

coefficients, it is possible to predict the peak particle velocity of the structure. This may then be compared with allowable peak particle velocity. This analysis is somewhat complex and should be carried out by specialists. In practice it usually will be the case that the response of the structure is already being monitored to measure the effect of the ground vibrations (see below) and thus its response to water shockwave will also be detected.

Human response to watershock varies widely, depending on the physiology of the person in the water, their state of health, position in the water and so on. There is also some difficulty experienced with the correlation of the different explosive parameters with physical injury. Primary shock, duration of pulse, impulse and energy have all been examined by various navies but as yet no reliable comprehensive guidance has been forthcoming. In view of this, and the extreme difficulty of predicting the shockwave characteristics at a particular point remote from the explosion, it is difficult to define safe distances for swimmers and divers.

The code of practice for the safe use of explosives in the construction industry, (BS 5607, 1988), suggests various minimum safe distances for swimmers from charges of TNT freely suspended in the water, in relation to the weight of charge. It also states that swimmers should never be close enough to an underwater blast so as to be exposed to a shockwave whose impulse at that point is greater than 14 Pa-s or whose peak overpressure is greater than 170 kPa. In view of the difficulty of predicting these parameters, in the authors' opinion specialist advice should always be sought on this subject. Marine animals can be killed or injured by underwater explosions if they are within around 200 metres of the explosion. Beyond this radius they seem to be less susceptible and are able to recover. Fish with well-developed air bladders are most at risk, whereas crustaceans can withstand much higher levels of shock. The importance of the effects of the shockwaves from blasting operations will depend on the size and confinement of the charges, the size of the fish populations and the rarity of the various species which would be affected.

Mitigating measures

Measures for mitigating the effect of the shockwave from an underwater explosion are of two types; those that reduce the intensity of the shockwave at source, and those that reduce the intensity of the emitted wave at the target.

In the first category two measures are possible. These are to reduce the instantaneous charge weights being detonated and to ensure maximum confinement of the charges. The former can be achieved by the use of delay detonators. In the latter case this will mean that charges placed in boreholes should not protrude above the level of the rock.

The intensity of the emitted wave may be altered during its passage to the target by the installation of a bubble curtain. A bubble curtain consists of one or preferably two perforated tubes, laid on the seabed, through which compressed air is fed (see Figure 14.10). The air escaping through the small perforations produces a curtain of bubbles from bed to water surface. When the shockwave hits the bubble curtain it is absorbed by the bubbles and transmitted in an altered form. There is no doubt that peak pressures are reduced by such a curtain but the impulse of the wave may not be much reduced. Records of the overall efficiency of the method are conflicting (Hicks, 1980).

Other mitigating methods have been tried, including foamed rubber bulkheads placed in front of the structure, but there is no evidence of their effectiveness.

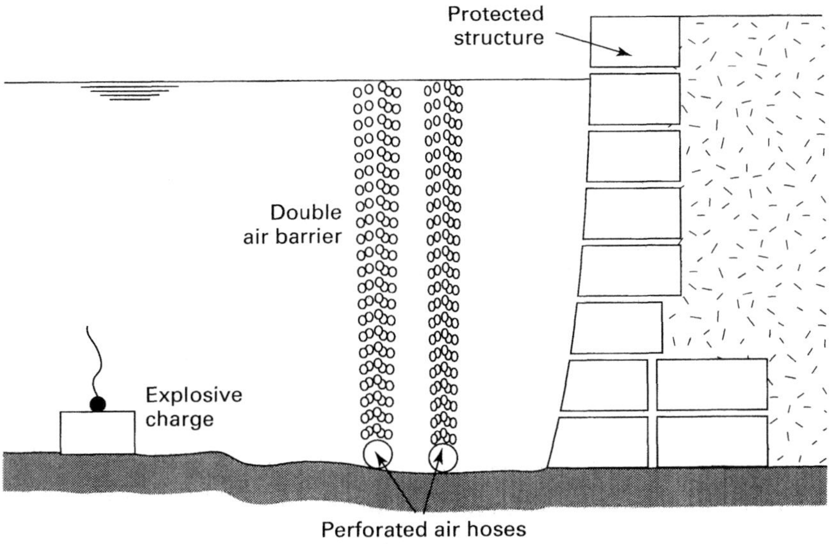

Figure 14.10 A bubble curtain for reducing the intensity of water shock waves

Measurement of shockwave

The measurement of water shockwave characteristics is difficult and currently there are no standard instruments which may be purchased for this purpose. Certain universities have devised laboratory methods for accurate measurements to be made during their research programmes and it is recommended that specialist organisations are contacted if it is thought necessary to make measurements of this kind.

14.7.3 GROUNDBORNE VIBRATION

Physical process

When an explosive charge detonates in a borehole a seismic wave propagates radially in the ground. Due to the inhomogeneity of the ground, and the interface between ground and water/air above, some portions of the wave will be reflected and/or refracted before they reach a point at which they can be measured. Any measurements made will therefore need to take account of the three components of movement, at right angles to one another, and the resultant of these at any particular moment.

The velocity of propagation of the shock wave, which is approximately the sonic velocity of the medium through which it is passing, will vary between 1000 and 7000 metres per second, the higher velocities occurring in dense and intact igneous and sedimentary rocks whilst the lower velocities will be found in less dense materials such as clays, weak sandstones and chalk or fragmented materials. The frequency of vibration, which has a considerable bearing on damage, will also vary, normally between 3 and 35 Hz.

The four parameters which characterise ground transmitted vibrations: amplitude, particle velocity, frequency and acceleration, are measurable on modern instrumentation and are linked by the standard formulae for sinusoidal vibration. More important, from the point of view of assessing the damage potential of a blast, is the relationship between these parameters and the site conditions for a particular weight of charge.

The general wave propagation equation relating to a site may be expressed in the following form:

$$V = K(R.W^{-A})^{-B} \qquad (14.12)$$

where:

V = Peak particle velocity (mm/s)
W = Instantaneous charge weight (kg)
R = Distance from blast (m)

K, A and B are site constants which are to be determined by measurement.

The index, A, is traditionally taken to be ⅓, ½ or ⅔. In the near field, that is within a radius of 30 metres from the explosion, it has been found that either ½ or ⅔ scaling provides more accurate prediction. However, ½ scaling is the most widely used for all distances.

The index, B, is usually found to be between 1.0 and 2.0, whilst the constant, K, may vary between 20 and 1000. A general formula published by the DuPont Company, based on a very large number of observations from many different sites, is:

$$V = 160(R.W^{-\frac{1}{2}})^{-1.6} \qquad (14.13)$$

where V, R and W denote the variables given in (14.12) above. However, this general formula may be unrepresentative of many sites and should only be adopted as a first approximation until actual site measurements have been carried out. Normally a data set of 30 or more readings of particle velocity for different charge weights and distances is sufficient to determine site characteristics. The measurements are plotted on log/log paper in the form: particle velocity versus scaled distance, where scaled distance is the expression:

$$\frac{R}{W^{\frac{1}{2}}}$$

An example of such a plot is shown in Figure 14.11. Regression analysis may subsequently be used to fit a line to the data points and the constant, K, and index, B, can then be determined from the position and slope of the line.

For the purposes of working on a site, an attenuation formula has to be developed. This is often the line which will ensure that, given a reasonably large number of blasts using the maximum allowable charge weight, 95 per cent of the peak particle velocities are less than a certain value. Maximum charge weights for various distances may then be based on this line. The adoption of this line also implies that some 5 per cent of the blasts, using maximum charge weights, will exceed the limiting value for peak particle velocity. This latter fact is one that sometimes escapes over-zealous supervisory engineers who wish to impose particle velocity limits that are never exceeded!

Effect on others

The effects of ground vibration on structures have been examined in some detail over the years and there is now a general similarity in the recommendations, although not in the absolute limiting values, given by the four most widely reported standards, namely the German (DIN 4150, 1986), Swedish, Swiss (SN 640 312, 1978) and US (Siskind *et al.*, 1980). At the present time there is no British equivalent and it is for this reason that The Building Research Establishment has produced a note on the subject in Digest 353 of 1990 (BRE, 1990). This note reviews the differences in the above quoted documents but cautions

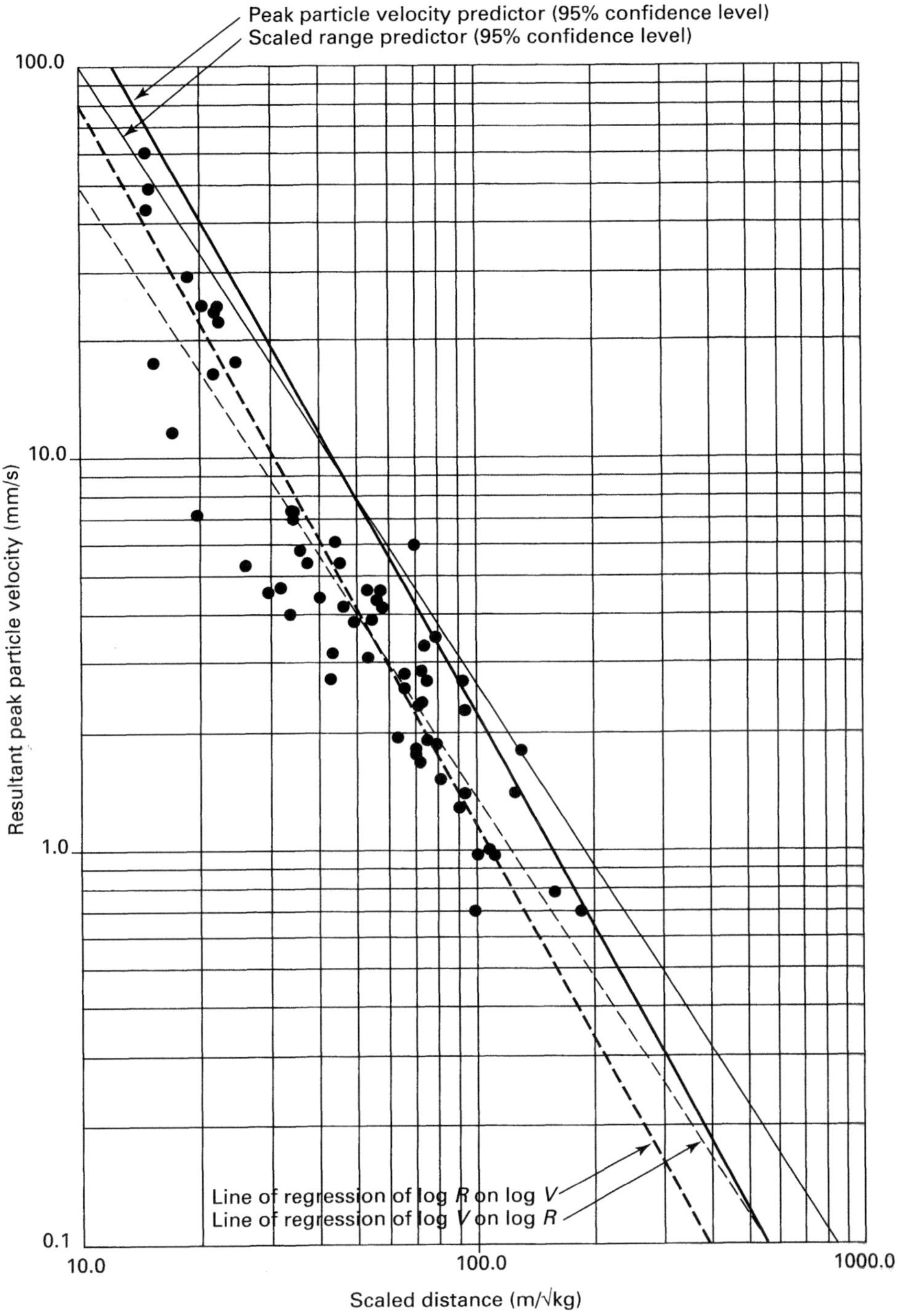

Figure 14.11 Graph of peak particle velocity versus scaled distance

that it is not strictly valid to apply any of the limits in the UK because the data on which they are based relate to different structural types and site characteristics.

To give an illustration of the level of vibration used in these standards and the manner in which they may be related to vibration frequency, the German guidelines are reproduced in Table 14.3. In this table the maximum peak particle velocity (i.e. the greatest of the *x*, *y* and *z* axes) is related to the predominant frequency, the type of structure being considered and the point of measurement. Although this illustration gives an indication of possible limiting values which could be used, it is recommended that specialist advice be sought on this subject. A good guide to American practice is given by the Blasting Guidance Manual published by the US Department of the Interior (Rosenthal and Morlock, 1987).

Table 14.3 Limiting groundborne vibrations as defined by the German Standard DIN 4150

	Vibration velocity v_i (mm/s)			
	Foundation at a frequency of			Plane of floor of uppermost full storey
Type of structure	less than 10 Hz	10–50 Hz	50–100 Hz*	Frequency mixture
Buildings used for commercial purposes, industrial buildings and buildings of similar design	20	20–40	40–50	40
Dwellings and buildings of similar design and/or use	5	5–15	15–20	15
Structures that, because of their particular sensitivity to vibration do not correspond to those listed above and are of great intrinsic value (e.g. buildings that are under a preservation order)	3	3–8	8–10	8

* For frequencies above 100 Hz, at least the values specific in this column shall be applied
v_i is the maximum value of the *x.y* and *z* axes vibrations

As far as the effect of vibration on humans is concerned, it is probably true to say that vibration from blasting is more often an irritant than a cause of acute discomfort or distress. The threshold of human perceptibility is a peak particle velocity of around 0.7 millimetres per second so the structural vibration limits illustrated above would be very noticeable. What is of greater importance is the nuisance value of the vibration and this is related to the time and frequency of blasting, the position of the human body and whether the person is busy or not at the time. The authoritative document on the subject in the UK is BS 6472: 1984 Guide to the evaluation of human exposure to vibration in buildings (1 Hz to 80 Hz).

Mitigation of effects

The effects of vibrations due to blasting may only be mitigated by reducing the intensity of the shockwave at source. Since it is unlikely that changing the type of explosive would be significant, or possible, the method of mitigation has to be to reduce the weight of the explosive being detonated at any moment. This is achieved by means of millisecond delays in the detonating system and usually results in individual shotholes, or small groups of shotholes, being detonated at intervals. The effect of this is to reduce the maximum vibration level to that which would result from the detonation of one or a small group of charges.

14.8 Safety and legislation

(Note: the explosives industry is constantly updating itself with respect to safety and legislation and readers in the UK are advised to contact the Institute of Explosives Engineers to ascertain how to obtain current recommendations.)

14.8.1 SAFETY

Great care needs to be exercised when using explosives for any application. In underwater blasting, additional complications arise due to the generally inhospitable environment in which the work is being performed. When divers are placing charges underwater they are normally working in a dark, pressurised and mobile medium which makes accuracy and checking much more difficult than when working on land. Operators loading shotholes from the sea surface are working remotely and blind. Electrical circuits have to be maintained without leakage, even when they pass through a conducting medium (salt water) and most of the operating platforms are composed predominantly of ferrous decking and structural members.There are many safety rules which should be applied to this type of work. Some of these are contained in BS 5607, Code of Practice for the Safe Use of Explosives in the Construction Industry. In general it is prudent to obtain specialist advice on this aspect of the work.

One particular potential danger often focussed on in this type of work is the possibility of accidental firing of electrical detonators due to currents induced in the detonator wires by radio transmissions. However, this should not generally be a problem. Research has shown that, for normal low power walkie-talkie and ship-to-shore transmissions, there is very little risk of significant currents being generated in this way, particularly if a few simple rules are observed with respect to the way detonators are handled. The DuPont Blaster's Handbook (Du Pont, 1980) gives advice on this subject, including safe distances for various types and powers of transmission. Some military transmissions are very powerful and it is recommended that when working in proximity to military zones, especially naval bases, discussions are held with the military to ascertain that the risk remains minimal.

14.8.2 LEGISLATION

There is no legal instrument which covers the way blasting is carried out in UK waters, although the use of explosives is covered in a general way by the Health and Safety at Work Act 1974, which not only covers an employer's own staff but also the safety of the general public. Where the law is specific is in the way in which the explosives are made, purchased, stored and transported, and a brief resume of the relevant legislation is given below.

- The Explosives Act of 1875 and subordinate SIs
  Controls the manufacture, purchase, storage and conveyance of explosives. Manufacture must be on licensed premises. Does not allow the mixing of two-part liquid explosives on site. Purchasers of explosives must have licensed storage facilities.

- Sale of Explosives Order 1928
  Allows the purchase and immediate use of explosives (but not storage) under an 'immediate use certificate'.
- Stores for Explosives Order 1951 (SI 1951/1163)
  Gives maximum quantities, minimum safety distances from other facilities and other safety aspects for explosives stores.
- Road Traffic (Carriage of Explosives) Regulations 1989
  Relates to transportation of explosives from place of storage to place of embarkation.
- Dangerous Substances in Harbour Areas Regulations 1987, (SI 1987/37)
  Controls the carriage, handling and storage of dangerous substances (which includes explosives) in harbours and harbour areas.
- Health and Safety at Work Act (Application outside Great Britain) Order 1977. (SI 1977/1232)
  Controls work on offshore installations, pipelines and certain activities (including diving) in territorial waters.
- Diving at Work Regulations 1981 (SI 1981/399)
  Controls the operations of all employed and self-employed divers.

Many other UK regulations exist, such as those relating to offshore installations and pipelines, as well as a number of international regulations covering conveyance of explosives on vessels. However, the former are of more relevance to offshore demolition companies and the latter to explosives manufacturers and shipping companies.

14.9 Dredging pre-treated rock

For a dredger to dredge rock after pre-treatment it is necessary for the rock to be adequately fragmented and bulked to a level well below the desired dredging level. Determination of the correct fragmentation and bulking will require a knowledge of the relative costs of dredging and pre-treatment, as well as the type of dredger to be used. An increase in dredgeability will usually only be achieved by an increase in the cost of pre-treatment. In rocks which are normally undredgeable, without pre-treatment, such as igneous or metamorphic, it is usual for pre-treatment to be carried out to a relatively small fragmentation size, because any unblasted rock or oversize fragments are extremely expensive to re-treat. In weaker rocks, which can just be dredged directly, complete effectiveness of pre-treatment is less critical. In the following descriptions it has been assumed that the rock is too strong to be dredged directly. Fragmentation size is the measurement of the largest dimension and should be assumed to be the D_{95} of the pre-treated material.

Trailing suction hopper dredger

Ideally fragmentation should be kept as small as possible, say 50 to 100 millimetres. The larger trailers will have suction pipes of between 800 and 1200 millimetres diameter, and the dragheads would normally be fitted with 300 to 450 millimetre grids to prevent large fragments passing through the pumping system. If fragmentation is too large, considerable time will be spent raising suction arms and removing blockages from dragheads. Good bulking of the rock is required, say over 25 per cent, and ripper teeth on the draghead may be beneficial if the bulking is too low.

Cutter suction dredger

The larger dredgers, of say 800 millimetres pipeline diameter and up, can handle fragments under 300 millimetres. All fragments must be able to pass between the cutterhead blades. The smaller the fragmentation the better for pumping. It is particularly important to ensure that all rock above dredged level has been fragmented since the cutterhead will be easily damaged on any projecting unblasted pinnacles. A bulking of over 10 per cent is required.

Bucket wheel dredger

Fragmentation should be similar to that required for the cutter suction dredger, and the dredger is even less able to deal with unbulked or unfragmented rock. Bulking of over 15 per cent is required.

Grab dredger

Generally, for grabs of five cubic metres and up, the fragmentation should be 400 millimetres or less, with a bulking factor of over 20 per cent. Grabs become less effective as the fragmentation gets larger and the bulking decreases. However, larger lumps can be removed if the tines of the grab bucket can penetrate and hold them. The maximum load is governed by the maximum line pull. In general it is preferable to use a large grab dredger rather than two or more smaller ones.

Backhoe dredger

With a bucket capacity of two to three cubic metres a maximum fragmentation of some 400 to 500 millimetres is desirable. Larger fragments can be recovered if the bucket size and installed power are sufficient. Bulking factor should be over ten per cent.

Dipper dredger

With a bucket size of five cubic metres or more, fragmentation should be under 800 millimetres with a bulking of 10 to 20 per cent for reasonable production. Much larger fragments than this can be dredged if bucket size and power allow.

Bucket dredger

If the dredger is fitted with rock buckets of 600 litres capacity, a fragmentation of less than 500 millimetres is required, with the majority of the material considerably smaller than this if possible. Bulking should be at least ten per cent. The bucket dredger can handle larger lumps, but the maximum size is restricted to that which can pass between the ladder hoist wires and through the well. It cannot easily pick up these larger lumps and tends to push them across the dredging area to one side.

References

Abrahams, J.L. (1974). Underwater drilling and blasting for rock dredging, Part 1, Paper 7728, pp 463–478, *Proceedings of the Institution of Civil Engineers*.

Building Research Establishment, 1990. Damage to structures from ground-borne vibration. *BRE Digest* 353, July.

Du Pont, 1980. *Blaster's Handbook*, (16th Edn). Explosives Products Division, E.I. Du Pont de Nemour & Co., (Inc), Wilmington, Delaware, USA.

German Standard DIN 4150, Part 3, 1986. Structural vibrations in buildings: Effects on structures.

Hicks, A.N. 1980. The effects of underwater explosions on structures. SECED informal discussion, Part 1, 68, August, pp 568–570, *Proceedings of the Institution of Civil Engineers*, London.

Langefors, U. & Kihlstrom, B. 1978. *The Modern Technique of Rock Blasting*, John Wiley,

MacGregor, K. 1967. *The Drilling of Rock*, C R Books.

Rosenthal, M.F. & Morlock, G.L. 1987. *Blasting Guidance Manual*, US Department of the Interior, Office of Surface Mining Reclamation and Enforcement.

Siskind, D.E., Stagg, M.S., Copp, J.W. & Dowding, C.H. 1980. Structure response and damage produced by ground vibration from surface mine blasting, *Report of Investigations* 8507, US Bureau of Mines.

Swansbourne, J.F.C. & Dudley, G. 1971. The development of Milford Haven, Supplement XV, Paper 7419 S, *Proceedings of the Institution of Civil Engineers*, London.

Swiss Standard SN 640 312:1978. Effects of vibrations on buildings.

APPENDIX

From: The London Convention Dredged Material Assessment Framework

Resolution LC.52(18) on a Dredged Material Assessment Framework

The eighteenth consultative meeting

1 RECALLING Article I of the Convention on the Prevention of Marine Pollution by Dumping of Wastes and Other Matter, 1972 (London Convention 1972), which provides that Contracting Parties shall individually and collectively promote the effective control of all sources of pollution of the marine environment;

2 RECOGNIZING the need for maintaining open shipping lanes and harbours for maritime transport and that undue burden should be avoided with regard to the interpretation and application of the provisions of the London Convention 1972;

3 RECOGNIZING ALSO that the major part of the sediments dredged from the waterways of the world is, by nature, similar to undisturbed sediments in inland waters, whereas a minor part is contaminated, mostly resulting from the emission of hazardous substances into internal waters, requiring application of major environmental constraints when depositing these sediments, and that problems will continue until such emissions are controlled at source;

4 RECALLING that the Tenth Consultative Meeting by resolution LDC.23(10) adopted Guidelines for the Application of the Annexes to the Disposal of Dredged Material with a view to assessing the suitability of dredged material for disposal at sea in accordance with the provisions of the London Convention 1972, and the agreement to review these Guidelines within five years time in light of experience gained by Contracting Parties, in particular with regard to the application of the terms 'trace contaminants', 'rapidly rendered harmless' and 'special care' as defined for disposal of dredged material at sea;

5 NOTING the experience with these Guidelines as reported by Contracting Parties;

6 RECALLING that the Fifteenth Consultative Meeting instructed the Scientific Group to carry out a full review of the Guidelines and that it considered the Waste Assessment Framework, which it had adopted on a provisional basis, to be an appropriate starting point for this review;

7 CONSIDERING that the Guidelines for the Application of the Annexes to the Disposal of Dredged Material (resolution LDC.23(10)) had primarily focused on Annex III, Part

A of the London Convention 1982, and that the review of these Guidelines would have to include, where appropriate to dredged material, a review of parts B and C of the Guidelines for the Implementation and Uniform Interpretation of Annex III to the London Dumping Convention as contained in resolution LDC.32(11);

8 NOTING the adoption of the Amendments to the Annexes to the Convention on the Prevention of Marine Pollution by Dumping of Wastes and Other Matter, 1972, Concerning Phasing out Sea Disposal of Industrial Waste and Concerning Disposal at Sea of Radioactive Wastes and Other Radioactive Matter, by resolutions LC.49(16) and LC.51(16) respectively, in particular with regard to the references contained therein to sea disposal of dredged material;

9 HAVING CONSIDERED the draft Dredged Material Assessment Framework prepared by the Scientific Group;

1 ADOPTS the Dredged Material Assessment Framework as set out at Annex hereto, thereby replacing the Guidelines for the Application of the Annexes to the Disposal of Dredged Material at Sea, as adopted by resolution LDC.23(10);

2 RESOLVES that Contracting Parties to the Convention when assessing the suitability of dredged material for disposal at sea shall take full account of the Dredged Material Assessment Framework;

3 AGREES to review the Dredged Material Assessment Framework within five years time in light of experience gained by Contracting Parties with it, and in light of relevant amendments to the London Convention 1972, adopted in accordance with resolution LC.48(16);

4 REQUESTS Contracting Parties to submit to the Organization for distribution to all Contracting Parties information on their experience gained with the Dredged Material Assessment Framework, including case studies;

5 CALLS UPON Contracting Parties to take all practicable steps at the source to prevent and reduce contamination of marine sediments.

Annex

Draft Dredged Material Assessment Framework

1 INTRODUCTION

1.1 Dredging is essential to maintain navigation in ports, harbours and inland waterways and for the development of port facilities. Much of the material removed during these necessary activities requires disposal at sea. The greater proportion of the total amount of material dredged world-wide is, by nature, similar to undisturbed sediments in inland and coastal waters. A smaller proportion of dredged material, however, is contaminated by human activity to an extent that major environmental constraints need to be applied when depositing these sediments.

1.2 Within the framework of the London Convention 1972, Contracting Parties have recognized that dredged material, due to its characteristics, can be managed separately from waste materials. In 1986, the Tenth Consultative Meeting adopted 'Guidelines for the Application of the Annexes to the Disposal of Dredged Material' (resolution LDC.23(10)). It was agreed that the guidelines should be kept under regular review to take into account developments in dredging technology and improved understanding of the environmental consequences of disposal at sea.

1.3 The Dredged Material Assessment Framework (DMAF) is a generic guideline for decision makers in the field of management of dredged material. It is derived from the Waste Assessment Framework and sets out the basic practical, though not necessarily detailed considerations required for determining the conditions under which dredged material might (or might not) be deposited at sea.

2 EVALUATION OF NEED FOR DREDGING AND DISPOSAL

2.1 There are a number of dredging activities which may give rise to the need to relocate or dispose of sediments. These include:

1 Capital dredging – for navigation, to enlarge or deepen existing channel and port areas or to create new ones; and for engineering purposes; e.g. trenches for pipes, cables, immersed tube tunnels, removal of material unsuitable for foundations, removal of overburden for aggregate extractions;
2 Maintenance dredging – to ensure that channels, berths or construction works are maintained at their designed dimensions; and
3 Clean-up dredging – deliberate removal of contaminated material for human health and environmental protection purposes.

2.2 Before beginning a full assessment of the material and the disposal options the question should be asked 'Is dredging necessary?'. In the event of a subsequent full assessment indicating no acceptable options for disposal it will be necessary to re-address this question in a broader context.

3 DREDGED MATERIAL CHARACTERIZATION

Physical characterization

3.1 Evaluation of the physical characteristics of sediments for disposal is necessary to determine potential environmental impact and the need for chemical and/or biological testing. The basic physical characteristics required are the amount of material, particle size distribution and specific gravity of solids.

Exemptions from detailed characterization

3.2 Dredged material may be exempted from the full characterization requested in paragraphs 3.3 to 3.9 below if it meets one of the criteria listed below:

1 dredged material is excavated from a site away from existing and historical sources of appreciable pollution, so as to provide reasonable assurance that the dredged material has not been contaminated, or
2 dredged material is composed predominantly of sand, gravel and/or rock, or
3 dredged material is composed of previously undisturbed geological materials.

Dredged material that does not meet one of these criteria will require a full characterization to assess its potential impact.

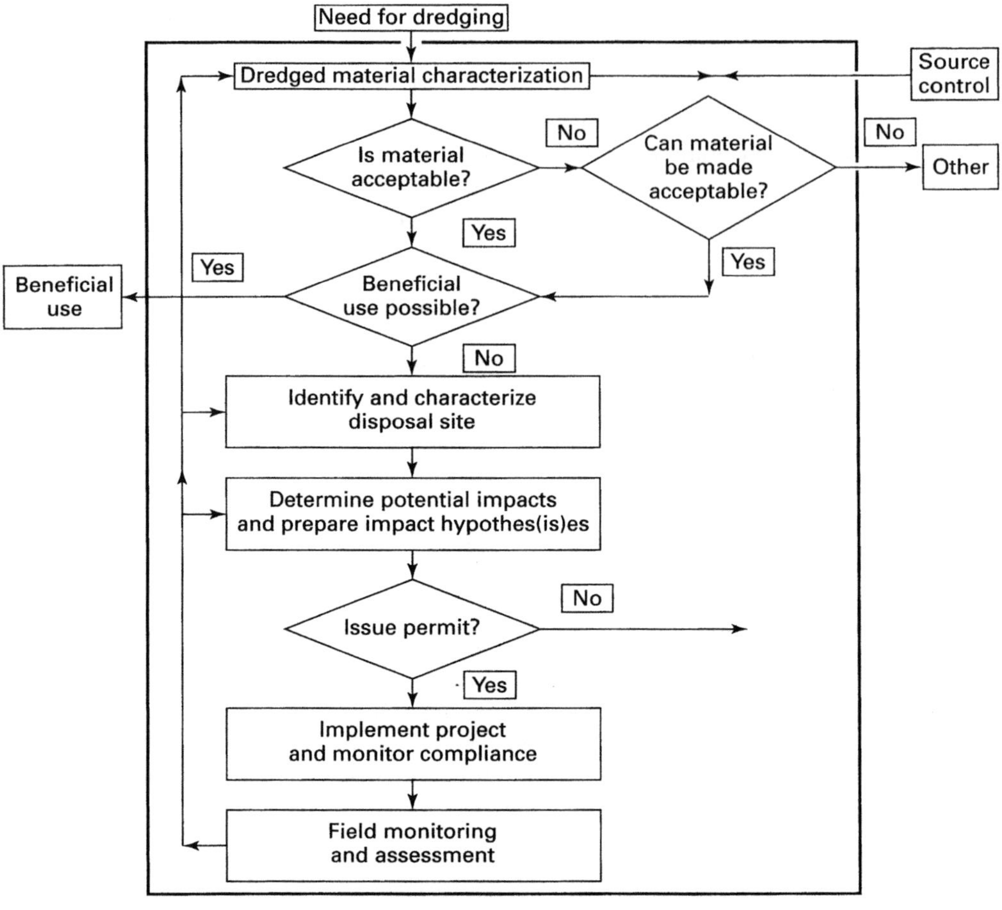

Representation of the jurisdictional boundary of the convention

Chemical characterization

3.3 Sufficient information for chemical characterization may be available from existing sources: in such cases new measurements may not be required of the potential impact of similar material at similar sites.

3.4 Considerations for additional chemical characterization of dredged material are as follows:

1 major geochemical characteristics of the sediment including redox status;
2 potential routes by which contaminants could reasonably have been introduced to the sediments;
3 data from previous sediment chemical characterization and other tests of the material or other similar material in the vicinity, provided this information is still reliable;
4 probability of contamination from agricultural and urban surface runoff;
5 spills of contaminants in the area to be dredged;
6 industrial and municipal waste discharges (past and present);

7 source and prior use of dredged materials (e.g. beach nourishment); and
8 substantial natural deposits of minerals and other natural substances.

3.5 Sampling of sediments from the proposed dredging site should represent the vertical and horizontal distribution and variability of properties of the materials to be dredged.

3.6 Further information may also be useful in interpreting the results of chemical testing, such as total organic carbon (TOC).

Biological characterization

3.7 If the potential impacts of the dredged material to be dumped cannot be assessed on the basis of the chemical and physical characterization and available biological information, biological testing should be conducted.

3.8 It is important to ascertain whether an adequate scientific basis exists on the characteristics and composition of the material to be dumped and on the potential impacts on marine life and human health. In this context, it is important to consider information about species known to occur in the area of the disposal site and the effects of the material to be dumped and of its constituents on organisms.

3.9 Biological tests should incorporate species that are considered appropriately sensitive and representative and should determine, where appropriate:

1 acute toxicity;
2 chronic toxicity such as long-term sub-lethal effects, covering an entire life cycle;
3 the potential for bioaccumulation; and
4 the potential for tainting.

Action List

3.10 The following is a screening mechanism for assessing properties and constituents of dredged material with a set of criteria for specific substances similar to that developed in the Waste Assessment Framework. These should reflect experience gained with published scientific research relating to the potential effects on human health or the marine environment. An Action List should be devised as a trigger mechanism for dredged material management decisions, including the identification and development of source control measures as described in paragraphs 3.13 to 3.15.

3.11 Action List levels[1] should be developed on a national or regional basis and might be set on the basis of concentration limits, biological responses, environmental quality standards, flux considerations or other reference values.

3.12 An Action List may include an upper and lower level giving these possible actions:

1 material which contains specified contaminants, or which causes biological responses, in excess of the relevant upper levels should generally be considered unsuitable for disposal at sea;

---

1 The Action List should, as a minimum, address the substances as currently contained in Annexes I and II to the Convention.

2 material which contains specified contaminants, or which causes biological responses, below the relevant lower levels should generally be considered of little environmental concern for disposal at sea; and
3 material of intermediate quality should require more detailed assessment before suitability for disposal at sea can be determined.

Contaminant Source Evaluation and Control

3.13 Contamination of estuarine and coastal marine sediments both as a consequence of historical and present day inputs presents a continuing problem for the management of dredged material. High priority should be given to the identification of sources, reduction and prevention of further contamination of sediments and should address both point and diffuse sources. Successful implementation of prevention strategies will require collaboration among agencies with responsibility for the control of point and diffuse sources of contamination.

3.14 In developing and implementing the source control strategy, appropriate agencies should take into account:

1 the continuing need for dredging;
2 the hazards posed by contaminants and the relative contributions of the individual sources to these hazards;
3 existing source control programmes and other regulations or legal requirements;
4 technical and economic feasibility;
5 the evaluation of the effectiveness of measures taken; and
6 consequences of not implementing contaminant reduction.

3.15 In cases where there has been historical contamination or where control measures are not fully effective in reducing contamination to acceptable levels, disposal management techniques, including the use of containment or treatment methods may be required.

4 EVALUATION OF DISPOSAL OPTIONS

4.1 The results of the physical/chemical/biological characterization will indicate whether the dredged material, in principle, is suitable for disposal at sea. Where sea disposal is identified as an acceptable option it is nonetheless important, recognizing the potential value of dredged material as a resource, to consider the availability of beneficial uses.

Beneficial Uses

4.2 There is a wide variety of beneficial uses depending on the physical and chemical characteristics of the material. Generally, a characterization carried out in accordance with chapter 3 of this Framework will be sufficient to match a material to possible uses such as:

1 engineering uses – land creation and improvement, beach nourishment, offshore berms, capping material and fill;
2 agricultural and product uses – aquaculture, construction material, liners; and
3 environmental enhancement – restoration and establishment of wetlands, upland habitats, nesting islands, and fisheries.

The technical aspects of beneficial uses are well-established and described in the literature.

Management Options

4.3 Where the characteristics of the dredged material are such that its disposal would not meet the requirements of the Convention, treatment or other management options should be considered. These options can be used to reduce or control impacts to a level that will not constitute an unacceptable risk to human health, or harm living resources, damage amenities or interfere with legitimate uses of the sea.

4.4 Treatment, such as separation of contaminated fractions, may make the material suitable for a beneficial use and should be considered before opting for sea disposal. Disposal management techniques may include placement on or burial in the sea floor followed by clean sediment capping, utilization of geochemical interactions and transformations of substances in dredged material when combined with sea water or bottom sediment, selection of special sites such as abiotic zones, or methods of containing dredged material in a stable manner.

5 SEA DISPOSAL SITE SELECTION[2]

5.1 The selection of a site for sea disposal involves not only considerations of an environmental nature but also economic and operational feasibility.

5.2 For the evaluation of a sea disposal site information should be obtained on the following, as appropriate:

1 the physical, geochemical and biological characteristics of the sea-bed (e.g. topography, redox status, benthic biota);
2 the physical, chemical and biological characteristics of the water column (e.g. currents, dissolved oxygen, pelagic species); and
3 proximity to:

- areas of natural beauty or significant cultural or historical importance;
- areas of special scientific or biological importance such as sanctuaries and critical habitats;
- recreational areas;
- subsistence, commercial and sport fishing areas;
- finfish and shellfish spawning, recruitment and nursery areas;
- migration routes of marine organisms;
- shipping lanes;
- military exclusion zones;
- engineering uses of the sea such as mining, undersea cables, water intakes, energy conversion sites, etc.

Such information can be obtained from existing sources complemented by field work where necessary.

2 Matters related to criteria for selection of sea disposal sites are addressed by the London Convention 1972 and are currently contained in Annex III thereto. These criteria should be considered in conjunction with this Framework.

5.3 The information on the characteristics of the sea disposal site referred to above is required to determine the probable fate and effects of the dumped material. The physical conditions in the vicinity of the sea disposal site will determine the transport and fate of the dredged material. The physico-chemical conditions can be used to assess the mobility and bioavailability of the chemical constituents of the material. The nature and distribution of the biological community and the proximity of the site of sea disposal to marine resources and amenities will, in turn, define the nature of the effects that are to be expected. Careful evaluation will then permit prediction of the consequences of dumping if it is authorized. It will also permit determination of environmental processes that may dominate the transport of material away from the sea disposal site. The influence of these processes may be reduced through the imposition of permit conditions.

5.4 In some cases, dumping can augment existing effects attributable to inputs of contaminants to coastal areas through land runoff and discharge, resource exploitation and maritime transport. These existing stresses on biological communities should be considered as part of the assessment of potential impacts caused by dumping. The proposed method of dumping and potential future uses of resources and amenities in the marine receiving area should also be taken into account.

6 *IMPACT ASSESSMENT*

6.1 Impact assessment should lead to a concise statement of the expected consequences of the disposal option (i.e. the Impact Hypothesis). Its purpose is to provide a basis for deciding whether to approve or reject the proposed disposal option and for defining environmental monitoring requirements.

6.2 Impact assessment comprises a summary of the potential effects on human health, living resources, amenities and other legitimate uses of the sea. It should define the nature, temporal and spatial scales and duration of expected impacts based on reasonably conservative assumptions.

6.3 For a retentive site, where the material deposited will remain within the vicinity of the site, the impact assessment should delineate the area that will be substantially altered by the presence of the deposited material and what the severity of these alterations might be. At the extreme, this may include an assumption that the immediate receiving area is entirely smothered. In such a case the likely timescale of recovery or recolonization should be projected after disposal operations have been completed as well as the likelihood that recolonization will be similar to, or different from, the existing benthic community structure. The impact assessment should specify the likelihood and scale of residual impacts outside the primary zone.

6.4 In the case of a dispersive site, the impact assessment should include a definition of the area likely to be altered in the shorter term by the proposed disposal operation (i.e. the near-field) and the severity of associated changes in that immediate receiving environment. It should also specify the likely extent of long-term transport of material from this area and what this flux represents in relation to existing transport fluxes in the area thereby permitting a statement regarding the likely scale and severity of effects in the long-term and far-field.

7 *PERMIT ISSUE*

7.1 If sea disposal is the selected option, then a permit authorizing sea disposal must be issued in advance. In granting a permit, the immediate impact of dredged material occurring within the boundaries of the disposal site such as alterations to the local, physical, chemical and biological environment is accepted by the permitting authority. Notwithstanding these consequences, the conditions under which a permit for sea disposal is issued should be such that environmental change beyond the boundaries of the disposal site are as far below the limits of allowable environmental change as practicable. The disposal operation should be permitted subject to conditions which further ensure that environmental disturbance and detriment are minimized and benefits maximized.

7.2 The permit is an important tool for managing sea disposal of dredged material and will contain the terms and conditions under which sea disposal may take place as well as provide a framework for assessing and ensuring compliance.

7.3 Permit conditions should be drafted in plain and unambiguous language and will be designed to ensure that:

1 only those materials which have been characterized and found acceptable for sea disposal, based on the impact assessment, are dumped;
2 the material is disposed of at the selected disposal site;
3 any necessary disposal management techniques identified during the impact analysis are carried out; and
4 any monitoring requirements are fulfilled and the results reported to the permitting authority.

7.4 Sufficient surveillance of sea disposal operations should assure the licensing authority that the permit conditions are met.

8 *MONITORING*

8.1 Monitoring in relation to disposal of dredged material is defined as measurements of compliance with permit requirements and of the condition and changes in condition of the receiving area to assess the Impact Hypothesis upon which the issue of a disposal permit was approved.

Specification of Baseline Conditions

8.2 It may usually be assumed that suitable specifications of existing (pre-disposal) conditions in the receiving area are already contained in the application for disposal. If the specification of such conditions is inadequate to permit the formulation of an Impact Hypothesis, additional information will be required by the licensing authority before any final decision on the permit application is made.

Post-Operational Monitoring

8.3 The Impact Hypothesis forms the basis for defining post-operational monitoring. The measurement programme should be designed to ascertain that changes in the receiving environment are within those predicted. In designing a monitoring programme the following questions must be answered:

1 What testable hypotheses can be derived from the Impact Hypothesis?
2 What measurements (type, location, frequency, performance requirements) are required to test these hypotheses?
3 How should the data be managed and interpreted?

8.4 The permitting authority is encouraged to take account of relevant research information in the design and modification of monitoring programmes. The measurements can be divided into two types – those within the zone of predicted impact and those outside.

8.5 Measurements should be designed to determine two things:

1 whether the zone of impact differs from that projected; and
2 whether the extent of change projected outside the zone of impact is within the scale predicted.

The first of these questions can be answered by designing a sequence of measurements in space and time that circumscribe the projected zone of impact to ensure that the projected spatial scale of change is not exceeded. The second question can be answered by the acquisition of measurements that provide information on the extent of change that occurs outside the zone of impact after the disposal operation. Frequently, this latter suite of measurements will only be able to be based on a null hypothesis – that no significant change can be detected.

Feedback

8.6 Information gained from field monitoring (or other related research studies) can be used to:

1 modify or terminate the field monitoring programme;
2 modify or revoke the permit; and
3 refine the basis on which applications to dump dredged material at sea are assessed.

Index

9780340545249